Algebraic Geometry and its Applications

Chandrajit L. Bajaj
Editor

Algebraic Geometry and its Applications

Collections of Papers from Shreeram S. Abhyankar's 60th Birthday Conference

Springer-Verlag
New York Berlin Heidelberg London
Paris Tokyo Hong Kong Barcelona Budapest

Chandrajit L. Bajaj
Department of Computer Science
Purdue University
West Lafayette, IN 47907
U.S.A.

Cover art and illustrations for the Part-opening pages were generated using the SHASTRA modelling environment developed by C. Bajaj.

Library of Congress Cataloging-in-Publication Data
Algebraic geometry & its applications : collection of papers from Shreeram Abhyankar's 60th birthday conference / [edited by] Chandrajit Bajaj.
p. cm.

ISBN-13: 978-1-4612-7614-2 e-ISBN-13: 978-1-4612-2628-4

DOI: 10.1007/978-1-4612-2628-4

1. Geometry, Algebraic–Conferences. I. Abhyankar, Shreeram Shankar. II. Bajaj, Chandrajit, 1958– . III. Title: Algebraic geometry and its applications.
QA564.A355 1994
516.3′5–dc20 93-33433

Printed on acid-free paper.

Softcover reprint of the hardcover 1st edition 1994

Production managed by Ken Dreyhaupt; manufacturing supervised by Vincent Scelta.
Photocomposed copy prepared from the authors' TeX files.

9 8 7 6 5 4 3 2 1

Shreeram S. Abhyankar

Shreeram Abhyankar received his Ph.D. from Harvard in 1955 under the guidance of Oscar Zariski. In his thesis, Abhyankar solved the problem of resolution of singularities of algebraic surfaces for nonzero characteristic. Ten years later, he solved the same problem for arithmetical surfaces and for three-dimensional varieties for nonzero characteristic. Since 1967, Abhyankar has been Distinguished Marshall Professor of Mathematics at Purdue University. In 1978, Abhyankar was awarded the Chauvenet Prize of the Mathematical Association of America for his article "Historical ramblings in algebraic geometry and related algebra" in the June 1976 issue of American Mathematical Monthly. *In that article, Abhyankar said that his father initiated him to mathematics and Sanskrit poetry by teaching him portions of Bhaskaracharya's treatise on algebra called Beejganita, which was written around AD 1150. In 1976, Abhynakar was also instrumental in founding the Bhaskaracharya Pratishthana, which is a research institute in Pune, India, named after this great mathematician.*

Preface

The "Algebraic Geometry and its Applications" conference was held in honor of Professor Shreeram Shankar Abhyankar and on the occasion of his 60th birthday.

A conference of renowned mathematicians, computer scientists, and engineers was held at Purdue University during the first four days of June 1990. There were over 150 research scientists and practitioners, a majority of which were students, collaborators, and colleagues of Professor Abhyankar during his illustrious career spanning 35 years so far. Professor Birkhoff gave the inaugural talk titled "Some Unorthodox Questions in Algebraic Geometry and its Applications" and Professor Masani a historical recollection of "Mathematics and India". The participants included a majority of the leading names in Algebraic Geometry—Abhyankar's main research field—Aroca, Berger, Buchweitz, Chiarli, Chow, Eisenbud, Hartshorne, Huneke, Igusa, Fulton, Mumford, Murthy, Nori, Sathaye, Seiler, Serre, Seshadri, Tessier, just to name a few. Artin and Nagata could not attend only due to last minute mishaps. The versatile nature of Abhyankar's talent was also evident in the applied geometry fields represented: Computer Algebra, Computational Geometry, Computational Real Algebraic Geometry, Computer Graphics and so on. Participants here included: Buchberger, Canny, Chazelle, Heintz, Kaltofen, Lazard, Roy, Sederberg, Warren, and many others. In addition to this, extensions related to Abhyankar's recent work in Group theory and Galois theory formed the substance of talks by Feit and by Harbater. The Deans of both the schools of Engineering and Science were also present and gave opening remarks on two of the four days.

In all there were over 150 scientists and engineers including twelve of Abhyankar's sixteen students. Lectures were held from 9 to 5:30 every day except that Sunday was a half day with lectures only in the morning. An interactive computer demonstration of computational algebraic geometry and computer graphics software was organized and run on Sunday afternoon, by the computer science department.

C. Bajaj
Purdue University

Contents

List of Contributors

Shreeram S. Abhyankar
Mathematics Department and Department of Computer Science
Purdue University
West Lafayette, IN 47907
USA

Chandrajit L. Bajaj
Department of Computer Science
Purdue University
West Lafayette, IN 47907
USA

Robert W. Berger
Fachbereich Mathematik
Universität des Saarlandes
D6600 Saarbrücken
Germany

Markus Brodmann
Mathematisches Institut der Universität Zürich
Rämistrasse 74
8001 Zürich
Switzerland

John Canny
Computer Science Division
University of California, Berkeley
Berkeley, CA 94720
USA

Geng-Zhe Chang
Department of Mathematics
University of Science and Technolgy of China
Hefei Anhui 230026
China

Bernard Chazelle
Department of Computer Science
Princeton University
Princeton, NJ 08544
USA

Nadia Chiarli
Dipartimento di Matematica
Politecnico di Torino
Italy

Walter Feit
Mathematics Department
Yale University
New Haven, CT 06520
USA

Sudhir R. Ghorpade
Department of Mathematics
Indian Institute of Technology
Powai, Bombay 400076
India

Angel Granja Baron
Dpto. de Matematicas
E.U. Graduados Sociales
Campus de Vegazana
Universidad de Leon
24007 Leon
Spain

Robin Hartshorne
Department of Mathematics
University of California-Berkeley
Berkeley, CA 94720
USA

Joos Heintz
Inst. Argentino de Matemática
CONICET
Viamonte 1636
(1055) Buenos Aires
Argentina

William Heinzer
Department of Mathematics
Purdue University
West Lafayette, IN 47907
USA

Craig Huneke
Mathematics Department
Purdue University
West Lafayette, IN 47907
USA

Jun-ichi Igusa
Department of Mathematics
Johns Hopkins University
Baltimore, MD 21218
USA

David Lantz
Department of Mathematics
Colgate University
Hamilton, NY 13346
USA

D. Lazard
Informatique, Université Paris VI
4 Place Jussieu
F-75252 Paris Cedex 05
France

Wei Li
Department of Mathematics and Statistics
McGill University
805 Sherbrooke St. West
Montreal, QU H3A 2K6
Canada

Anatoly S. Libgober
Department of Mathematics
University of Illinois at Chicago
Chicago, IL 60680
USA

P.R. Masani
University Professor Emeritus
Department of Mathematics and Statistics
University of Pittsburgh
Pittsburgh, PA 15260
USA

S.B. Mulay
Department of Mathematics
University of Tennessee
Knoxville, TN 37996-1300
USA

David Mumford
Department of Mathematics
Harvard University
Cambridge, MA 02138
USA

Masayoshi Nagata
Department of Applied Mathematics
Okayama University of Science
Okayama 700
Japan

Madhav Nori
Department of Mathematics
University of Chicago
Chicago, IL 60637
USA

Jack Ohm
Mathematics Department
Louisiana State University
Baton Rouge, LA 70803
USA

Leonidas Palios
Department of Computer Science
Princeton University
Princeton, NJ 08544
USA

Marie-Francoise Roy
IRMAR
Universite de Rennes I
35 042 Rennes Cedex
France

Peter Russell
Mathematics Department
McGill University
Montreal, QU
Canada

Judith D. Sally
Mathematics Department
Northwestern University
Evanston, IL 60201
USA

Avinash Sathaye
Mathematics Department
University of Kentucky
Lexington, KY 40506
USA

Thomas W. Sederberg
Department of Civil Engineering
Brigham Young University
Provo, UT 84601
USA

Wolfgang K. Seiler
Fakultä Fü Mathematick
Postfach 103462
6800 Mannheim
Germany

J-P. Serre
Chaire d'Algébre et Géométrie
Collège de France
Paris
France

Pablo Solerno
Inst. Argentino de Matemática
CONICET
Viamonte 1636
(1055) Buenos Aires
Argentina

Jon Stenerson
Mathematics Department
University of Kentucky
Lexington, KY 40506
USA

Shrinivas G. Udpikar
Department of Mathematics
S.P. College
Pune-411030
India

Sanjeevani V. Vaidya (Joshi)
Mathematics Department
Marist College
Poughkeepsie, NY 12601
USA

Joe Warren
Department of Computer Science
Rice University
USA

Roger Wiegand
Department of Mathematics and Statistics
University of Nebraska
Lincoln, NE 68588-0323
USA

Sylvia Wiegand
Department of Mathematics
University of Nebraska
Lincoln, NE 68588-0323
USA

Part I

Past, Present, and Future

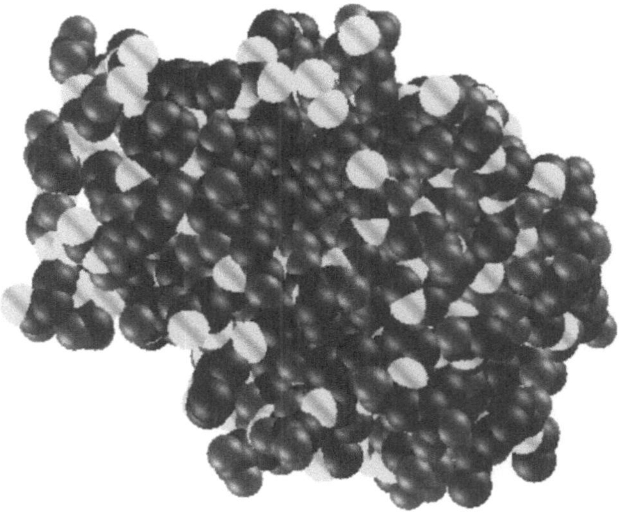

1
Mathematics and India

P.R. Masani

ABSTRACT Not being qualified to lecture on a technical subject appropriate to a conference on algebraic geometry, the organizers have been kind enough to let me articulate some thoughts on the role of mathematics in Indian culture instead. In the shaping of these thoughts my contacts with promising young students in India from 1949 to 1959 have been vital. First and foremost among these students was Shreeman Abhyankar, 1949–1951. It is therefore with much pleasure that I participate on the auspicious occasion of his 60th birthday. I will deal with "Mathematics and India" from a rather special angle that brings out the difference in the roles that mathematics has played in European and Indian civilization, and focuses on research initiation in India.

1.1 Mathematics in Europe

Mathematics occupied an exalted position in Greek philosophical thought. The great significance that the Pythagoreans attributed to it, became an important aspect of pre-Aristotelian Greek philosophy. Thus Plato wrote:

> ... The knowledge at which geometry aims is knowledge of *eternal being*, and not of that which at a particular time comes into being and perishes ([12], p. 287).

But the eternal being participates in the world, for Plato also tells us that

> ... in pitching a camp, or taking up a position, or closing or extending the lines of an army, or any other military maneuver, whether in actual battle or on a march, it will make all the difference whether a general is or is not a geometrician ([12], p. 287).

Thus, Plato realized that the study of the "eternal being" can significantly serve as a theoretical tool in bringing about changes in the world.

Latent in the Platonic view that a wooden triangle is a blurred and inaccurate copy of a purely ideal triangle in heaven is the principle that *raw data is a superposition of message (or pattern) and noise (or disturbance), and that to understand the data the message must be filtered and focused on, and the noise by and large ignored.* This is an important principle of scientific methodology, which is used more extensively today than ever before, cf. ([13], pp. 4–5).

Underlying this principle is the belief that the world is fundamentally harmonious. If you wish to understand the cosmos, you have to discover the hidden mathematical pattern. Every great mathematical physicist from Archimedes to Einstein has wittingly or unwittingly followed Plato's advice.

Most of the early fathers of the Christian church were neo-Platonists. some of them tried to explain eve the Incarnation, the mystery of how God can become a part of the world and yet retain his divinity, by the analogy of the participation of the ideal triangle in the wooden triangle. There was one important difference between the viewpoints of the church fathers and the Greeks. As the early congregations were largely made up of the dispossessed, there grew a strong Christian conviction in the *dignity of manual labor*. The Benedictine monks were obliged to spend much of their time on hard manual labor in the fields, and much of the rest on contemplation. Their monasteries became seats of both learning and craftsmanship. The Benedictines also ran schools in which Euclid was taught.

With the rise of the European towns and cities, the center of scholarship passed to the universities. Europe has had a long university tradition. The universities of Salerno and Bologna date back to the 11th century. Although in most of these the focus was on theology, philosophy and medicine, Euclidean geometry was never forgotten. Latin was the common medium in all universities. It thus happened that Copernicus, Stevinus, Kepler, and even Newton wrote in an Euclidean style in Latin. The professors were priests, often Dominican or Franciscan. The Dominicans leaned too heavily on Aristotle, and this retarded the growth of science, as we all know. The cry for more observation and experimentation, rather than deduction from vaguely-worded premises drawn from Aristotle, came from the Franciscans. The same Franciscans also wanted a movement towards mathematics. Foremost among them was Roger Bacon. Although he was an experimental philosopher and hardly a mathematician, here is what he wrote to the Poe in 1267:

> The neglect of mathematics for thirty or forty years has nearly destroyed the entire learning of Latin Christendom. for he who does not know mathematics cannot know any of the other sciences; what is more, he cannot discover his own ignorance or find its proper remedies. So it is that the knowledge of this science prepares the mind and elevates it to a well-authenticated knowledge of all things. For without mathematics neither what is antecedent nor consequent to it can be known; they perfect and regulate the former, and dispose and prepare the way for that which succeeds. (From a letter of Roger Bacon, Franciscan Friar and experimental philosopher, to Pope Clement IV,

c. 1267, cf. [12], pp. 55-56.)

This philosopher understood that raw data is a mixture of message and noise, and before one can experiment, one must know what pattern to look for. The *study of pattern* is mathematics. Thus, mathematics was demanded, not (à la Plato) in order to avoid experimentation, but in order to indulge in intelligent experimentation. This propadeutic role of mathematics became much clearer with Kant's *Critique of Pure Reason*, the appearance of which in 1781 marked a forward step in scientific methodology. Time does not allow speaking about Bacon's other Franciscan compatriots, the genius Duns Scotus, professor at Oxford at age 23, and William of Ockham, except to way two things.

1. A major issue that engaged their attention, and that of the great Dominicans such as St. Thomas Aquinas, to wit, the status of universals, had come from some extremely shrewd comments on Aristotle by the Islamic doctor Avicenna (Ibn Sina) who wrote in Bokhara around 1000 A.D., and

2. The Franciscan deliberations on Avicenna's principle has a rather profound influence on America's foremost philosopher, Charles Sanders Peirce (1839-1914). It lead him to *realistic* rather than nominalistic, conception of *pragmatism*[1]

1.2 Mathematics in India up to Ramanujan

When we turn from Greece and Europe to India, we read a very different story. Far from occupying an exalted place in Indian philosophy, mathematics (Ganita) remained a slave to commerce and astronomy, and astronomy meant in essence time-keeping. Mathematics, as in Babylonia, was a system of rules: do this, do that, without any demonstrations. There was no academic or university tradition in India. The only major complex that came close to being a university, was the conclave of Buddhist monasteries at Nalanda (near modern Patna), which flourished during the 5th, 6th and 7th centuries A.D. Here foreign as well as non-Buddhist scholars mingled. But the subjects discussed were grammer, rhetoric, poetry, logic, metaphysics, medicine and astronomy. Mathematics was not one of the main subjects. It is difficult to be specific since accurate historical sources are hard to come by, but there is no record of any mathematician associated with Nalanda. The scientific contribution of the university lay in the fields of *grammar*

[1]This "pragmaticism" of Peirce must be distinguished from the more popular and superficial pragmatism of John Dewey. Peirce, like Leibniz, was one of the creators of modern relation theory, and his realistic pragmaticism consolidates the Leibnizian bridge from matter to mind, cf. [10].

and *logic*. Its outstanding representatives were Dignaga (c. 480 A.D.) and his heir Dharmakirti (c. 650 A.D.).[2]

A major contribution of classical Indian mathematics was the *decimal place-value system of numeration* with a special symbol to indicate the absence of a unit. This system was well established in India by the 7th century A.D., but an early instance of it appears in the *Paulisa Siddhanta* (c. 400 A.D.). This Siddhanta, named after the Byzantine astrologer Paulus Alexandrus of the 3rd century A.D., suggests a possible Babylonian influence, transmitted to India during the Hellenic period. The salient facts in this regard may be stated as follows, cf. O. Neugebauer, ([6], pp. 5–28).

Old Babylonia (c. 1700 B.C.) had a sort of place-value system, but "had not yet developed a special sign for 'zero"' ([6], p. 16). In Babylonia in the Seleucid period (300 B.C.–300 A.D.), however, the absence of a unit was indicated initially by means of blank spaces, and in later papyri by the insertion in these spaces of special symbols such as "$\bar{0}$" and "$\bar{0}$". (The later Arabic "$\bar{0}$" was a simplification.) "But eve in the final phase of Babylonian writing we do not find any examples of zero signs at the end of numbers". The completed system came from Indian ([6], p. 20).

The place-value system of numeration is "one of the most fertile inventions of humanity" ([6], p 5). Its main beneficiaries are commerce, astronomy and other empirical fields, however. Pure mathematics of course gains, but to a limited extent. Archemedes was not appreciably hampered by his use of the Greek numerals, which have names only for integer multiples of 10, as with the Brahmi numerals of the Maurya period (3rd century B.C.). Nor must we forget the logical distinction between "0" as a symbol or punctuation to indicate the absence of a unit, and the number 0 as the neutral element for the binary operation of addition.

More important for mathematics as a science was Aryabhata's *use of variables* for unknowns, and Brahmagupta's assertion of the *rule of signs.* But this, and the original work done on the Pell equation by Brahmagupta and Bhaskara is overshadowed by the Greek contribution to number theory: the demonstration of the Euclidean algorithm, of the infinitude of the prime numbers, and the proof that every number of the form $2^{n-1}(2^n - 1)$, where $2^n - 1$ is a prime, is perfect. The diaphontine equation of greatest importance to mathematics as a whole is not Pell's equation, but rather the equation $y^2 = 2x^2$, that Pythagoras "solved" around 500 B.C. The latter raised the whole issue of incommensurability, and posed interesting problems such as the quadrature of the circle that brought Eudoxs (a student in Plat's Academy) close to the ideas of Dedekind, and brought Archimedes close to the discovery of the infinitesimal calculus. (As S. Bochner used to

[2]Contemporary with Nalanda was a smaller university at Vallabi in Kathiawar, and later on came universities at Vikramasali and Jagaddala and Odantapuri. I am very grateful to Professor G. Kallianpur for apprising me of the facts stated in the above paragraph and this footnote.

quip at tea-time in Princeton in the late 1940s, Archimedes missed only because "he had no graduate students who asked awkward questions".)

Responsible for this stunted development of mathematics (and also physics) in India was the institution not so much of a priesthood as of a hereditary priestly caste. Such castes are given to claiming exotic wisdom, to practicing secrecy, and to adopting a disdainful attitude towards sources of knowledge not their own. While the Moslems quickly made Greek thought their own, and people like Al-Biruni learned Sarikrit and translated Sanskrit treatises into Persian, there is no evidence of a Hindu scholar reciprocating in this manner. Actually, contacts with Greece were close in the post-Alexandrian period. Megasthenes was ambassador to India i Patna, c. 300 B.C. There were intermarriages at the royal level, and there was a noticeable Greek influence on Indian architecture. Indian mathematics began bearing the traces of Hellenic numerical methodology, but it bore none of the specifically Greek mathematics of Euclid and his followers.

How different the attitude was of the Hindu priesthood from the Christian (which is open to all ranks and in which celibacy is demanded) is clear from our earlier remarks on the Benedictines and Franciscans. To offer another illusration, the Franciscan friar Father Peirre Mersenne (1588–1648) not only encouraged scientists by acting as a clearing house between the great minds of his day, Descartes, Pascal, Fermat, Euler and others, but prevailed upon Cardinal Richleau to establish the French Academy of Sciences in 1635. It is hard to imagine a Brahmin priest stepping onto his shoes.

In the light of this record, it is fair to way that the first great mathematician from India is Srinivas Ramanujan (1887–1920). The dire circumstances faced by this boy genius, how at the age of 15 or so he began recovering some results of Euler, how his encounters with Carr's *Cynopsis* and Greenhill's book on elliptic functions led to his original researches before he was 20, his letter to G.H. Hardy of 1913, visit to Cambridge in 1914, subsequent stay at Trinity College, illness, return to India in 1919 and premature death in 1920 at age 32, are well recounted in Hardy's writings [8], [8]. However, it is only in recent decades that the full force of Ramanujan's previously unpublished "Notebooks" has come to light, thanks to the efforts of Professors Bruce Berndt, George Andrews and others, cf. [3], [4], [A2]. It transpires that of the more than 4,000 results in his Notebooks, almost two-thirds may turn out to be new theorems of high import. And the number of results that have been shown to be seriously flawed is embarrassingly small. What is even more remarkable is that the apparently incoherent arrangement of the results is deceptive, and that seemingly isolated results fall into integrated patterns when judged by the new vistas opened up since his death, in a wide variety of areas such as approximation theory, asymptotics, combinatorics, numerical analysis, algebraic number theory and orthogonal polynomials. There are strong indications that Ramanujan's ideas will bear on quite new areas of mathematical endeavor such as the recent Hopf algebras ("quantum groups").

Ramanujan then is India's gift to world mathematics. Professor E.H. Neville's description of him as "the only superlatively great mathematician whom India has produced in the last thousand years" stands even with the amendment "two thousand years".

1.3 Indology and Some Post-Ramanujan Development

To return to ancient India, its great contribution to the world civilization lay in the religious rather than in the scientific domain. It came as early as 500 B.C., from the Vedanta, the Buddha, the *Bhagavad Gita*, and from the spread of Buddhism in Asia. In India it brought into ascendence the most humane of all monarchs, Asoka Maurya (275–232 B.C.?), who used his pillars to spread a social gospel, which as D.D. Kosambi points out in his historical writings [7], [8], was articulated in Europe only in the 15th century A.D. by thinkers such as Erasmus of Rotterdam and St. Thomas More. (Thieves are not intrinsically bad fellows: most often they steal because they are hungry, and often they are hungry because they cannot find work.) Buddhist missionaries spread out not only into east Asia but into west Asia as well, almost to the shores of the Mediterranean. It seems that parts of the New Testament bear the imprint of Buddhist enlightenment.

These humane developments had a tempering effect on subsequent Indian governments, but did not directly affect west Asian and European history. The situation began to change in the late 18th century. The growing commercial intercourse, the rise of the sciences of *linguistics* and *anthropology*, and the study of *comparative mythology and religion* brought Indian culture into the European consciousness. Friedrich Max Muller, of German descent and professor of Oxford (1850–1875), translated the *Rig Veda* and edited 51 volumes of the *Sacred Books of the East.* There were other Indologists in Germany itself. There were for the most part influenced by the philosophy of Immanuel kant, who was himself influenced by the Romantic movement of the time. The availability of classical Indian literature has an influence on the German Romantic movement. The late philosophers Schoepenauer and Nitsche were influenced by Indian thought. It also affected the Romantic movement in German art. Goethe was influenced by Kalidas's *Shakuntalla.* Indian ideas penetrated even into German musical criticism. Thus, von Lens, in his *Critical Catalogue of Beethovan's Works*, writes that the best descriptions for the two movements in Beethovan's piano sonata Opus 111 are the words "Samsara" and "Nirvana". There was even a mild influence on German mathematical life in that the great mathematical pioneer Grassmann was a Sanskrit scholar to boot.

Buy a strange quirk, Indology, concerned as it is with the languages, literature and religions of ancient India, has had a bearing on post-Ramanujan mathematics in modern India. This happened in three odd ways.

First, in the 1920s and 1930s the word "Aryan" became an important term in the lexicon of the German National Socialist Party, and the party adopted as its emblem the Swastika, a Sankrit epigraph, that connotes well being or welfare. It was well known, especially in Germany, that the Aryans had their homeland 5,000 years ago in the northern regions of Eurasia. Southerly migrations began, which later bifurcated into an eastern branch that settled in India, and a western branch that settled in Iran and adjoining regions. It is generally agreed that such migrations did occur: the common etimology of Sanskrit and Avestan terms supports it. But the notion that the Indo- and Iranian Aryans and the present-day Nordic people are related is dubious, and that they constitute a superior race is quite absurd.[3] It is belied by our Nobel Laureates C.V. Ramam and S. Chandrasekhar, and by S. Ramanujan and S.S. Pillai (who solved Waring's problem) and by a host of others. It is not at all clear that these Dravidians (Brahmins all) are Aryans.[4]

The German folly led to the migration of German scientists, as we all know. India benefited. A German algebraist, a Privatdocent from Munich who had written a book on *Configuration Theory*, F.W. Levi, became Hardinge Professor of Mathematics at the University of Calcutta. In the mid-1930s he lectured on modern algebra, in particular *Galois theory.* Among his audience were two young Bengali statisticians, *R.C. Bose* and *S.N. Roy.* They are known the world over for their pioneering work in the application of Galois fields and finite geometries to *orthogonal Latin squares*, and thereby on the design of experiments — one of the major contributions of India to the field of mathematical statistics. A noteworthy offshoot of this development was the disproof of a conjecture of Euler in 1959 by Bose, S.S. Shrinkhande and E.T. Palmer. In 1947 Levi relinquished the Hardinge Chair and accepted a professorship at the Tata Institute of Fundamental Research in Bombay. He resigned in 1955 or so, to accept a professorship at the Free University in Berlin.

To continue with a second repercussion of Indology on Indian mathematics, one that led to contacts with American mathematics, we must turn to Harvard University and to Charles Rockwell LAnsman (1850–1941), who held the Chair in Sankrit at Harvard from 1880 to 1926. He edited the first 31 volumes of the *Harvard Oriental Series.* Now the Buddhist canonical

---

[3]Nevertheless, when as a young boy I was about to do something foolish, my father would retort in Gajarati "Hehi, unari". In Gajarati, "Nehi" meant "don't" but the word "unari" (un-Aryan) has the same connotation as the word "jungli" (from the jungle) viz. "stupid" or "ignoramus". Fortunately, such prejudices, ingrained in language, had no effect on our social intercourse.

[4]The issue of the ethnicity of the present-day Dravidians of south India is far from clear, cf. ([7], pp. 104–108).

literature is not in Sankrit but in a related language called *Pali* with a distinctive script of its own (*Brahmi*). Pali scholars are not easy to come by. There was a great man named Dharmanand Kosambi (1876–1947), a Saraswat Brahmin from Goa, with strong Buddhistic sympathies. After trampling (amid great hardship) over Buddhist lands, introducing Pali at the National College in Calcutta and serving as Professor of Pali at Calcutta University during 1907–1908, he was in the Bombay region in late 1908, endeavoring to introduce Pali studies in western India. In Bombay he gave lessons in Pali to a visitor from Harvard, *Dr. James H. Woods.* The latter persuaded Kosambi to join the Harvard staff in Indic philology. Kosambi's visits to Harvard were from 1910–1912, 1918–1922, 1926–1927 and 1931–1932.[5]

In the visit to Harvard in 1918, Kosambi was accompanied by his family. His son, Damodar Dharmanand Kosambi, left his school in Poona, and enrolled in the Cambridge Grammar School and Latin High School. In 1925 he entered Harvard to major in mathematics, and in 1929 received his B.S. in mathematics, Summa Cum Laude. Kosambi's interests were in *Lie Groups and Differential Geometry.* His teacher was Professor W.C. Graustein. But Kosambi had far wider interests that extended beyond mathematics, and among his professors at Harvard (and these included F.W. Osgood and M.H. Stone), the one he held in the highest esteem and affection was George David Birkhoff. Another blessing for him was the Widener Library, which he used profusely.

For reasons unclear, Kosambi did not get a Harvard post-graduate fellowship, and returned to India. Enroute, he took the opportunity to meet Elie Cartan in France and T. Levi Civita in Rome. He probably had introductory letters from Birkhoff. After a brief stay as lecturer at Benares Hindu University, he joined the Muslim University at Aligarh in 1931, thanks to Professor André Weil.

André Weil's visit to Aligarh at that time calls for a brief comment, for here we have another strange ramification of Indology on Indian mathematics. The Vice-Chancellor of the University, Dr. Ross Masood, a reformer, had hopes of establishing a Chair in French Civilization at Aligarh. In Paris in 1929 he met the great French Indologist Sylvian Levy, who introduced him to André Weil. Masood offered the Chair to Weil. Later, when the prospects of having this Chair evaporated, he changed the offer to one of a professorship in mathematics, which Weil accepted in January 1930.

---

[5]Dharmanand Kosambi (1876–1947), apart from being a great Buddhist scholar, was an exceedingly humane and self-abnegating individual. A non-violent socialist and disciple of Mahatma Gandhi, he rendered colossal social service to the laboring classes. When deteriorating health precluded any further activity, he undertook a traditional fast unto death. This took place in Mahatma Gandhi's Ashram, at the latter's request. He died in June 1947, only seven months before the Mahatma's assassination on January 30, 1948.

The mathematics department was being over-hauled. This led to the appointments of T. Vijayaraghavan and D.D. Kosambi to the mathematics faculty. There were hopes that S. Chowla could also be induced to join.[6] This would have meant the concentration at Aligarh of four relatively young men of great promise, a nucleus that could have grown into a promising mathematical research center in India. Alas, for reasons unclear, the group disintegrated within two year. Vijayaraghavan left for Dacca in 1930, Weil returned to Europe in 1932, and shortly thereafter Kosambi joined Fergusson College in Poona.

I met D.D. Kosambi in 1940 just after I received my Bachelor's degree in Bombay. My teachers included three Cambrigde mathematical tripos men, but Kosambi was the first individual I met who had a comprehension of modern mathematics and of mathematical research.[7] He was instrumental in propelling me to Harvard University, which I joined in the late fall of 1941. Whe I returned from the United States in 1948, I joined the Tata Institute of Fundamental Research (TIFR) as a Senior research fellow.[8] Both Kosambi and F.W. Levi were there as professors. Kosambi and H.J. Bhabha, the director, wanted this to be an interim arrangement until I could find a suitable position in the university. This suited me, for before my return I had been a part-time instructor at Princeton University, having confessed to Professor Marston Morse that full-time membership at the Institute for Advanced Study during the year before had been too much of a responsibility. I remained in the TIFR for almost two years, after which I was appointed Professor and Head of the Mathematics Department of the Institute of Science, the University of Bombay — the same institution where I had been an undergraduate. This was in December 1949.

One afternoon in early 1949, when the TIFR was still located in a pretty home (Kenilworth, on Peddar Road), a young lad, barefooted and shabbily dressed, dropped in my office. He had come to Bombay from Indore in central India, and was a junior at the Institute of Science. He had learned

---

[6]For the facts relating to Dharmanand Kosambi and Dr. Ross Masood, I am very grateful to Professors S.D. Deshmukh of Bombay University and André Weil of the Institute for Advanced Study, Princeton. The former is writing a biography of D.D. Kosambi in Marathi, hopefully with an English translation.

[7]D.D. Kosambi (1907–1966) lacked his father's humility but was also a great scholar. His mathematical contribution was small, but he has left a valuable and lasting imprint on Indology by his application of scientific tools. His contributions range from archeology, tribal anthropology, economic history, and classical poetry to linguistics, genetics and eugenics. His important place in Indian science was recognized in a posthumously published commemorative volume [10] in 1974, with a Foreword by V.V. Giri, then President of India.

[8]The TIFR is only the latest manifestation of the interest in the furtherance of scientific research shown by the industrial house of Tata & Sons (founded by Jamshedji N. Tata, 1839–1904). In 1907 it endowed the Indian Institute of Science in Bangalore, and in 1932 the Cancer Research Center in Bombay.

through the grapevine that I might succeed Professor K.R. Gunjikar (my own professor) who was about to retire. He expressed a strong desire to pursue mathematical research. A brief conversation convinced me that he was much more than the usual "good student". As he was living with some relatives in Malad, a suburb not too far from my home, I invited him to stop by my home at tea-time over the weekend.

After a few such evenings at my home, learned that this 18-year-old (who hardly needs to be identified) had discovered an interesting gap in the usual demonstrations in textbooks that a conic section

$$ax^2 + 2hxy + by^2 + 2gx + 2fy + c = 0$$

is the union of a pair of straight lines if and only if its discriminant vanishes:

$$det \begin{bmatrix} a & h & g \\ h & b & f \\ g & f & c \end{bmatrix} = 0.$$

For the reality of the lines the extra condition $h^2 - ab \geq 0$ has to be imposed. Left out in the textbooks is the crucial lemma that if quadratic polynomials S_1, S_2 on $\mathbb{C}^2$ have the same set of zeros, then they differ by a constant factor. He had found out that this result fails for cubic polynomials, but had succeeded in giving a generalization valid for m polynomials in n variables. I urged him to write a note on this in the *Mathematics Student*, cf. Vol. XIX (1951), p. 61–. This episode provides a glimpse of the early background alluded to in the following excerpt from a 1990 paper:

> First taking up the question as to what is Algebraic Geometry, long ago, to a major extent in my Father's time, and to a lesser extent in my own time, in high school and college, we used to learn the two subjects of Analytic Geometry and Theory of Equations. Analytic Geometry consists of studying geometric figures using algebraic equations. Theory of Equations, or High School Algebra, was manipulative in nature and dealt with simplifying expressions, solving equations, and factoring polynomials. These two subjects were later synthesized into and started being collectively called Algebraic Geometry. ([A1], p. 267)

The main thrust of my own lectures at the Elphinstone College and Institute of Science (sister institutions) in 1950 was the introduction of modern algebraic ideas. This could be doe only in the first and third years, since the other two involved university examinations in which what counted was cramming half-baked mathematics. I began with a course on the Theory of Equations for juniors. Abhyankar, who was by then a senior, could not attend.

When the TIFR moved into the quarters of the old Royal Bombay Yacht Club, it was only a short walk away from our Institute. I had introduced

Abhyankar to Kosambi and Levi, and from both of them he picked up a lot of mathematics. On a problem proposed to him by Levi, Abhyankar wrote a short "Note on the decomposition of the full symmetric group", which he sent to the PAMS, some months before his exams. The referee was able to provide a much shorter proof, however, and the paper did not appear.

It was clear to me that the TIFR was not a substitute for what our promising graduates needed (and there were some very good ones in physics as well), viz. excellent departments of postgraduate studies. In this regard the situation in India was the same as in 1940 when I met D.D. Kosambi. Thus, Abhyankar became the first of a line of my students who did their postgraduate work at Harvard and elsewhere in the United States. By way of indicating the difficulties involved, it will suffice to quote a few excerpts from the fifth of my letters to Professor Garrett Birkhoff in 1951, dated July 18, 1951:

> I must have written about 50 letters on Abhyankar's behalf, the majority of which no doubt must have quickly entered the waste basket. A few seem to have hit the spot, and believe me I am glad he is leaving India.

The most important of these letters was to the Honorable Manohar Singh Mehta, Education Minister, State of Madhya Bharat (Indore)—Abhyankar's state of domicile. This state generously awarded him a traveling fellowship of Rs. 12,000 ($2,400), just adequate to scrape by for one year at that time. Regarding the scholarship policy of the U.S. Educational Foundation, I wrote:

> Their people do not seem to know enough about our academic situation or about research. Many of their selected candidates are of the goody-goody school-teachery type, on whom we can ill afford to spend the few dollars alloted to us by the U.S. Government. It is the (somewhat rough) undergraduate who is passionately fond of his subject, that needs encouragement. Such students most often hail from poor families, and are in dire need of financial aid. It is annoying to find the Foundation not seeking the advice of competent people. ... Unfortunately Americans here usually come into contact with only wealthy Indians, who themselves have very hazy ideas about science and research.

1.4 Conclusion

This account of post-Ramanujan developments in Indian mathematics is colored by the nature of the occasion we are celebrating. The omissions are too numerous to list. In reference to the many fine contributions from India that have been left out, it will suffice to cite V.S. Varadharajan's paper [14] and merely add that today India is very much on the international mathematical map, and unlike the distant past, its value is appreciated in Indian intellectual circles. But as in the rest of the world, it is monetary greed and showmanship rather than intellectuality that reign. Working conditions are still unsatisfactory, especially at the crucila undergraduate level, and collegate staffs are poorly treated. At that level the conditions are in some ways worse than those in 1964 described in [9]. Fortunately, many of us, natives in India, have found a very congenial working environment and home in the United States and Canada. The same can be said of the Indian engineering and medical professions.

Perhaps the time has come to show our gratitude to our new land of domicile by bearing a bit of "the brown man's burden". One way would be to endow a school for American children, free from the quackery and social piffle that passes off for education in many an American public school Religion, India's main gift to mankind, must form an integral part of the curriculum. Discipline must be strict. But the motto should be "Thous shalt love thy God with all thy mind", and the sciences, especially mathematics (Euclidean demonstration) and physics, must be heavily stressed as well as history, geography, languages and literature. The clientele should be from the more deprived sections of American natives. To ensure this, and prevent parochialism, a "reverse discrimination" policy of denying scholarships to children of Indian descent, might be adopted.

1.5 References

[1] Abhyankar, S.S., Parametrization of curves and surfaces, *Proc. Soc. of Photo Optical Instrumentation Engineers*, **1251**, 1990, pp. 122-129.

[2] Andrews, G.E., et al., *Ramanujan Revisited*, Academic Press, Boston, 1988.

[3] Berndt, B. and Joshi, T., *Chapter 9 of Ramanujan's Notebook*, Contemporary Mathematics, American Mathematical Society, **23**, 1983.

[4] Berndt, B. and Joshi, T., *Ramanujan's Notebooks*, Part I, 1985, Part II, 1989, Springer-Verlag, New York.

[5] Hardy, G.H., Ramanujan, *Twelve Lectures on Subjects Suggested By His Life and Work*, Cambridge University Press, 1940.

[6] *Science and Human Progress* (Essays in honor of the late Professor D.D. Kosambi, Scientist Indololist and Humanist), Foreword by V.V. Furi, President of India, Popular Prakhashan, Bombay, 1974.

[7] Kosambi, D.D., *An Introduction to the Study of Indian History*, Popular Prakhashan, Bombay, 1956.

[8] Kosambi, D.D., *The Culture and Civilization of Ancient India*, Routledge & Kegan Paul, London, 1970.

[9] Masani, P., The basis of mathematical miseducation in the Indian universities, *Amer. Math. Monthly*, **71**, 1964, pp. 671–676.

[10] Masani, P., Norbert Wiener: The continuation of the tradition of Leibniz, Vico and Peirce, paper read at the Charles S. Peirce Sesquicentennial Congress, Harvard University, September 5–10, 1989, to appear.

[11] Neugebauer, O., *The Exact Sciences in Antiquity* (2nd ed.), Dover, New York, 1969.

[12] Plato, *The Republic* (translated by B. Jowett), Airmont Publishing Company, New York, 1968.

[13] Ramanujan, S., *Collected Papers*, edited by Seshu Aiyar, G.H. Hardy and B.M. Wilson (1927), Chelsea, New York, 1962.

[14] Varadharajan, V.S., Mathematics in and out of Indian universities, *Mathematical Intelligencer*, **5**, 1983, No. 1, pp. 38–42.

[15] Weisheipl, J.A., *The Development of Physical Theory in the Middle Ages*, University of Michigan Press, 1971.

[16] Whittaker, Sir Edmund, *From Euclid to Eddington*, Dover Publications, New York, 1958.

Part II

Algebraic Curves

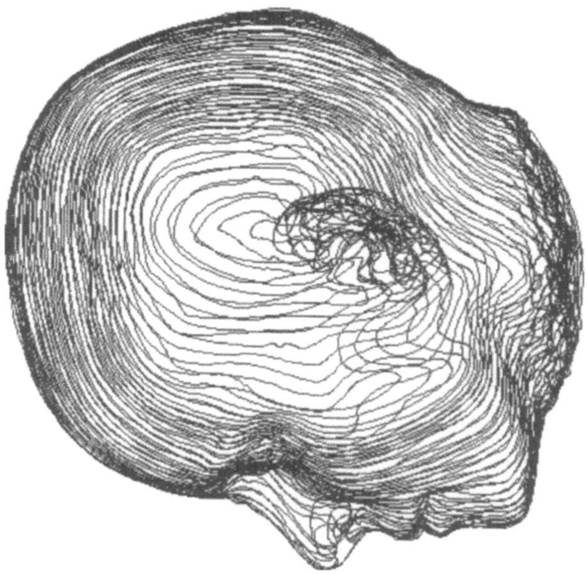

2

Square-root Parametrization of Plane Curves

Shreeram S. Abhyankar

2.1 Introduction

By calculating the genus of an irreducible algebraic plane curve of degree n in terms of its singularities, we see that, counted properly, the curve can have at most $\frac{(n-1)(n-2)}{2}$ double points, and it can be rationally parametrized iff this maximum is reached. If the maximum falls short by one or two, then the curve can still be parametrized by square-roots of rational functions. Such a square-root parametrization is used for factoring certain bivariate polynomials over a finite field.

This use of the square-root parametrization is similar to the school-book computations with quadratic surds, i.e., numbers of the form $\frac{a+b\sqrt{2}}{c+d\sqrt{2}}$ with integers a, b, c, d. The factorization also uses high-school ideas which say that integer roots of a monic integral polynomial are factors of its constant term, and if the polynomial is not monic then the factors of the leading coefficient suffice as denominators of the roots.

In this analogy, the writing of a rational number as a power product of prime numbers is to be replaced by the "divisor" of an algebraic function. While dealing with hyperelliptic functions, i.e., with functions admitting a square-root parametrization, by taking norms, computations with "divisors" are almost reduced to computations of rational functions. This is inspired by rationalization of the quadratic surd $\frac{a+b\sqrt{2}}{c+d\sqrt{2}}$ by multiplying the numerator and denominator by $c - d\sqrt{2}$ which is the "conjugate" of the denominator.

Another noteworthy feature of our calculations is the use of judiciously ad-hoc expansions which are somewhat modelled after "slow Taylor expansions" or "Taylor series with remainders". In other words, you don't have to be very pedantic in making your expansions. You should do them cleverly, and that could cut down a lot of tedious work. These ideas are applied in finding extensions of valuations and calculating the values of various functions at these extensions.[1]

Now the original bivariate polynomial is a univariate polynomial whose coefficients are hyperelliptic functions. After invoking the square-root parametrization, a suitable norm of the original polynomial becomes a genuine bivariate polynomial with coefficients in a finite field. I successfully used MACSYMA and REDUCE to actually find this new bivariate polynomial. However, both these computer algebra packages refused to factor a bivariate polynomial over a finite field. So I reverted to the original polynomial to hand-calculate in a hyperelliptic function field, using MACSYMA only to verify ordinary polynomial operations; see the Summary and the Note at the end of the Bar Polynomial Section of [4].

This factorization is employed in the calculation of Galois groups. The said calculation shows that $PSL(2,8)$ belongs to the algebraic fundamental group of the affine line in characteristic 7, i.e., $PSL(2,8)$ is the Galois group of a certain unramified covering of the affine line over an algebraically closed field of characteristic 7. To recapitulate, in my 1957 paper [A1], I wrote down the equation $Y^{p+t} - XY^p + 1 = 0$ giving an unramified covering of the affine line over an algebraically closed field of nonzero characteristic p, and suggested that its Galois group $\overline{G}_{p+t,p}$ be calculated; here t is a positive integer prime to p. Recently Serre[2] showed that if $t = 1$ then $\overline{G}_{p+t,p} = PSL(2,p)$. Inspired by this, in [4], by using CT,[3] I have shown that if $t > 1$ and $(p,t) \neq (7,2)$ then $\overline{G}_{p+t,p} =$ the alternating group A_{p+t} on $p+t$ letters. In the present paper, I complete the calculation by showing that in the remaining case of $(p,t) = (7,2)$ we have $\overline{G}_{9,7} = PSL(2,8)$.[4] Actually, I shall consider the slightly more general equation $Y^9 - XY^7 + e = 0$ where e is a nonzero element in a field k of characteristic 7, which is not assumed to be algebraically closed, and I shall show that its Galois group over $k(X)$ is $PSL(2,8)$ or $P\Gamma L(2,8)$ according as $4e$ does or does not have a cube root in k. Note that, for any power q of any prime number p, by $PSL(2,q)$ we denote the group of all fractional linear transformations $\frac{a'x+b'}{c'x+d'}$ with a', b', c', d' in $GF(q)$ such that $a'd' - b'c' = 1$, where $GF(q)$ is the field containing exactly q elements; moreover, $P\Gamma L(2,q)$ is the semidirect product of $PSL(2,q)$ with the cyclic group of order s where $q = p^s$.

All in all, this paper may be regarded as a huge exercise in the high-school art of factoring polynomials. But the high-school is to be mixed with a goodly dose of things like valuations of algebraic function fields from college algebra, and resolution of singularities of plane curves from

[1] In the same vein, resolution of singularities of plane curves by quadratic transformations is an implicit version of Newton–Puiseux expansion.

[2] Unpublished.

[3] CT = the Classification Theorem of finite simple groups.

[4] An alternative proof of this was recently communicated to me by Serre in his letter, dated 30 April 1991, which with his kind permission is being included in these Proceedings; see [8].

geometry, and so on.

At any rate, I enormously enjoyed working on this project which was a cliff-hanger to the end because I did not know whether the polynomial would factor or not. The deep concentration reached while doing it seemed to give a semblance of "savikalpa samadhi".

2.2 Hyperelliptic Curves

Consider a plane curve $C : F(X,Y) = 0$ where $F(X,Y)$ is an irreducible polynomial of degree n in indeterminates X, Y with coefficients in an algebraically closed field k^*. In case the curve is of genus zero, we can obtain a rational parametrization for it by means of adjoints of degree $n-2$. Referring to Lecture 19 of [A3][5] for details of this, let us proceed to show how, in case the genus is two, a square-root parametrization can be obtained by passing adjoints of degree $n-3$.

So, with notation as in the said Lecture 19, let $(P_i)_{i=1,2,\dots}$ be all the singular points of C, distinct as well as infinitely near, at finite distance as well as at ∞, and let ν_i be the multiplicity of C at P_i. Now

$$\text{genus of } C = \frac{(n-1)(n-2)}{2} - \delta(C) = \text{ a nonnegative integer}$$

where $\delta(C)$ is the number of double points of C counted properly, i.e.,

$$\delta(C) = \sum_i \frac{\nu_i(\nu_i - 1)}{2}.$$

Recall that a curve is adjoint to C if, for each i, its multiplicity at P_i is at least $\nu_i - 1$. Let $\mathrm{ADJ}_m(C)$ be the k^*-vector space of all polynomials $F^*(X,Y)$ of degree at most m in X, Y with coefficients in k^* such that the curve $F^*(X,Y) = 0$ is adjoint to C. By calculating the adjoint conditions imposed by the points P_i we get the vector space dimension estimate

$$[\mathrm{ADJ}_m(C) : k^*] \geq \frac{(m+1)(m+2)}{2} - \delta(C)$$

as on page 151 of [A3].

For a moment assume that C is of genus 2 and take $m = n-3$. Then

$$\frac{(m+1)(m+2)}{2} - \delta(C) = \frac{(n-1)(n-2)}{2} - \delta(C) = 2$$

and hence we can take two linearly independent members $F'(X,Y)$ and $F''(X,Y)$ of $\mathrm{ADJ}_{n-3}(C)$. Now upon letting

[5] A summary of this Lecture 19 may also be found in [5].

$\widehat{F}(X,Y) = F'(X,Y) + TF''(X,Y)$ we have that $\widehat{F}(X,Y) = 0$ is adjoint to C and for its intersection multiplicity with C at the singularities of C we have

$$\sum_i \nu_i(\nu_i - 1) = n(n-3) - 2$$

and hence, by Bezout's Theorem, it meets C in exactly 2 free points whose coordinates must be the roots of quadratic polynomials in T. Assuming the characteristic of k^* to be different from 2 and by completing the square, it follows that C can be parametrized by rational functions of T and the square-root of a polynomial in T whose degree, because of the Riemann–Hurwitz genus formula,[6] can be construed to be 5 or 6.

If the genus of C were 1, in a similar manner, but by using adjoints of degree $n-2$, we can get a square-root parametrization of C where the degree of the corresponding polynomial in T is 3 or 4.

Now curves of genus 1 are called elliptic curves, and curves admitting a square-root parametrization are called hyperelliptic curves. Thus a genus 2 curve is always hyperelliptic, but not conversely.

It may be noted that the complexity analysis of rational parametrization of genus zero curves given in [6] carries over to square-root parametrization of curves of genus one or two.

2.3 Deriving the Special Polynomial

We shall apply the square-root parametrization method to a plane curve which occurs in the calculation of Galois groups in the following manner.

By taking $(h, c_1, t) = (1, e, 2)$ in Proposition 1 of my 1957 paper [A1] we get the equation

$$(1^\bullet) \qquad x = y^2 + ey^{-p}$$

where x and y are variables over a field k of prime characteristic $p > 2$, and e is a nonzero element of k; for instance k could be the finite field k_p containing exactly p elements, and e could be 1. By changing y to Y and multiplying both sides by Y^p we get the polynomial

$$\Theta = \Theta(Y) = Y^{p+2} - xY^p + e$$

having y as a root. The Y-derivative of this is $2Y^{p+1}$ and by substituting its only root $Y = 0$ in the original polynomial we get e. Therefore, for every value of x, the equation $\Theta(Y) = 0$ has $p+2$ distinct roots. Consequently, this gives an unramified covering of the affine line over k, i.e., $x = \infty$ is the only

[6]For Bezout's Theorem and Riemann–Hurwitz formula see pages 146 and 58 of [A3] or pages 426 and 436 of [A2].

valuation of $k(x)/k$ which is ramified in $k(x,y) = k(y)$. We are interested in the Galois group $\overline{G}_{p+2,p} = \mathrm{Gal}(\Theta, k(x))$, which we regard as a permutation group of degree $p+2$, i.e., as a subgroup of the symmetric group S_{p+2}; note that $\Theta(Y)$ is clearly irreducible in the polynomial ring $k(x)[Y]$, and hence $\overline{G}_{p+2,p}$ is a transitive permutation group. In [4], I showed that if k is algebraically closed and $p \neq 7$ then $\overline{G}_{p+2,p} =$ the alternating group A_{p+2}; see the Bar Polynomial Section of [4] and especially the Summary at the end of that Section.[7] In [4], I also showed that if k is algebraically closed and $p = 7$ then $\overline{G}_{9,7} = PSL(2,8)$ or A_9 according as a certain *special polynomial* $\Psi(U)$ is reducible or not; see the Summary and the Note at the end of the Bar Polynomial Section of [4].[8] Here $PSL(2,8)$ denotes the group of all fractional linear transformations $\frac{a'x+b'}{c'x+d'}$ with a', b', c', d' in $GF(8)$ such that $a'd' - b'c' = 1$, where $GF(8)$ is the field containing exactly 8 elements. Let us now proceed to derive this polynomial $\Psi(U)$, and establish its reducibility when $p = 7$, thereby showing that if k is algebraically closed then $\overline{G}_{9,7} = PSL(2,8)$. The said special polynomial $\Psi(U)$ will essentially be obtained by successively "throwing away" three roots of $\Theta(Y)$. In establishing the reducibility of $\Psi(U)$ we shall not assume k to be algebraically closed and hence, without that assumption, from what we have said in the Bar Polynomial Section of [4] it will follow that $\overline{G}_{9,7} = PSL(2,8)$ or $P\Gamma L(2,8)$ where $P\Gamma L(2,8)$ is the semidirect product of $PSL(2,8)$ with the cyclic group of order 3.

So let $\Theta'(Y)$ be obtained by "throwing away" the root y of $\Theta(Y)$. Then

$$\begin{aligned}
&\Theta'(Y)\\
&= \tfrac{\Theta(Y+y)-\Theta(y)}{Y}\\
&= \tfrac{[(Y+y)^{p+2}-x(Y+y)^p+e]-[y^{p+2}-xy^p+e]}{Y}\\
&= \tfrac{[(Y^p+y^p)(Y^2+2yY+y^2)-x(Y^p+y^p)+e]-[y^{p+2}-xy^p+e]}{Y}\\
&= Y^{p+1} + 2yY^p + (y^2-x)Y^{p-1} + y^pY + 2y^{p+1}\\
&= Y^{p+1} + 2yY^p - ey^{-p}Y^{p-1} + y^pY + 2y^{p+1} \text{ because } x = y^2 + ey^{-p}.
\end{aligned}$$

Let $\Delta(Y)$ be obtained by "reciprocating" the roots of $\Theta'(Y)$. Then

$$\begin{aligned}
\Delta(Y) &= \frac{1 + 2yY - ey^{-p}Y^2 + y^pY^p + 2y^{p+1}Y^{p+1}}{2y^{p+1}}\\
&= Y^{p+1} + \frac{Y^p}{2y} - \frac{eY^2}{2y^{2p+1}} + \frac{Y}{y^p} + \frac{1}{2y^{p+1}}.
\end{aligned}$$

---

[7]Although in [4] we had taken $e = 1$, the calculations remain valid for any nonzero e.

[8]Again the calculations of [4] remain valid for any nonzero e.

Let $\Lambda(Z)$ be obtained by multiplying the roots of $\Delta(Z)$ by $2y$. Then

$$\begin{aligned}
&\Lambda(Z)\\
&= (2y)^{p+1}\Delta\left(\frac{Z}{2y}\right)\\
&= (2y)^{p+1}\left[\tfrac{Z^{p+1}}{(2y)^{p+1}} + \tfrac{Z^p}{2y(2y)^p} - \tfrac{eZ^2}{2y^{2p+1}(2y)^2} + \tfrac{Z}{y^p(2y)} + \tfrac{1}{2y^{p+1}}\right]\\
&= Z^{p+1} + Z^p - 2^{p-2}ey^{-p-2}Z^2 + 2^pZ + 2^p\\
&= (Z+2)^p(Z+1) - 2^{p-2}ey^{-p-2}Z^2.
\end{aligned}$$

Let $\Lambda^*(Z)$ be obtained from $\Lambda(Z)$ by writing y^* for $2^{p-2}y^{-p-2}$. In other words, let

$$y^* = 2^{p-2}ey^{-p-2} \tag{$2^\bullet$}$$

and

$$\Lambda^*(Z) = (Z+2)^p(Z+1) - y^*Z^2.$$

Upon letting z to be a root of $\Lambda^*(Z)$ we get

$$y^* = \frac{(z+2)^p(z+1)}{z^2}. \tag{$3^\bullet$}$$

Now let $\Lambda'(Z)$ be obtained by throwing away the root z of $\Lambda^*(Z)$. Then

$$\begin{aligned}
\Lambda'(Z) &= \tfrac{\Lambda^*(Z+z)-\Lambda^*(z)}{Z}\\
&= \tfrac{[(Z+z+2)^p(Z+z+1)-(z+2)^p(z+1)]-y^*[(Z+z)^2-z^2]}{Z}\\
&= [Z^p + (z+1)Z^{p-1} + (z+2)^p] - (z+1)(z+2)^pz^{-2}[Z+2z] \text{ by } (3^\bullet)\\
&= Z^p + (z+1)Z^{p-1} - (z+1)(z+2)^pz^{-2}Z - (z+2)^{p+1}z^{-1}.
\end{aligned}$$

Let $\Phi(W)$ be obtained by multiplying the roots of $\Lambda'(W)$ by z. Then $\Phi(W) = z^p\Lambda'\left(\frac{W}{z}\right)$ and hence

$$\Phi(W) = W^p + (z+1)zW^{p-1} - (z+1)(z+2)^pz^{p-3}W - (z+2)^{p+1}z^{p-1}$$

and upon letting ω to be a root of $\Phi(W)$ we have $\Phi(\omega) = 0$.

Finally let $\Psi(U)$ be obtained by throwing away the root ω of $\Phi(U)$. Then

$$\Psi(U) = \frac{\Phi(U+\omega) - \Phi(\omega)}{U} \quad \text{and} \quad \frac{(U+\omega)^p - \omega^p}{U} = U^{p-1}$$

and

$$\begin{aligned}
&\tfrac{(U+\omega)^{p-1}-\omega^{p-1}}{U}\\
&= \tfrac{\omega^{p-1}\left[-1+(1+\frac{U}{\omega})^p(1+\frac{U}{\omega})^{-1}\right]}{U}\\
&= \tfrac{\omega^{p-1}\left[-1+(1+\frac{U^p}{\omega^p})(1-\frac{U}{\omega}+\frac{U^2}{\omega^2}-\frac{U^3}{\omega^3}+\dots)\right]}{U}\\
&= \tfrac{\omega^{p-1}\left[-1+(1-\frac{U}{\omega}+\frac{U^2}{\omega^2}-\dots+\frac{U^{p-1}}{\omega^{p-1}})\right]}{U}\\
&= U^{p-2} - \omega U^{p-3} + w^2U^{p-4} - \dots + \omega^{p-3}U - w^{p-2}
\end{aligned}$$

and hence

$$(4^\bullet)\qquad \begin{aligned}\Psi(U) \quad = \quad & U^{p-1} \\ & +(z+1)z[U^{p-2} - \omega U^{p-3} + \omega^2 U^{p-4} - \ldots + \omega^{p-3}U] \\ & -(z+1)z[\omega^{p-2} + (z+2)^p z^{p-4}].\end{aligned}$$

By $(1^\bullet)$ we see that the valuation $x = \infty$ of $k(x)/k$ splits in $k(y)$ into the valuations $y = 0$ and $y = \infty$, and as we have said, this is the only valuation of $k(x)/k$ which is ramified in $k(y)$; therefore by $(2^\bullet)$ and $(3^\bullet)$ we see that $w_0^* : y^* = 0$ and $w_\infty^* : y^* = \infty$ are the only valuations of $k(y^*)/k$ which are ramified in $k(z)$, the valuation w_0^* splits into the valuations $\kappa_1 : z+1 = 0$ and $\kappa_2 : z+2 = 0$ of $k(z)/k$ with reduced ramification exponents $\bar{r}(\kappa_1 : w_0^*) = 1$ and $\bar{r}(\kappa_2 : w_0^*) = p$, and the valuation w_∞^* splits into the valuations $\kappa_0 : z = 0$ and $\kappa_\infty : z = \infty$ of $k(z)/k$ with reduced ramification exponents $\bar{r}(\kappa_0 : w_\infty^*) = 2$ and $\bar{r}(\kappa_\infty : w_\infty^*) = p - 1$. Thus, we have the *special ramification diagram* for the field extension $k(y^*) \subset k(z)$ where the square bracketed numbers are the reduced ramification exponents:

$$w_0^* : y^* = 0 \;—\!\!\left|\begin{array}{l} \to \kappa_1 : z+1 = 0[1] \\ \\ \to \kappa_2 : z+2 = 0[p] \end{array}\right.$$

$$w_\infty^* : y^* = \infty —\!\!\left|\begin{array}{l} \to \kappa_0 : z = 0[2] \\ \\ \to \kappa_\infty : z = \infty[p-1]. \end{array}\right.$$

Since the degree of $\Phi(W)$ is p, by the above diagram we see that $\Phi(W)$ is *irreducible* in $k(z)[W]$, the valuation κ_1 has a unique extension λ_1 to $k(z,\omega)$, and for its reduced ramification exponent we have $\bar{r}(\lambda_1 : \kappa_1) = p$. By the above diagram we also see that $\kappa_0, \kappa_1, \kappa_2$ are the only valuations of $k(z)/k$ which are ramified in $k(z,\omega)$.

As we have indicated in the beginning of this Section, for the Galois group $\overline{G}_{p+2,p}$ of the polynomial $Y^{p+2} - xY^p + 1$ over the field $k(x)$, in [4] it was shown that if k is algebraically closed and $p \neq 7$ then $\overline{G}_{p+2,p} = A_{p+2}$, whereas if k is algebraically closed and $p = 7$ then $\overline{G}_{9,7} = PSL(2,8)$ or A_9 according as the special polynomial $\Psi(U)$ is reducible in $k(z,\omega)[U]$ or not. Now the coefficients of $\Psi(U)$ are functions on the irreducible curve $\Phi = 0$ in the (z,ω)-plane, and so we shall first study this *auxiliary curve*. In particular we shall show that, in case k is algebraically closed and $p = 7$, its genus is 2, and this will enable us to find a square-root parametrization for it, which in turn will facilitate our investigation of the reducibility of $\Psi(U)$. In [4] it was also shown that, in case of $p = 7$, the polynomial $\Psi(U)$

is reducible in $k(z,\omega)[U]$ iff it factors into two irreducible polynomials of degree 3. We shall *indirectly* use this fact in investigating the reducibility of $\Psi(U)$. I say indirectly because, by trying out factors only of this type, we shall be able to find them!

2.4 Singularities of the Auxiliary Curve

Thus we want to consider the auxiliary curve given by

$$(1)\qquad \begin{aligned}\Phi(\omega) = 0 \text{ with } \Phi(W) = {} & W^p + (z+1)zW^{p-1} \\ & -(z+1)(z+2)^p z^{p-3} W \\ & -(z+2)^{p+1} z^{p-1}.\end{aligned}$$

Let us start by giving a direct proof of the fact that $\kappa_0, \kappa_1, \kappa_2$ are the only valuations of $k(z)/k$ which are ramified in $k(z,\omega)$. To this end, by calculating the W-derivative $\Phi_W(W)$ of $\Phi(W)$ we get

$$\Phi_W(W) = -(z+1)z\Phi'(W)$$

where

$$\Phi'(W) = W^{p-2} + (z+2)^p z^{p-4}$$

and clearly

$$\Phi(W) = [W^2 + (z+1)zW]\Phi'(W) - (z+2)^p z^{p-4}\Phi''(W)$$

where

$$\Phi''(W) = W^2 + 2(z+1)zW + (z+2)z^3$$

and obviously

$$\Phi''(W) = [W + z^2][W + (z+2)z]$$

and hence, upon letting Res_W and Disc_W stand for W-resultant and W-discriminant respectively, we have

$$\begin{aligned}&\mathrm{Disc}_W(\Phi(W)) \\ &= \mathrm{Res}_W(\Phi(W), \Phi_W(W)) \\ &= [(z+1)z]^{p-2}\mathrm{Res}_W(-(z+2)^p z^{p-4}\Phi''(W), \Phi_W(W)) \\ &= [(z+1)z]^{p-2}\mathrm{Res}_W(-(z+2)^p z^{p-4}\Phi''(W), -(z+1)z\Phi'(W)) \\ &= [(z+1)z]^p[-(z+2)^p z^{p-4}]^{p-2}\mathrm{Res}_W(\Phi''(W), \Phi'(W))\end{aligned}$$

and

$$\begin{aligned}&\mathrm{Res}_W(\Phi''(W), \Phi'(W)) \\ &= \Phi'(-z^2)\Phi'(-(z+2)z) \\ &= [(z+2)^p z^{p-4} - z^{2p-4}][(z+2)^p z^{p-4} - (z+2)^{p-2} z^{p-2}] \\ &= 8(z+2)^{p-2}(z+1)z^{2(p-4)}\end{aligned}$$

and therefore

(2) $$\mathrm{Disc}_W(\Phi(W)) = -8(z+2)^{(p+1)(p-2)}(z+1)^{p+1}z^{p(p-3)}.$$

Consequently, no valuation of $k(z)/k$, other than the valuations $\kappa_0, \kappa_1, \kappa_2, \kappa_\infty$, is ramified in $k(z,\omega)$.

By substituting $z+1=0$ in (1) we see that the unique extension λ_1 of $\kappa_1 : z+1=0$ to $k(z,\omega)$ corresponds to $\omega-1=0$, i.e., $\lambda_1(\omega-1)>0$.

To find the extensions of $\kappa_2 : z+2=0$ to $k(z,\omega)$, by substituting $z+2=0$ in (1) we see that there is an unramified extension λ_2 with $\lambda_2(\omega+2)>0$ and $\bar{r}(\lambda_2 : \kappa_2)=1$, and at the others the value of ω is positive and upon putting $\omega=(z+2)\hat{\omega}$ in (1) and dividing throughout by $(z+2)^{p-1}$ we get

$$(z+2)\hat{\omega}^p + (z+1)z\hat{\omega}^{p-1} - (z+1)(z+2)^2z^{p-3}\hat{\omega} - (z+2)^2z^{p-1} = 0.$$

Therefore $\hat{\epsilon}\hat{\omega}^{p-1} = \hat{\delta}(z+2)^2$ where $\hat{\epsilon} = (z+2)\hat{\omega} + (z+1)z$ and $\hat{\delta} = (z+1)z^{p-3}\hat{\omega} - z^{p-1}$ are units in the relevant ring,[9] and hence $z+2 = \pm\hat{\epsilon}'\hat{\omega}^{\frac{p-1}{2}}$ where $\hat{\epsilon}'$ is a unit in the completion of that ring with $\hat{\epsilon}'^2 = \hat{\epsilon}\hat{\delta}^{-1}$.[10] This gives extensions λ_{21} and λ_{22} of κ_2 to $k(z,\omega)$ with $\lambda_{21}(z+2) = \lambda_{22}(z+2) = \frac{p-1}{2}$ and $\lambda_{21}(\omega) = \lambda_{22}(\omega) = \frac{p+1}{2}$ and $\bar{r}(\lambda_{21} : \kappa_2) = \bar{r}(\lambda_{22} : \kappa_2) = \frac{p-1}{2}$.

By (1) we see that at an extension of $\kappa_0 : z=0$ to $k(z,\omega)$, either the value of $\frac{\omega}{z}$ is positive or the values of ω and $\frac{z}{\omega}$ are positive. In the first case, upon putting $\omega = z\tilde{\omega}$ in (1) and dividing throughout by z^{p-2} we get

$$z^2\tilde{\omega}^p + (z+1)z^2\tilde{\omega}^{p-1} - (z+1)(z+2)^p\tilde{\omega} - (z+2)^{p+1}z = 0$$

and hence $\epsilon\tilde{\omega} = \delta z$ where $\epsilon = (z+1)(z+2)^p$ and $\delta = z\tilde{\omega}^p + (z+1)z\tilde{\omega}^{p-1} - (z+2)^{p+1}$ are units in the relevant ring; this gives an extension λ_0 of κ_0 to $k(z,\omega)$ with $\lambda_0(z)=1$ and $\lambda_0(\omega)=2$ and $\bar{r}(\lambda_0 : \kappa_0)=1$. In the second case, upon putting $z=\tilde{z}\omega$ in (1) and dividing throughout by ω^{p-2} we get

$$\omega^2 + (z+1)\tilde{z}\omega^2 - (z+1)(z+2)^p\tilde{z}^{p-3} - (z+2)^{p+1}\tilde{z}^{p-1}\omega = 0$$

and hence $\tilde{\epsilon}\omega^2 = \tilde{\delta}\tilde{z}^{p-3}$ where $\tilde{\epsilon} = 1+(z+1)\tilde{z}$ and $\tilde{\delta} = (z+1)(z+2)^p + (z+2)^{p+1}\tilde{z}^2\omega$ are units in the relevant ring, and therefore $\omega = \pm\tilde{\epsilon}'\tilde{z}^{\frac{p-3}{2}}$ where $\tilde{\epsilon}'$ is a unit in the completion of that ring with $\tilde{\epsilon}'^2 = \tilde{\delta}\tilde{\epsilon}^{-1}$; this gives extensions λ_{01} and λ_{02} of κ_0 to $k(z,\omega)$ with $\lambda_{01}(\omega) = \lambda_{02}(\omega) = \frac{p-3}{2}$ and $\lambda_{01}(z) = \lambda_{02}(z) = \frac{p-1}{2}$ and $\bar{r}(\lambda_{01} : \kappa_0) = \bar{r}(\lambda_{02} : \kappa_0) = \frac{p-1}{2}$.

Thus the special ramification diagram gives rise to the following *extended ramification diagram*, with y^* as in ($3^\bullet$) of Section 3, where the square

---

[9]That is, units in the local ring of the relevant point of the relevant curve.

[10]Here, and in a few similar places, we assume k to be algebraically closed, but the final genus calculations etc. remain valid without this assumption.

bracketed numbers are the reduced ramification exponents:

$$
w_0^* : y^* = 0 \text{ —— }
\left|
\begin{array}{l}
\to \kappa_1 : z+1 = 0[1] \text{ —— } \left| \to \lambda_1 : \omega - 1 = 0[p] \right. \\[2ex]
\to \kappa_2 : z+2 = 0[p] \text{ —— } \left| \begin{array}{l} \to \lambda_2 : \omega + 2 = 0[1] \\ \to \lambda_{21} : \omega = 0[\frac{p-1}{2}] \\ \to \lambda_{22} : \omega = 0[\frac{p-1}{2}] \end{array} \right.
\end{array}
\right.
$$

$$
w_\infty^* : y^* = \infty \text{ —— }
\left|
\begin{array}{l}
\to \kappa_0 : z = 0[2] \text{ —— } \left| \begin{array}{l} \to \lambda_0(\omega) = 2[1] \\ \to \lambda_{01}(\omega) = \frac{p-3}{2}[\frac{p-1}{2}] \\ \to \lambda_{02}(\omega) = \frac{p-3}{2}[\frac{p-1}{2}] \end{array} \right. \\[2ex]
\to \kappa_\infty : z = \infty[p-1] \text{ —— } \left| \begin{array}{l} \text{all} \\ \text{extensions} \\ \text{unramified.} \end{array} \right.
\end{array}
\right.
$$

For giving a direct proof that κ_∞ is unramified in $k(z, \omega)$, first we note that

(3)
$$
\begin{aligned}
z^{-2p}\Phi(z^2W) = W^p + (\bar{z}+1)W^{p-1} - (\bar{z}+1)(2\bar{z}+1)^pW \\
-(2\bar{z}+1)^{p+1} \text{ where } \bar{z} = \tfrac{1}{z}
\end{aligned}
$$

and then by putting

(4)
$$z^* = 2\bar{z}+1 = \frac{z+2}{z} \quad \text{and} \quad \omega^* = \frac{\omega}{z^2} \quad \text{and} \quad \Phi^*(W) = z^{-2p}\Phi(z^2W)$$

we have

(5)
$$z^*+1 = \frac{2(z+1)}{z} \qquad \text{and} \qquad z^*-1 = \frac{2}{z}$$

and we get the *curve*

(6)
$$
\begin{aligned}
\Phi^*(\omega^*) = 0 \text{ with } \Phi^*(W) = W^p + \tfrac{1}{2}(z^*+1)W^{p-1} \\
-\tfrac{1}{2}(z^*+1)z^{*p}W - z^{*p+1}
\end{aligned}
$$

as a *birational transform* of the curve $\Phi = 0$. By (2) and (4) we have

$$
\begin{aligned}
\mathrm{Disc}_W(\Phi^*(W)) &= \frac{\mathrm{Disc}_W(\Phi(W))}{z^{2p(p-1)}} \\
&= -\left(\frac{z+2}{z}\right)^{(p+1)(p-2)} \left(\frac{2(z+1)}{z}\right)^{p+1} \left(\frac{2}{z}\right)^{p+1}
\end{aligned}
$$

and hence by (4) and (5) we get

(7) $$\mathrm{Disc}_W(\Phi^*(W)) = -z^{*(p+1)(p-2)}(z^*+1)^{p+1}(z^*-1)^{p+1}.$$

By (4) and (5) we see that $k(z^*,\omega^*) = k(z,\omega)$ and, as valuations of $k(z^*)/k$, the valuations $\kappa_1, \kappa_2, \kappa_0, \kappa_\infty$ are given by

(8) $$\kappa_1 : z^*+1=0, \quad \kappa_2 : z^*=0, \quad \kappa_0 : z^*=\infty, \quad \kappa_\infty : z^*-1=0.$$

Now $\kappa_\infty(\bar{z}) = 1$, and by (3) we have

$$\begin{aligned}\Phi^*(W) &= (W+1)(W^{p-1}-1) + [(W^{p-1}-1)-(W+1)]\bar{z}\\ &\quad -2(W+1)\bar{z}^p - 2(W+2)\bar{z}^{p+1}\end{aligned}$$

and clearly

$$(W^{p-1}-1) = (W+1)(W+2)\dots(W+p-1)$$

and hence we get an extension $\lambda_{\infty i}$ of κ_∞ to $k(z^*,\omega^*)$ with $\lambda_{\infty i}(\omega^*+i) = \bar{r}(\lambda_{\infty i} : \kappa_\infty) = 1$ for $2 \le i \le p-1$, and: ($1^\diamond$) *the center of* $\lambda_{\infty i}$ *on the curve* $\Phi^* = 0$ *is at the simple point* $(z^*,\omega^*) = (1,-i)$ *for* $2 \le i \le p-1$.

For the remaining extensions of κ_∞ to $k(z^*,\omega^*)$, the value of w^*+1 is positive, and obviously

$$W^{p-1} - 1 = W' + W'^2 + \dots + W'^{p-1} \quad \text{where} \quad W' = W+1$$

and hence

$$\begin{aligned}\Phi^*(W) &= W'^2 + W'^3 + \dots + W'^p + (W'^2 + W'^3 + \dots + W'^{p-1})\bar{z}\\ &\quad -2W'\bar{z}^p - 2(1+W')\bar{z}^{p+1}\end{aligned}$$

and therefore upon letting $\omega' = \omega^*+1$ and $z' = 2\bar{z} = z^*-1$ we have

(9) $$\epsilon'\omega'^2 = 2^{1-p}\omega' z'^p + 2^{-p}(1+\omega')z'^{p+1}$$

where

$$\begin{aligned}\epsilon' &= 1 + \omega' + \dots + \omega'^{p-2} + 2^{-1}z'(1+\omega'+\dots+\omega'^{p-3})\\ &= \text{a unit in the relevant ring}\end{aligned}$$

and hence $\bar{\omega}' = \frac{\omega'}{z'}$ is an element of positive value and we have

(10) $$\epsilon'\bar{\omega}'^2 = \delta' z'^{p-1}$$

where

$$\delta' = 2^{1-p} + 2\bar{\omega}' + 2^{-p}\bar{\omega}'z' = \text{ a unit in the relevant ring.}$$

Now (10) represents a *tacnode of index* $\frac{p-1}{2}$ *with horizontal tangent*, and hence remembering the equations

$$z' = z^*-1 \quad \text{and} \quad \omega' = \omega^*+1 \quad \text{and} \quad \bar{\omega}' = \frac{\omega'}{z'}$$

we see that the valuation κ_∞ has extensions λ_∞ and λ'_∞ to $k(z^*,\omega^*)$ such that $\lambda_\infty(\omega^*+1) = \lambda'_\infty(\omega^*+1) = \frac{p+1}{2}$ and $\bar{r}(\lambda_\infty : \kappa_\infty) = \bar{r}(\lambda'_\infty : \kappa_\infty) = 1$, and: ($2^\diamond$) *the common center of* λ_∞ *and* λ'_∞ *on the curve* $\Phi^* = 0$ *is at the double point* $(z^*,\omega^*) = (1,-1)$, *where the curve has a higher tacnode of index* $\frac{p+1}{2}$ *with horizontal tangent,* [11] *and hence in particular it has a double point in* $\frac{p-1}{2}$ *successive neighbourhoods of* $(1,-1)$ *which thus accounts for* $\frac{p+1}{2}$ *double points in the genus formula.*

To discuss the valuation λ_1, by (6) we get

$$\omega_1^p = \delta_1 z_1 \quad \text{where} \quad z_1 = z^*+1 \quad \text{and} \quad \omega_1 = \omega^* - 1$$

and where

$$\begin{aligned} \delta_1 &= 2 + \frac{1}{2}(\omega_1^2 - \omega_1^3 + \ldots + \omega_1^{p-1}) - \frac{1}{2} z_1^p(1+\omega_1) + z_1^{p-1} - z_1^p \\ &= \text{a unit in the relevant ring} \end{aligned}$$

and hence $\lambda_1(\omega^*-1) = 1$ and $\bar{r}(\lambda_1 : \kappa_1) = p$, and: ($3^\diamond$) *the center of* λ_1 *on the curve* $\Phi^* = 0$ *is at the simple point* $(z^*,\omega^*) = (-1,1)$ *with a vertical tangent whose intersection multiplicity with the curve is* p.

To revisit the extensions of κ_2, by (6) we have

$$(\omega^* + \frac{1}{2})\omega^{*p-1} + z^*(\frac{1}{2}\omega^{*p-1} - \frac{1}{2}z^{*p-1}\omega^* - \frac{1}{2}z^{*p}\omega^* - z^{*p}) = 0$$

and hence we get an extension λ_2 of κ_2 to $k(z^*,\omega^*)$ with $\lambda_2(\omega^*+\frac{1}{2}) = \bar{r}(\lambda_2 : \kappa_2) = 1$, and: ($4^\diamond$) *the center of* λ_2 *on the curve* $\Phi^* = 0$ *is at the simple point* $(z^*,\omega^*) = (0,-\frac{1}{2})$.

For the remaining extensions of κ_2 to $k(z^*,\omega^*)$, the value of ω^* is positive and

$$\epsilon_2 \omega^{*p-1} = z^{*p+1} + \frac{1}{2}z^{*p}\omega^* + \frac{1}{2}z^{*p+1}\omega^*$$

where

$$\epsilon_2 = \frac{1}{2} + \omega^* + \frac{1}{2}z^* = \text{ a unit in the relevant ring}$$

and hence

$$\epsilon_2 \bar{\omega}_2^{p-1} = \delta_2 z^{*2} \quad \text{where } \bar{\omega}_2 = \frac{\omega^*}{z^*}$$

and

$$\delta_2 = 1 + \frac{1}{2}\bar{\omega}_2 + \frac{1}{2}z^*\bar{\omega}_2 = \text{ a unit in the relevant ring.}$$

[11] That is, near the origin in the (z',ω')–plane, the curve (9) looks like the union of the two *flat* parabolas $\omega' = z'^d$ and $\omega' = -z'^d$ where $d = \frac{p+1}{2}$. At least this is so if d is even. For odd d you may want to think of it as a union of two "flat flexes". Note that a higher tacnode of index 4 (resp: 2) is called a *tacnode* (resp: *node*).

Therefore the valuation κ_2 has extensions λ_{21} and λ_{22} to $k(z^*,\omega^*)$ such that $\lambda_{21}(\omega^*) = \lambda_{22}(\omega^*) = \frac{p+1}{2}$ and $\bar{r}(\lambda_{21} : \kappa_2) = \bar{r}(\lambda_{22} : \kappa_2) = \frac{p-1}{2}$, and: ($5^\diamond$) *the common center of* λ_{21} *and* λ_{22} *on the curve* $\Phi^* = 0$ *is at the* $(p-1)$*-fold point* $(z^*,\omega^*) = (0,0)$, *where the curve has a higher tac-cusp of index* $\frac{p+1}{2}$ *with horizontal tangent,* [12] *and hence in particular it has a double point in* $\frac{p-1}{2}$ *successive neighbourhoods of the origin which thus accounts for* $\frac{(p-1)(p-2)}{2} + \frac{p-1}{2} = \frac{(p-1)^2}{2}$ *double points in the genus formula.*

Finally, to revisit the extensions of κ_0 to $k(z^*,\omega^*)$, their centers on the curve $\Phi^* = 0$ are precisely the points at infinity of this curve. By looking at the highest degree terms of Φ^*, we find these to be the point $(z^*, w^*, 1) = (1,0,0)$, i.e., the point at infinity along the z^*-axis in the (z^*,ω^*)-plane, and the point $(z^*, w^*, 1) = (0,1,0)$, i.e., the point at infinity along the ω^*-axis in the (z^*,ω^*)-plane. By introducing a homogenizing variable in Φ^*, and then on the one hand by putting $z^* = 1$ we see that $(1,0,0)$ is a simple point of the curve $\Phi^* = 0$, and on the other hand by putting $\omega^* = 1$ we see that $(0,1,0)$ is a double point of the curve $\Phi^* = 0$ where it has a higher tacnode of index $\frac{p+1}{2}$.

In greater detail, by (6) we have

$$\omega_0^p + \frac{1}{2}(z_0+1)z_0\omega_0^{p-1} - \frac{1}{2}(z_0+1)z_0^{p-3}\omega_0 - z_0^{p-1} = 0$$

where

$$z_0 = \frac{1}{z^*} \quad \text{and} \quad \omega_0 = \frac{\omega^*}{z^{*2}}$$

and hence at an extension of κ_0 to $k(z^*,\omega^*)$, either the value of $\frac{\omega_0}{z_0}$ is positive or the values of ω_0 and $\frac{z_0}{\omega_0}$ are positive. In the first case, by putting $\omega_0 = z_0\bar{\omega}_0$ we get $\bar{\omega}_0 = \delta_0 z_0$ where $\delta_0 = -2 - \bar{\omega}_0 + z_0\bar{\omega}_0^{p-1} + z_0^2\bar{\omega}_0^{p-1} + 2z_0\bar{\omega}_0^p$ = a unit in the relevant ring; this gives an extension λ_0 of κ_0 to $k(z^*,\omega^*)$ with $\lambda_0(\omega^*+2) = \lambda_0(\bar{\omega}_0) = \lambda_0(z_0) = 1$ and $\bar{r}(\lambda_0 : \kappa_0) = 1$, and: ($6^\diamond$) *the center of* λ_0 *on the curve* $\Phi^* = 0$ *is at the simple point* $(z^*,\omega^*,1) = (1,0,0)$.

In the second case, by putting $z_0 = \omega_0\bar{z}_0$ we have $\bar{\epsilon}_0\omega_0^2 = \bar{\delta}_0\bar{z}_0^{p-3}$ where $\bar{\epsilon}_0 = 2 + \bar{z}_0^2\omega_0 + \bar{z}_0$ and $\bar{\delta}_0 = 1 + \bar{z}_0\omega_0 + 2\bar{z}_0^2\omega_0$ are units in the relevant ring, and hence we get extensions λ_{01} and λ_{02} of κ_0 to $k(z^*,\omega^*)$ with $\lambda_{01}(\omega^*) = \lambda_{02}(w^*) = -\frac{p+1}{2}$ and $\bar{r}(\lambda_{01} : \kappa_0) = \bar{r}(\lambda_{02} : \kappa_0) = \frac{p-1}{2}$, and ($7^\diamond$) *the common center of* λ_{01} *and* λ_{02} *on the curve* $\Phi^* = 0$ *is at the double point* $(z^*,\omega^*,1) = (0,1,0)$, *where the curve has a higher tacnode of index* $\frac{p+1}{2}$, *and hence in particular it has a double point in* $\frac{p-1}{2}$ *successive*

---

[12]That is, near the origin in the (z^*,ω^*)-plane, the curve $\Phi^* = 0$ looks like the union of the two higher cusps $w^{*d-1} = z^{*d}$ and $w^{*d-1} = -z^{*d}$ where $d = \frac{p+1}{2}$. Note that, after one quadratic transformation, a higher tac-cusp of index d is transformed into a higher tacnode of index $d-1$. Also note that a *higher cusp* of index d, i.e., a singularity of the type $w^{*d-1} = z^{*d}$, is a $(d-1)$-fold point which is resolved by one quadratic transformation. Finally note that a *cusp* is a higher cusp of index 3.

neighbourhoods of $(0,1,0)$ *which thus account for* $\frac{p+1}{2}$ *double points in the genus formula.*

Thus we have the following *final ramification diagram* where the square bracketed numbers again denote ramification exponents:

$w_0^* : y^* = 0$ —
- $\to \kappa_1 : z^* + 1 = 0[1]$ —
  - $\to \lambda_1(\omega^* - 1) = 1[p]$
- $\to \kappa_2 : z^* = 0[p]$ —
  - $\to \lambda_2(\omega^* + \frac{1}{2}) = 1[1]$
  - $\to \lambda_{21}(\omega^*) = \frac{p+1}{2}[\frac{p-1}{2}]$
  - $\to \lambda_{22}(\omega^*) = \frac{p+1}{2}[\frac{p-1}{2}]$

$w_\infty^* : y^* = \infty$ —
- $\to \kappa_0 : z^* = \infty[2]$ —
  - $\to \lambda_0(\omega^* + 2) = 1[1]$
  - $\to \lambda_{01}(\omega^*) = -\frac{p+1}{2}[\frac{p-1}{2}]$
  - $\to \lambda_{02}(\omega^*) = -\frac{p+1}{2}[\frac{p-1}{2}]$
- $\to \kappa_\infty : z^* - 1 = 0[p-1]$ —
  - $\to \lambda_\infty(\omega^* + 1) = \frac{p+1}{2}[1]$
  - $\to \lambda'_\infty(\omega^* + 1) = \frac{p+1}{2}[1]$
  - $\to \lambda_{\infty i}(\omega^* + i) = 1[1]$ for $2 \le i \le p-1$.

At any rate, $\kappa_1, \kappa_2, \kappa_0$ are the only valuations of $k(z^*)/k$ which are ramified in $k(z^*, \omega^*)$, and in view of the LCM Theorem (Proposition 7 of [A1]) we see that $\lambda_1, \lambda_2, \lambda_0$ are the only valuations of $k(z^*, \omega^*)/k$ which are ramified in a splitting field of $\Psi(U)$ over $k(z^*, \omega^*)$, and for any extension of any one of these three valuations to the said splitting field, the reduced ramification exponent is $\frac{p-1}{2}$. Note that a splitting field of $\Psi(U)$ over $k(z^*, \omega^*)$ is the same as a splitting field of $\Psi^*(U)$ over $\kappa(z^*, \omega^*)$ where, upon letting $\Psi^*(U) = z^{-2(p-1)}\Psi(z^2U)$, in view of (4) and (5), by $(4^\bullet)$ of Section 3 we have

$$\Psi^*(U) = U^{p-1} + \tfrac{1}{2}(z^*+1)[U^{p-2} - \omega^* U^{p-3} + \omega^{*2}U^{p-4} - \ldots + \omega^{*p-3}U] - \tfrac{1}{2}(z^*+1)[\omega^{*p-2} + z^{*p}].$$

The three singularities $(z^*, \omega^*, 1) = (1, -1, 1)$, $(0,0,1)$, $(0,1,0)$ of the curve $\Phi^* = 0$ of degree $p+2$ account for $\frac{p+1}{2} + \frac{(p-1)^2}{2} + \frac{p+1}{2} = \frac{p^2+3}{2}$ double points and hence its genus is $\frac{(p+1)p}{2} - \frac{p^2+3}{2}$, and thus

$$\text{(11)} \qquad (\text{genus of } \Phi^* = 0) = \frac{p-3}{2}$$

and hence in particular: ($8^\diamond$) if $p = 7$ then (genus of $\Phi^* = 0$) = 2.

If we project a curve of degree n from an m-fold point then its degree in one of the variables becomes $n - m$. So, to simplify matters, let us project the curve $\Phi^* = 0$ of degree $p + 2$ from the $(p - 1)$-fold point $(z^*, \omega^*, 1) = (0, 0, 1)$. In other words, let us arrange things so that this point becomes the point $(\xi, \eta, 1) = (0, 1, 0)$ by putting

$$(12) \qquad \xi = \frac{z^*}{\omega^*} \quad \text{and} \quad \eta = \frac{1}{\omega^*}$$

and

$$\begin{cases} \phi^*(X, Y) = 2Y^{p+2} \text{ times } \Phi^*(W) \\ \text{with } \left(\frac{X}{Y}, \frac{1}{Y}\right) \text{ replacing } (z^*, W) \end{cases}$$

to get

$$(13) \quad \phi^*(\xi, \eta) = 0 \text{ with } \phi^*(X, Y) = Y^3 + (X+2)Y^2 - (2X+1)X^pY - X^{p+1}.$$

Now $k(\xi, \eta) = k(z^*, \omega^*)$, and the three singularities of the curve $\phi^* = 0$ are: a double point at $(\xi, \eta, 1) = (-1, -1, 1)$ where the curve has a higher tacnode of index $\frac{p+1}{2}$; a $(p - 1)$-fold point at $(\xi, \eta, 1) = (0, 1, 0)$ where the curve has a higher tac-cusp of index $\frac{p+1}{2}$; and a double point at $(\xi, \eta, 1) = (0, 0, 1)$ where the curve has a higher tacnode of index $\frac{p+1}{2}$. At the valuations $\lambda_1, \lambda_2, \lambda_0$ of $k(\xi, \eta)/k$ we have: $\lambda_1(\xi + 1) = \lambda_1(\eta - 1) = 1$, and $\lambda_2(\xi) = \lambda_2(\eta + 2) = 1$, and $\lambda_0(\xi) = -1$ and $\lambda_0(2\eta + 1) = 1$. Also, these are the only three valuations of $k(\xi, \eta)/k$ which are ramified in a splitting field of $\Psi^*(U)$ over $k(\xi, \eta)$. Moreover, for any extension of any one of these three valuations to the said splitting field, the reduced ramification exponent is $\frac{p-1}{2}$. Finally note that a splitting field of $\Psi^*(U)$ over $k(\xi, \eta)$ is the same as a splitting field of $\psi^*(\xi, \eta, Z)$ over $k(\xi, \eta)$ where, upon letting

$$\begin{cases} \psi^*(X, Y, Z) = Y^{2(p-1)} \text{ times } \Psi^*(U) \\ \text{with } \left(\frac{X}{Y}, \frac{1}{Y}, \frac{Z}{Y^2}\right) \text{ replacing } (z^*, \omega^*, U) \end{cases}$$

we have

$$(14) \quad \begin{aligned} &\psi^*(X, Y, Z) \\ &= Z^{p-1} + \tfrac{1}{2}(X + Y)Y[Z^{p-2} - YZ^{p-3} + Y^2Z^{p-4} - \ldots + Y^{p-3}Z] \\ &\quad -\tfrac{1}{2}(X + Y)Y^{p-3}[Y^2 + X^p]. \end{aligned}$$

We know that if $p = 7$ then the genus of $\phi^* = 0$ is 2 and using this fact we want to factor $\psi^*(\xi, \eta, Z)$ in $k(\xi, \eta)[Z]$ into two monic factors of degree 3 in Z.

2.5 Parametrizing the Auxiliary Curve in Characteristic Seven

Henceforth assuming the characteristic of the ground field k to be 7, we have the curve

(1′) $\phi^*(\xi,\eta)=0$ with $\phi^*(X,Y)=Y^3+(X+2)Y^2-(2X+1)X^7Y-X^8$

of genus 2 with the three singularities: a double point at $(\xi,\eta,1)$ $(-1,-1,1)$ where the curve has a higher tacnode of index 4; a 6-fold point at $(\xi,\eta,1)=(0,1,0)$ where the curve has a higher tac-cusp of index 4; and a double point at $(\xi,\eta,1)=(0,0,1)$ where the curve has a higher tacnode of index 4. We are considering the polynomial

$$(2') \qquad \begin{aligned}&\psi^*(X,Y,Z)\\ &=Z^6-3(X+Y)Y[Z^5-YZ^4+Y^2Z^3-Y^3Z^2+Y^4Z]\\ &\qquad +3(X+Y)Y^4(Y^2+X^7).\end{aligned}$$

We want to factor $\psi^*(\xi,\eta,Z)$ in $k(\xi,\eta)[Z]$ into two monic factors of degree 3 in Z. To facilitate the said factorization, in this Section we shall obtain a square-root parametrization for the curve $\phi^*=0$.

To simplify matters we make the birational transformation

(3′) $\xi=\hat{\xi}^{-1}$ and $\eta=\hat{\xi}^{-4}\hat{\eta}$ and $\hat{\phi}(X,Y)=X^{12}\phi^*(X^{-1},X^{-4}Y)$

to get

(4′) $\hat{\phi}(\hat{\xi},\hat{\eta})=0$ with $\hat{\phi}(X,Y)=Y^3+(2X+1)X^3Y^2-(X+2)Y-X^4.$

Since the curve (1′) has no singularities outside $\xi=0,\infty$, except a double point at $(\xi,\eta,1)=(-1,-1,1)$ where it has a higher tacnode of index 4, by (3′) it follows that the curve (4′) has no singularities outside $\hat{\xi}=0,\infty$, except a double point at $(\hat{\xi},\hat{\eta},1)=(-1,-1,1)$ where it has a higher tacnode of index 4. By putting $X=0$ in $\hat{\phi}(X,Y)$ we get $Y(Y^2-2)$ which has no multiple factors and hence the curve (4′) has no singularities with $\hat{\xi}=0$.

To simplify matters further, we make the translation

(5′) $\hat{\xi}=x-1$ and $\hat{\eta}=y-1$ and $\phi'(X,Y)=\hat{\phi}(X-1,Y-1)$

to get

$$(6') \qquad \begin{aligned}\phi'(x,y)=0 \text{ with } \phi'(X,Y)=Y^3&+(2X^4+2X^2+2X+5)Y^2\\ &+(3X^4+3X^2+2X)Y\\ &+(X^4+4X^3+3X^2)\end{aligned}$$

which is a sextic (= curve of degree 6) having no singularities outside $x=\infty$, except a double point at $(x,y,1)=(0,0,1)$ where it has a higher tacnode of index 4. To examine the points of this sextic with $x=\infty$, by looking at the highest degree terms of ϕ', we find these to be the triple point $(x,y,1)=(0,1,0)$ where the curve has a higher cusp of index 4, and the double point $(x,y,1)=(1,0,0)$ where the curve has a node. So the three singularities $(x, y, 1) = (0, 0, 1)$, $(0, 1, 0)$, $(1, 0, 0)$ account for $4 + 3 + 1 = 8$ double points of the sextic, thus verifying that its genus is $\frac{(6-1)(6-2)}{2} - 8 = 2$.

To square-root parametrize this sextic, we consider its cubic adjoints. Such an adjoint must pass through the double points $(0, 0, 1)$ and $(1, 0, 0)$ of the sextic, and it must have at least a double point at the triple point $(0, 1, 0)$ of the sextic, and hence the equation of such a cubic must be of the form

$$(aX^2 + bX + c)Y - AX^2 - BX = 0 \tag{7'}$$

where a, b, c, A, B are constants to be determined. Since the sextic has a higher tacnode of index 4 at (0,0,1), the intersection multiplicity of the above cubic adjoint with the sextic at this point must be at least 8. Since the sextic is monic in Y, the said intersection multiplicity is less equal the X-order of the Y-resultant $r(X)$ of the cubic and the sextic, and hence

$$\mathrm{ord}_X r(X) \geq 8. \tag{8'}$$

By solving the cubic we get $Y = \frac{AX^2+BX}{aX^2+bX+c}$, and now $r(X)$ is found by substituting this root in the sextic and then multiplying throughout by the cube of the denominator $aX^2 + bX + c$. Therefore

$$\begin{aligned} r(X) = {} & (AX^2 + BX)^3 \\ & +(2X^4 + 2X^2 + 2X + 5)(AX^2 + BX)^2(aX^2 + bX + c) \\ & +(3X^4 + 3X^2 + 2X)(AX^2 + BX)(aX^2 + bX + c)^2 \\ & +(X^4 + 4X^3 + 3X^2)(aX^2 + bX + c)^3 \\ = {} & 3c(2B - c)^2X^2 + \text{ terms of degree } > 2 \text{ in } X \end{aligned} \tag{9'}$$

and hence by $(8')$ we see that either $c = 0$ or $c = 2B$. If $c = 0$ then the cubic consists of the line $X = 0$ and the conic $(aX + b)Y - AX - B = 0$; since the intersection multiplicity of the line $X = 0$ with the sextic at $(0, 0, 1)$ is clearly 2 and since the intersection multiplicity of the cubic with the sextic at $(0, 0, 1)$ is at least 8, the conic must pass through $(0, 0, 1)$, i.e., we must have $B = 0$. Thus always

$$c = 2B. \tag{10'}$$

Substituting $(10')$ in $(9')$ we get

$$\begin{aligned} r(X) = {} & (AX^2 + BX)^3 \\ & +(2X^4 + 2X^2 + 2X + 5)(AX^2 + BX)^2(aX^2 + bX + 2B) \\ & +(3X^4 + 3X^2 + 2X)(AX^2 + BX)(aX^2 + bX + 2B)^2 \\ & +(X^4 + 4X^3 + 3X^2)(aX^2 + bX + 2B)^3 \end{aligned}$$

and by collecting coefficients of like powers of X we have

$$r(X) = \sum_{i=2}^{10} r_i X^i \tag{11'}$$

where

$$r_2 = 3B^3 + B^3 + 3B^3 = 0 \tag{12'}$$

and

$$r_3 = B^3 + (5bB^2 + 6AB^2 + 4B^3) + (bB^2 + AB^2 + 5B^3) + (bB^2 + 4B^3) = 0 \tag{13'}$$

and

$$r_4 = 6b^2B + 4bAB + 6bB^2 + 3A^2B + 2AB^2 + 5B^3 = 6B(b - 2A - 3B)^2 \tag{14'}$$

and

$$\begin{aligned} r_5 &= 3b^3 + 5abB + 2b^2A + 6b^2B \\ &\qquad +4aAB + 6aB^2 + 5bA^2 + 2bAB + AB^2 + 5B^3 \\ &= 3b(b - 2A - 3B)^2 + 3B(4a + b + B)(b - 2A - 3B) \end{aligned} \tag{15'}$$

and

$$\begin{aligned} r_6 &= 2ab^2 + 4b^3 + 6a^2B + 4abA + 5abB + 3b^2A + 6b^2B \\ &\qquad +5aA^2 + 2aAB + 2bA^2 + 4bAB + 5bB^2 \\ &\qquad +A^3 + 4A^2B + 5AB^2 + 4B^3 \\ &= (b - 2A - 3B)(4b^2 + 2ab + 4bA + 4bB + aA + 4aB \\ &\qquad\qquad +3A^2 + 3AB + 3B^2) \\ &\qquad +6B(a - 3A - 6B)^2 \end{aligned} \tag{16'}$$

and

$$\begin{aligned} r_7 &= 2a^2b + 5ab^2 + b^3 + 2a^2A + 6a^2B + 6abA + 5abB + 3b^2B \\ &\qquad +2aA^2 + 4aAB + 5aB^2 + 2bA^2 + 5bAB + 2bB^2 + AB^2 \\ &= (b - 2A - 3B)(b^2 + 5ab + 2bA + 6bB + 2a^2 + 2aA + 6aB) \\ &\qquad +(b - 2A - 3B)(6A^2 + 2AB + 6B^2) \\ &\qquad +(a - 3A - 6B)[6A(a - 3A) + 5aB + 3AB + 4B^2] \end{aligned} \tag{17'}$$

and

$$\begin{aligned} r_8 &= 3a^3 + 5a^2b + 3ab^2 + 3a^2A + 6a^2B + 6abB + 3b^2A \\ &\qquad +2aA^2 + 5aAB + 2aB^2 + 4bAB + 4A^2B \\ &= (b - 2A - 3B)(3ab + 3bA + 5a^2 + 6aA + aB + 6A^2 + 6AB) \\ &\qquad +(a - 3A - 6B)(3a^2 + aA + 4aB + 3A^2 + AB + B^2) \\ &\qquad +6A^2B + 6AB^2 + 6B^3 \end{aligned} \tag{18'}$$

and

$$\begin{aligned} r_9 &= 4a^3 + 3a^2b + 3a^2B + 6abA + 4aAB + 2bA^2 \\ &= (b - 2A - 3B)(3a^2 + 6aA + 2A^2) \\ &\qquad +(a - 3A - 6B)(4a^2 + 4aA + aB + 3A^2 + 6B^2) \\ &\qquad +6A^3 + 3A^2B + 4AB^2 + B^3 \end{aligned} \tag{19'}$$

and

$$(20')\qquad \begin{aligned} r_{10} &= a^3 + 3a^2A + 2aA^2 \\ &= (a - 3A - 6B)(a^2 + 6aA + 6aB + 6A^2 + 5AB + B^2) \\ &\quad +4A^3 + 2A^2B + 5AB^2 + 6B^3. \end{aligned}$$

By $(8')$ and $(14')$ we see that either $B = 0$ or $b = 2A + 3B$. If $B = 0$ then by $(8')$ and $(15')$ we see that either $b = 0$ or $b = 2A + 3B$. For a moment suppose that $B = 0$ and $b = 0$; then by $(10')$ we have $c = 0$, and hence the cubic consists of the line $X = 0$ counted twice together with the line $aY = A$; by substituting in $(6')$ we see that the total intersection multiplicity of these three with the sextic at (0,0,1) is 4 or 6 according as $A \neq 0$ or $A = 0$; this contradicts the observation that the intersection multiplicity of cubic adjoint with the sextic at (0,0,1) is at least 8. Thus we must always have

$$(21')\qquad b = 2A + 3B.$$

By $(8')$, $(16')$ and $(21')$ we see that either $B = 0$ or $a = 3A+6B$. By $(8')$, $(17')$ and $(21')$ we see that if $B = 0 \neq A$ then $a = 3A + 6B$. If $B = 0 = A$ then by $(21')$ we get $b = 0$ and, as we have seen in the above paragraph, this leads to a contradiction. Thus we always have

$$(22')\qquad a = 3A + 6B.$$

In view of $(10')$, $(21')$ and $(22')$, by $(7')$ we get

$$(23')\qquad [(3A + 6B)X^2 + (2A + 3B)X + 2B]Y - AX^2 - BX = 0$$

as the pencil of cubic adjoints, and by $(11')$ to $(20')$ we get

$$(24')\qquad \begin{aligned} r(X) = X^8[&(4A^3 + 2A^2B + 5AB^2 + 6B^3)X^2 \\ &+(6A^3 + 3A^2B + 4AB^2 + B^3)X \\ &+(6A^2B + 6AB^2 + 6B^3)] \end{aligned}$$

as its Y-resultant with the sextic. Upon discarding the factor X^8 which gives the intersection at (0,0,1), and upon putting $A = 1$, $B = T$, $X = x$, and $Y = y$, by $(23')$ and $(24')$ we get

$$(25')\qquad \begin{aligned} &g'(T, x, y) = 0 \\ &\text{with } g'(T, X, Y) = [(6T + 3)X^2 + (3T + 2)X + 2T]Y \\ &\qquad\qquad\qquad -X^2 - TX \end{aligned}$$

and

$$(26')\qquad \begin{aligned} &f'(T, x) = 0 \\ &\text{with } f'(T, X) = (6T^3 + 5T^2 + 2T + 4)X^2 \\ &\qquad\qquad +(T^3 + 4T^2 + 3T + 6)X + (6T^3 + 6T^2 + 6T) \end{aligned}$$

as a square-root parametrization of the sextic. In other words, by solving (25′) we get

$$T = \frac{(4x^2 + 5x)y + x^2}{(6x^2 + 3x + 2)y + 6x} \in k(x, y) \tag{27′}$$

and x satisfies the quadratic equation (26′) over $k(T)$, and y satisfies the linear equation (25′) over $k(T, x)$. [13]

Now first in view of (5′), upon letting

$$\hat{g}(T, X, Y) = g'(T, X + 1, Y + 1) \quad \text{and} \quad \hat{f}(T, X) = f'(T, X + 1)$$

we get

$$\begin{aligned}\hat{g}(T, \hat{\xi}, \hat{\eta}) = 0 \text{ with } \hat{g}(T, X, Y) = {} & [(6X^2 + X + 4)T + (3X^2 + X + 5)]Y \\ & -[(X^2 + 4)T + (5X^2 + X + 3)]\end{aligned}$$

and

$$\begin{aligned}\hat{f}(T, \hat{\xi}) = 0 \text{ with } \hat{f}(T, X) = {} & (6X^2 + 6X + 6)T^3 \\ & +(5X^2 + 1)T^2 + (2X^2 + 4)T + (4X^2 + 3)\end{aligned}$$

as a square-root parametrization of (4′) and then in view of (3′), upon letting

$$g(T, X, Y) = 2X^6 \hat{g}(T, X^{-1}, X^{-4}Y) \quad \text{and} \quad f(T, X) = 6X^2 \hat{f}(T, X^{-1})$$

we get

$$\begin{aligned}& g(T, \xi, \eta) = 0 \\ & \text{with } g(T, X, Y) = [(X^2 + 2X + 5)T + (3X^2 + 2X + 6)]Y \\ & \qquad -[(X^6 + 2X^4)T + (6X^6 + 2X^5 + 3X^4)]\end{aligned} \tag{28′}$$

and

$$\begin{aligned}f(T, \xi) = 0 \text{ with } f(T, X) = {} & (T^3 + 6T^2 + 3T + 4)X^2 \\ & +T^3 X + (T^3 + 2T^2 + 5T + 3)\end{aligned} \tag{29′}$$

as a square-root parametrization of the nonic [14] (1′) where by solving (28′) we have

$$T = \frac{(4\xi^2 + 5\xi + 1)\eta + (6\xi^6 + 2\xi^5 + 3\xi^4)}{(\xi^2 + 2\xi + 5)\eta + (6\xi^6 + 5\xi^4)} \in k(\xi, \eta). \tag{30′}$$

---

[13] Having found the parametrization, we can directly verify its validity and forget about adjoints and such. In other words, starting with (25′) which defines T, substitute in (6′) the value of y obtained from (25′) and then clear the denominator and open the parenthesis and so on, as is high-school. This will simply reproduce the RHS of (24′) equated to zero with $(A, B, X) = (1, T, x)$, and now dividing out by x^8 we would get (26′).

[14] That is, a curve of degree nine.

To explicitly bring out the square-root aspect of this parametrization, first by letting

$$(31') \quad P = E\xi \text{ with } E = T^3 + 6T^2 + 3T + 4 = (T+2)(T+5)(T+6)$$

we get

$$(32') \qquad \begin{aligned} P^2 = 6T^3P + 6EL \text{ with } L &= T^3 + 2T^2 + 5T + 3 \\ &= (T+2)(T+3)(T+4) \end{aligned}$$

and then by letting

$$(33') \qquad \begin{aligned} S^* = P + 4T^3 \text{ and } D^* &= 2T^6 - EL \\ &= T^6 + 6T^5 + T^4 + 6T^3 + T^2 + 6T + 2 \end{aligned}$$

we get

$$(34') \quad S^{*2} = D^* \text{ with } D^* = 2 + T(T+2)(T+3)(T+4)(T+5)(T+6).$$

Note that now

$$(35') \qquad k(\xi,\eta) = k(T,\xi) = k(T,P) = k(T,S^*)$$

and by (28′) we have

$$(36') \quad [(\xi^2+2\xi+5)T + (3\xi^2+2\xi+6)]\eta = (\xi^6+2\xi^4)T + (6\xi^6+2\xi^5+3\xi^4).$$

To express η in terms of S^* and T, in view of (31′), multiplying both sides of (36′) by E^6 we get

$$\begin{aligned} &E^4[(P^2+2EP+5E^2)T + (3P^2+2EP+6E^2)]\eta \\ &= P^4[(T+6)P^2 + 2EP + 2(T+5)E^2] \end{aligned}$$

and hence by repeatedly substituting the value of P^2 given in (32′) we get

$$E^4(\mu P + \nu)\eta = \mu' P + \nu'$$

where

$$\mu = 2(T+1)E + 6T^3(T+3) \quad \text{and} \quad \nu = 5(T+4)E^2 + 6(T+3)EL$$

and

$$\begin{aligned} \mu' = {} & 2E^3L^2 + 4T^3(T+5)E^3L + 4T^3(T+6)E^2L^2 + T^6E^2L \\ & +5T^9(T+5)E^2 + 4T^9(T+6)EL + 2T^{12}E + 6T^{15}(T+6) \end{aligned}$$

and

$$\begin{aligned} \nu' = {} & 2(T+5)E^4L^2 + 6(T+6)E^3L^3 + 3T^3E^3L^2 \\ & +5T^6(T+5)E^3L + 3T^6(T+6)E^2L^2 + 2T^9E^2L + 6T^{12}(T+6)EL \end{aligned}$$

By solving the above equation for η we get

$$(37') \qquad \eta = \frac{\mu' P + \nu'}{E^4(\mu P + \nu)}.$$

To clear the denominator, we want to find the inverse of $\mu P + \nu$. By (32′) we have the equation $P^2 + T^3 P + EL = 0$. Dividing the corresponding polynomial $Z^2 + T^3 Z + EL$ in indeterminate Z by the polynomial $\mu Z + \nu$ we have

$$\mu^2(Z^2 + T^3 Z + EL) = (\mu Z + \nu)(\mu Z + T^3\mu + 6\nu) - (6EL\mu^2 + T^3\mu\nu + 6\nu^2)$$

and hence substituting P for Z we get

$$(\mu P + \nu)(\mu P + T^3\mu + 6\nu) = (6EL\mu^2 + T^3\mu\nu + 6\nu^2).$$

Thus, upto a multiplicative constant, $\mu P + T^3\mu + 6\nu$ is an inverse of $\mu P + \nu$ and upon multiplying the numerator and denominator of the RHS of (37′) by it we get

$$\eta = \frac{(\mu P + T^3\mu + 6\nu)(\mu' P + \nu')}{E^4(6\mu^2 EL + T^3\mu\nu + 6\nu^2)}$$

and hence by (32′) we have

$$\eta = \frac{(\mu\nu' + 6\nu\mu')P + (6EL\mu\mu' + T^3\mu\nu' + 6\nu\nu')}{E^4(6\mu^2 EL + T^3\mu\nu + 6\nu^2)}$$

and therefore by (33′) we get

$$(38') \qquad \eta = \frac{(\mu\nu' + 6\nu\mu')S^* + [3T^3(\mu\nu' + 6\nu\mu') + (6EL\mu\mu' + T^3\mu\nu' + 6\nu\nu')]}{E^4(6\mu^2 EL + T^3\mu\nu + 6\nu^2)}.$$

By direct calculation with polynomials in T we get

$$
\begin{aligned}
&E^4(6\mu^2EL + T^3\mu\nu + 6\nu^2)\\
&= E^4[\{3(T+1)^2E^3L + 4T^3(T+1)(T+3)E^2L + 6T^6(T+3)^2EL\}\\
&\qquad +\{3T^3(T+1)(T+4)E^3 + 5T^3(T+1)(T+3)E^2L\\
&\qquad\qquad +2T^6(T+3)(T+4)E^2 + T^6(T+3)^2EL\}\\
&\qquad +\{3(T+4)^2E^4 + 3(T+3)(T+4)E^3L + 6(T+3)^2E^2L^2\}]\\
&= E^6[3(T+4)^2E^2 + 6(T^2+T+3)EL + 3T^3(T+1)(T+4)E\\
&\qquad +6(T+3)^2L^2 + 2T^3(T+1)(T+3)L + 2T^6(T+3)(T+4)]\\
&= E^6[3(T+2)^2(T+4)^2(T+5)^2(T+6)^2\\
&\qquad +6(T+2)^2(T+3)(T+4)(T+5)(T+6)(T^2+T+3)\\
&\qquad +3T^3(T+1)(T+2)(T+4)(T+5)(T+6)\\
&\qquad +6(T+2)^2(T+3)^4(T+4)^2\\
&\qquad +2T^3(T+1)(T+2)(T+3)^2(T+4) + 2T^6(T+3)(T+4)]\\
&= E^6[(T+2)^2(T+4)(T^2+4T+2)\times\\
&\qquad\qquad \{3(T+4)(T^2+4T+2) + 6(T+3)(T^2+T+3)\}\\
&\qquad +6(T+2)^2(T+3)^4(T+4)^2\\
&\qquad +T^3(T+1)(T+2)(T+4)\{3(T^2+4T+2) + 2(T^2+6T+2)\}\\
&\qquad +2T^6(T+3)(T+4)]\\
&= E^6[(T+2)^2(T+4)\times\\
&\qquad\qquad \{(T^2+4T+2)(2T^3+6T^2+6T+1) + 6(T+3)^4(T+4)\}\\
&\qquad +T^3(T+4)\{(T^2+3T+2)(5T^2+3T+3) + 2T^3(T+3)\}]\\
&= E^6[(T+2)^2(T+4)(T^5+5T^4+2T^3)\\
&\qquad +T^3(T+4)(3T^3+T^2+T+6)]\\
&= T^3(T+4)E^6[(T^2+4T+4)(T^2+5T+2) + (3T^3+T^2+T+6)]\\
&= T^3(T+4)E^6[(T^4+2T^3+5T^2+1) + (3T^3+T^2+T+6)]\\
&= T^3(T+4)E^6(T^4+5T^3+6T^2+T)\\
&= T^4(T+4)^4E^6
\end{aligned}
$$

and therefore

$$E^4(6\mu^2EL + T^3\mu\nu + 6\nu^2) = T^4(T+4)^4E^6. \tag{39'}$$

Likewise

$$
\begin{aligned}
&\mu\nu' + 6\nu\mu' \\
&= [4(T+1)(T+5) + 4(T+4)]E^5L^2 + T^3(T+4)(T+5)E^5L \\
&\quad +[5(T+1)(T+6) + 2(T+3)]E^4L^3 \\
&\quad +[6T^3(T+1) + 5T^3(T+3)(T+5) + T^3(T+4)(T+6) \\
&\qquad\quad +4T^3(T+3)(T+5)]E^4L^2 \\
&\quad +[3T^6(T+1)(T+5) + 2T^6(T+4)]E^4L + 3T^9(T+4)(T+5)E^4 \\
&\quad +[T^3(T+3)(T+6) + 4T^3(T+3)(T+6)]E^3L^3 \\
&\quad +[6T^6(T+1)(T+6) + 4T^6(T+3) + T^6(T+3)]E^3L^2 \\
&\quad +[4T^9(T+1) + 2T^9(T+3)(T+5) + T^9(T+4)(T+6) \\
&\qquad\quad +5T^9(T+3)(T+5)]E^3L \\
&\quad +4T^{12}(T+4)E^3 \\
&\quad +[4T^9(T+3)(T+6) + 4T^9(T+3)(T+6)]E^2L^2 \\
&\quad +[5T^{12}(T+1)(T+6) + 5T^{12}(T+3) + 2T^{12}(T+3)]E^2L \\
&\quad +5T^{15}(T+4)(T+6)E^2 \\
&\quad +[T^{15}(T+3)(T+6) + 6T^{15}(T+3)(T+6)]EL \\
&= (4T^2+1)E^5L^2 + T^3(T+4)(T+5)E^5L + (5T^2+2T+1)E^4L^3 \\
&\quad +T^3(3T^2+4T+4)E^4L^2 + T^6(3T^2+6T+2)E^4L \\
&\quad +3T^9(T+4)(T+5)E^4 \\
&\quad +5T^3(T+3)(T+6)E^3L^3 + T^6(6T^2+5T+2)E^3L^2 \\
&\quad +T^{11}E^3L + 4T^{12}(T+4)E^3 \\
&\quad +T^9(T+3)(T+6)E^2L^2 + 5T^{12}(T+1)(T+6)E^2L \\
&\quad +5T^{15}(T+4)(T+6)E^2 \\
&= [(4T^2+1)E + (5T^2+2T+1)L]E^4L^2 \text{ (by regrouping)} \\
&\quad +T^3[(T+4)(T+5)E^2 + (3T^2+4T+4)EL \\
&\qquad\quad +5(T+3)(T+6)L^2]E^3L \\
&\quad +T^6[(3T^2+6T+2)L + 3T^3(T+4)(T+5)]E^4 \\
&\quad +T^6[(6T^2+5T+2)E + T^3(T+3)(T+6)]E^2L^2 \\
&\quad +T^{11}[L + 4T(T+4)]E^3 + 5T^{12}(T+6)[(T+1)L + T^3(T+4)]E^2 \\
&= 2T^3(T+2)^2E^4L^2 \\
&\quad +T^3(T+2)^2(T+4)(T+6)(2T^4+2T^3+T^2+5T+6)E^3L \\
&\quad +T^6(T+4)^2(6T^3+5T^2+2T+3)E^4 \\
&\quad +T^6(T+6)^2(T^2+1)E^2L^2 \\
&\quad +T^{11}(T+4)^2(T+5)E^3 \\
&\quad +3T^{12}(T+4)^2(T+6)(T^2+6T+6)E^2 \\
&= 4T^5(T+2)^2(T+4)^2(T+6)^2E^3L \text{ (by adding consecutive pairs)} \\
&\quad +T^7(T+2)^2(T+4)^2(T+6)^2(6T^4+3T^3+5T^2+4T+2)E^2 \\
&\quad +4T^{11}(T+4)^2(T+6)^2(T^2+5T+5)E^2 \\
&= 4T^5(T+2)^2(T+4)^2(T+6)^2E^3L \\
&\quad +T^7(T+4)^3(T+6)^3(3T^4+3T^3+T^2+3T+5)E^2 \\
&= 4T^5(T+1)(T+4)^5(T+6)^5E^2
\end{aligned}
$$

and hence

$$
(40') \qquad \mu\nu' + 6\nu\mu' = 4T^5(T+1)(T+4)^5(T+6)^5E^2.
$$

By direct calculation with polynomials in T we also get

$$
\begin{aligned}
&6EL\mu\mu' + T^3\mu\nu' + 6\nu\nu' \\
&= 4(T+4)(T+5)E^6L^2 \\
&\quad +[3(T+1) + 5(T+4)(T+6) + 2(T+3)(T+5)]E^5L^3 \\
&\quad +[6T^3(T+1)(T+5) + 4T^3(T+1)(T+5) + 6T^3(T+4)]E^5L^2 \\
&\quad +3T^6(T+4)(T+5)E^5L + 6(T+3)(T+6)E^4L^4 \\
&\quad +[6T^3(T+1)(T+6) + 5T^3(T+1)(T+6) \\
&\qquad +2T^3(T+3) + 3T^3(T+3)]E^4L^3 \\
&\quad +[5T^6(T+1) + 6T^6(T+1) + 6T^6(T+4)(T+6) \\
&\qquad +4T^6(T+3)(T+5) + 5T^6(T+3)(T+5) \\
&\qquad +5T^6(T+3)(T+5)]E^4L^2 \\
&\quad +[4T^9(T+1)(T+5) + 3T^9(T+1)(T+5) + 4T^9(T+4)]E^4L \\
&\quad +[4T^6(T+3)(T+6) + T^6(T+3)(T+6) + 3T^6(T+3)(T+6)]E^3L^3 \\
&\quad +[6T^9(T+1)(T+6) + 6T^9(T+1)(T+6) \\
&\qquad +T^9(T+3) + 4T^9(T+3) + 2T^9(T+3)]E^3L^2 \\
&\quad +[3T^{12}(T+1) + 4T^{12}(T+1) + 5T^{12}(T+4)(T+6) \\
&\qquad +5T^{12}(T+3)(T+5) + 2T^{12}(T+3)(T+5)]E^3L \\
&\quad +[4T^{12}(T+3)(T+6) + 4T^{12}(T+3)(T+6) \\
&\qquad +6T^{12}(T+3)(T+6)]E^2L^2 \\
&\quad +[2T^{15}(T+1)(T+6) + 5T^{15}(T+1)(T+6) \\
&\qquad +2T^{15}(T+3) + 5T^{15}(T+3)]E^2L \\
&\quad +[6T^{18}(T+3)(T+6) + T^{18}(T+3)(T+6)]EL \\
&= 4(T+4)(T+5)E^6L^2 + 6(T+1)E^5L^3 + T^3(3T^2+3T+4)E^5L^2 \\
&\quad +3T^6(T+4)(T+5)E^5L + 6(T+3)(T+6)E^4L^4 \\
&\quad +T^3(4T^2+5T+4)E^4L^3 \\
&\quad +T^6(6T^2+T+1)E^4L^2 + 4T^9(T+4)E^4L + T^6(T+3)(T+6)E^3L^3 \\
&\quad +5T^9(T+1)(T+6)E^3L^2 + 5T^{12}(T+4)(T+6)E^3L \\
&= [4(T+4)(T+5)E^2 + 6(T+1)EL + 6(T+3)(T+6)L^2]E^4L^2 \\
&\quad +T^3[(3T^2+3T+4)E + (4T^2+5T+4)L]E^4L^2 \text{ (by regrouping)} \\
&\quad +T^6[3(T+4)(T+5)E^2 + (6T^2+T+1)EL + (T+3)(T+6)L^2]E^3L \\
&\quad +T^9[4(T+4)E + 5(T+1)(T+6)L]E^3L + 5T^{12}(T+4)(T+6)E^3L \\
&= 3T^4(T+2)^2(T+4)(T+6)E^4L^2 + 6T^4(T+1)(T+2)^2E^4L^2 + \\
&\quad +T^8(T+2)^2(T+4)(T+6)(3T^2+6T+6)E^3L \\
&\quad +5T^{10}(T+2)^2(T+4)(T+6)E^3L + 5T^{12}(T+4)(T+6)E^3L \\
&= 3T^4(T+2)^2(T+4)(T+6)E^4L^2 \\
&\quad +T^4(T+4)^2(T+6)^2(6T^6+5T^5+5T^4+4T^3+5T^2+5T+4)E^3L
\end{aligned}
$$

and hence

$$
\begin{aligned}
&6EL\mu\mu' + T^3\mu\nu' + 6\nu\nu' \\
&= T^4(T+4)^3(T+6)^3(2T^4+5T^3+6T^2+2T+1)E^3L
\end{aligned}
$$

and therefore by (40′) we have

$$\begin{aligned}
&3T^3(\mu\nu' + 6\nu\mu') + (6EL\mu\mu' + T^3\mu\nu' + 6\nu\nu') \\
&= [5T^8(T+1)(T+4)^5(T+6)^5]E^2 \\
&\quad +[T^4(T+2)^2(T+3)(T+4)^4(T+5)(T+6)^4\times \\
&\qquad (2T^4+5T^3+6T^2+2T+1)]E^2 \\
&= T^4(T+4)^4(T+6)^4(2T^8+6T^7+6T^6+4T^5+2T^4+2T+4)E^2
\end{aligned}$$

and hence in view of (38′) to (40′) we get

$$(41') \qquad \eta = \frac{[4T(T+1)(T+4)(T+6)]S^* + (2T^8+6T^7+6T^6+4T^5+2T^4+2T+4)}{(T+2)^4(T+5)^4}$$

By (31′) and (33′) we also have

$$(42') \qquad \xi = \frac{S^* + 3T^3}{(T+2)(T+5)(T+6)}.$$

Thus ξ and η have "poles" at $T+2$ and $T+5$. To simplify matters we send one of these, say $T+2$, to infinity by making the fractional linear substitution

$$(43') \qquad R = \frac{T}{T+2}$$

and we note that then

$$(44') \quad T = \frac{R}{3R+4} \quad \text{and} \quad \frac{T+1}{T+2} = 4(R+1) \quad \text{and} \quad \frac{1}{T+2} = 3(R+6)$$

and

$$(45') \qquad \frac{T+3}{T+2} = 3(R+4) \quad \text{and} \quad \frac{T+4}{T+2} = 6(R+5)$$

and

$$(46') \qquad \frac{T+5}{T+2} = 2(R+3) \quad \text{and} \quad \frac{T+6}{T+2} = 5(R+2)$$

and hence upon letting

$$(47') \qquad S = \frac{5S^*}{(T+2)^3} \quad \text{and} \quad D = \frac{4D^*}{(T+2)^6}$$

and upon multiplying (34′) throughout by $\frac{4}{(T+2)^6}$ we get

$$\begin{aligned}
S^2 = D &= \frac{4D^*}{(T+2)^6} \\
&= \frac{1+4T(T+2)(T+3)(T+4)(T+5)(T+6)}{(T+2)^6} \\
&= 3^6(R+6)^6 + 4R[3(R+4)][6(R+5)][2(R+3)][5(R+2)] \\
&= R^6 + R^4 + R^2 + 1 \\
&= (R^2+1)(R^2+3R+1)(R^2+4R+1)
\end{aligned}$$

and hence we have

$$(48')\qquad S^2 = D \quad \text{and} \quad \begin{aligned} D &= R^6 + R^4 + R^2 + 1 \\ &= (R^2+1)(R^2+3R+1)(R^2+4R+1). \end{aligned}$$

Now upon multiplying the numerator and denominator of the RHS of (41′) by $\frac{4}{(T+2)^8}$ we get

$$(49')\qquad \eta = \frac{L_\eta S + M_\eta}{(R+3)^4}$$

where

$$L_\eta = 4R(R+1)(R+2)(R+5)(R+6)$$

and

$$M_\eta = [R^4 + 6R^2(R+6)^2 + R(R+6)^3 + 3(R+6)^4]\times \\ [R^4 + 2R^3(R+6) + 6R(R+6)^3 + 6(R+6)^4]$$

and hence

$$(50')\qquad L_\eta = 4R^5 + R^3 + 2R \quad \text{and} \quad M_\eta = 4R^8 + 3R^6 + 4R^4 + 4.$$

Likewise, upon multiplying the numerator and denominator of the RHS of (42′) by $\frac{5}{(T+2)^3}$ we get

$$(51')\qquad \xi = \frac{S + R^3}{(R+2)(R+3)}.$$

At the same time we simplify ψ^* by putting

$$\psi(Z) = \eta^{-6}\psi^*(\xi, \eta, -\eta Z)$$

to get

$$(52')\qquad \psi(Z) = Z^6 + J[Z^5 + Z^4 + Z^3 + Z^2 + Z] + K$$

with

$$(53')\qquad J = 3(\xi + \eta) \quad \text{and} \quad K = 3(\xi + \eta)(\eta^2 + \xi^7)\eta^{-2}$$

and we note that what we want to do is to factor $\psi(Z)$ in $k(\xi, \eta)[Z] = k(R, S)[Z]$ into two monic factors of degree 3 in Z.

To express J in terms of R and S, by (49′) to (53′) we have

$$(54')\qquad J = \frac{L_J S + M_J}{(R+2)(R+3)^4}$$

where

$$(55')\qquad \begin{aligned} L_J &= 3(R+3)^3 + 3(4R^5 + R^3 + 2R)(R+2) \\ &= 5R^6 + 3R^5 + 3R^4 + 2R^3 + 5R^2 + 2R + 4 \end{aligned}$$

and

$$(56') \qquad \begin{aligned} M_J &= 3R^3(R+3)^3 + 3(4R^8 + 3R^6 + 4R^4 + 4)(R+2) \\ &= 5R^9 + 3R^8 + 2R^7 + 4R^5 + 4R^3 + 5R + 3 \end{aligned}$$

To find the "zeroes" of J we consider its "norm"

$$(57') \qquad N_J = \frac{(L_J S + M_J)(-L_J S + M_J)}{(R+2)^2(R+3)^8} = \frac{M_J^2 - S^2 L_J^2}{(R+2)^2(R+3)^8} = \frac{M_J^2 - DL_J^2}{(R+2)^2(R+3)^8}$$

and by (48′), (55′) and (56′) we get

$$(58') \qquad \begin{aligned} M_J^2 - DL_J^2 &= 6R^{13} + 3R^{12} + 6R^{11} + 4R^{10} + R^9 + R^8 + 3R^7 \\ &= 6R^7(R+2)(R+3)^4(R+4). \end{aligned}$$

Likewise, to find the "zeroes" of η we consider its norm

$$(59') \qquad N_\eta = \frac{(L_\eta S + M_\eta)(-L_\eta S + M_\eta)}{(R+3)^8} = \frac{M_\eta^2 - S^2 L_\eta^2}{(R+3)^8} = \frac{M_\eta^2 - DL_\eta^2}{(R+3)^8}$$

and by (48′) to (50′) we have

$$(60') \qquad M_\eta^2 = 2R^{16} + 3R^{14} + 6R^{12} + 3R^{10} + 6R^8 + 3R^6 + 4R^4 + 2$$

and

$$(61') \qquad DL_\eta^2 = 2R^{16} + 3R^{14} + 6R^{12} + 3R^{10} + 5R^8 + 4R^6 + R^4 + 4R^2$$

and hence we get

$$(62') \qquad M_\eta^2 - DL_\eta^2 = R^8 + 6R^6 + 3R^4 + 3R^2 + 2 = (R+3)^4(R+4)^4.$$

To express K in terms of R and S, by (49′) to (51′) we have

$$(\eta^2 + \xi^7)\eta^{-2} = \frac{(R+2)^7(L_\eta S + M_\eta)^2 + (R+3)(S+R^3)^7}{(R+2)^7(L_\eta S + M_\eta)^2}$$

and multiplying the numerator and denominator by $(-L_\eta S + M_\eta)^2$ we get

$$(\eta^2 + \xi^7)\eta^{-2} = \frac{(R+2)^7(M_\eta^2 - S^2L_\eta^2)^2 + (R+3)(S^7 + R^{21})(-L_\eta S + M_\eta)^2}{(R+2)^7(M_\eta^2 - S^2L_\eta^2)^2}$$

and hence, in view of (48′), substituting D for S^2 we have

$$(63') \qquad (\eta^2 + \xi^7)\eta^{-2} = \frac{(R+2)^7(M_\eta^2 - DL_\eta^2)^2 + (R+3)(L'S + M')}{(R+2)^7(M_\eta^2 - DL_\eta^2)^2}$$

where

$$L' = D^3(M_\eta^2 + DL_\eta^2) - 2R^{21}L_\eta M_\eta$$

and

$$M' = -2D^4 L_\eta M_\eta + R^{21}(M_\eta^2 + DL_\eta^2).$$

By (50′) we have

$$2L_\eta M_\eta = 4R^{13} + 4R^{11} + 5R^9 + 6R^7 + 6R^5 + R^3 + 2R$$

and hence in view of (60′) and (61′) we get

$$\begin{aligned} L' &= (R^6 + R^4 + R^2 + 1)^3 \times \\ &\qquad (4R^{16} + 6R^{14} + 5R^{12} + 6R^{10} + 4R^8 + 5R^4 + 4R^2 + 2) \\ &\quad +R^{21}(3R^{13} + 3R^{11} + 2R^9 + R^7 + R^5 + 6R^3 + 5R) \\ &= 2R^{20} + 3R^{18} + 2R^{16} + 6R^{14} + 3R^6 + R^4 + 3R^2 + 2 \\ &= (R+3)^7(R+4)^7(2R^6 + 3R^4 + 2R^2 + 6) \end{aligned}$$

and

$$\begin{aligned} M' &= (R^6 + R^4 + R^2 + 1)^4 \times \\ &\qquad (3R^{13} + 3R^{11} + 2R^9 + R^7 + R^5 + 6R^3 + 5R) \\ &\quad +R^{21}(4R^{16} + 6R^{14} + 5R^{12} + 6R^{10} + 4R^8 + 5R^4 + 4R^2 + 2) \\ &= 5R^{23} + 5R^{21} + R^{19} + R^{17} + R^{15} + 4R^9 \\ &\quad +4R^7 + 5R^5 + 5R^3 + 5R \\ &= (R+3)^7(R+4)^7(5R^9 + 5R^7 + R^5 + R^3 + R) \end{aligned}$$

and obviously we have

$$(R+2)^7(R+4)+(5R^9+5R^7+R^5+R^3+R) = 5R^9+R^8+2R^7+R^5+R^3+3R+1$$

and hence in view of (62′) and (63′) we get

$$(64') \qquad (\eta^2 + \xi^7)\eta^{-2} = \frac{L^*S + M^*}{(R+2)^7(R+4)}$$

where

$$(65') \qquad L^* = 2R^6 + 3R^4 + 2R^2 + 6 \text{ and } M^* = 5R^9 + R^8 + 2R^7 + R^5 + R^3 + 3R + 1$$

and for the corresponding product of conjugates we have

$$(66') \qquad \begin{aligned} (L^*S + M^*)(-L^*S + M^*) &= M^{*2} - DL^{*2} \\ &= 3R^{17} + 5R^{16} + 4R^{15} + 2R^{14} \\ &\quad +2R^{13} + R^{12} + 2R^{11} + 2R^9 \\ &\quad +4R^8 + 4R^7 + 4R^6 + 2R^5 \\ &\quad +4R^4 + 2R^3 + 5R^2 + 6R \\ &= 3R(R+2)^7(R+3)^4(R+4)^5. \end{aligned}$$

Now by (48′), (53′), (54′) and (64″) we have

$$K = \frac{(L_J M^* + M_J L^*)S + (DL_J L^* + M_J M^*)}{(R+2)^8(R+3)^4(R+4)}$$

and, in view of (48′), (55′), (56′) and (65′), by direct calculation with polynomials in R we get

$$\begin{aligned} L_J M^* + M_J L^* &= (5R^6 + 3R^5 + 3R^4 + 2R^3 + 5R^2 + 2R + 4)\times \\ &\qquad (5R^9 + R^8 + 2R^7 + R^5 + R^3 + 3R + 1) \\ &\quad +(5R^9 + 3R^8 + 2R^7 + 4R^5 + 4R^3 + 5R + 3)\times \\ &\qquad (2R^6 + 3R^4 + 2R^2 + 6) \\ &= 5R^{14} + 5R^{13} + 6R^{11} + 3R^9 + 3R^8 + 3R^7 + 3R^6 \\ &\quad +5R^5 + 6R^4 + 6R^3 + 3R^2 + 2R + 1 \\ &= (R+4)(5R^{13} + 6R^{12} + 4R^{11} + 4R^{10} + 5R^9 \\ &\qquad +4R^8 + R^7 + 6R^6 + 5R^4 + 6R^2 + 2) \end{aligned}$$

and

$$\begin{aligned} DL_J L^* + M_J M^* &= (R^6 + R^4 + R^2 + 1)\times \\ &\quad (5R^6 + 3R^5 + 3R^4 + 2R^3 + 5R^2 + 2R + 4)\times \\ &\quad (2R^6 + 3R^4 + 2R^2 + 6) \\ &\quad +(5R^9 + 3R^8 + 2R^7 + 4R^5 + 4R^3 + 5R + 3)\times \\ &\quad (5R^9 + R^8 + 2R^7 + R^5 + R^3 + 3R + 1) \\ &= 5R^{17} + 5R^{16} + 6R^{15} + 5R^{14} \\ &\quad +2R^9 + 3R^7 + 5R + 6 \\ &= (R+4)(5R^{16} + 6R^{15} + 3R^{14} + 2R^8 + 6R^7 + 5) \end{aligned}$$

and hence we have

$$K = \frac{L_K S + M_K}{(R+2)^8 (R+3)^4} \tag{67′}$$

where

$$L_K = 5R^{13}+6R^{12}+4R^{11}+4R^{10}+5R^9+4R^8+R^7+6R^6+5R^4+6R^2+2 \tag{68′}$$

and

$$M_K = 5R^{16} + 6R^{15} + 3R^{14} + 2R^8 + 6R^7 + 5 \tag{69′}$$

and for the norm of K we have

$$N_K = \frac{(L_K S+M_K)(-L_K S+M_K)}{(R+2)^{16}(R+3)^8} = \frac{M_K^2 - S^2 L_K^2}{(R+2)^{16}(R+3)^8} = \frac{M_K^2 - DL_K^2}{(R+2)^{16}(R+3)^8} \tag{70′}$$

and, in view of (53′), (62″) and (66″), we get

$$\begin{aligned} M_K^2 - DL_K^2 &= (R+4)^{-2}(M_J^2 - DL_J^2)(M^{*2} - DL^{*2}) \\ &= 4R^8(R+2)^8(R+3)^8(R+4)^4. \end{aligned} \tag{71′}$$

Note that now $k(\xi,\eta) = k(T,S^*) = k(R,S)$, a splitting field of $\psi^*(\xi,\eta,Z)$ over $k(\xi,\eta)$ is the same thing as a splitting field of $\psi(Z)$ over $k(R,S)$, and concerning the valuations $\lambda_1, \lambda_2, \lambda_0$ of $k(\xi,\eta)/k = k(R,S)/k$, by what we have said at the end of the last Section, we see that: these are the only three valuations of $k(R,S)/k$ which are ramified in a splitting field of $\psi(Z)$ over $k(R,S)$; for any extension of any one of these three valuations to the said splitting field, the reduced ramification exponent is 3; and, in terms of ξ and η, these valuations are given by $\lambda_1(\xi+1)=\lambda_1(\eta-1)=1$, and $\lambda_2(\xi)=\lambda_2(\eta+2)=1$, and $\lambda_0(\xi)=-1$ and $\lambda_0(2\eta+1)=1$. Looking at (30′) to (34′) we see that, in terms of T and S^*, the said valuations are given by $\lambda_1(T)=\lambda_1(S^*+4)=1$, $\lambda_2(T+4)=\lambda_2(S^*+4)=1$, and $\lambda_0(T+6)=\lambda_0(S^*+4)=1$. Therefore, because of (43′) to (47′), in terms of R and S, they are given by $\lambda_1(R)=\lambda_1(S+6)=1$, and $\lambda_2(R+5)=\lambda_2(S+1)=1$, and $\lambda_0(R+2)=\lambda_0(S+1)=1$.

2.6 Finding the factors

By referring to items (48′), (52′) to (58′), and (67′) to (71′) of the previous Section, we are assuming the characteristic of the ground field k to be 7, and we are considering the univariate algebraic function field $k(R,S)$ given by

(1*) $$S^2=D \quad \text{with} \quad \begin{aligned} D&=R^6+R^4+R^2+1\\ &=(R^2+1)(R^2+3R+1)(R^2+4R+1)\end{aligned}$$

and we want to factor the polynomial

(2*) $\psi(Z)=Z^6+J[Z^5+Z^4+Z^3+Z^2+Z]+K$ with J and K in $k(R,S)$

into two monic factors of degree 3 in $k(R,S)[Z]$, where the coefficient J is given by

(3*) $$J=\frac{L_JS+M_J}{(R+2)(R+3)^4}$$

with

(4*) $$L_J=5R^6+3R^5+3R^4+2R^3+5R^2+2R+4$$

and

(5*) $$M_J=5R^9+3R^8+2R^7+4R^5+4R^3+5R+3$$

and its norm is given by

(6*) $$N_J=\frac{(L_JS+M_J)(-L_JS+M_J)}{(R+2)^2(R+3)^8}=\frac{M_J^2-S^2L_J^2}{(R+2)^2(R+3)^8}=\frac{M_J^2-DL_J^2}{(R+2)^2(R+3)^8}$$

with

(7*) $$M_J^2-DL_J^2=6R^7(R+2)(R+3)^4(R+4)$$

and the coefficient K is given by

$$K = \frac{L_K S + M_K}{(R+2)^8 (R+3)^4} \tag{8*}$$

with

$$L_K = 5R^{13} + 6R^{12} + 4R^{11} + 4R^{10} + 5R^9 + 4R^8 + R^7 + 6R^6 + 5R^4 + 6R^2 + 2 \tag{9*}$$

and

$$M_K = 5R^{16} + 6R^{15} + 3R^{14} + 2R^8 + 6R^7 + 5 \tag{10*}$$

and its norm is given by

$$N_K = \frac{(L_K S + M_K)(-L_K S + M_K)}{(R+2)^{16}(R+3)^8} = \frac{M_K^2 - S^2 L_K^2}{(R+2)^{16}(R+3)^8} = \frac{M_K^2 - D L_K^2}{(R+2)^{16}(R+3)^8} \tag{11*}$$

with

$$M_K^2 - D L_K^2 = 4R^8 (R+2)^8 (R+3)^8 (R+4)^4. \tag{12*}$$

The "finding" of the factors of $\psi(Z)$ uses high-school ideas which say that integer roots of a monic integral polynomial are factors of its constant term, and if the polynomial is not monic then the factors of the leading coefficient suffice as denominators of the roots. In this analogy, the writing of a rational number as a power product of prime numbers is to be replaced by the "divisor" of an algebraic function. While dealing with hyperelliptic functions, i.e., with functions admitting a square-root parametrization, by taking norms, computations with "divisors" are almost reduced to computations of rational functions. This is inspired by rationalization of the quadratic surd $\frac{a+b\sqrt{2}}{c+d\sqrt{2}}$ by multiplying the numerator and denominator by $c - d\sqrt{2}$ which is the "conjugate" of the denominator.

Thus the "finding" of the factors of $\psi(Z)$ hinges on the "divisors " of J and K which themselves will be read off from their norms. By (6*), (7*), (11*) and (12*), these norms are given by

$$N_J = \frac{6R^7 (R+4)}{(R+2)(R+3)^4} \tag{13*}$$

and

$$N_K = \frac{4R^8 (R+4)^4}{(R+2)^8}. \tag{14*}$$

Now the linear factors $R, (R+6), (R+5), (R+4), (R+3), (R+2), (R+1)$ correspond to the values $R = 0, 1, 2, 3, 4, 5, 6$ respectively, and upon substituting these values in the expressions (1*), (4*), (5*), (9*) and (10*), we get the corresponding values of D, L_J, M_J, L_K, M_K, S, L_JS+M_J and L_KS+M_K as described in the following table.

R	0	1	2	3	4	5	6
L_J	4	3	4	1	5	3	3
M_J	3	5	5	6	5	4	0
L_K	2	6	0	5	2	4	4
M_K	5	6	4	5	2	3	3
D	1	4	1	1	1	1	4
S	1,6	2,5	1,6	1,6	1,6	1,6	2,5
L_JS+M_J	0,6	4,6	2,1	0,5	3,0	0,1	6,1
L_KS+M_K	0,3	4,1	4,4	3,0	4,0	0,6	4,2

This gives valuations V_{i6} and V_{i1} of $k(R,S)/k$ with

(15*) $$V_{i6}(R+i)=1\le V_{i6}(S+6) \text{ and } V_{i1}(R+i)=1\le V_{i1}(S+1) \quad \text{for } i=0,5,4,3,2$$

and valuations V_{i5} and V_{i2} of $k(R,S)/k$ with

(16*)
$$V_{i5}(R+i)=1\le V_{i5}(S+5) \text{ and } V_{i2}(R+i)=1\le V_{i2}(S+2) \text{ for } i=6,1$$

and for these valuations we have

(17*) $$V_{ij}(L_JS+M_J)>0 \text{ for } (i,j)=(0,6),(4,6),(3,1),(2,6)$$

and

(18*) $$V_{ij}(L_JS+M_J)=0 \text{ for } (i,j)=(0,1),(6,5),(6,2),(5,6),(5,1),(4,1),(3,6),(2,1),(1,5),(1,2)$$

and

(19*) $$V_{ij}(L_KS+M_K)>0 \text{ for } (i,j)=(0,6),(4,1),(3,1),(2,6)$$

and

(20*) $$V_{ij}(L_KS+M_K)=0 \text{ for } (i,j)=(0,1),(6,5),(6,2),(5,6),(5,1),(4,6),(3,6),(2,1),(1,5),(1,2).$$

In view of (1*) we also get valuations $V_{\infty 6}$ and $V_{\infty 1}$ of $k(R,S)/k$ with

(21*) $$\begin{cases} V_{\infty 6}\left(\frac{1}{R}\right)=1\le V_{\infty 6}\left(\frac{S}{R^3}+6\right) \\ \text{and } V_{\infty 1}\left(\frac{1}{R}\right)=1\le V_{\infty 1}\left(\frac{S}{R^3}+1\right)\end{cases}$$

and by (4*) and (5*) we see that

(22*) $$V_{\infty j}\left(\frac{L_J}{R^6}+2\right)>0<V_{\infty j}\left(\frac{M_J}{R^9}+2\right) \quad \text{for } j=6,1$$

and hence

$$(23^*) \qquad V_{\infty 1}\left(\frac{L_J S + M_J}{R^9}\right) > 0 = V_{\infty 6}\left(\frac{L_J S + M_J}{R^9}\right)$$

and by (9*) and (10*) we see that

$$(24^*) \qquad V_{\infty j}\left(\frac{L_K}{R^{13}} + 2\right) > 0 < V_{\infty j}\left(\frac{M_K}{R^{16}} + 2\right) \quad \text{for } j = 6, 1$$

and hence

$$(25^*) \qquad V_{\infty 1}\left(\frac{L_K S + M_K}{R^{16}}\right) > 0 = V_{\infty 6}\left(\frac{L_K S + M_K}{R^{16}}\right).$$

In view of (7*) and (12*), by (15*) to (25*), for the divisors of $L_J S + M_J$ and $L_K S + M_K$ we have

$$(26^*) \qquad (L_J S + M_J) = 7V_{06} + V_{26} + 4V_{31} + V_{46} - 4V_{\infty 1} - 9V_{\infty 6}$$

and

$$(27^*) \qquad (L_K S + M_K) = 8V_{06} + 8V_{26} + 8V_{31} + 4V_{41} - 12V_{\infty 1} - 16V_{\infty 6}$$

i.e.,

$$V(L_J S + M_J) = \begin{cases} 7, 1, 4, 1, -4, -9 & \text{if } V = V_{06}, V_{26}, V_{31}, V_{46}, V_{\infty 1}, V_{\infty 6} \\ 0 & \text{if } V \text{ is any other val. of } k(R,S)/k \end{cases}$$

and

$$V(L_K S + M_K) = \begin{cases} 8, 8, 8, 4, -12, -16 & \text{if } V = V_{06}, V_{26}, V_{31}, V_{41}, V_{\infty 1}, V_{\infty 6} \\ 0 & \text{if } V \text{ is any other val. of } k(R,S)/k \end{cases}$$

and more suggestively we may express this by writing

$$(28^*) \qquad (L_J S + M_J) = \frac{R^{*7}(R+2)^*(R+3)'^4(R+4)^*}{R'^4_\infty R^{*9}_\infty}$$

and

$$(29^*) \qquad (L_K S + M_K) = \frac{R^{*8}(R+2)^{*8}(R+3)'^8(R+4)'^4}{R'^{12}_\infty R^{*16}_\infty}$$

where

$$\text{a factor refers to +ve value of} \begin{cases} S+1 & \text{if primed \& unsubscripted} \\ S+6 & \text{if starred \& unsubscripted} \\ \frac{S}{R^3}+1 & \text{if primed \& } \infty\text{-subscripted} \\ \frac{S}{R^3}+6 & \text{if starred \& } \infty\text{-subscripted} \end{cases}$$

and hence, in view of (3*) and (8*), for the divisors of J and K we have

(30*) $$(J) = 7V_{06} + V_{46} + V_{\infty 1} - V_{21} - 4V_{36} - 4V_{\infty 6}$$

and

(31*) $$(K) = 8V_{06} + 4V_{31} + 4V_{41} - 8V_{21} - 4V_{36} - 4V_{\infty 6}$$

i.e.,

$$V(J) = \begin{cases} 7, 1, 1, -1, -4, -4 & \text{if } V = V_{06}, V_{46}, V_{\infty 1}, V_{21}, V_{36}, V_{\infty 6} \\ 0 & \text{if } V \text{ is any other valuation of } k(R,S)/k \end{cases}$$

and

$$V(K) = \begin{cases} 8, 4, 4, -8, -4, -4 & \text{if } V = V_{06}, V_{31}, V_{41}, V_{21}, V_{36}, V_{\infty 6} \\ 0 & \text{if } V \text{ is any other valuation of } k(R,S)/k \end{cases}$$

and again more suggestively we may express this by writing

(32*) $$(J) = \frac{R^{*7}(R+4)^{*}R'_{\infty}}{(R+2)'(R+3)^{*4}R^{*4}_{\infty}}$$

and

(33*) $$(K) = \frac{R^{*8}(R+3)'^{4}(R+4)'^{4}}{(R+2)'^{8}(R+3)^{*4}R^{*4}_{\infty}}.$$

Let

(□) $$\psi(Z) = (Z^3 + A'Z^2 + B'Z + C')(Z^3 + A''Z^2 + B''Z + C'')$$

with $A', B', C', A'', B'', C''$ in $k(R,S)$ be a possible factorization. By (1*) we see that D is square-free in $k[R]$, and hence the ring $k[R]S + k[R]$ is normal, i.e., it is integrally closed in its quotient field $k(R,S)$. Moreover, by (3*) and (8*) we see that

$$(R+2)^{12}(R+3)^{24}\psi((R+2)^{-2}(R+3)^{-4}Z)$$

is a monic polynomial of degree 6 in Z with coefficients in $k[R]S + k[R]$, and hence

(34*) $$\begin{cases} \text{the quantities } (R+2)^2(R+3)^4A', \\ (R+2)^4(R+3)^8B',\ (R+2)^6(R+3)^{12}C', \\ \text{as well as the quantities} (R+2)^2(R+3)^4A'', \\ (R+2)^4(R+3)^8B'',\ (R+2)^6(R+3)^{12}C'', \\ \text{all belong to } k[R]S + k[R]. \end{cases}$$

To decrease the above exponents, let us prove the following

Elementary Lemma. *Let V be a valuation* [15] *of a field Ω and let $f(Z) = f'(Z)f''(Z)$ be monic polynomials of positive degrees n, n', n'' in Z with coefficients in a field Ω, so that $f(Z) = Z^n + \sum_{i=1}^{n} a_i Z^{n-i}$ and $f'(Z) = Z^{n'} + \sum_{i=1}^{n'} a'_i Z^{n'-i}$ and $f''(Z) = Z^{n''} + \sum_{i=1}^{n''} a''_i Z^{n''-i}$ with a_i, a'_i, a''_i in Ω. Then we have the following.*

(35) If d is a positive integer such that $V(a_1) = -d \le V(a_i)$ for $2 \le i \le n$ then either $V(a'_i) \ge 0$ for $1 \le i \le n'$ and $V(a''_1) = -d \le V(a''_i)$ for $2 \le i \le n''$, or $V(a''_i) \ge 0$ for $1 \le i \le n''$ and $V(a'_1) = -d \le V(a'_i)$ for $2 \le i \le n'$.*

(36) If d is a rational number such that $V(a_i) \ge id$ for $1 \le i \le n$ then $V(a'_i) \ge [id]$ for $1 \le i \le n'$ and $V(a''_i) \ge [id]$ for $1 \le i \le n''$ where $[id]$ is the smallest integer $\ge id$.*

To see this, take $e \in \Omega$ with $V(e) = 1$.

To prove (35*), for a moment assume that d is a positive integer such that $V(a_1) = -d \le V(a_i)$ for $2 \le i \le n$. Let $g(Z) = e^{nd} f(e^{-d}Z)$ and $g'(Z) = e^{n'd} f'(e^{-d}Z)$ and $g''(Z) = e^{n''d} f''(e^{-d}Z)$. Then $g(Z) = g'(Z)g''(Z)$. Also $g(Z) = Z^n + \sum_{i=1}^{n} b_i Z^{n-i}$ and $g'(Z) = Z^{n'} + \sum_{i=1}^{n'} b'_i Z^{n'-i}$ and $g''(Z) = Z^{n''} + \sum_{i=1}^{n''} b''_i Z^{n''-i}$ with $b_i = e^{id} a_i$ and $b'_i = e^{id} a'_i$ and $b''_i = e^{id} a''_i$. It follows that $V(b_1) = 0$ and $V(b_i) \ge (i-1)d \ge 0$ for $2 \le i \le n$, and hence $V(b'_i) \ge 0$ for $1 \le i \le n'$ and $V(b''_i) \ge 0$ for $1 \le i \le n''$. By comparing the coefficients of Z^{n-1} in the equation $g(Z) = g'(Z)g''(Z)$ we get $b_1 = b'_1 + b''_1$, and hence either $V(b'_1) = 0$ or $V(b''_1) = 0$. Therefore by suitably permuting the factors $(f'(Z), f''(Z))$ of $f(Z)$ we may assume that $V(b''_1) = 0$. We can uniquely extend V to a valuation of $\Omega(Z)$, which we again denote by V, so that for every $0 \ne F(Z) = \sum_{i \ge 0} \alpha_i Z^i \in \Omega[Z]$ with $\alpha_i \in \Omega$ we have $V(F(Z)) = \min\{V(\alpha_i) : i \ge 0\}$. Let $h(Z) = g(e^d Z)$ and $h'(Z) = g'(e^d Z)$ and $h''(Z) = g''(e^d Z)$. Then $h(Z) = h'(Z)h''(Z)$ and hence $V(h(Z)) = V(h'(Z)) + V(h''(Z))$. Also $h(Z) = e^{(n-1)d}[e^d Z^n + \sum_{i=1}^{n} c_i Z^{n-i}]$ where $c_i = e^{(1-i)d} b_i$, and hence $V(c_1) = 0 \le V(c_i)$ for $2 \le i \le n$, and therefore $V(h(Z)) = (n-1)d$. Moreover $h'(Z) = \sum_{i=0}^{n'} c'_i Z^{n'-i}$ where $c'_0 = e^{n'd}$ and $c'_i = e^{(n'-i)d} b'_i$ for $1 \le i \le n'$, and hence $V(c'_0) = n'd$ and $V(c'_i) \ge (n'-i)d$ for $1 \le i \le n'$, and therefore

$$V(h'(Z)) = \min\{V(c'_i) : 0 \le i \le n'\} \le n'd.$$

Finally $h''(Z) = e^{n''d} Z^{n''} + \sum_{i=1}^{n''} c''_i Z^{n''-i}$ where $c''_i = e^{(n''-i)d} b''_i$, and hence $V(c''_1) = (n''-1)d$ and $V(c''_i) \ge (n''-i)d$ for $2 \le i \le n''$, and therefore

$$V(h''(Z)) = \min\{V(c''_i) : 1 \le i \le n''\} \le (n''-1)d.$$

Since $V(h'(Z)) + V(h''(Z)) = V(h(Z)) = (n-1)d = n'd + (n''-1)d$, by

[15] All valuations in this paper are assumed to be real discrete, i.e., they map the nonzero elements of the relevant field onto the set of all ordinary integers.

the above two displayed relations for $V(h'(Z))$ and $V(h''(Z))$ we conclude that

$$\min\{V(c'_i) : 0 \le i \le n'\} = n'd \text{ and } \min\{V(c''_i) : 1 \le i \le n''\} = (n''-1)d.$$

Therefore $V(c'_0) = n'd \le V(c'_i)$ for $1 \le i \le n'$ and $V(c''_1) = (n''-1)d \le V(c''_i)$ for $2 \le i \le n''$. Consequently $V(a'_i) \ge 0$ for $1 \le i \le n'$ and $V(a''_1) = -d \le V(a''_i)$ for $2 \le i \le n''$.

To prove (36*), for a moment assume that d is a rational number such that $V(a_i) \ge id$ for $1 \le i \le n$. We can first take a positive integer m such that $d = \frac{\delta}{m}$ where δ is an integer, and then we can take an element ϵ in an overfield of Ω such that $\epsilon^m = e$, and finally we can take an extension V^* of V to $\Omega^* = \Omega(\epsilon)$. Now $V^*(\alpha) = mV(\alpha)$ for all $\alpha \in \Omega$, and hence $V^*(a_i) \ge i\delta$ for $1 \le i \le n$. Let $f^*(Z) = \epsilon^{-n\delta} f(\epsilon^\delta Z)$ and $f'^*(Z) = \epsilon^{-n'\delta} f'(\epsilon^\delta Z)$ and $f''^*(Z) = \epsilon^{-n''\delta} f''(\epsilon^\delta Z)$. Then $f^*(Z) = f'^*(Z) f''^*(Z)$. Also $f^*(Z) = Z^n + \sum_{i=1}^n a_i^* Z^{n-i}$ and $f'^*(Z) = Z^{n'} + \sum_{i=1}^{n'} a_i'^* Z^{n'-i}$ and $f''^*(Z) = Z^{n''} + \sum_{i=1}^{n''} a_i''^* Z^{n''-i}$ with $a_i^* = \epsilon^{-i\delta} a_i$ and $a_i'^* = \epsilon^{-i\delta} a'_i$ and $a_i''^* = \epsilon^{-i\delta} a''_i$. It follows that $V^*(a_i^*) \ge 0$ for $1 \le i \le n$, and hence $V^*(a_i'^*) \ge 0$ for $1 \le i \le n'$ and $V^*(a_i''^*) \ge 0$ for $1 \le i \le n''$. Therefore $V^*(a'_i) \ge i\delta$ for $1 \le i \le n'$ and $V^*(a''_i) \ge i\delta$ for $1 \le i \le n''$, and hence $V(a'_i) \ge [id]$ for $1 \le i \le n'$ and $V(a''_i) \ge [id]$ for $1 \le i \le n''$.

In view of (2*), (32*), (33*) and (35*), by suitably permuting the primed and double primed factors we can arrange matters so that

$$\begin{cases} V_{36}(A') \ge 0, \ V_{36}(B') \ge 0, \ V_{36}(C') \ge 0, \\ V_{36}(A'') = -4, \ V_{36}(B'') \ge -4, \ V_{36}(C'') \ge -4. \end{cases}$$

Now in view of (2*), (32*), (33*) and (36*), we see that

$$\begin{cases} V_{31}(A') \ge 0, \ V_{31}(B') \ge 0, \ V_{31}(C') \ge 0, \\ V_{31}(A'') \ge 0, \ V_{31}(B'') \ge 0, \ V_{31}(C'') \ge 0 \end{cases}$$

and

$$\begin{cases} V_{26}(A') \ge 0, \ V_{26}(B') \ge 0, \ V_{26}(C') \ge 0, \\ V_{21}(A') \ge -1, \ V_{21}(B') \ge -2, \ V_{21}(C') \ge -4 \end{cases}$$

and

$$\begin{cases} V_{26}(A'') \ge 0, \ V_{26}(B'') \ge 0, \ V_{26}(C'') \ge 0, \\ V_{21}(A'') \ge -1, \ V_{21}(B'') \ge -2, \ V_{21}(C'') \ge -4 \end{cases}$$

and

$$\begin{cases} V_{06}(A') \ge 2, \ V_{06}(B') \ge 3, \ V_{06}(C') \ge 4, \\ V_{06}(A'') \ge 2, \ V_{06}(B'') \ge 3, \ V_{06}(C'') \ge 4. \end{cases}$$

In view of (34*), by the material in the above five displayed braces concerning the valuations V_{36}, V_{31}, V_{26}, V_{21} and V_{06}, we conclude that

$$(37^*) \qquad A' = \tfrac{L_{A'}S + M_{A'}}{(R+2)}, \quad B' = \tfrac{L_{B'}S + M_{B'}}{(R+2)^2}, \quad C' = \tfrac{L_{C'}S + M_{C'}}{(R+2)^4}$$

and

$$(38^*) \qquad A'' = \tfrac{L_{A''}S+M_{A''}}{(R+2)(R+3)^4}, \quad B'' = \tfrac{L_{B''}S+M_{B''}}{(R+2)^2(R+3)^4}, \quad C'' = \tfrac{L_{C''}S+M_{C''}}{(R+2)^4(R+3)^4}$$

where

$$(39^*) \qquad \begin{cases} L_{A'}, M_{A'}, L_{B'}, M_{B'}, L_{C'}, M_{C'}, \\ L_{A''}, M_{A''}, L_{B''}, M_{B''}, L_{C''}, M_{C''} \\ \text{are elements in } k[R] \end{cases}$$

such that

$$(40^*) \qquad \begin{cases} V_{31}(L_{A''}S + M_{A''}) \geq 4, \;\; V_{31}(L_{B''}S + M_{B''}) \geq 4, \\ V_{31}(L_{C''}S + M_{C''}) \geq 4 \end{cases}$$

and

$$(41^*) \; V_{26}(L_{A'}S+M_{A'}) \geq 1, \;\; V_{26}(L_{B'}S+M_{B'}) \geq 2, \;\; V_{26}(L_{C'}S+M_{C'}) \geq 4$$

and

$$(42^*) \qquad \begin{cases} V_{26}(L_{A''}S + M_{A''}) \geq 1, \;\; V_{26}(L_{B''}S + M_{B''}) \geq 2, \\ V_{26}(L_{C''}S + M_{C''}) \geq 4 \end{cases}$$

and

$$(43^*) \; V_{06}(L_{A'}S+M_{A'}) \geq 2, \;\; V_{06}(L_{B'}S+M_{B'}) \geq 3, \;\; V_{06}(L_{C'}S+M_{C'}) \geq 4$$

and

$$(44^*) \qquad \begin{cases} V_{06}(L_{A''}S + M_{A''}) \geq 2, \;\; V_{06}(L_{B''}S + M_{B''}) \geq 3, \\ V_{06}(L_{C''}S + M_{C''}) \geq 4. \end{cases}$$

By (2*), (32*), (33*) and (36*), we see that

$$\begin{cases} V_{\infty 1}(A') \geq 0, \; V_{\infty 1}(B') \geq 0, \; V_{\infty 1}(C') \geq 0, \\ V_{\infty 1}(A'') \geq 0, \; V_{\infty 1}(B'') \geq 0, \; V_{\infty 1}(C'') \geq 0 \end{cases}$$

and hence by (37*) and (38*) we get

$$(45^*) \qquad \begin{cases} V_{\infty 1}(L_{A'}S + M_{A'}) \geq -1, \\ V_{\infty 1}(L_{B'}S + M_{B'}) \geq -2, \\ V_{\infty 1}(L_{C'}S + M_{C'}) \geq -4 \end{cases}$$

and

$$(46^*) \qquad \begin{cases} V_{\infty 1}(L_{A''}S + M_{A''}) \geq -5, \\ V_{\infty 1}(L_{B''}S + M_{B''}) \geq -6, \\ V_{\infty 1}(L_{C''}S + M_{C''}) \geq -8. \end{cases}$$

Since $V_{\infty 6}(\frac{S}{R^3}+6) > 0 < V_{\infty 1}(\frac{S}{R^3}+1)$, we have

$$(\triangle) \qquad \overline{L}, \overline{M} \in k[R] \Rightarrow \begin{cases} \max(3+\deg \overline{L}, \deg \overline{M}) \\ = \max(-V_{\infty 6}(\overline{L}S+\overline{M}), -V_{\infty 1}(\overline{L}S+\overline{M})). \end{cases}$$

where deg denotes degree in R, and where we use the obvious conventions about infinity. Finally, by (2*), (32*), (33*) and (35*), we see that either

$$\begin{cases} V_{\infty 6}(A') \geq 0, \ V_{\infty 6}(B') \geq 0, \ V_{\infty 6}(C') \geq 0, \\ V_{\infty 6}(A'') = -4, \ V_{\infty 6}(B'') \geq -4, \ V_{\infty 6}(C'') \geq -4 \end{cases}$$

or

$$\begin{cases} V_{\infty 6}(A'') \geq 0, \ V_{\infty 6}(B'') \geq 0, \ V_{\infty 6}(C'') \geq 0, \\ V_{\infty 6}(A') = -4, \ V_{\infty 6}(B') \geq -4, \ V_{\infty 6}(C') \geq -4 \end{cases}$$

and hence by (37*) and (38*) we conclude that either

$$(\dagger) \qquad \begin{cases} V_{\infty 6}(L_{A'}S+M_{A'}) \geq -1, & V_{\infty 6}(L_{A''}S+M_{A''}) = -9, \\ V_{\infty 6}(L_{B'}S+M_{B'}) \geq -2, & V_{\infty 6}(L_{B''}S+M_{B''}) \geq -10, \\ V_{\infty 6}(L_{C'}S+M_{C'}) \geq -4, & V_{\infty 6}(L_{C''}S+M_{C''}) \geq -12 \end{cases}$$

or

$$(\dagger\dagger) \qquad \begin{cases} V_{\infty 6}(L_{A'}S+M_{A'}) = -5, & V_{\infty 6}(L_{A''}S+M_{A''}) \geq -5, \\ V_{\infty 6}(L_{B'}S+M_{B'}) \geq -6, & V_{\infty 6}(L_{B''}S+M_{B''}) \geq -6, \\ V_{\infty 6}(L_{C'}S+M_{C'}) \geq -8, & V_{\infty 6}(L_{C''}S+M_{C''}) \geq -8. \end{cases}$$

Suppose if possible that we have (†). Then by (45*) and (△) we get

$$L_{A'} = L_{B'} = 0 \text{ and } \deg M_{A'} \leq 1 \text{ and } \deg M_{B'} \leq 2$$

and hence by (43*) we see that $M_{A'} = M_{B'} = 0$, and therefore by (37*) we have $A' = B' = 0$. Consequently by (2*) and (□) we get

$$(\dagger^*) \ Z^6+J[Z^5+Z^4+Z^3+Z^2+Z]+K = (Z^3+C')(Z^3+A''Z^2+B''Z+C'').$$

Now

$$\begin{aligned} A'' &= \text{coefficient of } Z^5 \text{ in RHS of } (\dagger^*) \\ &= \text{coefficient of } Z^5 \text{ in LHS of } (\dagger^*) \\ &= J \\ &= \text{coefficient of } Z^2 \text{ in LHS of } (\dagger^*) \\ &= \text{coefficient of } Z^2 \text{ in RHS of } (\dagger^*) \\ &= C'A'' \end{aligned}$$

and hence, because $J \neq 0$, we must have $C' = 1$, and so we get

$$\begin{aligned} C'' = C'C'' &= \text{coefficient of } Z^0 \text{ in RHS of } (\dagger^*) \\ &= \text{coefficient of } Z^0 \text{ in LHS of } (\dagger^*) \\ &= K \end{aligned}$$

and therefore

$$\begin{aligned}1+K = C'+C'' &= \text{coefficient of } Z^3 \text{ in RHS of } (\dagger^*)\\ &= \text{coefficient of } Z^3 \text{ in LHS of } (\dagger^*)\\ &= J\end{aligned}$$

and hence by (3*) and (8*) we conclude that

$$L_K = (R+2)^7 L_J$$

which is a contradiction by (4*) and (9*).

Thus we must have

$$(47^*)\qquad \begin{cases} V_{\infty 6}(L_{A'}S+M_{A'}) = -5, & V_{\infty 6}(L_{A''}S+M_{A''}) \geq -5,\\ V_{\infty 6}(L_{B'}S+M_{B'}) \geq -6, & V_{\infty 6}(L_{B''}S+M_{B''}) \geq -6,\\ V_{\infty 6}(L_{C'}S+M_{C'}) \geq -8, & V_{\infty 6}(L_{C''}S+M_{C''}) \geq -8.\end{cases}$$

and hence, in view of ($\triangle$), by (45*) and (46*) there exist elements

$$(48^*)\qquad A'_i, \overline{A}'_i, B'_i, \overline{B}'_i, C'_i, \overline{C}'_i, A''_i, \overline{A}''_i, B''_i, \overline{B}''_i, C''_i, \overline{C}''_i \text{ in } k$$

such that

$$(49^*)\qquad \begin{cases} L_{A'} = \sum_{i=0}^{2} A'_i R^i, & L_{B'} = \sum_{i=0}^{3} B'_i R^i, & L_{C'} = \sum_{i=0}^{5} C'_i R^i,\\ M_{A'} = \sum_{i=0}^{5} \overline{A}'_i R^i, & M_{B'} = \sum_{i=0}^{6} \overline{B}'_i R^i, & M_{C'} = \sum_{i=0}^{8} \overline{C}'_i R^i\end{cases}$$

and

$$(50^*)\qquad \begin{cases} L_{A''} = \sum_{i=0}^{2} A''_i R^i, \quad L_{B''} = \sum_{i=0}^{3} B''_i R^i,\\ L_{C''} = \sum_{i=0}^{5} C''_i R^i,\\ M_{A''} = \sum_{i=0}^{5} \overline{A}''_i R^i, \quad M_{B''} = \sum_{i=0}^{6} \overline{B}''_i R^i,\\ M_{C''} = \sum_{i=0}^{8} \overline{C}''_i R^i\end{cases}$$

and

$$(51^*)\qquad \text{either } A'_2 \neq 0 \text{ or } \overline{A}'_5 \neq 0$$

and

$$(52^*)\qquad \text{either } A''_2 \neq 0 \text{ or } \overline{A}''_5 \neq 0.$$

Let us now convert the valuation conditions (40*) to (45*) into linear conditions on the elements (48*). So let us consider the values of an element of the form

$$H = (H_0 + H_1R + \ldots + H_qR^q)S + (\overline{H}_0 + \overline{H}_1R + \ldots + \overline{H}_{q+3}R^{q+3})$$

where q is a positive integer and $H_0, H_1, \ldots, H_q, \overline{H}_0, \overline{H}_1, \ldots, \overline{H}_{q+3}$ are elements in k.

For the valuation $V_{\infty 1}$ we have $V_{\infty 1}(\frac{S}{R^3}+1)>0$ and by (1*) we get

$$\overline{S}^2 = 1+\overline{R}^2+\overline{R}^4+\overline{R}^6 \quad \text{where} \quad \overline{R}=\frac{1}{R} \quad \text{and} \quad \overline{S}=\frac{S}{R^3}$$

and hence

$$\overline{S}'(5+\overline{S}') = \overline{R}^2+\overline{R}^4+\overline{R}^6 \quad \text{where} \quad \overline{S}=6+\overline{S}' \quad \text{with} \quad V_{\infty 1}(\overline{S}')>0$$

and therefore

$$\overline{S}' = 3\overline{R}^2+\overline{R}^2\overline{S}'' \quad \text{with} \quad V_{\infty 1}(\overline{S}'')>0$$

and by substituting this value of $\overline{S}'$ in the previous equation we get

$$(3\overline{R}^2+\overline{R}^2\overline{S}'')(5+3\overline{R}^2+\overline{R}^2\overline{S}'') = \overline{R}^2+\overline{R}^4+\overline{R}^6$$

and hence $\overline{S}'' = \overline{R}^2\overline{S}'''$ with $V_{\infty 1}(\overline{S}''')\geq 0$ and therefore

$$\overline{S} = 6+3\overline{R}^2+\overline{R}^4\overline{S}''' \quad \text{with} \quad V_{\infty 1}(\overline{S}''')\geq 0$$

and by substituting this value of $\overline{S}$ in the defining equation of H we get

$$\begin{aligned}
\tfrac{H}{R^{q+3}} &= (H_q+H_{q-1}\overline{R}+\ldots+H_0\overline{R}^q)(6+3\overline{R}^2+\overline{R}^4\overline{S}''') \\
&\quad +(\overline{H}_{q+3}+\overline{H}_{q+2}\overline{R}+\ldots+\overline{H}_0\overline{R}^{q+3}) \\
&= (6H_q+\overline{H}_{q+3})+(6H_{q-1}+\overline{H}_{q+2})\overline{R} \\
&\quad +(3H_q+6H_{q-2}+\overline{H}_{q+1})\overline{R}^2+(3H_{q-1}+6H_{q-3}+\overline{H}_q)\overline{R}^3 \\
&\quad +(\text{terms with } V_{\infty 1}\text{-value } \geq 4)
\end{aligned}$$

with the understanding that

$$H_{-1}=H_{-2}=\ldots=0$$

and hence

$$V_{\infty 1}(H)\geq 1-q \Leftrightarrow \begin{cases} \overline{H}_{q+3}=H_q \text{ and } \overline{H}_{q+2}=H_{q-1} \text{ and} \\ \overline{H}_{q+1}=4H_q+H_{q-2} \text{ and } \overline{H}_q=4H_{q-1}+H_{q-3}. \end{cases}$$

In view of (45*) and (49*), by taking $(q,H_i,\overline{H}_i)=(2,A_i',\overline{A}_i')$, $(3,B_i',\overline{B}_i')$, $(5,C_i',\overline{C}_i')$ in the above implication we conclude that

$$(53^*)\quad \begin{cases} \overline{A}_5'=A_2', & \overline{A}_4'=A_1', & \overline{A}_3'=4A_2'+A_0', & \overline{A}_2'=4A_1', \\ \overline{B}_6'=B_3', & \overline{B}_5'=B_2', & \overline{B}_4'=4B_3'+B_1', & \overline{B}_3'=4B_2'+B_0', \\ \overline{C}_8'=C_5', & \overline{C}_7'=C_4', & \overline{C}_6'=4C_5'+C_3', & \overline{C}_5'=4C_4'+C_2'. \end{cases}$$

For the valuation V_{06} we have $V_{06}(R)=1\leq V_{06}(S+6)$ and hence by (1*) we get

$$S_0(2+S_0)=R^2+R^4+R^6 \quad \text{where} \quad S=1+S_0 \quad \text{with} \quad V_{06}(S_0)>0$$

and hence

$$S_0 = 4R^2 + R^2 S_0' \quad \text{with} \quad V_{06}(S_0') > 0$$

and by substituting this value of S_0 in the previous equation we get

$$(4R^2 + R^2 S_0')(2 + 4R^2 + R^2 S_0') = R^2 + R^4 + R^6$$

and therefore $S_0' = R^2 S_0''$ with $V_{06}(S_0'') \geq 0$ and hence

$$S = 1 + 4R^2 + R^4 S_0'' \quad \text{with} \quad V_{06}(S_0'') \geq 0$$

and by substituting this value of S in the defining equation of H we get

$$\begin{aligned} H &= (H_0 + H_1 R + \ldots + H_q R^q)(1 + 4R^2 + R^4 S_0'') \\ &\quad +(\overline{H}_0 + \overline{H}_1 R + \ldots + \overline{H}_{q+3} R^{q+3}) \\ &= (H_0 + \overline{H}_0) + (H_1 + \overline{H}_1)R + (4H_0 + H_2 + \overline{H}_2)R^2 \\ &\quad +(4H_1 + H_3 + \overline{H}_3)R^3 + (\text{terms with } V_{06}\text{-value} \geq 4). \end{aligned}$$

In view of (43*), (44*), (49*) and (50*), by taking $(q, H_i, \overline{H}_i) = (2, A_i', \overline{A}_i')$, $(3, B_i', \overline{B}_i')$, $(5, C_i', \overline{C}_i')$, $(2, A_i'', \overline{A}_i'')$, $(3, B_i'', \overline{B}_i'')$, $(5, C_i'', \overline{C}_i'')$ in the above equation we conclude that

$$(54^*) \quad \begin{cases} \overline{A}_0' = 6A_0', & \overline{A}_1' = 6A_1', \\ \overline{B}_0' = 6B_0', & \overline{B}_1' = 6B_1', \quad \overline{B}_2' = 3B_0' + 6B_2', \\ \overline{C}_0' = 6C_0', & \overline{C}_1' = 6C_1', \quad \overline{C}_2' = 3C_0' + 6C_2', \quad \overline{C}_3' = 3C_1' + 6C_3' \end{cases}$$

and

$$(55^*) \quad \begin{cases} \overline{A}_0'' = 6A_0'', & \overline{A}_1'' = 6A_1'', \\ \overline{B}_0'' = 6B_0'', & \overline{B}_1'' = 6B_1'', \quad \overline{B}_2'' = 3B_0'' + 6B_2'', \\ \overline{C}_0'' = 6C_0'', & \overline{C}_1'' = 6C_1'', \quad \overline{C}_2'' = 3C_0'' + 6C_2'', \\ \quad \overline{C}_3'' = 3C_1'' + 6C_3''. \end{cases}$$

For the valuation V_{26} we have

$$R = 5 + R_2 \quad \text{and} \quad S = 1 + S_2 \quad \text{with} \quad V_{26}(R_2) = 1 \leq V_{26}(S_2)$$

and by (1*) we get

$$S_2(2 + S_2) = S^2 + 6 = R^2 + R^4 + R^6 = 3R_2 + 6R_2^2 + 5R_2^4 + 2R_2^5 + R_2^6$$

and hence

$$S_2 = 5R_2 + R_2 S_2' \quad \text{with} \quad V_{26}(S_2') > 0$$

and by substituting this value of S_2 in the previous equation we get

$$(5R_2 + R_2 S_2')(2 + 5R_2 + R_2 S_2') = 3R_2 + 6R_2^2 + 5R_2^4 + 2R_2^5 + R_2^6$$

and subtracting $3R_2 + 4R_2^2$ from both sides and then dividing throughout by $2R_2$ we get

$$S_2' + 5R_2S_2' + 4R_2S_2'^2 = R_2 + 6R_2^3 + R_2^4 + 4R_2^5$$

and therefore

$$S_2' = R_2 + R_2S_2'' \quad \text{with} \quad V_{26}(S_2'') > 0$$

and by substituting this value of S_2' in the previous equation and subtracting $R_2 + 5R_2^2 + 4R_2^3$ from both sides and then dividing throughout by R_2 we get

$$S_2'' + 5R_2S_2'' + R_2^2S_2'' + 4R_2^2S_2''^2 = 2R_2 + 2R_2^2 + R_2^3 + 4R_2^4$$

and hence

$$S_2'' = 2R_2 + R_2^2S_2''' \quad \text{with} \quad V_{26}(S_2''') \geq 0$$

and therefore

$$S = 1 + 5R_2 + R_2^2 + 2R_2^3 + R_2^4S_2''' \quad \text{with} \quad V_{26}(S_2''') \geq 0$$

and by substituting this value of S in the defining equation of H we get

$$\begin{aligned}
H &= (H_0 + H_1R + \ldots + H_qR^q)(1 + 5R_2 + R_2^2 + 2R_2^3 + R_2^4S_2''') \\
&\quad +(\overline{H}_0 + \overline{H}_1R + \ldots + \overline{H}_{q+3}R^{q+3}) \\
&= [H_0 + H_1(5 + R_2) + \ldots + H_q(5 + R_2)^q](1 + 5R_2 + R_2^2 + 2R_2^3) \\
&\quad +[\overline{H}_0 + \overline{H}_1(5 + R_2) + \ldots + \overline{H}_{q+3}(5 + R_2)^{q+3}] \\
&\quad +(\text{terms with } V_{26}\text{-value} \geq 4) \\
&= \left[\textstyle\sum_{i=0}^{q} 5^iH_i \; + \; \sum_{i=0}^{q+3} 5^i\overline{H}_i\right] \\
&\quad + \left[\textstyle\sum_{i=0}^{q}(5^{i+1} + i5^{i-1})H_i \; + \; \sum_{i=0}^{q+3} i5^{i-1}\overline{H}_i\right] R_2 \\
&\quad + \left[\textstyle\sum_{i=0}^{q} \left(5^i + i5^i + \binom{i}{2} 5^{i-2}\right) H_i \right.\\
&\qquad\qquad \left. + \textstyle\sum_{i=0}^{q+3} \binom{i}{2} 5^{i-2}\overline{H}_i\right] R_2^2 \\
&\quad + \left[\textstyle\sum_{i=0}^{q} \left(2(5^i) + i5^{i-1} + \binom{i}{2} 5^{i-1} + \binom{i}{3} 5^{i-3}\right) H_i \right.\\
&\qquad\qquad \left. + \textstyle\sum_{i=0}^{q+3} \binom{i}{3} 5^{i-3}\overline{H}_i\right] R_2^3 \\
&\quad +(\text{terms with } V_{26}\text{-value} \geq 4)
\end{aligned}$$

and upon simplifying the last expression we get

$$\begin{aligned}
H &= \left[\textstyle\sum_{i=0}^{q+3} 5^i\overline{H}_i \; - \; \sum_{i=0}^{q}(6)5^iH_i\right] \\
&\quad + \left[\textstyle\sum_{i=0}^{q+3}(3i)5^i\overline{H}_i \; - \; \sum_{i=0}^{q}(4i + 2)5^iH_i\right] R_2 \\
&\quad + \left[\textstyle\sum_{i=0}^{q+3} i(i-1)5^i\overline{H}_i \; - \; \sum_{i=0}^{q}(6i^2 + 6)5^iH_i\right] R_2^2 \\
&\quad + \left[\textstyle\sum_{i=0}^{q+3} i(i-1)(i-2)5^i\overline{H}_i \; - \; \sum_{i=0}^{q}(6i^3 + 5i^2 + 5)5^iH_i\right] R_2^3 \\
&\quad +(\text{terms with } V_{26}\text{-value} \geq 4).
\end{aligned}$$

In view of (41*), (42*), (49*) and (50*), by taking $(q, H_i, \overline{H}_i) = (2, A'_i, \overline{A}'_i)$, $(3, B'_i, \overline{B}'_i)$, $(5, C'_i, \overline{C}'_i)$, $(2, A''_i, \overline{A}''_i)$, $(3, B''_i, \overline{B}''_i)$, $(5, C''_i, \overline{C}''_i)$ in the above equation we conclude that

$$(56^*)\qquad \begin{cases} \sum_{i=0}^{5} 5^i \overline{A}'_i = \sum_{i=0}^{2} (6) 5^i A'_i, \\ \sum_{i=0}^{6} 5^i \overline{B}'_i = \sum_{i=0}^{3} (6) 5^i B'_i, \\ \sum_{i=0}^{6} (3i) 5^i \overline{B}'_i = \sum_{i=0}^{3} (4i+2) 5^i B'_i, \\ \sum_{i=0}^{8} 5^i \overline{C}'_i = \sum_{i=0}^{5} (6) 5^i C'_i, \\ \sum_{i=0}^{8} (3i) 5^i \overline{C}'_i = \sum_{i=0}^{5} (4i+2) 5^i C'_i, \\ \sum_{i=0}^{8} i(i-1) 5^i \overline{C}'_i = \sum_{i=0}^{5} (6i^2+6) 5^i C'_i, \\ \sum_{i=0}^{8} i(i-1)(i-2) 5^i \overline{C}'_i = \sum_{i=0}^{5} (6i^3+5i^2+5) 5^i C'_i \end{cases}$$

and

$$(57^*)\qquad \begin{cases} \sum_{i=0}^{5} 5^i \overline{A}''_i = \sum_{i=0}^{2} (6) 5^i A''_i, \\ \sum_{i=0}^{6} 5^i \overline{B}''_i = \sum_{i=0}^{3} (6) 5^i B''_i, \\ \sum_{i=0}^{6} (3i) 5^i \overline{B}''_i = \sum_{i=0}^{3} (4i+2) 5^i B''_i, \\ \sum_{i=0}^{8} 5^i \overline{C}''_i = \sum_{i=0}^{5} (6) 5^i C''_i, \\ \sum_{i=0}^{8} (3i) 5^i \overline{C}''_i = \sum_{i=0}^{5} (4i+2) 5^i C''_i, \\ \sum_{i=0}^{8} i(i-1) 5^i \overline{C}''_i = \sum_{i=0}^{5} (6i^2+6) 5^i C''_i, \\ \sum_{i=0}^{8} i(i-1)(i-2) 5^i \overline{C}''_i = \sum_{i=0}^{5} (6i^3+5i^2+5) 5^i C''_i. \end{cases}$$

By substituting (54*) in (56*) and dividing by suitable powers of 5 we get

$$\begin{cases} \sum_{i=2}^{5} 5^{i-2} \overline{A}'_i = 6A'_2, \\ \sum_{i=3}^{6} 5^{i-3} \overline{B}'_i = 5B'_0 + 6B'_3, \\ \sum_{i=3}^{6} (3i) 5^{i-3} \overline{B}'_i = 4B'_1 + 6B'_2, \\ \sum_{i=4}^{8} 5^{i-4} \overline{C}'_i = C'_0 + 5C'_1 + 6C'_4 + 2C'_5, \\ \sum_{i=4}^{8} (3i) 5^{i-4} \overline{C}'_i = C'_1 + 4C'_2 + 6C'_3 + 4C'_4 + 5C'_5, \\ \sum_{i=4}^{8} i(i-1) 5^{i-4} \overline{C}'_i = 5C'_0 + 4C'_1 + C'_2 + 2C'_3 + 4C'_4 + 3C'_5, \\ \sum_{i=4}^{8} i(i-1)(i-2) 5^{i-4} \overline{C}'_i = 6C'_0 + 6C'_2 + 3C'_3 + 4C'_5 \end{cases}$$

and separately simplifying these equations we get

$$\begin{cases} \overline{A}'_2 + 5\overline{A}'_3 + 4\overline{A}'_4 + 6\overline{A}'_5 = 6A'_2, \\ \overline{B}'_3 + 5\overline{B}'_4 + 4\overline{B}'_5 + 6\overline{B}'_6 = 5B'_0 + 6B'_3, \\ \overline{B}'_3 + 2\overline{B}'_4 + 2\overline{B}'_5 + 5\overline{B}'_6 = 2B'_1 + 3B'_2, \\ \overline{C}'_4 + 5\overline{C}'_5 + 4\overline{C}'_6 + 6\overline{C}'_7 + 2\overline{C}'_8 = C'_0 + 5C'_1 + 6C'_4 + 2C'_5, \\ \overline{C}'_4 + \overline{C}'_5 + 6\overline{C}'_6 + 4\overline{C}'_8 = 3C'_1 + 5C'_2 + 4C'_3 + 5C'_4 + C'_5, \\ \overline{C}'_4 + 6\overline{C}'_5 + 3\overline{C}'_6 = C'_0 + 5C'_1 + 3C'_2 + 6C'_3 + 5C'_4 + 2C'_5, \\ \overline{C}'_4 + 2\overline{C}'_5 + 6\overline{C}'_6 = 2C'_0 + 2C'_2 + C'_3 + 6C'_5 \end{cases}$$

and simultaneously solving them we get

$$
(58^*)\qquad \begin{cases}
\overline{A}'_2 + 5\overline{A}'_3 + 4\overline{A}'_4 + 6\overline{A}'_5 = 6A'_2, \\
\overline{B}'_4 + 3\overline{B}'_5 + 5\overline{B}'_6 = 4B'_0 + 4B'_1 + 6B'_2 + 2B'_3, \\
\overline{B}'_3 + 2\overline{B}'_4 + 2\overline{B}'_5 + 5\overline{B}'_6 = 2B'_1 + 3B'_2, \\
\overline{C}'_7 + 6\overline{C}'_8 = C'_0 + 2C'_1 + 4C'_2 + C'_3 + 5C'_4 + 3C'_5, \\
\overline{C}'_6 + 4\overline{C}'_8 = 3C'_0 + 6C'_1 + 5C'_2 + 6C'_3 + C'_4 + C'_5, \\
\overline{C}'_5 + \overline{C}'_6 = 5C'_0 + 3C'_1 + 2C'_2 + 3C'_3 + 3C'_4 + 6C'_5, \\
\overline{C}'_4 + 2\overline{C}'_5 + 6\overline{C}'_6 = 2C'_0 + 2C'_2 + C'_3 + 6C'_5.
\end{cases}
$$

Likewise, by substituting (55*) in (57*) and simultaneously solving the resulting equations we get (58*) with double primes replacing single primes, i.e., we get

$$
(59^*)\qquad \begin{cases}
\overline{A}''_2 + 5\overline{A}''_3 + 4\overline{A}''_4 + 6\overline{A}''_5 = 6A''_2, \\
\overline{B}''_4 + 3\overline{B}''_5 + 5\overline{B}''_6 = 4B''_0 + 4B''_1 + 6B''_2 + 2B''_3, \\
\overline{B}''_3 + 2\overline{B}''_4 + 2\overline{B}''_5 + 5\overline{B}''_6 = 2B''_1 + 3B''_2, \\
\overline{C}''_7 + 6\overline{C}''_8 = C''_0 + 2C''_1 + 4C''_2 + C''_3 + 5C''_4 + 3C''_5, \\
\overline{C}''_6 + 4\overline{C}''_8 = 3C''_0 + 6C''_1 + 5C''_2 + 6C''_3 + C''_4 + C''_5, \\
\overline{C}''_5 + \overline{C}''_6 = 5C''_0 + 3C''_1 + 2C''_2 + 3C''_3 + 3C''_4 + 6C''_5, \\
\overline{C}''_4 + 2\overline{C}''_5 + 6\overline{C}''_6 = 2C''_0 + 2C''_2 + C''_3 + 6C''_5.
\end{cases}
$$

By substituting (53*) in (58*) we get

$$
\begin{cases}
5A'_0 + A'_1 + 5A'_2 = 6A'_2, \\
B'_1 + 3B'_2 + 2B'_3 = 4B'_0 + 4B'_1 + 6B'_2 + 2B'_3, \\
B'_0 + 2B'_1 + 6B'_2 + 6B'_3 = 2B'_1 + 3B'_2, \\
C'_4 + 6C'_5 = C'_0 + 2C'_1 + 4C'_2 + C'_3 + 5C'_4 + 3C'_5, \\
C'_3 + C'_5 = 3C'_0 + 6C'_1 + 5C'_2 + 6C'_3 + C'_4 + C'_5, \\
C'_2 + C'_3 + 4C'_4 + 4C'_5 = 5C'_0 + 3C'_1 + 2C'_2 + 3C'_3 + 3C'_4 + 6C'_5, \\
\overline{C}'_4 + 2C'_2 + 6C'_3 + C'_4 + 3C'_5 = 2C'_0 + 2C'_2 + C'_3 + 6C'_5
\end{cases}
$$

and simplifying these equations we get

$$
\begin{cases}
A'_2 = 5A'_0 + A'_1, \\
4B'_2 = 4B'_0 + 3B'_1, \\
3B'_2 + 6B'_3 = 6B'_0, \\
6C'_3 + 3C'_4 + 3C'_5 = C'_0 + 2C'_1 + 4C'_2, \\
2C'_3 + 6C'_4 = 3C'_0 + 6C'_1 + 5C'_2, \\
5C'_3 + C'_4 + 5C'_5 = 5C'_0 + 3C'_1 + C'_2, \\
\overline{C}'_4 = 2C'_0 + 2C'_3 + 6C'_4 + 3C'_5
\end{cases}
$$

and simultaneously solving them we get

$$(60^*)\qquad \begin{cases} A'_2 = 5A'_0 + A'_1, \\ B'_2 = B'_0 + 6B'_1, \quad B'_3 = 4B'_0 + 4B'_1, \\ C'_3 = 3C'_0 + 6C'_1, \quad C'_4 = 3C'_0 + 6C'_1 + 2C'_2, \\ C'_5 = 3C'_0 + 6C'_1 + 4C'_2, \\ \overline{C}'_4 = 3C'_1 + 3C'_2. \end{cases}$$

In view of (60*), by (54*) we get

$$(61^*)\quad \begin{cases} \overline{A}'_0 = 6A'_0, \quad \overline{A}'_1 = 6A'_1, \\ \overline{B}'_0 = 6B'_0, \quad \overline{B}'_1 = 6B'_1, \quad \overline{B}'_2 = 2B'_0 + B'_1, \\ \overline{C}'_0 = 6C'_0, \quad \overline{C}'_1 = 6C'_1, \quad \overline{C}'_2 = 3C'_0 + 6C'_2, \quad \overline{C}'_3 = 4C'_0 + 4C'_1 \end{cases}$$

and by (53*) we get

$$(62^*)\qquad \begin{cases} \overline{A}'_5 = 5A'_0 + A'_1, \quad \overline{A}'_4 = A'_1, \quad \overline{A}'_3 = 4A'_1, \quad \overline{A}'_2 = 4A'_1, \\ \overline{B}'_6 = 4B'_0 + 4B'_1, \quad \overline{B}'_5 = B'_0 + 6B'_1, \\ \overline{B}'_4 = 2B'_0 + 3B'_1, \quad \overline{B}'_3 = 5B'_0 + 3B'_1, \\ \overline{C}'_8 = 3C'_0 + 6C'_1 + 4C'_2, \quad \overline{C}'_7 = 3C'_0 + 6C'_1 + 2C'_2, \\ \overline{C}'_6 = C'_0 + 2C'_1 + 2C'_2, \quad \overline{C}'_5 = 5C'_0 + 3C'_1 + 2C'_2. \end{cases}$$

For the valuation V_{41} we have

$$R = 3 + R_4 \quad \text{and} \quad S = 6 + S_4 \quad \text{with} \quad V_{41}(R_4) = 1 \le V_{41}(S_4)$$

and by (1*) we get

$$S_4(5+S_4) = S^2 + 6 = R^2 + R^4 + R^6 = 4R_4 + 3R_4^2 + 6R_4^3 + 3R_4^4 + 4R_4^5 + R_4^6$$

and hence

$$S_4 = 5R_4 + R_4S'_4 \quad \text{with} \quad V_{41}(S'_4) > 0$$

and by substituting this value of S_4 in the previous equation we get

$$(5R_4 + R_4S'_4)(5 + 5R_4 + R_4S'_4) = 4R_4 + 3R_4^2 + 6R_4^3 + 3R_4^4 + 4R_4^5 + R_4^6$$

and subtracting $4R_4 + 4R_4^2$ from both sides and then dividing throughout by $5R_4$ we get

$$S'_4 + 2R_4S'_4 + 3R_4S_4'^2 = 4R_4 + 4R_4^2 + 2R_4^3 + 5R_4^4 + 3R_4^5$$

and therefore

$$S'_4 = 4R_4 + R_4S''_4 \quad \text{with} \quad V_{41}(S''_4) > 0$$

and by substituting this value of S'_4 in the previous equation and subtracting $4R_4 + R_4^2 + 6R_4^3$ from both sides and then dividing throughout by R_4 we get

$$S''_4 + 2R_4S''_4 + 3R_4^2S''_4 + 3R_4^2S_4''^2 = 3R_4 + 3R_4^2 + 5R_4^3 + 3R_4^4$$

and hence

$$S_4'' = 3R_4 + R_4^2 S_4''' \quad \text{with} \quad V_{41}(S_4''') \geq 0$$

and therefore

$$S = 6 + 5R_4 + 4R_4^2 + 3R_4^3 + R_4^4 S_4''' \quad \text{with} \quad V_{41}(S_4''') \geq 0$$

and by substituting this value of S in the defining equation of H we get

$$\begin{aligned}
H &= (H_0 + H_1 R + \ldots + H_q R^q)(6 + 5R_4 + 4R_4^2 + 3R_4^3 + R_4^4 S_4''') \\
&\quad + (\overline{H}_0 + \overline{H}_1 R + \ldots + \overline{H}_{q+3} R^{q+3}) \\
&= [H_0 + H_1(3 + R_4) + \ldots + H_q(3 + R_4)^q](6 + 5R_4 + 4R_4^2 + 3R_4^3) \\
&\quad + [\overline{H}_0 + \overline{H}_1(3 + R_4) + \ldots + \overline{H}_{q+3}(3 + R_4)^{q+3}] \\
&\quad + (\text{terms with } V_{41}\text{-value} \geq 4) \\
&= \left[\textstyle\sum_{i=0}^q (6)3^i H_i \; + \; \sum_{i=0}^{q+3} 3^i \overline{H}_i\right] \\
&\quad + \left[\textstyle\sum_{i=0}^q \left((5)3^i + 6i3^{i-1}\right) H_i \; + \; \sum_{i=0}^{q+3} i3^{i-1} \overline{H}_i\right] R_4 \\
&\quad + \left[\textstyle\sum_{i=0}^q \left((4)3^i + 5i3^{i-1} + 6\binom{i}{2} 3^{i-2}\right) H_i \right. \\
&\qquad\qquad \left. + \textstyle\sum_{i=0}^{q+3} \binom{i}{2} 3^{i-2} \overline{H}_i\right] R_4^2 \\
&\quad + \left[\textstyle\sum_{i=0}^q \left(3^{i+1} + 4i3^{i-1} + 5\binom{i}{2} 3^{i-2} + 6\binom{i}{3} 3^{i-3}\right) H_i \right. \\
&\qquad\qquad \left. + \textstyle\sum_{i=0}^{q+3} \binom{i}{3} 3^{i-3} \overline{H}_i\right] R_4^3 \\
&\quad + (\text{terms with } V_{41}\text{-value} \geq 4)
\end{aligned}$$

and upon simplifying the last expression we get

$$\begin{aligned}
H &= \left[\textstyle\sum_{i=0}^{q+3} 3^i \overline{H}_i \; - \; \sum_{i=0}^q 3^i H_i\right] \\
&\quad + \left[\textstyle\sum_{i=0}^{q+3} (5i)3^i \overline{H}_i \; - \; \sum_{i=0}^q (5i+2)3^i H_i\right] R_4 \\
&\quad + \left[\textstyle\sum_{i=0}^{q+3} 2i(i-1)3^i \overline{H}_i \; - \; \sum_{i=0}^q (2i^2 + i + 3)3^i H_i\right] R_4^2 \\
&\quad + \left[\textstyle\sum_{i=0}^{q+3} i(i-1)(i-2)3^i \overline{H}_i \; - \; \sum_{i=0}^q (i^3 + i^2 + 6i + 4)3^i H_i\right] R_4^3 \\
&\quad + (\text{terms with } V_{41}\text{-value} \geq 4).
\end{aligned}$$

In view of (2*), (3*), (37*) and (38*), by equating the coefficients of Z^0 and Z on both sides of (□) and then multiplying throughout by $(R+2)^8(R+3)^4$ and $(R+2)^6(R+3)^4$ respectively we get

$$(\square_0) \qquad L_K S + M_K = (L_{C'} S + M_{C'})(L_{C''} S + M_{C''})$$

and

$$(\square_1) \qquad \begin{aligned}(L_J S + M_J)(R+2)^5 &= (L_{B'} S + M_{B'})(L_{C''} S + M_{C''}) \\ &\quad + (L_{B''} S + M_{B''})(L_{C'} S + M_{C'})\end{aligned}$$

and hence by (28*) and (29*) we see that either

$$(\ddagger) \qquad V_{41}(L_{C'}S + M_{C'}) = 4 = 4 + V_{41}(L_{C''}S + M_{C''})$$

or

$$(\ddagger\ddagger) \qquad V_{41}(L_{C''}S + M_{C''}) = 4 = 4 + V_{41}(L_{C'}S + M_{C'})$$

and by taking $(q, H_i, \overline{H}_i) = (5, C'_i, \overline{C}'_i)$ or $(5, C''_i, \overline{C}''_i)$ in the above simplified expression of H we conclude that if ($\ddagger$) then

$$(\ddagger') \qquad \begin{cases} \sum_{i=0}^{8} 3^i \overline{C}'_i = \sum_{i=0}^{5} 3^i C'_i, \\ \sum_{i=0}^{8} (5i) 3^i \overline{C}'_i = \sum_{i=0}^{5} (5i+2) 3^i C'_i, \\ \sum_{i=0}^{8} 2i(i-1) 3^i \overline{C}'_i = \sum_{i=0}^{5} (2i^2 + i + 3) 3^i C'_i, \\ \sum_{i=0}^{8} i(i-1)(i-2) 3^i \overline{C}'_i = \sum_{i=0}^{5} (i^3 + i^2 + 6i + 4) 3^i C'_i \end{cases}$$

whereas if ($\ddagger\ddagger$) then

$$(\ddagger\ddagger') \qquad \begin{cases} \sum_{i=0}^{8} 3^i \overline{C}''_i = \sum_{i=0}^{5} 3^i C''_i, \\ \sum_{i=0}^{8} (5i) 3^i \overline{C}''_i = \sum_{i=0}^{5} (5i+2) 3^i C''_i, \\ \sum_{i=0}^{8} 2i(i-1) 3^i \overline{C}''_i = \sum_{i=0}^{5} (2i^2 + i + 3) 3^i C''_i, \\ \sum_{i=0}^{8} i(i-1)(i-2) 3^i \overline{C}''_i = \sum_{i=0}^{5} (i^3 + i^2 + 6i + 4) 3^i C''_i. \end{cases}$$

If ($\ddagger'$) then by substituting (54*) in ($\ddagger'$) and dividing by suitable powers of 3 we would get

$$\begin{cases} \sum_{i=4}^{8} 3^{i-4} \overline{C}'_i = 6C'_0 + 4C'_1 + C'_2 + 3C'_3 + C'_4 + 3C'_5, \\ \sum_{i=4}^{8} (5i) 3^{i-4} \overline{C}'_i = 3C'_0 + C'_1 + 4C'_2 + 6C'_3 + C'_4 + 4C'_5, \\ \sum_{i=4}^{8} 2i(i-1) 3^{i-4} \overline{C}'_i = 3C'_1 + 5C'_2 + 5C'_3 + 4C'_4 + 6C'_5, \\ \sum_{i=4}^{8} i(i-1)(i-2) 3^{i-4} \overline{C}'_i = C'_0 + 3C'_1 + 5C'_3 + 3C'_4 + 6C'_5 \end{cases}$$

and separately simplifying these equations we would get

$$\begin{cases} \overline{C}'_4 + 3\overline{C}'_5 + 2\overline{C}'_6 + 6\overline{C}'_7 + 4\overline{C}'_8 = 6C'_0 + 4C'_1 + C'_2 + 3C'_3 + C'_4 + 3C'_5, \\ \overline{C}'_4 + 2\overline{C}'_5 + 3\overline{C}'_6 + \overline{C}'_8 = 4C'_0 + 6C'_1 + 3C'_2 + C'_3 + 6C'_4 + 3C'_5, \\ \overline{C}'_4 + 5\overline{C}'_5 + 5\overline{C}'_6 = C'_1 + 4C'_2 + 4C'_3 + 6C'_4 + 2C'_5, \\ \overline{C}'_4 + 4\overline{C}'_5 + 3\overline{C}'_6 = 5C'_0 + C'_1 + 4C'_3 + C'_4 + 2C'_5 \end{cases}$$

and simultaneously solving them we would get

$$(\ddagger'') \qquad \begin{cases} \overline{C}'_7 + 5\overline{C}'_8 + 5C'_3 + 3C'_4 + 6C'_5 = 3C'_0 + 3C'_2, \\ \overline{C}'_6 + 2\overline{C}'_8 + 6C'_3 + 5C'_4 + 5C'_5 = 6C'_0 + 3C'_1 + C'_2, \\ \overline{C}'_5 + 2\overline{C}'_6 + 2C'_4 = 2C'_0 + 4C'_2 \\ \overline{C}'_4 + 4\overline{C}'_5 + 3\overline{C}'_6 + 3C'_3 + 6C'_4 + 5C'_5 = 5C'_0 + C'_1. \end{cases}$$

Likewise, if (‡‡′) then by substituting (55*) in (‡‡′) and simultaneously solving the resulting equations we would get (‡″) with double primes replacing single primes, i.e., we would get

$$(\ddagger\ddagger'') \qquad \begin{cases} \overline{C}''_7 + 5\overline{C}''_8 + 5C''_3 + 3C''_4 + 6C''_5 = 3C''_0 + 3C''_2, \\ \overline{C}''_6 + 2\overline{C}''_8 + 6C''_3 + 5C''_4 + 5C''_5 = 6C''_0 + 3C''_1 + C''_2, \\ \overline{C}''_5 + 2\overline{C}''_6 + 2C''_4 = 2C''_0 + 4C''_2 \\ \overline{C}''_4 + 4\overline{C}''_5 + 3\overline{C}''_6 + 3C''_3 + 6C''_4 + 5C''_5 = 5C''_0 + C''_1. \end{cases}$$

By (60*) to (62*) we get

$$\begin{cases} \overline{C}'_7 + 5\overline{C}'_8 + 5C'_3 + 3C'_4 + 6C'_5 = 4C'_0 + C'_1 + 3C'_2, \\ \overline{C}'_6 + 2\overline{C}'_8 + 6C'_3 + 5C'_4 + 5C'_5 = 6C'_0 + 5C'_1 + 5C'_2, \\ \overline{C}'_5 + 2\overline{C}'_6 + 2C'_4 = 6C'_0 + 5C'_1 + 3C'_2, \\ \overline{C}'_4 + 4\overline{C}'_5 + 3\overline{C}'_6 + 3C'_3 + 6C'_4 + 5C'_5 = 2C'_0 \end{cases}$$

and since the LHS of these equations coincides with the LHS of the equations (‡″), we see that if (‡″) then the corresponding RHS would coincide, i.e., we would get

$$\begin{cases} 4C'_0 + C'_1 + 3C'_2 = 3C'_0 + 3C'_2, \\ 6C'_0 + 5C'_1 + 5C'_2 = 6C'_0 + 3C'_1 + C'_2, \\ 6C'_0 + 5C'_1 + 3C'_2 = 2C'_0 + 4C'_2, \\ 2C'_0 = 5C'_0 + C'_1 \end{cases}$$

and simultaneously solving the above equations we would get $C'_0 = C'_1 = C'_2 = 0$ and by (60*) to (62*) this would yield $C'_i = 0 = \overline{C}'_j$ for $0 \le i \le 5$ and $0 \le j \le 8$ and hence by (37*) and (49*) we would get $C' = 0$ and this by (2*) and (□) we would yield $K = 0$ which would be a contradiction by (8*) to (10*). Therefore we must have (‡‡) and (‡‡″), i.e., we must have

$$(63^*) \qquad V_{41}(L_{C''}S + M_{C''}) = 4 = 4 + V_{41}(L_{C'}S + M_{C'})$$

and

$$(64^*) \qquad \begin{cases} \overline{C}''_7 + 5\overline{C}''_8 + 5C''_3 + 3C''_4 + 6C''_5 = 3C''_0 + 3C''_2, \\ \overline{C}''_6 + 2\overline{C}''_8 + 6C''_3 + 5C''_4 + 5C''_5 = 6C''_0 + 3C''_1 + C''_2, \\ \overline{C}''_5 + 2\overline{C}''_6 + 2C''_4 = 2C''_0 + 4C''_2 \\ \overline{C}''_4 + 4\overline{C}''_5 + 3\overline{C}''_6 + 3C''_3 + 6C''_4 + 5C''_5 = 5C''_0 + C''_1. \end{cases}$$

Since the degree of the divisor of a function is zero, in view of (29*) and ($\square_0$), by (63*) and (39*) to (47*) we conclude that

$$(65^*) \qquad V_{31}(L_{C'}S + M_{C'}) = 4 = V_{31}(L_{C''}S + M_{C''})$$

and

$$(66^*) \qquad \begin{cases} (L_{C'}S + M_{C'}) = \frac{R^{*4}(R+2)^{*4}(R+3)'^4}{R'^4_\infty R^{*8}_\infty} \quad \text{and} \\ (L_{C''}S + M_{C''}) = \frac{R^{*4}(R+2)^{*4}(R+3)'^4(R+4)'^4}{R'^8_\infty R^{*8}_\infty} \end{cases}$$

and hence by (37*) and (38*) we get

$$(67^*) \qquad (C') = \frac{R^{*4}(R+3)'^4}{(R+2)'^4 R^{*4}_\infty} \quad \text{and} \quad (C'') = \frac{R^{*4}(R+4)'^4}{(R+2)'^4(R+3)^{*4}}.$$

For the valuation V_{31} we have

$$R = 4 + R_3 \quad \text{and} \quad S = 6 + S_3 \quad \text{with} \quad V_{31}(R_3) = 1 \le V_{31}(S_3)$$

and by (1*) we get

$$S_3(5+S_3) = S^2 + 6 = R^2 + R^4 + R^6 = 3R_3 + 3R_3^2 + R_3^3 + 3R_3^4 + 3R_3^5 + R_3^6$$

and hence

$$S_3 = 2R_3 + R_3 S_3' \quad \text{with} \quad V_{31}(S_3') > 0$$

and by substituting this value of S_3 in the previous equation we get

$$(2R_3 + R_3S_3')(5 + 2R_3 + R_3S_3') = 3R_3 + 3R_3^2 + R_3^3 + 3R_3^4 + 3R_3^5 + R_3^6$$

and subtracting $3R_3 + 4R_3^2$ from both sides and then dividing throughout by $5R_3$ we get

$$S_3' + 5R_3S_3' + 3R_3S_3'^2 = 4R_3 + 3R_3^2 + 2R_3^3 + 2R_3^4 + 3R_3^5$$

and therefore

$$S_3' = 4R_3 + R_3S_3'' \quad \text{with} \quad V_{31}(S_3'') > 0$$

and by substituting this value of S_3' in the previous equation and subtracting $4R_3 + 6R_3^2 + 6R_3^3$ from both sides and then dividing throughout by R_3 we get

$$S_3'' + 5R_3S_3'' + 3R_3^2S_3'' + 3R_3^2S_3''^2 = 4R_3 + 3R_3^2 + 2R_3^3 + 3R_3^4$$

and hence

$$S_3'' = 4R_3 + R_3^2S_3''' \quad \text{with} \quad V_{31}(S_3''') \ge 0$$

and therefore

$$S = 6 + 2R_3 + 4R_3^2 + 4R_3^3 + R_3^4S_3''' \quad \text{with} \quad V_{31}(S_3''') \ge 0$$

and by substituting this value of S in the defining equation of H we get

$$\begin{aligned}
H &= (H_0 + H_1R + \ldots + H_qR^q)(6 + 2R_3 + 4R_3^2 + 4R_3^3 + R_3^4S_3'''') \\
&\quad +(\overline{H}_0 + \overline{H}_1R + \ldots + \overline{H}_{q+3}R^{q+3}) \\
&= [H_0 + H_1(4 + R_3) + \ldots + H_q(4 + R_3)^q](6 + 2R_3 + 4R_3^2 + 4R_3^3) \\
&\quad +[\overline{H}_0 + \overline{H}_1(4 + R_3) + \ldots + \overline{H}_{q+3}(4 + R_3)^{q+3}] \\
&\quad +(\text{terms with } V_{31}\text{-value} \geq 4) \\
&= \left[\textstyle\sum_{i=0}^{q}(6)4^iH_i \;+\; \sum_{i=0}^{q+3} 4^i\overline{H}_i\right] \\
&\quad + \left[\textstyle\sum_{i=0}^{q} \left((2)4^i + 6i4^{i-1}\right) H_i \;+\; \sum_{i=0}^{q+3} i4^{i-1}\overline{H}_i\right] R_3 \\
&\quad + \left[\textstyle\sum_{i=0}^{q} \left(4^{i+1} + 2i4^{i-1} + 6\begin{pmatrix} i \\ 2 \end{pmatrix} 4^{i-2}\right) H_i \right. \\
&\qquad\qquad \left. + \textstyle\sum_{i=0}^{q+3} \begin{pmatrix} i \\ 2 \end{pmatrix} 4^{i-2}\overline{H}_i\right] R_3^2 \\
&\quad + \left[\textstyle\sum_{i=0}^{q} \left(4^{i+1} + i4^i + 2\begin{pmatrix} i \\ 2 \end{pmatrix} 4^{i-2} + 6\begin{pmatrix} i \\ 3 \end{pmatrix} 4^{i-3}\right) H_i \right. \\
&\qquad\qquad \left. + \textstyle\sum_{i=0}^{q+3} \begin{pmatrix} i \\ 3 \end{pmatrix} 4^{i-3}\overline{H}_i\right] R_3^3 \\
&\quad +(\text{terms with } V_{31}\text{-value} \geq 4)
\end{aligned}$$

and upon simplifying the last expression we get

$$\begin{aligned}
H &= \left[\textstyle\sum_{i=0}^{q+3} 4^i\overline{H}_i \;-\; \sum_{i=0}^{q} 4^iH_i\right] \\
&\quad + \left[\textstyle\sum_{i=0}^{q+3}(2i)4^i\overline{H}_i \;-\; \sum_{i=0}^{q}(2i + 5)4^iH_i\right] R_3 \\
&\quad + \left[\textstyle\sum_{i=0}^{q+3} 2i(i-1)4^i\overline{H}_i \;-\; \sum_{i=0}^{q}(2i^2 + i + 3)4^iH_i\right] R_3^2 \\
&\quad + \left[\textstyle\sum_{i=0}^{q+3} 6i(i-1)(i-2)4^i\overline{H}_i \right. \\
&\qquad\qquad \left. - \textstyle\sum_{i=0}^{q}(6i^3 + 6i^2 + i + 3)4^iH_i\right] R_3^3 \\
&\quad +(\text{terms with } V_{31}\text{–value} \geq 4).
\end{aligned}$$

In view of (40*), (49*), (50*) and (65*), by taking $(q, H_i, \overline{H}_i) = (2, A_i'', \overline{A}_i'')$,

$(3, B_i'', \overline{B}_i'')$, $(5, C_i'', \overline{C}_i'')$, $(5, C_i', \overline{C}_i')$ in the above equation we conclude that

$$(68^*)\quad \begin{cases} \sum_{i=0}^{5} 4^i \overline{A}_i'' = \sum_{i=0}^{2} 4^i A_i'', \\ \sum_{i=0}^{5} (2i)4^i \overline{A}_i'' = \sum_{i=0}^{2} (2i+5)4^i A_i'', \\ \sum_{i=0}^{5} 2i(i-1)4^i \overline{A}_i'' = \sum_{i=0}^{2} (2i^2+i+3)4^i A_i'', \\ \sum_{i=0}^{5} 6i(i-1)(i-2)4^i \overline{A}_i'' = \sum_{i=0}^{2} (6i^3+6i^2+i+3)4^i A_i'', \\ \sum_{i=0}^{6} 4^i \overline{B}_i'' = \sum_{i=0}^{3} 4^i B_i'', \\ \sum_{i=0}^{6} (2i)4^i \overline{B}_i'' = \sum_{i=0}^{3} (2i+5)4^i B_i'', \\ \sum_{i=0}^{6} 2i(i-1)4^i \overline{B}_i'' = \sum_{i=0}^{3} (2i^2+i+3)4^i B_i'', \\ \sum_{i=0}^{6} 6i(i-1)(i-2)4^i \overline{B}_i'' = \sum_{i=0}^{3} (6i^3+6i^2+i+3)4^i B_i'', \\ \sum_{i=0}^{8} 4^i \overline{C}_i'' = \sum_{i=0}^{5} 4^i C_i'', \\ \sum_{i=0}^{8} (2i)4^i \overline{C}_i'' = \sum_{i=0}^{5} (2i+5)4^i C_i'', \\ \sum_{i=0}^{8} 2i(i-1)4^i \overline{C}_i'' = \sum_{i=0}^{5} (2i^2+i+3)4^i C_i'', \\ \sum_{i=0}^{8} 6i(i-1)(i-2)4^i \overline{C}_i'' = \sum_{i=0}^{5} (6i^3+6i^2+i+3)4^i C_i'' \end{cases}$$

and

$$(69^*)\quad \begin{cases} \sum_{i=0}^{8} 4^i \overline{C}_i' = \sum_{i=0}^{5} 4^i C_i', \\ \sum_{i=0}^{8} (2i)4^i \overline{C}_i' = \sum_{i=0}^{5} (2i+5)4^i C_i', \\ \sum_{i=0}^{8} 2i(i-1)4^i \overline{C}_i' = \sum_{i=0}^{5} (2i^2+i+3)4^i C_i', \\ \sum_{i=0}^{8} 6i(i-1)(i-2)4^i \overline{C}_i' = \sum_{i=0}^{5} (6i^3+6i^2+i+3)4^i C_i'. \end{cases}$$

By substituting (55*) in (68*) and dividing by suitable powers of 4 we get

$$\begin{cases} \sum_{i=2}^{5} 4^{i-2} \overline{A}_i'' = A_0'' + 4A_1'' + A_2'', \\ \sum_{i=2}^{5} (2i)4^{i-2} \overline{A}_i'' = 6A_0'' + 4A_1'' + 2A_2'', \\ \sum_{i=2}^{5} 2i(i-1)4^{i-2} \overline{A}_i'' = 5A_0'' + 5A_1'' + 6A_2'', \\ \sum_{i=2}^{5} 6i(i-1)(i-2)4^{i-2} \overline{A}_i'' = 5A_0'' + 4A_1'', \\ \sum_{i=3}^{6} 4^{i-3} \overline{B}_i'' = 3B_0'' + B_1'' + 4B_2'' + B_3'', \\ \sum_{i=3}^{6} (2i)4^{i-3} \overline{B}_i'' = 2B_0'' + B_1'' + 5B_2'' + 4B_3'', \\ \sum_{i=3}^{6} 2i(i-1)4^{i-3} \overline{B}_i'' = 3B_1'' + 6B_2'' + 3B_3'', \\ \sum_{i=3}^{6} 6i(i-1)(i-2)4^{i-3} \overline{B}_i'' = 3B_0'' + B_1'' + 5B_3'' \end{cases}$$

and

$$\begin{cases} \sum_{i=4}^{8} 4^{i-4} \overline{C}_i'' = 6C_0'' + 3C_1'' + C_2'' + 4C_3'' + C_4'' + 4C_5'', \\ \sum_{i=4}^{8} (2i)4^{i-4} \overline{C}_i'' = 4C_0'' + C_1'' + 3C_2'' + 6C_3'' + 6C_4'' + 4C_5'', \\ \sum_{i=4}^{8} 2i(i-1)4^{i-4} \overline{C}_i'' = 4C_1'' + 5C_2'' + 2C_3'' + 4C_4'' + C_5'', \\ \sum_{i=4}^{8} 6i(i-1)(i-2)4^{i-4} \overline{C}_i'' = 6C_0'' + 3C_1'' + 5C_3'' + 4C_4'' + 6C_5'' \end{cases}$$

and separately simplifying these equations we get

$$\begin{cases} \overline{A}_2'' + 4\overline{A}_3'' + 2\overline{A}_4'' + \overline{A}_5'' = A_0'' + 4A_1'' + A_2'', \\ \overline{A}_2'' + 6\overline{A}_3'' + 4\overline{A}_4'' + 6\overline{A}_5'' = 5A_0'' + A_1'' + 4A_2'', \\ \overline{A}_2'' + 5\overline{A}_3'' + 5\overline{A}_4'' + 3\overline{A}_5'' = 3A_0'' + 3A_1'' + 5A_2'', \\ \overline{A}_3'' + 2\overline{A}_4'' + 6\overline{A}_5'' = 3A_0'' + A_1'', \\ \overline{B}_3'' + 4\overline{B}_4'' + 2\overline{B}_5'' + \overline{B}_6'' = 3B_0'' + B_1'' + 4B_2'' + B_3'', \\ \overline{B}_3'' + 3\overline{B}_4'' + \overline{B}_5'' + 2\overline{B}_6'' = 5B_0'' + 6B_1'' + 2B_2'' + 3B_3'', \\ \overline{B}_3'' + \overline{B}_4'' + 2\overline{B}_5'' + 5\overline{B}_6'' = 2B_1'' + 4B_2'' + 2B_3'', \\ \overline{B}_3'' + 2\overline{B}_4'' + 6\overline{B}_5'' + 6\overline{B}_6'' = 3B_0'' + B_1'' + 5B_3'' \end{cases}$$

and

$$\begin{cases} \overline{C}_4'' + 4\overline{C}_5'' + 2\overline{C}_6'' + \overline{C}_7'' + 4\overline{C}_8'' = 6C_0'' + 3C_1'' + C_2'' + 4C_3'' + C_4'' + 4C_5'', \\ \overline{C}_4'' + 5\overline{C}_5'' + 3\overline{C}_6'' + \overline{C}_8'' = 4C_0'' + C_1'' + 3C_2'' + 6C_3'' + 6C_4'' + 4C_5'', \\ \overline{C}_4'' + 2\overline{C}_5'' + 5\overline{C}_6'' = 6C_1'' + 4C_2'' + 3C_3'' + 6C_4'' + 5C_5'', \\ \overline{C}_4'' + 3\overline{C}_5'' + 3\overline{C}_6'' = 5C_0'' + 6C_1'' + 3C_3'' + C_4'' + 5C_5'' \end{cases}$$

and simultaneously solving them we get

$$\begin{cases} \overline{A}_5'' = 4A_0'' + 2A_1'' + 3A_2'', \\ \overline{A}_4'' = A_0'' + 6A_1'' + 2A_2'', \\ \overline{A}_3'' + 2\overline{A}_4'' + 6\overline{A}_5'' = 3A_0'' + A_1'', \\ \overline{A}_2'' + 4\overline{A}_3'' + 2\overline{A}_4'' + \overline{A}_5'' = A_0'' + 4A_1'' + A_2'', \\ \overline{B}_6'' = 4B_1'' + 2B_2'' + 3B_3'', \\ \overline{B}_5'' + 6\overline{B}_6'' = 4B_0'' + 4B_1'' + 4B_2'' + 6B_3'', \\ \overline{B}_4'' + 4\overline{B}_5'' + \overline{B}_6'' = 3B_0'' + 6B_1'' + 3B_2'' + 3B_3'', \\ \overline{B}_3'' + 2\overline{B}_4'' + 6\overline{B}_5'' + 6\overline{B}_6'' = 3B_0'' + B_1'' + 5B_3'' \end{cases}$$

and

$$\begin{cases} \overline{C}_7'' + 2\overline{C}_8'' = 4C_0'' + 4C_2'' + 2C_3'' + 3C_4'' + C_5'', \\ \overline{C}_6'' + 2\overline{C}_8'' = 6C_0'' + 4C_1'' + C_2'' + 6C_3'' + 2C_4'' + 5C_5'', \\ \overline{C}_5'' + 5\overline{C}_6'' = 5C_0'' + 3C_2'' + 2C_4'', \\ \overline{C}_4'' + 3\overline{C}_5'' + 3\overline{C}_6'' = 5C_0'' + 6C_1'' + 3C_3'' + C_4'' + 5C_5'' \end{cases}$$

and further simultaneously solving these we get

$$(70^*)\qquad \begin{cases} \overline{A}_5'' = 4A_0'' + 2A_1'' + 3A_2'', \\ \overline{A}_4'' = A_0'' + 6A_1'' + 2A_2'', \\ \overline{A}_3'' = 5A_0'' + 5A_1'' + 6A_2'', \\ \overline{A}_2'' = 3A_0'' + 5A_1'' + 5A_2'', \\ \overline{B}_6'' = 4B_1'' + 2B_2'' + 3B_3'', \\ \overline{B}_5'' = 4B_0'' + B_1'' + 6B_2'' + 2B_3'', \\ \overline{B}_4'' = B_0'' + 5B_1'' + 5B_2'' + 6B_3'', \\ \overline{B}_3'' = 5B_0'' + 3B_1'' + 5B_2'' + 5B_3'' \end{cases}$$

and

$$(71^*)\qquad \begin{cases} \overline{C}''_8 + 5\overline{C}''_4 = 2C''_0 + 4C''_1 + C''_2 + 4C''_3 + 4C''_4 + 3C''_5, \\ \overline{C}''_7 + 4\overline{C}''_4 = 6C''_1 + 2C''_2 + C''_3 + 2C''_4 + 2C''_5, \\ \overline{C}''_6 + 4\overline{C}''_4 = 2C''_0 + 3C''_1 + 6C''_2 + 5C''_3 + C''_4 + 6C''_5, \\ \overline{C}''_5 + \overline{C}''_4 = 2C''_0 + 6C''_1 + C''_2 + 3C''_3 + 4C''_4 + 5C''_5. \end{cases}$$

Likewise, by substituting (54*) in (69*) and simultaneously solving the resulting equations we get (71*) with single primes replacing double primes, i.e., we get

$$(72^*)\qquad \begin{cases} \overline{C}'_8 + 5\overline{C}'_4 = 2C'_0 + 4C'_1 + C'_2 + 4C'_3 + 4C'_4 + 3C'_5, \\ \overline{C}'_7 + 4\overline{C}'_4 = 6C'_1 + 2C'_2 + C'_3 + 2C'_4 + 2C'_5, \\ \overline{C}'_6 + 4\overline{C}'_4 = 2C'_0 + 3C'_1 + 6C'_2 + 5C'_3 + C'_4 + 6C'_5, \\ \overline{C}'_5 + \overline{C}'_4 = 2C'_0 + 6C'_1 + C'_2 + 3C'_3 + 4C'_4 + 5C'_5. \end{cases}$$

Substituting the last two equations of (71*) in the last equation of (64*) we get

$$\overline{C}''_4 = 2C''_0 + 4C''_1 + C''_2 + 2C''_3 + 4C''_4 + C''_5$$

and substituting this value in (71*) we get

$$(73^*)\qquad \begin{cases} \overline{C}''_8 = 6C''_0 + 5C''_1 + 3C''_2 + C''_3 + 5C''_4 + 5C''_5, \\ \overline{C}''_7 = 6C''_0 + 4C''_1 + 5C''_2 + 5C''_5, \\ \overline{C}''_6 = C''_0 + C''_1 + 2C''_2 + 4C''_3 + 6C''_4 + 2C''_5, \\ \overline{C}''_5 = 2C''_1 + C''_3 + 4C''_5, \\ \overline{C}''_4 = 2C''_0 + 4C''_1 + C''_2 + 2C''_3 + 4C''_4 + C''_5 \end{cases}$$

and substituting these values in (64*) and the last four equations of (59*) we get

$$\begin{cases} 5C''_0 + C''_1 + 3C''_2 + 3C''_3 + C''_5 = 0, \\ C''_1 + 5C''_3 + 3C''_5 = 0, \\ 4C''_1 + 2C''_3 + C''_5 = 0 \\ 0 = 0 \\ 6C''_0 + 4C''_1 + 5C''_2 + 5C''_3 + 4C''_4 + 4C''_5 = 0, \\ C''_0 + C''_1 + 2C''_2 + 2C''_3 + 4C''_4 = 0, \\ 3C''_0 + 2C''_3 + 3C''_4 = 0, \\ 6C''_0 + 4C''_2 + 6C''_3 + 5C''_4 + C''_5 = 0. \end{cases}$$

and sending the first equation to the end and deleting the trivial equation $0 = 0$, and then simultaneously solving the last five equations we get

$$\begin{cases} C''_1 + 5C''_3 + 3C''_5 = 0, \\ 4C''_1 + 2C''_3 + C''_5 = 0 \\ 4C''_4 = 0, \\ C''_4 + C''_5 = 5C''_0, \\ C''_3 = 2C''_0 + 2C''_4, \\ C''_2 = 2C''_0 + 2C''_3 + 4C''_4 + 5C''_5 \\ C''_1 = 2C''_0 + 4C''_2 + 4C''_3 + 6C''_5 \end{cases}$$

and further simultaneously solving these equations we get

$$C_5'' = 5C_0'', \quad C_4'' = 0, \quad C_3'' = 2C_0'', \quad C_2'' = 3C_0'', \quad C_1'' = 3C_0''$$

and substituting these values in (55*) and (73*) we get

$$(74^*) \quad \begin{cases} C_5'' = 5C_0'', \quad C_4'' = 0, \quad C_3'' = 2C_0'', \quad C_2'' = 3C_0'', \quad C_1'' = 3C_0'', \\ \overline{C}_8'' = C_0'', \quad \overline{C}_7'' = 2C_0'', \quad \overline{C}_6'' = 0, \quad \overline{C}_5'' = 0, \\ \overline{C}_4'' = 5C_0'', \quad \overline{C}_3'' = 0, \quad \overline{C}_2'' = 0, \quad \overline{C}_1'' = 4C_0'', \quad \overline{C}_0'' = 6C_0''. \end{cases}$$

Substituting (70*) in the first three equations of (59*) we get

$$\begin{cases} 3A_1'' + 6A_2'' = 0, \\ 2B_0'' + 3B_1'' + 6B_2'' + 4B_3'' = 0, \\ B_0'' + 5B_1'' + 6B_2'' + B_3'' = 0 \end{cases}$$

and simultaneously solving these we get

$$A_2'' = 3A_1'', \quad B_3'' = 2B_0'' + 3B_1'', \quad B_2'' = 3B_0'' + B_1''$$

and substituting these values in (55*) and (70*) we get

$$(75^*) \quad \begin{cases} A_2'' = 3A_1'', \quad B_3'' = 2B_0'' + 3B_1'', \quad B_2'' = 3B_0'' + B_1'', \\ \overline{A}_5'' = 4A_0'' + 4A_1'', \quad \overline{A}_4'' = A_0'' + 5A_1'', \quad \overline{A}_3'' = 5A_0'' + 2A_1'', \\ \overline{A}_2'' = 3A_0'' + 6A_1'', \quad \overline{A}_1'' = 6A_1'', \quad \overline{A}_0'' = 6A_0'', \\ \overline{B}_6'' = 5B_0'' + B_1'', \quad \overline{B}_5'' = 5B_0'' + 6B_1'', \quad \overline{B}_4'' = 0, \\ \overline{B}_3'' = 2B_0'' + 2B_1'', \quad \overline{B}_2'' = 6B_1'', \quad \overline{B}_1'' = 6B_1'', \quad \overline{B}_0'' = 6B_0''. \end{cases}$$

Substituting (60*) to (62*) in (72*) we get

$$\begin{cases} 3C_0' + 5C_2' = 0, \\ 2C_0' + 3C_1' = 0, \\ 2C_0' + 5C_1' + 4C_2' = 0, \\ 2C_0' + 5C_1' + 4C_2' = 0 \end{cases}$$

and simultaneously solving these we get

$$C_2' = 5C_0' \quad \text{and} \quad C_1' = 4C_0'$$

and substituting these values in the relevant part of (60*) to (62*) we get

$$(76^*) \quad \begin{cases} C_5' = 5C_0', \quad C_4' = 2C_0', \quad C_3' = 6C_0', \quad C_2' = 5C_0', \quad C_1' = 4C_0', \\ \overline{C}_8' = 5C_0', \quad \overline{C}_7' = 2C_0', \quad \overline{C}_6' = 5C_0', \quad \overline{C}_5' = 6C_0', \\ \overline{C}_4' = 6C_0', \quad \overline{C}_3' = 6C_0', \quad \overline{C}_2' = 5C_0', \quad \overline{C}_1' = 3C_0', \quad \overline{C}_0' = 6C_0' \end{cases}$$

and by the remaining part of (60*) to (62*) we have

$$
(77^*)\quad \begin{cases}
A'_2 = 5A'_0 + A'_1, \quad B'_3 = 4B'_0 + 4B'_1, \quad B'_2 = B'_0 + 6B'_1, \\
\overline{A}'_5 = 5A'_0 + A'_1, \quad \overline{A}'_4 = A'_1, \quad \overline{A}'_3 = 4A'_1, \\
\overline{A}'_2 = 4A'_1, \quad \overline{A}'_1 = 6A'_1, \quad \overline{A}'_0 = 6A'_0, \\
\overline{B}'_6 = 4B'_0 + 4B'_1, \quad \overline{B}'_5 = B'_0 + 6B'_1, \\
\overline{B}'_4 = 2B'_0 + 3B'_1, \quad \overline{B}'_3 = 5B'_0 + 3B'_1, \\
\overline{B}'_2 = 2B'_0 + B'_1, \quad \overline{B}'_1 = 6B'_1, \quad \overline{B}'_0 = 6B'_0.
\end{cases}
$$

By (49*), (50*) and (74*) to (77*) we get

$$
(78^*)\quad \begin{cases}
L_{A'} = (5R^2 + 1)A'_0 + (R^2 + R)A'_1, \\
M_{A'} = (5R^5 + 6)A'_0 + (R^5 + R^4 + 4R^3 + 4R^2 + 6R)A'_1, \\
L_{A''} = A''_0 + (3R^2 + R)A''_1, \\
M_{A''} = (4R^5 + R^4 + 5R^3 + 3R^2 + 6)A''_0 \\
\qquad\qquad +(4R^5 + 5R^4 + 2R^3 + 6R^2 + 6R)A''_1
\end{cases}
$$

and

$$
(79^*)\quad \begin{cases}
L_{B'} = (4R^3 + R^2 + 1)B'_0 + (4R^3 + 6R^2 + R)B'_1, \\
M_{B'} = (4R^6 + R^5 + 2R^4 + 5R^3 + 2R^2 + 6)B'_0 \\
\qquad\qquad +(4R^6 + 6R^5 + 3R^4 + 3R^3 + R^2 + 6R)B'_1, \\
L_{B''} = (2R^3 + 3R^2 + 1)B''_0 + (3R^3 + R^2 + R)B''_1, \\
M_{B''} = (5R^6 + 5R^5 + 2R^3 + 6)B''_0 \\
\qquad\qquad +(R^6 + 6R^5 + 2R^3 + 6R^2 + 6R)B''_1
\end{cases}
$$

and

$$
(80^*)\quad \begin{cases}
L_{C'} = (5R^5 + 2R^4 + 6R^3 + 5R^2 + 4R + 1)C'_0, \\
M_{C'} = (5R^8 + 2R^7 + 5R^6 + 6R^5 + 6R^4 \\
\qquad\qquad +6R^3 + 5R^2 + 3R + 6)C'_0, \\
L_{C''} = (5R^5 + 2R^3 + 3R^2 + 3R + 1)C''_0, \\
M_{C''} = (R^8 + 2R^7 + 5R^4 + 4R + 6)C''_0.
\end{cases}
$$

In view of (2*), (3*), (8*), (37*) and (38*), by equating the coefficients of Z^3, Z^4 and Z^5 on both sides of (□) and then multiplying throughout by $(R+2)^4(R+3)^4$, $(R+2)^2(R+3)^4$ and $(R+2)(R+3)^4$ respectively we get

$$
(\square_3)\quad \begin{aligned}
(L_J S + M_J)(R+2)^3 = {} & (L_{C''}S + M_{C''}) \\
& +(L_{C'}S + M_{C'})(R+3)^4 \\
& +(L_{A'}S + M_{A'})(L_{B''}S + M_{B''})(R+2) \\
& +(L_{A''}S + M_{A''})(L_{B'}S + M_{B'})(R+2)
\end{aligned}
$$

and

$$
(\square_4)\quad \begin{aligned}
(L_J S + M_J)(R+2) = {} & (L_{B''}S + M_{B''}) \\
& +(L_{B'}S + M_{B'})(R+3)^4 \\
& +(L_{A'}S + M_{A'})(L_{A''}S + M_{A''})
\end{aligned}
$$

and

$$(\square_5) \qquad (L_J S + M_J) = (L_{A''} S + M_{A''}) + (L_{A'} S + M_{A'})(R+3)^4.$$

By (4*) we have

$$\begin{aligned} \text{LHS of } (\square_5) &= (5R^6 + 3R^5 + 3R^4 + 2R^3 + 5R^2 + 2R + 4)S + M_J \\ &\quad \text{with } M_J \in k[R] \end{aligned}$$

and by (78*) we get

$$\begin{aligned} \text{RHS of } (\square_5) &= [(L_{A''} + L_{A'}(R+3)^4]S + Y_5 \\ &= [A''_0 + (3R^2 + R)A''_1 \\ &\quad +(5R^6 + 4R^5 + 5R^4 + 6R^3 + 4R^2 + 3R + 4)A'_0 \\ &\quad +(R^6 + 6R^5 + 3R^4 + R^3 + 4R)A'_1]S + Y_5 \\ &\quad \text{with } Y_5 = M_{A''} + M_{A'}(R+3)^4 \in k[R] \end{aligned}$$

and hence by equating the coefficients of S, RS, R^2S, R^3S on the two sides of $(\square_5)$ we get

$$\begin{cases} A''_0 + 4A'_0 = 4, \\ A''_1 + 3A'_0 + 4A'_1 = 2, \\ 3A''_1 + 4A'_0 = 5, \\ 6A'_0 + A'_1 = 2 \end{cases}$$

and simultaneously solving these equations we get

$$A'_0 = 4, \quad A'_1 = 6, \quad A''_0 = 2, \quad A''_1 = 1$$

and substituting these values in (78*) we get

$$(81^*) \qquad \begin{cases} L_{A'} = 5R^2 + 6R + 4, \\ M_{A'} = 5R^5 + 6R^4 + 3R^3 + 3R^2 + R + 3, \\ L_{A''} = 3R^2 + R + 2, \\ M_{A''} = 5R^5 + 5R^3 + 5R^2 + 6R + 5. \end{cases}$$

By (4*) we have

$$\begin{aligned} \text{LHS of } (\square_4) &= L_J(R+2)S + X_4 \\ &= (5R^7 + 6R^6 + 2R^5 + R^4 + 2R^3 \\ &\qquad +5R^2 + R + 1)S + X_4 \\ &\quad \text{with } X_4 = M_J(R+2) \in k[R] \end{aligned}$$

and by (1*), (79*) and (81*) we get

$$\begin{aligned} \text{RHS of } (\square_4) &= [(L_{B''} + L_{B'}(R+3)^4 + L_{A'}M_{A''} + M_{A'}L_{A''}]S + Y_4 \\ &= [(2R^3 + 3R^2 + 1)B''_0 + (3R^3 + R^2 + R)B''_1 \\ &\quad +(4R^7 + 4R^5 + 4R^4 + 3R^3 + 2R^2 + 3R + 4)B'_0 \\ &\quad +(4R^7 + 5R^6 + 2R^5 + 5R^4 + 4R^3 + 6R^2 + 4R)B'_1 \\ &\quad +(5R^7 + 4R^6 + 2R^4 + R^3 + 6R^2 + 3R + 5)]S + Y_4 \\ &\quad \text{with } Y_4 = M_{B''} + M_{B'}(R+3)^4 \\ &\qquad\qquad +L_{A'}L_{A''}D + M_{A'}M_{A''} \in k[R] \end{aligned}$$

and hence by equating the coefficients of S, RS, R^2S, R^3S on the two sides of $(\square_4)$ we get

$$\begin{cases} B_0'' + 4B_0' + 5 = 1, \\ B_1'' + 3B_0' + 4B_1' + 3 = 1, \\ 3B_0'' + B_1'' + 2B_0' + 6B_1' + 6 = 5, \\ 2B_0'' + 3B_1'' + 3B_0' + 4B_1' + 1 = 2 \end{cases}$$

and simultaneously solving these equations we get

$$B_0' = 1, \quad B_1' = 6, \quad B_0'' = 6, \quad B_1'' = 6$$

and substituting these values in (79*) we get

$$(82^*) \qquad \begin{cases} L_{B'} = 2R^2 + 6R + 1, \\ M_{B'} = 2R^5 + 6R^4 + 2R^3 + R^2 + R + 6, \\ L_{B''} = 2R^3 + 3R^2 + 6R + 6, \\ M_{B''} = R^6 + 3R^5 + 3R^3 + R^2 + R + 1. \end{cases}$$

By (4*) we have

$$\begin{aligned} \text{LHS of } (\square_3) &= L_J(R+2)^3S + X_3 \\ &= (5R^9 + 5R^8 + 4R^7 + 5R^6 \\ &\qquad +3R^4 + R^3 + 4R^2 + R + 4)S + X_3 \\ &\quad \text{with } X_3 = M_J(R+2)^3 \in k[R] \end{aligned}$$

and by (1*), (80*), (81*) and (82*) we get

$$\begin{aligned} \text{RHS of } (\square_3) &= [(L_{C''} + L_{C'}(R+3)^4 \\ &\quad +(L_{A'}M_{B''} + M_{A'}L_{B''} + L_{A''}M_{B'} \\ &\qquad +M_{A''}L_{B'})(R+2)]S + Y_3 \\ &= [(5R^5 + 2R^3 + 3R^2 + 3R + 1)C_0'' \\ &\quad +(5R^9 + 6R^8 + 6R^7 + 4R^6 \\ &\qquad +R^5 + 2R^4 + R^3 + 2R^2 + 5R + 4)C_0' \\ &\quad +(R^9 + 3R^8 + 2R^7 + 6R^6 \\ &\qquad +3R^5 + 2R^3 + 3R^2 + 6R + 1)]S + Y_3 \end{aligned}$$

with

$$\begin{aligned} Y_3 = {} & M_{C''} + M_{C'}(R+3)^4 \\ & +(L_{A'}L_{B''}D + M_{A'}M_{B''} + L_{A''}L_{B'}D + M_{A''}M_{B'})(R+2) \in k[R] \end{aligned}$$

and hence by equating the coefficients of S, RS, R^2S on the two sides of $(\square_3)$ we get

$$\begin{cases} C_0'' + 4C_0' + 1 = 4, \\ 3C_0'' + 5C_0' + 6 = 1, \\ 3C_0'' + 2C_0' + 3 = 4 \end{cases}$$

and simultaneously solving these equations we get

$$C_0' = 5, \quad C_0'' = 4$$

and substituting these values in (80*) we get

$$(83^*) \quad \begin{cases} L_{C'} = 4R^5 + 3R^4 + 2R^3 + 4R^2 + 6R + 5, \\ M_{C'} = 4R^8 + 3R^7 + 4R^6 + 2R^5 + 2R^4 + 2R^3 + 4R^2 + R + 2, \\ L_{C''} = 6R^5 + R^3 + 5R^2 + 5R + 4 \\ M_{C''} = 4R^8 + R^7 + 6R^4 + 2R + 3. \end{cases}$$

2.7 The factorization

By referring to (1*) to (10*), (□), (37*), (38*), (49*), (50*), (81*), (82*) and (83*) of the previous Section, we are assuming the characteristic of the ground field k to be 7, and we are considering the univariate algebraic function field $k(R, S)$ given by

$$(1^{**}) \qquad S^2 = D \quad \text{and} \quad D = R^6 + R^4 + R^2 + 1$$

and we are considering the polynomial

$$(2^{**}) \quad \psi(Z) = Z^6 + J[Z^5 + Z^4 + Z^3 + Z^2 + Z] + K \text{ with } J \text{ and } K \text{ in } k(R, S)$$

where the coefficient J is given by

$$(3^{**}) \qquad J = \frac{L_J S + M_J}{(R+2)(R+3)^4}$$

with

$$(4^{**}) \qquad L_J = 5R^6 + 3R^5 + 3R^4 + 2R^3 + 5R^2 + 2R + 4$$

and

$$(5^{**}) \qquad M_J = 5R^9 + 3R^8 + 2R^7 + 4R^5 + 4R^3 + 5R + 3$$

and the coefficient K is given by

$$(6^{**}) \qquad K = \frac{L_K S + M_K}{(R+2)^8 (R+3)^4}$$

with

$$(7^{**}) \qquad \begin{aligned} L_K = 5R^{13} + 6R^{12} + 4R^{11} + 4R^{10} + 5R^9 \\ +4R^8 + R^7 + 6R^6 + 5R^4 + 6R^2 + 2 \end{aligned}$$

and

$$(8^{**}) \qquad M_K = 5R^{16} + 6R^{15} + 3R^{14} + 2R^8 + 6R^7 + 5$$

and we have shown that if $\psi(Z)$ factors into two monic factors of degree 3 in Z with coefficients in $k(R,S)$ then these factors must be of the form

$$Z^3 + A'Z^2 + B'Z + C' \quad \text{and} \quad Z^3 + A''Z^2 + B''Z + C''$$

with

$$(9^{**}) \qquad A' = \frac{L_{A'}S+M_{A'}}{(R+2)}, \quad B' = \frac{L_{B'}S+M_{B'}}{(R+2)^2}, \quad C' = \frac{L_{C'}S+M_{C'}}{(R+2)^4}$$

and

$$(10^{**}) \qquad A'' = \frac{L_{A''}S+M_{A''}}{(R+2)(R+3)^4}, \quad B'' = \frac{L_{B''}S+M_{B''}}{(R+2)^2(R+3)^4}, \quad C'' = \frac{L_{C''}S+M_{C''}}{(R+2)^4(R+3)^4}$$

where

$$(11^{**}) \qquad \begin{cases} L_{A'} = 5R^2 + 6R + 4, \\ M_{A'} = 5R^5 + 6R^4 + 3R^3 + 3R^2 + R + 3, \\ L_{A''} = 3R^2 + R + 2, \\ M_{A''} = 5R^5 + 5R^3 + 5R^2 + 6R + 5 \end{cases}$$

and

$$(12^{**}) \qquad \begin{cases} L_{B'} = 2R^2 + 6R + 1, \\ M_{B'} = 2R^5 + 6R^4 + 2R^3 + R^2 + R + 6, \\ L_{B''} = 2R^3 + 3R^2 + 6R + 6, \\ M_{B''} = R^6 + 3R^5 + 3R^3 + R^2 + R + 1 \end{cases}$$

and

$$(13^{**}) \qquad \begin{cases} L_{C'} = 4R^5 + 3R^4 + 2R^3 + 4R^2 + 6R + 5, \\ M_{C'} = 4R^8 + 3R^7 + 4R^6 + 2R^5 + 2R^4 + 2R^3 + 4R^2 + R + 2, \\ L_{C''} = 6R^5 + R^3 + 5R^2 + 5R + 4 \\ M_{C''} = 4R^8 + R^7 + 6R^4 + 2R + 3. \end{cases}$$

Comparing coefficients of Z^0, Z, Z^2, Z^3, Z^4, Z^5, and clearing denominators, we see that

$$(\square) \qquad \psi(Z) = (Z^3 + A'Z^2 + B'Z + C')(Z^3 + A''Z^2 + B''Z + C'')$$

if and only if

$$(\square_0) \qquad L_K S + M_K = (L_{C'}S + M_{C'})(L_{C''}S + M_{C''})$$

and

$$(\square_1) \qquad \begin{aligned} (L_J S + M_J)(R+2)^5 = {} & (L_{B'}S + M_{B'})(L_{C''}S + M_{C''}) \\ & +(L_{B''}S + M_{B''})(L_{C'}S + M_{C'}) \end{aligned}$$

and

$$(\square_2) \qquad \begin{aligned} (L_J S + M_J)(R+2)^4 = {} & (L_{A'}S + M_{A'})(L_{C''}S + M_{C''}) \\ & +(L_{A''}S + M_{A''})(L_{C'}S + M_{C'}) \\ & +(L_{B'}S + M_{B'})(L_{B''}S + M_{B''})(R+2) \end{aligned}$$

and

$$(\square_3)\qquad \begin{aligned}(L_J S + M_J)(R+2)^3 = (L_{C''}S + M_{C''}) &+ (L_{C'}S + M_{C'})(R+3)^4\\ &+(L_{A'}S + M_{A'})(L_{B''}S + M_{B''})(R+2)\\ &+(L_{A''}S + M_{A''})(L_{B'}S + M_{B'})(R+2)\end{aligned}$$

and

$$(\square_4)\qquad \begin{aligned}(L_J S + M_J)(R+2) = (L_{B''}S + M_{B''}) &+ (L_{B'}S + M_{B'})(R+3)^4\\ &+(L_{A'}S + M_{A'})(L_{A''}S + M_{A''})\end{aligned}$$

and

$$(\square_5)\qquad (L_J S + M_J) = (L_{A''}S + M_{A''}) + (L_{A'}S + M_{A'})(R+3)^4.$$

In view of (1**), equation ($\square_0$) is equivalent to the two equations

$$(\square_1')\qquad L_K = L_{C'}M_{C''} + M_{C'}L_{C''}$$

and

$$(\square_2')\qquad M_K = L_{C'}L_{C''}D + M_{C'}M_{C''}$$

and equation ($\square_1$) is equivalent to the two equations

$$(\square_3')\qquad L_J(R+2)^5 = L_{B'}M_{C''} + M_{B'}L_{C''} + L_{B''}M_{C'} + M_{B''}L_{C'}$$

and

$$(\square_4')\qquad M_J(R+2)^5 = L_{B'}L_{C''}D + M_{B'}M_{C''} + L_{B''}L_{C'}D + M_{B''}M_{C'}$$

and equation ($\square_2$) is equivalent to the two equations

$$(\square_5')\qquad \begin{aligned}L_J(R+2)^4 = {}&L_{A'}M_{C''} + M_{A'}L_{C''} + L_{A''}M_{C'} + M_{A''}L_{C'}\\ &+(L_{B'}M_{B''} + M_{B'}L_{B''})(R+2)\end{aligned}$$

and

$$(\square_6')\qquad \begin{aligned}M_J(R+2)^4 = {}&L_{A'}L_{C''}D + M_{A'}M_{C''} + L_{A''}L_{C'}D + M_{A''}M_{C'}\\ &+(L_{B'}L_{B''}D + M_{B'}M_{B''})(R+2)\end{aligned}$$

and equation ($\square_3$) is equivalent to the two equations

$$(\square_7')\qquad \begin{aligned}L_J(R+2)^3 = {}&L_{C''} + L_{C'}(R+3)^4\\ &+(L_{A'}M_{B''} + M_{A'}L_{B''}\\ &\quad +L_{A''}M_{B'} + M_{A''}L_{B'})(R+2)\end{aligned}$$

and

$$(\square_8')\qquad \begin{aligned}M_J(R+2)^3 = {}&M_{C''} + M_{C'}(R+3)^4\\ &+(L_{A'}L_{B''}D + M_{A'}M_{B''}\\ &\quad +L_{A''}L_{B'}D + M_{A''}M_{B'})(R+2)\end{aligned}$$

and equation $(\square_4)$ is equivalent to the two equations

$$(\square_9') \qquad L_J(R+2) = L_{B''} + L_{B'}(R+3)^4 + L_{A'}M_{A''} + M_{A'}L_{A''}$$

and

$$(\square_{10}') \qquad M_J(R+2) = M_{B''} + M_{B'}(R+3)^4 + L_{A'}L_{A''}D + M_{A'}M_{A''}$$

and equation $(\square_5)$ is equivalent to the two equations

$$(\square_{11}') \qquad L_J = L_{A''} + L_{A'}(R+3)^4$$

and

$$(\square_{12}') \qquad M_J = M_{A''} + M_{A'}(R+3)^4.$$

By direct calculation with polynomials in R we see that both sides of $(\square_1')$ are equal to

$$5R^{13} + 6R^{12} + 4R^{11} + 4R^{10} + 5R^9 + 4R^8 + R^7 + 6R^6 + 5R^4 + 6R^2 + 2$$

and both sides of $(\square_2')$ are equal to

$$5R^{16} + 6R^{15} + 3R^{14} + 2R^8 + 6R^7 + 5$$

and both sides of $(\square_3')$ are equal to

$$5R^{11} + 4R^{10} + 2R^9 + 6R^8 + R^7 + 2R^6 + 6R^5 + 6R^4 + 3R^2 + 6R + 2$$

and both sides of $(\square_4')$ are equal to

$$5R^{14} + 4R^{13} + R^{12} + R^{11} + 3R^{10} + 5R^9 + 4R^7$$
$$+2R^6 + 4R^5 + 4R^4 + 4R^3 + 3R^2 + R + 5$$

and both sides of $(\square_5')$ are equal to

$$5R^{10} + R^9 + 6R^7 + 3R^6 + 3R^5 + 6R^3 + 2R^2 + 6R + 1$$

and both sides of $(\square_6')$ are equal to

$$5R^{13} + R^{12} + 6R^{11} + 3R^{10} + 4R^9 + 4R^8 + 6R^7$$
$$+6R^6 + 4R^5 + 3R^4 + 5R^3 + R^2 + R + 6$$

and both sides of $(\square_7')$ are equal to

$$5R^9 + 5R^8 + 4R^7 + 5R^6 + 3R^4 + R^3 + 4R^2 + R + 4$$

and both sides of $(\square_8')$ are equal to

$$5R^{12} + 5R^{11} + 3R^{10} + 4R^9 + 3R^8 + 5R^7 + 3R^6 + 4R^4 + 2R^3 + R^2 + 6R + 3$$

and both sides of $(\square_9')$ are equal to

$$5R^7+6R^6+2R^5+R^4+2R^3+5R^2+R+1$$

and both sides of $(\square_{10}')$ are equal to

$$5R^{10}+6R^9+R^8+4R^7+4R^6+R^5+4R^4+R^3+5R^2+6R+6$$

and both sides of $(\square_{11}')$ are equal to

$$5R^6+3R^5+3R^4+2R^3+5R^2+2R+4$$

and both sides of $(\square_{12}')$ are equal to

$$5R^9+3R^8+2R^7+4R^5+4R^3+5R+3.$$

Therefore $(\square)$ is true with the values given in (9**) to (13**). This completes the proof that, with $p=7$ and with Θ as in the beginning of Section 3, for the Galois group $\overline{G}_{9,7}=\mathrm{Gal}(\Theta,k(x))$ we have: $\overline{G}_{9,7}=PSL(2,8)$ if we assume k to be algebraically closed, and $\overline{G}_{9,7}=PSL(2,8)$ or $P\Gamma L(2,8)$ without that assumption.

2.8 Galois groups

Thus we are considering the Galois group $\overline{G}_{9,7}$ of the polynomial

$$\Theta=Y^9-xY^7+e$$

over $k(x)$ where e is a nonzero element in the field k of characteristic 7. Now essentially by throwing away the three roots y,z,ω of Θ we obtained the polynomial

$$\psi(Z)=Z^6+J[Z^5+Z^4+Z^3+Z^2+Z]+K$$

cited in (52′) with ξ,η,J,K as in (49′) to (53′) where $k(z,\omega)=k(R,S)$.
By (48′) we have

$$S^2=D=1+R^2+R^4+R^6$$

and hence the R-adic valuation of $k(R)/k$ splits into two valuations V_{06} and V_{01} of $k(R,S)$ given by

$$V_{06}(R)=1\le V_{06}(S+6)\quad\text{and}\quad V_{01}(R)=1\le V_{01}(S+1).$$

Modulo R^8 we have

$$\begin{aligned}D^{1/2}&\equiv 1+(1/2)(R^2+R^4+R^6)+(1/2)(-1/2)(1/2)(R^2+R^4+R^6)^2\\&\quad+(1/2)(-1/2)(-3/2)(1/2)(1/3)(R^2+R^4+R^6)^3\\&\equiv 1+4R^2+3R^4+6R^6\end{aligned}$$

and hence

$$(1'') \qquad S \equiv 1 + 4R^2 + 3R^4 + 6R^6 \text{ modulo } M(V_{06})^8$$

where $M(V_{06})$ is the maximal ideal in the valuation ring of V_{06}. Therefore by (55′) and (56′), modulo $M(V_{06})^8$ we have

$$\begin{aligned} L_J S + M_J &\equiv (5R^6 + 3R^5 + 3R^4 + 2R^3 + 5R^2 + 2R + 4)\times \\ &\qquad (1 + 4R^2 + 3R^4 + 6R^6) \\ &\quad +(5R^9 + 3R^8 + 2R^7 + 4R^5 + 4R^3 + 5R + 3) \\ &\equiv 4R^7 \end{aligned}$$

and hence by (54′) we get

$$(2'') \qquad J \equiv 4R^7 \text{ modulo } M(V_{06})^8.$$

In view of (1″), by (65′) we get

$$L^* S + M^* \equiv 3R \text{ modulo } M(V_{06})^2$$

and hence by (64′) we have

$$(3'') \qquad (\eta^2 + \xi^7)\eta^{-2} \equiv 3R \text{ modulo } M(V_{06})^2.$$

In view of (2″) and (3″), by (53′) we get

$$(4'') \qquad K \equiv 5R^8 \text{ modulo } M(V_{06})^9$$

and

$$(5'') \qquad (\xi + \eta)\xi^7\eta^{-2} \equiv R^7 \text{ modulo } M(V_{06})^8.$$

In view of (1″), by (49′) to (51′) we see that

$$\xi \equiv 6 \text{ modulo } M(V_{06}) \quad \text{and} \quad \eta \equiv 1 \text{ modulo } M(V_{06})$$

and hence

$$\xi - \eta \equiv 5 \quad \text{modulo } M(V_{06})$$

and therefore

$$(6'') \qquad (\xi - \eta)^{-6} \equiv 1 \text{ modulo } M(V_{06}).$$

Upon letting

$$u = (\xi + \eta)\xi^7\eta^{-2}(\xi - \eta)^{-6}$$

by (5″) and (6″) we get

$$(7'') \qquad u \equiv R^7 \text{ modulo } M(V_{06})^8.$$

By (5) and (12) we have

$$z = \frac{2}{z^* - 1} \quad \text{and} \quad z^* = \frac{\xi}{\eta}$$

and hence

$$z = \frac{2\eta}{\xi - \eta} \quad \text{and} \quad z + 1 = \frac{\xi + \eta}{\xi - \eta} \quad \text{and} \quad z + 2 = \frac{2\xi}{\xi - \eta}$$

and by ($2^\bullet$) and ($3^\bullet$) we also have

$$y^{-9} = \frac{(z+2)^7(z+1)}{4ez^2}$$

and hence we get

$$(8'') \qquad y^{-9} = ue^{-1}.$$

Let

$$y' = y^{-4}R^{-3} \quad \text{and} \quad u' = u^4e^{-4}R^{-27}.$$

Then obviously $k(y, R, S) = k(y', R, S)$ and by ($7''$) and ($8''$) we get

$$(9'') \qquad y'^9 = u'$$

and

$$(10'') \qquad u' \equiv e^{-4}R \text{ modulo } M(V_{06})^2$$

and hence V_{06} has a unique extension V'_{06} to $k(y', R, S)$, and V'_{06} is totally ramified over V_{06}, i.e., the ramification exponent of V'_{06} over V_{06} is 9, and V'_{06} is residually rational over V_{06}, and hence the residue field of V'_{06} is k. Upon letting

$$\begin{aligned}\psi'(Z) &= y'^{-72}\psi(y'^{12}Z)\\ &= Z^6 + a_1Z^5 + a_2Z^4 + a_3Z^3 + a_4Z^2 + a_5Z + a_6 \in k(y', R, S)[Z]\end{aligned}$$

by ($2''$), ($4''$), ($9''$) and ($10''$) we see that $V'_{06}(a_i) > 0$ for $1 \le i \le 5$, and $V'_{06}(a_6) = 0$, and $a_6 \equiv -2e^{32}$ modulo $M(V'_{06})$.

Therefore, by what we have said at the end of the previous Section, we see that: the polynomial $\psi(Z)$ factors completely (into linear factors) in $k(y, z, \omega)[Z] \Leftrightarrow$ the polynomial $\psi(Z)$ factors completely in $k(y', R, S))[Z] \Leftrightarrow$ the polynomial $Z^6 - 2e^{32}$ factors completely in $k[Z]$. Consequently, $G = PSL(2, 8)$ or $P\Gamma L(2, 8)$ according as k does or does not contain a cube root of $4e$. Hence, in particular, if $k = GF(7)$ then $G = PSL(2, 8)$ or $P\Gamma L(2, 8)$ according as $e \in \{2, 5\}$ or $e \in \{1, 3, 4, 6\}$.

2.9 References

[1] S. S. Abhyankar, *Coverings of algebraic curves,* American Journal of Mathematics, vol 79, 1957, pages 825-856.

[2] S. S. Abhyankar, *Historical ramblings in algebraic geometry and related algebra,* American Mathematical Monthly, vol 83, 1976, pages 409-448.

[3] S. S. Abhyankar, *Algebraic Geometry for Scientists and Engineers,* American Mathematical Society, Providence, 1990.

[4] S. S. Abhyankar, *Galois theory on the line in nonzero characteristic,* Bulletin of the American Mathematical Society, (to appear).

[5] S. S. Abhyankar and C. Bajaj, *Automatic parametrization of curves and surfaces III: Algebraic plane curves,* Computer Aided Geometric Design, vol 5, 1988, pages 309-321.

[6] S. S. Abhyankar and C. Bajaj, *Computations with algebraic curves,* Springer Lecture Notes In Computer Science, vol 358, 1989, pages 274-284.

[7] J-P. Serre, *A letter as an appendix to the square-root paper of Abhyankar,* These Proceedings.

3

A Letter as an Appendix to the Square-Root Parameterization Paper Of Abhyankar

J-P. Serre

The Letter

Dear Abhyankar,

I received two days ago your preprint "Square–root ..." containing the $SL(2,8)$–extension in char. 7.

As I told you in Paris, I suspected from the beginning that this extension could be constructed from a similar extension in char. 0. I have now checked (with the help of Mestre and of old papers by Fricke and Goursat) that this is indeed the case. Let me explain how this is done:

Notation: $G = SL(2,8)$, the unique simple group of order 504; I put $G.3 = \mathrm{Aut}(G)$, as in the ATLAS.

The group $G.3$ has a natural embedding in the alternating group A_9 (action on the points of the projective line over F_8 by fractional transformations composed with Galois conjugation).

One can generate G by three elements x, y, z with the relations

$$xyz = 1, \quad x^2 = y^3 = z^7 = 1.$$

As elements of A_9, x is of type 2^4 (product of 4 disjoint transpositions), y of type 3^3 and z of course of type 7. Conversely, any three elements of A_9 of such types, with product 1, generate a subgroup isomorphic to G (this can be checked e.g. on character tables, as explained in my Harvard notes).

The G–covering of the line in char. 0

Because of the above relations, we know that there is a covering of the projective line with group G, which is ramified at $1, 0, \infty$, with ramification groups generated by x, y, z as above. It turns out that this covering has been written explicitly by Fricke (Math. Ann. 52 (1899), 321-339). The curve corresponding to the Galois extension has genus 7, and one finds nice equations for it in a paper of Macbeath (Proc. London Math. Soc. 15 (1965), 527-542, – see also Wohlfahrt, Glasgow Math. J. 27 (1985), 239-247). However, it is more convenient to use the degree 9 covering corresponding to the embedding of G in A_9. It turns out (genus formula!)

that this covering has genus 0. Hence we can view it as given by a rational function $t = F(u)$ of degree 9. Because of the very demanding ramification properties of our covering, the function F has the following properties (for a suitable normalisation of u):

a) $F(u)$ is of the form $p_9(u)/p_2(u)$, where p_i means a polynomial of degree i;

b) p_9 is a cube: $p_9 = p_3^3$, where $p_3 = u^3 + au^2 = bu + c$; moreover: $p_9 - p_2 = up_4(u)^2$, where $p_4 = u^4 + du^3 + \ldots + g$.

(These conditions express the ramification at $\infty, 0$ and 1 respectively.)

If one writes p_2 as $hu^2 + iu + j$, one gets a system of identities on the coefficients $a, b, c \ldots, j$ which **determines** them, up to a homogeneity factor. The computation was done for me by J-F. Mestre (but it had been done 103 years earlier by Goursat, according to what Fricke says). The result is:

$$a = 4, \quad b = 10, \quad c = 6, \quad d = 6, \quad e = 21,$$

and

$$f = 35, \quad g = 63/2, \quad h = 27, \quad i = 351/4, \quad j = 216.$$

(You will have fun to check this on your computer.)

The conclusion of all this is that *the 9–degree equation*

$$p_3(u)^3 - tp_2(u) = 0$$

defines a degree 9 extension of $\mathbf{C}(t)$ *with Galois group* $G \subset A_9$.

Notice however that the coefficients of the equation belong to $\mathbf{Q}$ (could this be proved a priori ? I suspect it could). Hence one can play the game over $\mathbf{Q}$ and ask what the Galois group is over $\mathbf{Q}(t)$. Answer: it is not G, but the slightly bigger group $G.3$; moreover, the "3" (i.e. the constant field extension inside the Galois closure) is the cubic extension of $\mathbf{Q}$ given by $2cos(2\pi/7)$, i.e. by the root z of the equation

$$z^3 + z^2 - 2z - 1 = 0.$$

(Sketch of proof: the new Galois group contains G as a normal subgroup; but the normalizer of G in A_9 is $G.3$; hence the Galois group we want is either G or $G.3$. But an easy ramification argument shows that the Galois field we have contains $\mathbf{Q}(z)$. Hence the result.)

From characteristic 0 to characteristic 7

Let me explain the strategy first. I shall make some rescaling on the variables t and u in such a way that F "reduces well" mod 7, with the important difference that the ramification points of order 2 and 7 will be "fused" in one point. Hence we shall get a G–extension of the projective line in char. 7 which is ramified at *two points* only: one of them (infinity, say) with inertia group the dihedral group D_7 of order 14, and the other one with tame ramification of order 3 (at 0, say). By using your lemma, one makes the tame ramification disappear. Hence a covering of the affine line with Galois group G. By writing down explicit equations, one finds that it is isomorphic to yours, i.e., given by $Y^9 - XY^7 + 1 = 0$.

Here are the details of the computation:

Rescaling: The first step is to replace the variable u by v, with $v = u - 1$. The new coefficients of the p_i polynomials become:

$p_2 \rightarrow 27$, $567/4$, $1323/4$ (note that $567 = 3^4 \cdot 7$ and $1323 = 3^3 \cdot 7^2$)

$p_3 \rightarrow 1, 7, 21, 21$

p_4 does not matter.

The essential point is the divisibility by 7 of the coefficients of p_2 and p_3. To take advantage of this, one makes a further change of variable

$$v = w\lambda \quad \text{where} \quad \lambda = 7^{1/3}.$$

We have thus

$p_2 = 27\lambda^2(w^2 + \ldots)$

$p_3 = \lambda^3(w^3 + \ldots + 3)$,

where the unwritten coefficients are all divisible by λ. If we put

$$T = \lambda^{-7}t,$$

we have

$$T = (w^3 + \ldots + 3)^3/27(w^2 + \ldots).$$

We can now safely reduce mod λ, i.e. go to char. 7. We then find that the rational function T is given by

$$T = -(w^3 + 3)^3/w^2 \qquad \text{(note that } 27 = -1 \text{ in char. 7).}$$

The upshot of this is that the 9–degree equation

$$(w^3 + 3)^3 + Tw^2 = 0$$

has Galois group (in char. 7) the group G (a priori we could get a subgroup of G, but by looking at the ramificaiton, we see that it is the full group G – moreover, this works over the ground field F_7 itself).

More changes of variables

Let $p = (w^3 + 3)/w$; one checks that $w = -T/p^3$ and the equation connecting T and the new variable p is:

$$3p^9 + Tp^7 - T^3 = 0.$$

Put $T = X^3$ and $Y = p/X$.

We then have:

$3Y^9 + XY^7 - 1 = 0.$

This is not quite your equation, but becomes so by replacying Y by WY and X by $3W^2X$ where W is an element in the ground field (which is assumed to be algebraically closed). [1]

[1] Here I am correcting a computational mistake in Serre's original letter – Abhyankar.

I suspect (but I have not proved) that there is a "modular" interpretation for all these little miracles. It should be something like the following:

Let $K = \mathbf{Q}(z)$ be the cubic field considered above. Let D be the quaternion algebra with center K which is ramified at two of the three real places of K, and nowhere else. Let $O(D)$ be a maximal order of D (there should be only one, up to conjugation). Let Γ be the group of units of $O(D)$ of reduced norm 1. The group $\Gamma/(\pm 1)$ can be viewed as a discrete subgroup with compact quotient of $PSL_2(\mathbf{R})$ and should be the triangle group $(2,3,7)$ (this must have been checked by K. Takeuchi, J. Math. Soc. Japan 29 (1977), 91-106 – see also Fricke-Klein, Vorl. autom. Funkt., Bd.I, 610-611). The congruence subgroup $\Gamma(2)$ of Γ (mod 2) is such that $\Gamma/\Gamma(2) = SL(2,8)$ (notice that the residue field of K at the prime 2 is F_8). The Riemann surface associated with Γ (resp. with $\Gamma(2)$) should be the projective line (resp. the genus 7 curve studied by Frick and Macbeath) and the covering given by the inclusion $\Gamma(2) \to \Gamma$ should be our G–Galois covering. Hence we may view this covering as a "**Shimura curve**" of level 2, relative to K and the quaternoin algebra D. The rather surprising good reduction at 7 of this curve can probably be obtained from general theorems.

Even if all this turns out to be correct, there will still be one miracle left unexplained: *why is this* 9 *degree equation so simple in characteristic* 7?

With best regards,
Yours

J-P.Serre

4
Equisingularity Invariants of Plane Curves

Angel Granja

4.1 Introduction

First we want to state some questions about the classification of singularities of plane curves.

Let k be a field and denoted by L the algebraic closure of k. We call plane curve over k to a reduced ring $A = k[[X,Y]]/(f)$.

If k is an algebraically closed field ($k = L$) then Zariski's equisingularity theory classifies the singularity of A and there are a lot of invariants for this classification ([5], [6], [15]).

For k a non algebraically closed field ($k \neq L$) we can consider $A_L = (A \otimes_k L)^* = L[[X,Y]]/(f)$. (Where $(A \otimes_k L)^*$ is the completion of $A \otimes_k L$).

If k is a zero characteristic field then A_L is also a plane curve and the singularity type of A_L is the same for each coefficient field k of A ([2], [16]). This idea has been used by Zariski to introduce the concept of singularity of dimensional type one. etc).

On the other hand, for positive characteristic fields, Abhyankar has given examples showing that A_L need not be a plane curve and also that even being A_L an irreducible plane curve (that is A_L is a domain) the singularity type may change when we change the coefficient field of A [2].

So we can put two questions:

1. When is A_L useful to classify?
2. How to classify the singularity of A when the coefficient field is not algebraically closed?

If A is a domain the answer to Question 1 is the following:

We can use A_L if and only if $A'/M(A')$ is a separable extnsion of $A/M(A)$, where A' is the integral closure of A and $M(-)$ denotes the maximal ideal [9].

To answer Question2 we are going to consider a more general situation by studying Abhyankar' curves, that is pairs (R, j) where R is a regular noetherian local ring of Krull dimension two, and J is a non zero principal ideal of R. $(0 \neq J = fR)$, [4].

If R is a complete ring, using Abhyankar's initial form for these curves one can generalize Zariski's idea of tangentially stable pairing and one can get to the same Zriski's equivalent definitions of a)–equivalence, b)–equivalence and formal equivalence [11].

At this point we must find invariants (as in the classical case) to characterize the singularity type.

First we will give some examples to see how some classical invariants work for our curves. We will consider the mutliplicity sequence and the semigroup of values.

Examples. Let R be the ring $R = \mathbb{R}[[X,Y]]$, where $\mathbb{R}$ is the set of real numbers.

Consider the irreducible curves $(R,(X^2+Y^2)R)$, $(R,(X^4+Y^2)R)$ and $(R,(X^3+Y^2)R)$.

The desingularization processes are the following

a) $(R,(X^2+Y^2)R)$, $((R[Y/X])_{(X,(Y/X)^2+1)},$
$(1+(Y/X)^2)(R[Y/X])+(X,(Y/X)^2+1))$.

b) $(R,(X^4+y^2)R)$, $((R[Y/X])_{(X,Y/X)},(X^2+(Y/X)^2)(R[Y/X])_{(X,Y/X)})$,
$(R_1[Y/X^2])_{(X,(Y/X^2)^2+1)},(1+(Y/X^2)^2)(R_1[Y/X^2])_{(X,(Y/X^2)^2+1)})$
where $R_1 = (R[Y/X])_{(X,Y/X)}$

c) $(R,(X^3+Y^2)R)$, $((R[Y/X])_{(X,Y/X)},(X^1+(Y/X)^2)(R[Y/X])_{(X,Y/X)})$.

We have

Curve	Multiplicity Sequcne	Semigroup of Values
$(R,(X^2+Y^2)R)$	2,1	$2\mathbb{Z}_+ = \{0,2,4,6,\ldots\}$
$(R,(X^2+Y^4)R)$	2,2,1	$2\mathbb{Z}_+ = \{0,2,4,6,\ldots\}$
$(R,(X^2+Y^3)R)$	2,2,1	$\mathbb{Z}_+ - \{1\} = \{0,2,3,4,5,\ldots\}$.

(Here the semigroup of values of a curve (R,J) is the set

$$S(R,J) = \{L_R(R/gR+J),\ g \in R-J\},$$

where L_R denotes the length fuction of R-modules).

So we must find some alternatives to those invariants. Instead of the multiplicity sequence we will consider the equiresolutin class and instead of the semigroup of values we will consider a special set of generators of this semigroup — the Apery Basis. The rest of the paper is devoted to study this set of generators.

In the following we will denote by $M(A)$ the maximal ideal of a local ring A and we assume that (R,J) is an irreducible plane curve. That is, R is a complete regular noetherian local ring of Krull dimension two and J is a non zero principal ideal of R. We will denote by $m(R, J)$ the multiplicity of the curve, that is, $m(R, J)$ is the non negative integer d such that $J \subset (M(R))^d$ and $J \not\subset (M(R))^{d+1}$.

4.2 Equiresolution Class

By a formal quadratic transform of R we mean a ring R' such that R' is the completion of a ring $(R[z^{-1}M(R)])_Q$, where $z \in M(R)$, $z \notin (M(R))^2$ and Q is a prime ideal of $R[z^{-1}M(R)]$. The ring $(R[z^{-1}M(R)]_Q$ is a quadratic transform of R (see [4]).

By a formal quadratic transform of (R, J) we mean a curve (R', J') such that R' is a formal quadratic transform of R and J' is the unique ideal of R' such that $J'(M(R))^{m(R,J)}R' = JR'$.

As (R, J) is an irreducible plane curve there is one and only one formal quadratic transform (R', J') of (R, J) such that $R' \neq J'$ and (R', J') is also an irreducible plane curve. So we have a sequence $(R, J) = (R_0, J_0)$, $(R_1, J_1), \ldots, (R_n, J_n), \ldots$ such that (R_i, J_i) is the unique formal quadratic transform of (R_{i-1}, J_{i-1}) with $R_i \neq J_i$, $i = 1, 2, \ldots$.

On the other hand there is a non negative integer N such that $m(R_n, J_n) = 1$ for each $n > N$ and $m(R_n, J_n) > 1$ for each $n < N$. (This is an easy consequence of the Proper Desingularization Theorem of [4]).

We call resolution process of (R, J) to the finite sequence of irreducible plane curves $(R, J) = (R_0, J_0), (R_1, J_1), \ldots, (R_N, J_N)$. And by the equiresolution class of (R, J) we mean the set

$$E(R, J) = \{(n_i, s_i)\}_{0 \leq i \leq N},$$

where

$$n_i = m(R_i, J_i), \qquad 0 \leq i \leq N,$$
$$s_i = [R_{i+1}/M(R_{i+1}) : R_i/M(R_i)], \qquad 0 \leq i \leq N-1$$

and $s_N = 1$. Note that R_N/J_N is the integral closure of R/J.

4.3 Apery Basis Relative to a Parameter

We will denote by L_R the length function of R-modules. If $g \notin J$ then we have

$$L_R(R/gR + J) = [R_N/M(R_N) : R/M(R)] L_{R_N}(R_N/gR_N + J_n)$$

([13] Th. 13, p. 168).

Let (x, y) be a basis of $M(R)$ and assume that $x \notin J$ and $p = L_R(R/xR + J)$, hence $J = (y^p + xf)R$.

4.3.1 Lemma

The set

$$V_i(R, J, x) = \{L_R(R/gR + J); L_R(R/gR + xR) = i\},$$

is finite $0 \leq i < p$. (Note that if $L_R(R/gR + xR) = i < p$ then $g \notin J$.)

Proof. We must only see that $V_{p-1}(R, J, x)$ is finite.

$$(V_{i+1}(R, J, x) \supset L_R(R/yR + J) + V_i(R, J, x)).$$

If $V_{p-1}(R, J, x)$ is not finite then there are $g_1, g_2, \ldots, g_3, \ldots$ such that $L_R(R/g_jR + xR) = p - 1$ and

$$L_R(R/g_jR + J) > [(R_N/M(R_N)) : (R/M(R))]j, \qquad j \geq 1.$$

So

$$L_{R_N}(R_N/gR_N + J_N) = j$$

and

$$\eta_{R/J}(g_j) = \eta_{R_N/J_N}(g_j) \in (M(R_N/J_N))_j$$

where

$$\eta_{R/J} : R \to R/J$$

and

$$\eta_{R_N/J_N} : R_N \to R_N/J_N$$

are the canonical epimorphisms.

On the other hand as $(M(R_N/J_N))^{h_0} \subset M(R/J)$ for some h_0, if $j > ph_0$ then

$$\eta_{R/J}(g_j) \in (M(R_N/J_N))^j \subset (M(R_N/J_N))^{ph_0} \subset (M(R/J))^p,$$

so

$$g_j = \sum_{i+j+p} a_{ij}x^iy^j + g_j'(y^p + xf),$$

with $g_j' \in R$, $L_R(R/g_jR + J) \geq p$ and we get to a contradiction.

4.3.2 Definition

We will write

$$a_i(R, J, x) = Sup(V_i(R, J, x))$$

for $0 \leq i < p$, and we will say that

$$A(R, J, x) = \{0 = a_0(R, J, x) < a_1(R, J, x)$$

$$< a_2(R, J, x) < \ldots < a_{p-1}(R, J, x)\}$$

is the x-Apery basis of (R, J).

4.3.3 Remark

We point out that

$$a_i(R, J, x) \geq a_j(R, J, x) + a_k(R, J, x),$$

for each $j, k \geq 1$, with $j + k = i$, and also we note that if $A(R, J, x) = \{0\}$ then $m(R, J) = 1$.

4.3.4 Lemma

With the above notations, for $1 \leq i < p$ there is $g_i \in R$ such that

$$g_i = y^i + \sum_{(j,k)\in Q_1} g^i_{jk} x^j y^k$$

and

$$L_R(R/g_iR + J) = a_i(R, J, x),$$

where Q_i is a finite subset of

$$\{(i, j) \in \mathcal{Z}_+ \times \mathcal{Z}_+; \qquad k < i, j > 0$$

and $g^i_{jk} \notin M(R)$ for each $(j, k) \in Q_i$, $1 \leq i < p$.

Proof. We can assume that there is

$$g'_i = y^i + g'_i x,$$

such that

$$L_R(R/g'_iR + J) = a_i(R, J, x)$$

and

$$L_R(R/g_i' + J) = i.$$

So we can write

$$g_i' = \sum_{0 \le j+k \le a_i(R,J,x)} g_{jk}^i x^j y^k + g_i^{**},$$

with

$$g_i^{**} \in (M(r))^{a_i(R,J,x)+1}$$

and

$$g_{jk}^i \notin M(R), \qquad 0 \le j+k \le a_i(R,J,x).$$

Note that if $g_i^{**} \neq 0$ then $L_R(R/g_i^{**}R + J) > a_i(R,J,x)$. So

$$g_i' = uy^i + x(\sum_{\substack{0 \le j+k \le a_i(R,J,x) \\ k<i}} g_{jk}^i x^j y^k + g_i^{**}),$$

where

$$u = 1 + x(\sum_{\substack{0 \le j+k \le a_i(R,J,x) \\ k<i}} g_{jk}^i x^j y^{k-1}).$$

We must only take $g_i = u^{-1}(g_i' - xg_i^{**})$ to have the lemma.

4.3.5 Definition

We will say that a non negative integer r is an x-index of contact of (R, J) if $r = 0, 1$ or $r > 1$ and $a_r(R,J,x) > a_j(R,J,x) + a_k(R,J,x)$, for each $r > j, k \ge 0$ with $j + k = r$.

If

$$0 + r_0(R,J,x) < 1 = r_1(R,J,x) < r_2(R,J,x) < \ldots < r_{g(R,J,x)}(R,J,x)$$

are the x-indices of contact of (R, J) we say that $g(R,J,x)$ is the x-genre of (R, J) and we will write

$$v_0(*R,J,x) = p$$

and

$$v_i(R,J,x) = a_{r_i(R,J,x)}(R,J,x), 1 \le i \le g(R,J,x).$$

We will say that

$$\{p = v_0(R, J, x), v_1(R, J, x), \ldots, v_{g(R,J,x)}(R, J, x)\}$$

is the x-sequence of (R, J).

4.3.6 Lemma

With the above notations, if $1 \leq i < p$ and

$$i = \sum_{1 \leq j \leq g(R,J,x)} d_j r_j(R, J, x)$$

then

$$a_i(R, J, x) \geq \sum_{1 \leq j \leq g(R,J,x)} d_j v_j(R, J, x).$$

Proof. For $i = 1$ we certainly have $a_1(R, J, x) \geq v_1(R, J, x)$. Now assume the result for each j with $1 \leq j < i (i > 1)$. So there is $d_k \neq 0$ and we can consider

$$1 = \sum_{l \leq j \leq g(R,J,x), j \neq k} d_j r_j(R, J, x) + (d_k - 1) r_k(R, J, x).$$

By the induction hypothesis one has that

$$a_i(R, J, x) \geq a_1(R, J, x) + a_{r_k}(R, J, x) = a_1(R, J, x) + v_k(R, J, x)$$

$$\geq \textstyle\sum_{1 \leq j \leq g(R,J,x), j \neq k} d_j v_j(R, J, x) + (d_k - 1) v_k(R, J, x) + v_k(R, J, x)$$

$$= \textstyle\sum_{1 \leq j \leq g(R,J,x)} d_j v_j(R, J, x).$$

4.3.7 Remark

We want to point out that for $1 \leq i < p$ there are $d_1, \ldots, d_{g(R,J,x)}$ such that

$$i = \sum_{1 \leq j \leq g(R,J,x)} d_j r_j(R, J, x)$$

and

$$a_i(R, J, x) = \sum_{1 \leq j \leq g(R,J,x)} d_j v_j(R, J, x).$$

Also note that if $r_i(R,J,x)$ is an x-index of contact and

$$r_i(R,J,x) = \sum_{1\le j\le i-1} d_j r_j(R,J,x),$$

then

$$v_i(R,J,x) > \sum_{1\le j\le i-1} d_j v_j(R,J,x).$$

(If

$$v_i(R,J,x) = \sum_{1\le j\le i-1} d_j v_j(R,J,x),$$

as $d_k \neq 0$ for some $1 \le j < i$, then

$$v_i(R,J,x) \le v_k(R,J,x) + a_1(R,J,x),$$

with

$$1 = \sum_{\substack{1\le j\le i-1\\ j\neq k}} d_j r_j(R,J,x) + (d_k - 1) r_k(R,J,x)$$

and $r_i(R,J,x) = 1 + r_k(R,J,x)$, and this is a contradiction).

4.4 Technical Lemmas

Let (x,y) be a basis of $M(R)$ and let p,q be two non negative rational numbers ($p \neq 0, q \neq 0$). We will write

$$\begin{aligned} N_{p,q} = p\mathcal{Z}_+ \times q\mathcal{Z}_+ \quad &= \{ip + jq; (i,j) \in \mathcal{Z}_+ \times \mathcal{Z}_+\} \\ &= \{0 = N_0 < N_1 < N_2 < \ldots\}. \end{aligned}$$

Note that, for each $d \in \mathcal{Z}_+$, there is $k \in \mathcal{Z}_+$ such that

$$(M(R))^d \supset \{x^i y^j; \qquad ip + jp \ge N_k\}$$

4.4.1 Lemma

Let $f \in R$ be such that

$$f = f_{i_0 j_0} x^{i_0} y^{j_0} + \sum_{(i,j)\in Q} f_{ij} x^i y^j,$$

with $i_0 \ge 1$, $j_0 \ge 1$,

$$Q \subset \{(i,j) \in \mathcal{Z}_+ \times \mathcal{Z}_+; \quad ip + jq > i_0p + j_0q\},$$

$f_{ij} \in R$ and $f_{i_0j_0} \notin M(R)$. Then there are $g_1, g_2 \in R$ such that $f = g_1g_2$.

Proof. We can assume that $f_{i_0j_0} = 1$.

Consider $g_1^0 = x^{i_0}$ and $g_2^0 = y^{j_0}$. Then

$$f - g_1^0g_2^0 = \sum_{(i,j)\in Q_0} f_{ij}^0 x^i y^j,$$

with $Q_0 = Q$ and $f_{ij}^0 = f_{ij}$ for each $(i,j) \in Q_0$.

If $Q_0 = \phi$ we take $g_1 = g_1^0$ and $g_2 = g_2^0$.

If $Q_0 \neq \phi$ we take $r_1 = \min\{ip + jq; \quad (i,j) \in Q_0\}$ $(i_0p + j_0q < r_1)$.

Now consider

$$A_1^0 = \{(i,j) \in Q_0; \quad j \geq j_0\}$$

and

$$A_2^0 = \{(i,j) \in Q_0; \quad j \geq j_0\}.$$

If $(i,j) \in A_2^0$, as $ip + jq \geq r_1 > i_0p + j_0q$, then $i > i_0$.

We define

$$g_1^1 = g_1^0 + \sum_{(i,j)\in A_1^0} f_{ij}^0 x^i y^{j-j_0}$$

and

$$g_2^1 = g_2^0 + \sum_{(i,j)\in A_2^0} f_{ij}^0 x^{i-i_0} y^j.$$

We have

$$f - g_1^1g_2^1 = \sum_{(i,j)\in Q_1} f_{ij}^1 x^i y^j,$$

with

$$Q_1 \subset \{(i,j) \in \mathcal{Z}_+ \times \mathcal{Z}_+; \quad ip + jq > r_1\}$$

and $g_1^1 - g_1^0$ and $g_2^1 - g_2^0$ are in the ideal generated by $\{x^iy^j; \quad ip + jq \geq r_1\}$.

If $Q_1 = \phi$ we take $g_1 = g_1^1$ and $g_2 = g_2^1$.

If $q_1 \neq \phi$ we take $r_2 = \min\{ip + jq; \quad (i,j) \in Q_1\}$ $(r_1 < r_2)$. Now we can repeat the above process.

Then either we get to a step n with $q_n = \phi$ or we have a sequence of rational numbers $i_0p + j_0q = r_0 < r_1 < \ldots < r_n < \ldots$ with $r_n \in N_{p,q}$ and sequence of elements of $R\{g_j^n\}_{n\in\mathcal{Z}_+}$, $1 \leq j \leq 2$ such that both $g_j^{n+1} - g_j^n$ and $f - g_1^n g_2^n$ are in the ideal generated by $\{x^i y^j;\quad ip + jq > r_n\}$. As R is complete then $\{g_j^n\}_{n\in\mathcal{Z}_+}$ converge to g_j, $1 \leq j \leq 2$ and so $f = g_1 g_2$.

4.4.2 Lemma

Assume that $L_R(R/xR + J) = p$, $L_4(R/yR + J) = q$ and $e = gcd(p, q)$. If $r > 1$ is an x-index of contat, then p/e divides r. Moreover, there is

$$g = y^{dp/e} + f_{d-1} y^{(d-1)p/e} x^{q/e} + \ldots + f_0 x^{dq/e} + \sum_{(i,j)\in Q} f_{ij} x^i y^j \in R,$$

such that

$$L_R(R/gR + J) = a_r(R, J, x)$$

and $f_0 \notin M(R)$, where $dp/e = r$ and

$$Q \subset \{(i,j) \in \mathcal{Z}_+ \times \mathcal{Z}_+; \qquad ip + jq > rq\}.$$

Proof. By 4.2.4 we can take $g \in R$ with

$$L_R(R/gR + J) = a_r(R, J, x)$$

and

$$g = y^r + \sum_{(i,j)\in Q'} f_{ij} x^i y^j,$$

where Q' is a finite subset of

$$\{(i,j) \in \mathcal{Z}_+ \times \mathcal{Z}_+; \qquad i \geq 1, \quad j < r\}$$

and $f_{ij} \notin M(R)$ for each $(i,j) \in Q'$. As r is an x-index of contact then $Q' \neq \phi$.

Now consider $d_0 = \min\{ip + jq; \quad (i,j) \in Q'\}$. We have two possibilities.

1. $\mathbf{d_0 < rq}$. In this case let $(i_0, j_0) \in \{(i,j) \in Q';\ ip + jq = d_0\}$ be such that $j_0 \geq j$ for each $(i,j) \in Q'$ with $ip + jq = d_0$. We have $i_0 \geq 1$, $j_0 < r$, $f_{i_0 j_0} \notin M(R)$, $j_0 < d_0/q$ and $j_0 > 0$. (If $j_0 = 0$ then $L_R(R/gR + j) = i_0 p < rq < a_r(R, J, x)$).

   If $d_1 = \min\{ip + jq;\ (i,j) \in Q',\ ip + jq > d_0\}$ and β is a rational number such that $(d_0/q) < \beta < (d_1/q)$, then it is easy to check that

$$g = f_{i_0 j_0} x^{i_0} y^{j_0} \sum_{(i,j)\in Q} f_{ij} x^i y^j,$$

where $i_0 > 0$, $j_0 > 0$,

$$Q \subset \{(i,j) \in \mathcal{Z}_+ \times \mathcal{Z}_+; \quad ip' + jq' > c_0\},$$

$$p' = \beta - j_0,$$

$$q' = i_0,$$

$$c_0 = i_0\beta = p'i_0 + q'j_0$$

and

$$f_{i_0 j_0} \notin M(R).$$

So by 4.3.1, there exist $g_1, g_2 \in R$ such that $g = g_1 g_2$. Then

$$a_r(R, J, x) = L_R(R/g_1R + J) + L_R(R/g_2R + J),$$
$$r = L_R(R/gR + xR) = L_R(R/g_1R + xR) + L_R(R/g_2R + xR),$$
$$L_R(R/g_1R + xR) = k_1,$$
$$L_R(R/g_2R_xR) = k_2$$

and

$$a_r(R, J, x) \leq a_{k_1}(R, J, x) + a_{k_2}(R, J, x)$$

which is a contracition.

2. $\mathbf{d_0 = rq}$. In this case let $(i_0, j_0) \in \{(i,j) \in Q'; ip + jq = d.\}$ be such that $j_0 \leq j$ for each $(i,j) \in Q'$ with $ip + jq = d_0$. We also have two possibilities. (Note that if $\{(i,j) \in Q'; \quad ip + jq = d_0\} = \phi$ then $L_R(R/gR + J) = rq < a_r(R, J, x)$.)

3. (i) $j_0 > 0$. consider

$$d_1 = \min\{ip + jq; \quad (i,j) \in Q',\ ip + jq > d_0 = rq\}$$

and let β be a rational number such that $(d_0/q) < \beta < (d_1/q)$, $(\beta > i_0)$.

It is easy to check that

$$g = f_{i_0 j_0} x^{i_0} y^{j_0} + \sum_{(i,j)\in Q} f_{ij} x^i y^j,$$

where $i_0 > 0$, $j_0 > 0$, $Q \subset \{(i,j) \in \mathcal{Z}_+ \times \mathcal{Z}_+; \ ip' + jq' > c_0\}$, $p' = \beta - j_0$, $q' = i_0$, $c_0 = i_0\beta = p'i_0 + q'j_0$ and $f_{i_0 j_0} \notin M(R)$. Now we can repeat the above argument to get to a contradiction.

4. (ii) $j_0 = 0$. In this case $i_0 p = rq$, $r = dp/e$, $i_0 = dq/e$ and $f_{dq/e,0} \notin M(R)$.

4.4.3 Remark

We can take $g = y + g_0 x^d$ such that $g_0 \notin M(R)$, $L_R(R/gR+j) = a_1(R,J,x)$ and $(p/q) \leq d \leq (p/q)+1$.

4.5 Inversion

Let (x, y) be a basis of $M(R)$ and assume that $p = L_R(R/xR + J)$ and $q = L_R(R/yR + J)$. Let

$$0 = r_0(R,J,x) < 1 = r_1(R,J,x) < \ldots < r_{g(R,J,x)}(R,J,x)$$

resp.

$$0 = r_0(R,J,y) < 1 = r_1(R,J,y) < \ldots < r_{g(R,J,y)}(R,J,y)$$

be the x-indices) (resp. y-indices) of contact, let

$$\{p = v_0(R,J,x), v_1(R,J,x), \ldots, v_{g(R,J,x)}(R,J,x)\}$$

resp.

$$\{q = v_0(R,J,y), v_1(R,J,y), \ldots, v_{g(R,J,y)}(R,J,y)\}$$

be the x-sequence (resp. y-sequence) and finally let

$$A(R,J,x) = \{0 = a_0(R,J,x) < a_1(R,J,x) < \ldots < a_{p-1}(R,J,x)\}$$

resp.

$$A(R,J,y) = \{0 = a_0(R,J,y) < a_1(R,J,y) < \ldots < a_{q-1}(R,J,y)\}$$

be the x-Apery (resp. y-Apery) basis of (R, J). By 4.3.2 we can take

$$f_i^x = y^{b_i p/e} + f_{b_i-1}^{ix} y^{(b_i-1)p/e} x^{q/e} + \ldots + f_0^{ix} x^{b_i q/e} + \sum_{(k,\ell)\in Q_\ell^x} f_{k1}^x x^k y^1$$

and

$$f_j^y = x^{c_j q/e} + f_{c_j-1}^{jy} x^{(c_j-1)q/e} y^{p/e} + \ldots + f_0^{jy} y^{c_j p/e} + \sum_{(k,\ell)\in Q_j^y} f_{k\ell}^y x^k y^\ell,$$

where $e = gcd(p, q)$,

$$b_i p/e = r_i(R, J, x),$$

$$c_j q/e = r_j(R, J, y),$$

$$f_0^{ix}, f_0^{jy} \notin M(R),$$

$$L_R(R/f_i^x R + J) = v_i(R, J, x)$$

and

$$L_R(R/f_j^y R + J) = v_j(R, J, y), \qquad 2 \leq i \leq g(R, J, x),\ 2 \leq j \leq g(R, J, y).$$

Moreover, by 4.3.3 we can take $f_1^x = y + f_0^{1x} x^{b_1}$ and $f_1^y = x + f_0^{ly} y^{c_1}$, with f_0^{lx}, $f_0^{1y} \notin M(R)$, $L_R(R/f_1^x R + J) = v_1(R, J, x)$ and $L_R(R/f_1^y R + J) = v_1(R, J, y)$.

Then we have

$$v_1(R, J, x) \leq a_{b_1}(R, J, y),$$

$$v_1(R, J, y) \leq a_{c_1}(R, J, x),$$

$$v_i(R, J, x) \leq a_{(q/p) r_i(R,J,x)}(R, J, y)$$

and

$$v_j(R, J, y) \leq a_{(p/q)} r_j(R, J, y), \qquad 2 \leq i, \quad j \leq \min(g(R, J, x),\ g(R, J, y)).$$

In fact, if $r_j(R, J, y) = (q/p) r_i(R, J, x)$ for i and j, then $v_j(R, J, y) = v_i(R, J, x)$.

The main objective of this section is to prove the following

4.5.1 Inversion Theorem

With the above notations we have:

If $p = q$ then

$$\begin{aligned} g(R, J, y) &= g(R, J, x), \\ r_j(R, J, y) &= (q/p) r_j((R, J, x) \quad \text{and} \\ v_j(R, J, y) &= v_j(R, J, x), \qquad 0 \leq i \leq g(R, J, x). \end{aligned} \tag{5.1.1}$$

If $p < q < v_1(R, J, x)$ then

$$\begin{aligned} g(R,J,y) &= g(R,J,x)+1, \\ r_0(R,J,x) &= 0, \\ v_0(R,J,x) &= q, \\ r_1(R,J,y) &= 1, \\ v_1(R,J,y) &= p, \\ r_{j+1}(R,J,y) &= (q/p)r_j(R,J,x), \quad \text{and} \\ v_{j+1}(R,J,y) &= v_j(R,J,x), \quad 1 \le j \le g(R,J,x). \end{aligned} \tag{5.1.2}$$

If $p < q = v_1(R,J,x)$ then

$$\begin{aligned} g(R,J,y) &= g(R,J,x), \\ r_0(R,J,x) &= 0, \\ v_0(R,J,x) &= q, \\ r_1(R,J,y) &= 1, \\ v_1(R,J,y) &= p, \\ r_j(R,J,y) &= (q/p)r_j(R,J,x), \quad \text{and} \\ v_j(R,J,y) &= v_j(R,J,x), \quad 2 \le j \le g(R,J,x). \end{aligned} \tag{5.1.3}$$

If $p < q$ and $r_2(R,J,x) = p/q$ then

$$\begin{aligned} g(R,J,y) &= g(R,J,x)-1, \\ r_0(R,J,x) &= 0, \\ v_0(R,J,x) &= q, \\ r_j(R,J,y) &= (q/p)r_{j+1}(R,J,x), \text{and} \\ v_j(R,J,y) &= v_{j+1}(R,J,x), \quad 1 \le j \le g(R,J,x)-1. \end{aligned} \tag{5.1.4}$$

If $q < p$ and $\gamma_2(R,J,x) > p/q$ then

$$\begin{aligned} g(R,J,y) &= g(R,J,x), \\ r_0(R,J,x) &= 0, \\ v_0(R,J,x) &= q, \\ r_j(R,J,y) &= 1, \\ v_1(R,J,y) &= p, \\ \gamma_j(R,J,y) &= (q/p)r_j(R,J,x), \quad \text{and} \\ v_j(R,J,y) &= v_j(R,J,x), \quad 2 \le j \le g(R,J,x). \end{aligned} \tag{5.1.5}$$

Proof. We will prove (5.1.1). The proof of the remaining cases need only minor modifications.

For (R,J) a regular curve one has $p=1$ or $q=1$. In our case $p=q=1$. Thus $g(R,J,x) = g(R,J,y) = 0$ and $v_0(R,J,x) = v_0(R,J,y) = 1$.

Assume that (R,J) is not regular. Then $p=q>1$ and $p=q=gcd(p,q)$.

On the other hand, if $p=q=v_1(R,J,x)$ then $v_1(R,J,x) = v_1(R,J,y)$. (Note that if $q < v_1(R,J,y)$ then $c_1 = 1$ and $v_1(R,J,y) \le v_1(R,J,x)$.) If $p=q<v_1(R,J,x)$ then $b_1 = 1$, $p=q<v_1(R,J,x) \le v_1(R,J,y)$, $c_1 = 1$ and $v_1(R,J,y) \le v_1(R,J,x)$. So in any case $v_1(R,J,y) = v_1(R,J,x)$.

Now assume that $r_j(R,J,y) = r_j(R,J,x)$ and that $v_j(R,J,y) = v_j(R,J,x)$, $1 \le j \le i < \min(g(R,J,x), g(R,J,y))$.

If $r_{i+1}(R,J,x) < r_{i+1}(R,J,y)$ then there exist $d_j \in \mathcal{Z}_+$, $1 \le j \le i$, such that

$$r_{i+1}(R,J,x) = \sum_{1\le j\le i} d_j r_j(R,J,y),$$

and

$$a_{r_{i+1}(R,J,x)}(R,J,y) = \sum_{1\le j\le i} d_j v_j(R,J,y).$$

So

$$V_{i+1}(R,J,x) \le a_{r_{i+1}(R,J,x)}(R,J,y) = \sum_{1\le j\le i} d_j v_j(R,J,y) < v_{i+1}(R,J,x),$$

which is a contradiction. Thus $r_{i+1}(R,J,x) \ge r_{i+1}(R,J,y)$.

In the same way we get to $r_{i+1}(R,J,y) \le r_{i+1}(R,J,x)$. Thus $r_{i+1}(R,J,y) = r_{i+1}(R,J,x)$ and so $v_{i+1}(R,J,y) = v_{i+1}(R,J,x)$.

To finish the proof assume that $r_j(R,J,y) = r_j(R,J,x)$ and that $v_j(R,J,y) = v_j(R,J,x)$, $1 \le j \le \min(g(R,J,x), f(R,J,y))$. Then we have $g(R,J,y) = g(R,J,x)$.

If $g(R,J,x) < g(R,J,y)$ then $r_{g(R,J,x)+1}(R,J,y) = c_{g(R,J,x)+1} < q = p$ and so then there exist $d_j \in \mathcal{Z}_+$, $1 \le j \le g(R,J,x)$ such that

$$\begin{aligned} r_{g(R,J,x)+1}(R,J,y) &= \textstyle\sum_{1\le j\le g(R,J,x)} d_j r_j(R,J,x) \quad \text{and} \\ a_{r_{g(R,J,x)+1}(R,J,y)}(R,J,x) &= \textstyle\sum_{1\le j\le g(R,J,x)} d_j v_j(R,J,x). \end{aligned}$$

So

$$v_{g(R,j,x)+1}(R,J,y) \le a_{r_{g(R,J,x)+1}(R,J,y)}(R,J,x)$$

$$= \textstyle\sum_{1\le j\le g(R,J,x)} d_j v_j(R,J,x) < v_{g(R,J,x)+1}(R,J,y),$$

which is a contradiction. Thus $g(R,J,x) \ge g(R,J,y)$.

In the same way we get to $g(R,J,x) \le g(R,J,y)$ and we have $g(R,J,y) = g(R,J,x)$.

4.5.2 Remark

Note the similarity between the above Theorem and Theorem 1 of [1].

4.5.3 Corollary

Let $x, y \in R$ be such that there are z, $z' \in R$ with $xR + zR = yR + z'R = M(R)$. If $L_R(R/xR_J) = L_R(R/yR + J)$ then $A(R, J, x) = A(R, J, y)$. In particular $A(R, J, x)$ depends only on $L_R(R/xR + J)$.

Proof. If $xR + yR = M(R)$ then $A(R, J, x) = A(R, J, y)$ by (5.1.1).

If $xR+yR \neq M(R)$, as $zR+xR = z'R+yR = M(R)$, then $y = \beta_1 z+\beta_2 x$ and $x = \delta_1 z' + \delta_2 y$, with β_1, $\delta_1 \in M(R)$.

Now, if $\beta_2 \in M(R)$ then $x \in (M(R))^2$ and $zR + yR = M(R)$. On the other hand, if $\beta_2 \notin M(R)$ then $zR + yR = M(R)$ again. In both cases $zR + xR = zR + yR = M(R)$. The Inversion Theorem 4.1 can be applied to (z, x) and (z, y) and so we get to $A(R, J, x) = A(R, J, y)$.

4.5.4 Definition

Let p be a non negative integer such that there are $x, y \in M(R)$ with $xR + yR = M(R)$ and $L_R(R/xR + J) = p$. We define the Apery basis of (R, J) relative to p as

$$\begin{aligned} A(R,p) = A(R,J,x) = \{0 = a_0(R,p) = a_0(R,J,x) < a_1(R,p) \\ = a_1(R,J,x) < \ldots < a_{p-1}(R,p) = a_{p-1}(R,J,x)\}. \end{aligned}$$

We will say that r is a p-index of contact of (R, J) if r is an x-index of contact. We will write $g(R, J, p) = g(R, J, x)$ and $v_i(R, J, p) = v_i(R, J, x)$, $0 \leq i \leq g(R, J, p)$ and we will say that $g(R, J, p)$ is the genre of (R, J) releative to p and that $\{v_0(R, J, p), v_1(R, J, p), \ldots, v_{g(R,J,p)}(R, J, p)\}$ is the p-sequence of (R, J).

4.5.5 Definition

We will say that $x \in M(R)$ is a transversal parameter of (R, J) if $L_R(R/xR+J) \leq L_R(R/zR+J)$ for each $z \in M(R)$. ((R, J) being an irreducible curve then there is always a transversal parameter of (R, J)).

4.5.6 Lemma

If x is a transversal parameter of (R, J) then $m(R, J) = L_R(R/xR + J)$.

Proof. As x is transversal then there is $y \in M(R)$ such that $xR + yR = M(R)$ and we can write

$$J = \left(\sum_{i+j=m(R,J)} a_{ij} x^i y_j \right) R.$$

If $m(R,J) < L_R(R/xR+J)$ then $a_{0m(R,J)} \in M(R)$ and $J = xR$. (J is prime.) So $m(R,J) = L_R(R/yR+J)$ and x is not transversal.

Next we will make use of the apery basis $A(R,J,m(R,J))$ of (R,J) relative to $m(R,J)$.

4.6 Apery Basis and Formal Quadratic Transformations

Let x be a transversal parameter of (R,J) and set $p = L_R(R/xR+J) = m(R,J)$. Let y be a parameter of R such that $xR+yR = M(R)$ and let (R_1,J_1) be the formal quadratic transform of (R,J). Then R_1 is the completion of a ring $(R[y/x])_Q$ and $(y^p+xf) = J$.

On the other hand for each non negative integer s, we can write

$$h^1 = \sum_{0\le j<p} h_j^1(y/x)^j + h_s^1 + h^{1*},$$

with

$$h_s^1 \in (M(R_1))^s - (M(R_1))^{s+1}, \quad h^{1*} \in J_1$$

and

$$h_j^1 \in R, \quad 0 \le j < p.$$

Let $A(R,J,x) = \{0 = a_0(R,J,x) < a_1(R,J,x) < \ldots < a_{p-1}(R,J,x)\}$ be the x-Apery basis of (R,J), let

$$r_0(R,J,x) = 0, r_1(R,J,x) = 1, r_2(R,J,x), \ldots, r_{g(R,J,x)}(R,J,x)$$

be the x-index of contact and finally let

$$\{p = v_0(R,J,x), v_1(R,J,x), \ldots, v_{g(R,J,x)}(R,J,x)\}$$

be the x-sequence.

We can assume that $L_R(R/yR+J) = v_1(R,J,x) = a_1(R,J,x)$.

4.6.1 Lemma

If $s_0 = [(R_1/M(R_1) : (R/M(R))]$ then s_0 divides $r_i(R,J,x)$, $2 \le i \le g(R,J,x)$.

Proof. We have that $p = s_0 L_{R_1}(R_1/xR_1+J_1)$. If $s_0 = 1$ then the lemma is obviously true.

Assume that $s_0 > 1$. Then $(y/x) \in R_1 - M(R_1)$ and $p = v_1(R,J,x)$. By 4.2.4 we can take

$$g_i = y^{\gamma_i(R,J,x)} + g^i_{\gamma_1(R,J,x)-1} y^{r_i(R,J,x)-1} x + \ldots + g^i_0 x^{r_i(R,J,x)}$$

such that

$$L_R(R/g_iR) = v_i(R,J,x) > r_i(R,J,x)p, \quad 2 \le i \le g(R,J,x).$$

(Note that $g_i/x^{r_i(R,J,x)} \in M(R_1)$.)

Let $P(T)$ be a monic irreducible polynomial of $(R/M(R))[T]$ subh that $P(\eta_{R_1}(y/x)) = 0$, where $\eta_{R_1} : R_1 \to R_1/M(R_1)$ is the canonical epimorphism. There are $h_{s_0-1}, \dots, h_0 \in R$ such that

$$P(T) = T^{s_0} + \eta_R(h_{s_0-1})T^{s_0-1} + \ldots + \eta_R(g_0^i),$$

where $\eta_R : R \to R/M(R)$ is again the canonical epimorphism.

Now consider in $(R/M(R))[T]$ the monic polynomial $Q_i(T) = T^{r_i(R,J,x)} + \eta_R(g_{r_1(R,x)-1})T^{r_i(R,J,x)-1} + \ldots + \eta_R(g_0^i)$, $2 \le i \le g(R,J,x)$. As $Q_i(\eta_{R_1}(y/x)) = 0$, then we have $Q_1(T) = H_i(T)P(T)^{k_i}$, with $k_i \ge 1$ and being H_i and P coprime monic polynomials for every, $2 \le i \le g(R,J,x)$.

By Hensel's Lemma there are in $R[T]$ monic polynomials

$$H_i'(T) = T^{d_i} + b_{d_i-1}T^{d_i-1} + \ldots + b_0$$

and

$$G_i'(T) = t^{e_i} + c_{e_i-1}T^{e_i-1} + \ldots + c_0$$

such that

$$Q_i'(T) = H_i'(T)G_i'(T),$$

where

$$\begin{aligned} H_i(T) &= T^{d_i} + \eta_R(b_{d_i-1})T^{d_i-1} + \ldots + \eta_R(b_0), \\ P(T)^{k_i} &= T^{e_i} + \eta_R(C_{e_i-1})T^{e_i-1} + \ldots + \eta_R(c_0), \\ Q_i'(T) &= T^{r_i(R,J,x)} + g^i_{r_1(R,J,x)-1}T^{r_i(R,J,x)-1} + \ldots + g_0^i, \quad \text{and} \\ e_i &= s_0k_i, \qquad 2 \le i \le g(R,J,x). \end{aligned}$$

We have $g_i = x^{d_i}H_i'(y/x)x^{e_i}G_i'(y/x)$. So

$$L_R(R/g_iR+J) = L_R(R/x^{e_i}G_i'(y/x)R+J) + L_R(R/x^{d_i}H_i'(y/x)R+J), \quad d_i = 0$$

and

$$r_i(R,J,x) = k_is_0, \quad 2 \le i \le g(R,J,x).$$

(If $d_i \le 1$ then $v_i(R,J,x) \le a_{d_i}(R,J,x) + a_{e_i}(R,J,x)$ and we get to a contradiction.)

4.6.2 Remark

With the above notations, if $s_0 > 1$ then $r_2(R,J,x) = s_0$. (Note that $L_{R_1}(R_1/hR_1 + J_1) > 0$ with $h = (y/x)^{s_0} + h_{s_0-1}(y/x)^{s_0-1} + \ldots + h_0$, hence $L_R(R/x^{s_0}hR + J) > s_0 p$ and $s_0 \leq r_2(R,J,x)$.) Moreover $P(T) = Q_2(T)$, $xR_1 + (g_2/x^{r_2(R,J,x)})R_1 = M(R_1)$ and $g_i - g_2^{r_i(R,J,x)/r_2(R,J,x)} \in M(R)^{r_i(R,J,x)+1}$, $2 \leq i \leq g(R,J,x)$.

4.6.3 Theorem

With the above notations we have that the x-Apery basis of (R_1, J_1) is $A(R_1, J_1, X) = \{(1/s_0)(a_i(R,J,x) - ip); 1 \leq i < p\}$.

Proof. If $p = 1$ then (R,J) and (R_1,J_1) are regular curves and $L_{R_1}(R_1/xR_1) = L_R(R/xR) = 1$. So we can assume $p > 1$.

If $e = gcd(p, a_1(R,J,x))$ then there is

$$g_i = y^{k_i p/e} + f^i_{k_1-1} y^{(k_i-1)p/e} x^{e_1(R,J,x)/e} + \ldots + f^i_0 x^{k_i e_1(R,J,x)/e} + \sum_{(k,\ell)\notin Q_1} f^i_{k,\ell} x^k y^\ell,$$

with

$$\begin{aligned} L_R(R/g_i + J) &= v_i(R,J,x), \\ k_i p/e &= r_i(R,J,x) \quad \text{and} \\ f^i_0 \notin M(R), \quad 2 \leq i &\leq g(R,J,x). \end{aligned}$$

We can also take $g_1 = y$.

If $s_0 = 1$ then $a_1(R,J,x) > p$, $xR_1 + (y/x)R_1 = M(R_1)$ and

$$L_{R_1}(R_1/(g_i/x^{r_i(R,J,x)})R_1 + J_1) = r_i(R,J,x)/s_0, \quad 2 \leq i \leq g(R,J,x).$$

(Note that $g_i/x^{r_i(R,J,x)} - (y/x)^{r_i(R,J,x)} \in xR_1$.)

If $s_0 > 1$ then $xR_1 + (g_2/x^{r_2(R,J,x)})R_1 = M(R_1) r_2(R,J,x) = s_0$,

$$g_i - g_2^{r_i(R,J,x)/r_2(R,J,x)} \in (M(R))^{r_i(R,J,x)+1}$$

and

$$L_{R_1}(R_1/(g_i/x^{r_i(R,J,x)})R_1 + J_1) = r_i(R,J,x)/s_0, \quad 2 \leq i \leq g(R,J,x).$$

Note that we can write

$$g_i = g_2^{r_i(R,J,x)/r_2(R,J,x)} + \sum_{k<r_2(R,x),\, j+k+1r_2(R,x)=r_1(R,x)+1} g^i_{jk1} x^j y^k (g_2)^1,$$

with $g^i_{jk\ell} \in R$ for each $(j,k,1)$ and

$$g_i/x^{ri(R,J,x)} - (g_2/x^{r_2(R,J,x)})^{r_i(R,J,x)/r_2(R,J,x)} \in xR_1.$$

On the other hand, for $1 \leq i < p$ we have that

$$a_i(R,J,x) = \sum_{1\leq j\leq g(R,J,x)} b_j v_j(R,J,x),$$

with

$$i = \sum_{1\leq j=leqg(R,J,x)} b_j r_j(R,J,x)$$

and

$$b_1 < r_2(R,J,x).$$

So if

$$f_i = \prod_{1\leq j\leq g(R,J,x)} (g_j)^{b_i},$$

then

$$L_r(R/f_iR + J) = a_i(R,J,x),$$
$$L_{R_1}(R_1/(f_i/x^i)R_1 + xR_1) = k_i,$$

with $i = k_i s_0 + e_i$, $0 \leq e_i < s_0$.

Now let $h^i \in R_1$ be such that $L_{R_1}(R_1/(h^i/x^i)R_1 + J_1) = i$, with $1 \leq i < p'$ and $p' = L_R(R_1/xR_1 + xR)$, $(p = p's_0)$.

We can write

$$h^i h_{p-1}(y/x)^{p-1} + \ldots + h_{(i+1)s_0-1}(y/x)^{(i+1)s_0-1} + \ldots + h_{is_0}(y/x)^{is_0} + \ldots + h_0 + h_s^1 + h^{1*},$$

with

$$h_s^1 \in (M(R_1))^s - (M(R_1))^{s+1},\ h^{1*} \in J_1,\ h_j^1 \in R$$

and

$$s > (1/s_0)(a_{is_0}(R,J,x) - is_0 p), \quad 0 \leq j < p.$$

So we can also write

$$\begin{aligned} h^i = h'_{p-1}(f_{p-1}/x^{p-1}) + \ldots + h'_{(i+1)s_0-1}(f_{(i+1)s_0-1}/x^{(i+1)s_0-1}) \\ + \ldots + h'_{is_0}(f_{is_0}/x^{is_0}) + \ldots + h'_0 + h_s^1 + h^{1*} \end{aligned}$$

with $h'_j \in R$, $1 \leq j < p$. In fact, as $f_{is_0+j} = y^j f_{is_0}$, $0 \leq j < s_0$, then

$$\sum_{0\leq j<s_0} h'_{is_0+j}(y/x)^j \notin M(R_1)$$

and so

$$h'_{is_0+j} \notin M(R)$$

for some j. Consider j_0 such that $h'_{is_0+j} \in M(R)$ if $j_0 < j < s_0$ and $h'_{is_0+j_0} \notin M(R)$.

If

$$L_{R_1}(R_1/h_i r_1 + J_1) > (1/s_0)(a_{is_0}(R,J,x) - is_0 p)$$

we have

$$L_{R_1}(R_1/h^i_* R_1) > (1/s_0)(a_{is_0}(R,J,x) - is_0 p),$$

with

$$h^i_* = h'_{is_0+j_0}(f_{is_0+j_0}/x^{is_0+j_0})^{is_0+j_0} + \ldots + h'_0.$$

So

$$L_R(R/x^{is_0+j_0}h^i_* R + J) > a_{is_0}(R,J,x) + j_0 p = a_{is_0+j_0}(R,J,x),$$

$$L_R(R/x^{is_0+j_0}h^i_* R + xR) = is_0 + j_0$$

and we get to a contradiction. Thus $a_i(R_1,x) = (1/s_0)(a_{is_0}(R,J,x) - is_0 p)$, for $1 \leq i \leq p' = p/s_0$, and, as $a_{is_0+j}(R,J,x) = a_{is_0}(R,J,x) + jp$, $0 \leq j < s_0$, the Theorem is proved.

4.6.4 Remark

Note that $A(R_1, J_1, m(R,J)/s_0) = \{(1/s_0)(a_i(R,J,m(R)) - ip;\ 1 \leq i < p\}$ and that, by 4.4.1, we have $A(R_1, J_1, m(R_1, J_1))$.

On the other hand, if we know $A(R_1, J_1, m(R_1))$ and $(m(R,J), s_0)$ then by 4.4.1 we know $a(R_1, J_1, m(R,J)/s_0)$, and we have

$$A(R,J,m(R)) = \{(s_0(a_i(R_1,J_1,m(R,J)/s_0) + (is_0 + j)p;$$

$$1 \leq i < m(R,J)/s_0, \quad 0 \leq j < s_0\}.$$

4.6.5 Theorem

Let $E(R < J) = \{(n_i, s_i);\ 0 \leq i \leq N\}$ be the equiresolution class of (R,J). Then the following data are equivalent for (R,J).

1. The equiresolution class $E(R,J)$.
2. The Apery basis $A(R, m(R,J))$ of (R,J) relative to $m(R,J) = n_0$.

Proof. By induction on N the result follows as an easy consequence of 4.5.3 and 4.5.4.

4.7 Algebroid Case

Let K be an algebraically closed field, let R be the ring $K[[X,Y]]$ and let $f \in R$ be an irreducible power series.

Assume that $f \notin YR$ and set $n = L_R(R/yR + fR)$. If $S(R, fR)$ denotes the semigroup of values of (R, fR) (see [5], 1.0.1 and [6], 4.3.1), then the Apery basis of $S(R, fR)$ relative to n is defined as

$$A_n = \{0 = a_0 < a_1 < \ldots < a_{n-1}\} = \{\min(S(R, fR) \cap (k + n\mathcal{Z}_+);$$

$$0 \leq k \leq n-1\}$$

and the n-sequence in $S(R, fR)$, $\{v_0, \ldots, v_r\}$ is defined as $v_0 = n$, and

$$v_i = \min\{v \in S(R, fR);$$

$$gcd(v_0, v_1, \ldots, v_{i-1}) > gcd(v_0, v_1, \ldots, v_{i-1,v},)\},$$

$$1 \leq i \leq r.$$

We have the following properties of the Apery basis.

4.7.1 Proposition

If $M_j = K[[Y]] + XK[[Y]] + \ldots + X^j K[[Y]]$, $0 \leq j \leq n-1$, then:

$$\{a_j\} = v(M_{j-1} + X^j) - v(M_{j-1}), \quad 1 \leq j \leq n-1. \tag{7.1.1}$$

$$v(M_j) = \cup_{0 \leq i \leq j}(a_i + n\mathcal{Z}_+), \quad 0 \leq j \leq n-1. \tag{7.1.2}$$

$$a_i + a_j \leq a_{i+j}, \quad 0 \leq i+j \leq n-1. \tag{7.1.3}$$

Where $v(M_j) = \{L_R(R/gR + fR);\ g \in M_j - \{0\}\}$, $0 \leq j \leq n-1$ and

$$v(M_{j-1} + X^j) = \{L_R(R/(g + X^j)R + fR);\ g \in M_{j-1}\}, \quad 1 \leq j \leq n-1.$$

Proof. See [5], Satz 3 and [14], Proposition 2.

4.7.2 Remark

Note that $a_j = \max(v(M_{j-1} + X^j) - v(M_{j-1}))$, $1 \leq j \leq n-1$.

4.7.3 Proposition

$a_i = a_i(R, n)$, $0 \leq i < n$.

Proof. Note that if $L_R(R/hR + yR) = j$ then, by the Weierstrass Preparation Theorem, there is $h' \notin M(R)$ such that $h'h \in M_{j-1} + X^j$.

On the other hand, if $h \in M_{j-1} + X^j$ then $L_R(R/hR + yR) \leq j$, $1 \leq j \leq n-1$. So $a_i = a_i(R, fR, n)$, $1 \leq i \leq n-1$.

4.8 References

[1] Abhyankar, S.S., Inversion and Invariance of Characteristic Paris, *Amer. J. Math.*, **89**, (1967), 363–372.

[2] Abhyankar, S.S., Note on Coefficient Fields, *Amer. J. Math.*, **90**, (1968), 346–354.

[3] Abhyankar, S.S., *Lectures on Expansion Techniques in Algebraic Geometric*, Tata Institute of Fundamental Research, Bombay, 1977.

[4] Abhyankar, S.S., Desingularization of Plane Curves, *Proc. Sympos. Pure Math.*, **40**, Part 1 (1983), 1–45.

[5] Angermuler, G., Die Wertehalgruppe Einer Ebenen Irreduziblen Algebroiden Kurve, *Math. A.*, **153**, (1977), 267–282.

[6] Campillo, A., Algebroid Curves in Positive Characgeristic, *Lecture Notes in Math.*, **813**, Springer Verlag, 1980.

[7] Eprahim, R., Special Polars and Curves With One Place at Infinity, *Proc. Sympos. Pure Math.*, **40**, Part 1 (1983), 3553–359.

[8] Granja, A., Apery Basis End Polar Invariants of Plane Curve Singularities, *Pacific J. of Math.*, **140**, 1 (1989), 85–95.

[9] Granja, A., Coefficient Fields for Plane Curves and Equisingularity, *Communications in Algebra*, **18**, 1, (1990), 193–208.

[10] Granja, A., *Apery Basis of Plane Curves*, preprint, Leon, (1990).

[11] Granja, A., and T. Sanchex-Giralda, *Enriques Graphs of Plane Curves*, preprint, Valladolid, (1990).

[12] Nagata, M., *Local Rings*, Interscience Publishers, New York, (1962).

[13] Northcott, D.G., *Lessions on Rings Modules and Multiplicities*, Cambridge University Press, (1968).

[14] Pinkham, H., Courbes Planes Ayant une Seule Place a l'infini, *Publications de Centre de Mathematiques*, Ecole Polytechnique, F. 91128 Palaiseu, 1977–1978.

[15] Zariski, O., Studies in Equisingularity I, *Amer. J. Math.*, **87**, (1965), 507–536.

[16] Zariski, O., Studies in Equisingularity II, *Amer. J. Math.*, **87**, (1965), 972–1006.

5

Classification of Algebraic Space Curves, III

Robin Hartshorne

5.1 Introduction

Over a fixed algebraically closed ground field k, we consider irreducible non-singular projective curves in the projective space $\mathbf{P}^3_k$. The original classification problem for algebraic space curves could be described as finding all such curves, giving their numerical invariants, and determining the algebraic families they belong to. This was the problem tackled in the great treatises of M. Noether [10] and G. Halphen [5] over a hundred years ago, and has been the subject of numerous investigations since then. The determination of which pairs $< d, g >$ can occur as the degree and genus of a non-singular space curve was stated by Halphen, but only properly proved within the last decade by Gruson and Peskine [4]. Now it is reasonable to ask for finer numerical data, for example one can ask for the postulation of all possible curves, which means for each n, the number of conditions for a hypersurface of degree n to contain the curve.

To fix the notation, let C be a curve in $\mathbf{P}^3$, with structure sheaf $\mathcal{O}_C$ and ideal sheaf $\mathcal{I}_C$. Then the *postulation* of C is the function φ which to each $n \geq 0$ assigns the integer

$$\varphi(n) = h^0(\mathcal{O}_{\mathbf{P}^3}(n)) - h^0(\mathcal{I}_C(n)).$$

There is an exact sequence of cohomology groups

$$0 \to H^0(\mathcal{I}_C(n)) \to H^0(\mathcal{O}_{\mathbf{P}^3}(n)) \to H^0(\mathcal{O}_C(n)) \to H^1(\mathcal{I}_C(n)) \to 0,$$

so to know the postulation it is equivalent to know the integers $h^0(\mathcal{O}_C(n))$ and $h^1(\mathcal{I}_C(n))$. The former can be calculated from the Riemann–Roch theorem on C,

$$h^0(\mathcal{O}_C(n)) = nd + 1 - g + h^1(\mathcal{O}_C(n))$$

if one knows $h^1(\mathcal{O}_C(n))$—and in any case this vanishes for n large enough.

The latter, $h^1(\mathcal{I}_C(n))$ is also zero for large n, but to get exact information for all n, this leads us to consider the sum

$$M = \bigoplus_{n \in \mathbf{Z}} H^1(\mathcal{I}_C(n))$$

which can be regarded as a graded module over the homogeneous coordinate ring $S = k[x_0, x_1, x_2, x_3]$ of $\mathbf{P}^3$. One sees easily that M is a graded S-module of finite length. We call it the *H^1-module of C*, and regard it as an important invariant of the embedding of C in $\mathbf{P}^3$.

This leads us to expand our wishful thinking further, and we can ask to find for all space curves not only their postulation but also their H^1-modules.

In this talk I will concentrate my attention on recent work in this direction, even though there are also many other fascinating directions in the study of space curves that one could pursue.

5.2 The Classification

An essential tool is the technique of residual intersections, which in its simplest form is a method of obtaining new curves from old: given a curve C, let F, G be two surfaces containing C, and from this obtain a new curve C' as the closure of $F \cap G - C$. In its modern formulation it is called liaison (linkage) and is defined as follows. First we expand the class of curves we consider to include all one-dimensional closed subschemes of $\mathbf{P}^3$ with no embedded points or isolated points (henceforth called simply "curves"). Two curves C, C', which may even have irreducible components in common, or could even have the same support, are said to be *lié* (linked) if there exist surfaces F, G (locally principal closed subschemes of $\mathbf{P}^3$), whose scheme-theoretic intersection $X = F \cap G$ has support equal to the support of $C \cup C'$, and such that furthermore

$$\begin{aligned} \mathcal{I}_{C,X} &\cong \operatorname{Hom}(\mathcal{O}_{C'}, \mathcal{O}_X) \\ &\text{and} \\ \mathcal{I}_{C',X} &\cong \operatorname{Hom}(\mathcal{O}_C, \mathcal{O}_X). \end{aligned}$$

Two curves C, C' are in the same *liaison equivalence class* (respthe same *even liaison equivalence class*) if there exists a sequence $C = C_0, C_1, C_2, \ldots$, $C_n = C'$ (respwith n even) such that each C_i is linked to C_{i+1} as above by two surfaces.

Working with these equivalence classes separates the classification problem into two parts. One is to characterize the set of equivalence classes; the other is to study the structure of curves within a given equivalence class. One should note also that even if one's primary interest is the classification of irreducible non-singular curves, it is natural, and perhaps even essential to deal with this larger class of curves defined as general one-dimensional subschemes of $\mathbf{P}^3$, because in the process of liaison, starting from a non-singular curve contained in two surfaces, the residual intersection may not be irreducible or even reduced.

The technical aspects of the theory of liaison were established in the 1970's in papers of Peskine and Szpiro [11], Gruson and Peskine [3], and

Ellingsrud [2], who proved among other things the modern version of the theorem of Apéry and Gaeta. We say a curve is *arithmetically Cohen–Macaulay* (*ACM* for short) if its associated H^1-module is zero. This is the analogue of the property arithmetically normal, or projectively normal for non-singular curves. Then these results can be summarized in

Theorem 1 *A curve C is arithmetically Cohen–Macaulay if and only if it is in the even liaison equivalence class of a line. Furthermore, there are then formulae giving the degree, genus, postulation, irreducible components and dimensions of the Hilbert scheme for all such curves.*

As an aid in explaining the generalizations to arbitrary curves later in this talk, I would like to give the exact method of computing degree, genus, and postulation for ACM curves alluded to in Theorem 1. Gruson and Peskine did this using something they called the *numerical character* of the ACM curve. We will state the result here instead using the *postulation character* of Martin–Deschamps and Perrin [9] which lends itself better to generalization.

We defined above the postulation φ of a curve C, as a numerical function from $\mathbf{Z}$ to $\mathbf{Z}$. For any numerical function f, define its difference function ∂f by $\partial f(n) = f(n) - f(n-1)$. Then we define the *postulation character* of the curve C to be $\gamma_C = -\partial^3 \varphi$.

The postulation character has the following properties which one can easily verify:

1. $\gamma_C(n) = 0$ for $n < 0$
2. $\gamma_C(n) = 0$ for $n \gg 0$
3. $\sum_{n \in \mathbf{Z}} \gamma_C(n) = 0$ (hence the word "character").

In addition, we note

a) One can recover the postulation φ by numerical integration, since φ is normalized by $\varphi(n) = 0$ for all $n < 0$.

b) One can compute the degree d and arithmetic genus p_a of C easily from the character as follows:

$$\begin{aligned} d &= \textstyle\sum_{n>0} n \cdot \gamma(n) \\ p_a &= \textstyle\sum_{n>0} \frac{1}{2}(n-1)(n-2)\gamma(n). \end{aligned}$$

c) The postulation character γ_C of any curve C has the following form:

$$\ldots\ 0\ -1\ -1\ -1\ \ldots\ -1\ a_0\ a_1\ \ldots\ a_r\ 0\ 0\ \ldots$$

with $a_0 \geq 0$. Namely, there is an integer s_0 (the least degree of a surface containing C) such that

$$\begin{aligned} &\gamma_C(n) = -1 \text{ for } 0 \leq n < s_0 \\ &\gamma_C(s_0) \geq 0 \\ &\gamma_C(s_0 + i) = a_i \text{ arbitrary integers for } i > 0, \end{aligned}$$

eventually all zero.

Now using a) and b) above we can determine all possible degree, genus, and postulation of ACM curves from the following theorem (due essentially to Gruson and Peskine, but in the formulation of Deschamps and Perrin).

Theorem 2 *A numerical function γ satisfying properties* 1), 2), 3) *and* c) *above is the numerical character of an* ACM *curve in* $\mathbf{P}^3$ *if and only if all $a_i \geq 0$ (in which case we say γ is positive).*

In this statement is is essential that we understand "curve" in the large sense of 1-dimensional closed subscheme of $\mathbf{P}^3$. There will be an irreducible non-singular curve with the given character if and only if γ is also *connected* meaning $\{n \in \mathbf{Z} \mid \gamma_C(n) > 0\}$ is a connected set of integers. When we generalize to non-ACM curves, the questions of irreducibility and smoothness become very delicate, so we will deal only with curves in the broader sense.

Now let us consider the generalization of these results to general (non-ACM) curves in $\mathbf{P}^3$. The first result is that the H^1-module characterizes the even liaison classes.

Theorem 3 (Rao [12]) *Two curves C, C' in $\mathbf{P}^3$ are in the same even liaison class if and only if their H^1-modules M_C and $M_{C'}$ are isomorphic up to a shift in degrees of their grading. Furthermore, given any finite length graded S-module M, there exists an irreducible non-singular curve C with $M_C = M$ up to shift in degrees.*

In view of this result, we recognize the study of ACM curves to be the first example of an analysis of an even liaison class of curves, and it is natural to ask for analogous results for the even liaison class associated to any other finite length graded S-module M.

A first step in this direction was the remarkable result of Lazarsfeld and Rao [8], who showed that for given genus g, for degree d sufficientldy large with respect to g, and for a sufficiently general curve C_0 of degree d and genus g, the following hold:

1. C_0 is the only curve of degree d in its even liaison equivalence class;
2. all other curves in the even liaison class of C_0 have degree $> d$;
3. all other curves in the even liaison class can be obtained from C_0 by a rather specific method of "basic double linkages", (which we do not describe in detail) and deformations. In particular, it is possible to determine all possible degrees and genera of such curves.

Now this result has been extended to all even liaison equivalence classes by the independent work of Ballico, Bolondi, and Migliore [1] in Italy and Martin–Deschamps and Perrin [9] in France, with two provisos: one must allow curves in the general scheme sense described above, and the curve of minimal degree is no longer unique, but only unique up to deformation. Here is the result.

Theorem (Ballico, Bolondi, Migliore, Martin–Deschamps, Perrin) *Given a finite length graded S-module M, let $L(M)$ denote the even liaison class of curves associated to M. Then*

1. *There is a minimum degree d_0 of curves in $L(M)$.*
2. *The set of all curvees of degree d_0 in $L(M)$ is an irreducible flat family with constant postulation.*
3. *Any curve C' can be obtained from one C_0 of minimal degree by basic double linkages and deformation.*
4. *There are formulae to give all possible degrees, genera, and postulations for curves $C' \in L(M)$ once one has the corresponding data for a minimal curve C_0 in $L(M)$, and this in turn can be calculated from a resolution of the module M.*

Using the postulation character γ of a curve, defined above, we can make this last statement of the theorem precise as follows. First we define a partial ordering among numerical characters as follows. Let γ, γ' be numerical characters, i.e. functions $\mathbf{Z} \to \mathbf{Z}$ stisfying properties 1), 2), 3) and c) listed earlier. We say $\gamma \leq \gamma'$ if

i) $s_0(\gamma) \leq s_0(\gamma')$

ii) there exists an integer $h \geq 0$ such that $\gamma'(n) \geq 0$ for $s_0(\gamma') \leq n \leq s_0(\gamma) + h$, and for all $n \geq s_0(\gamma) + h$ we have $\gamma'(n) \geq \gamma(n-h)$.

Now the last statement of the theorem says more exactly that if γ_0 is the postulation character of a minimal curve $C_0 \in L(M)$, then another numerical character γ is the postulation character of a curve $C' \in L(M)$ if and only if $\gamma \geq \gamma_0$. From this we obtain all possible degrees, genera, and postulations of curves in $L(M)$, using the properties a) and b) of postulation characters.

5.3 Examples

Now let us give some examples.

1. A plane curve of degree d is an ACM curve. Its module is 0. Its postulation character is

$$-1\ 0\ 0\ \ldots\ 0\ 1\ 0\ \ldots$$

   with $\gamma(0) = -1$ and $\gamma(d) = 1$.

2. The twisted cubic curve in $\mathbf{P}^3$ is ACM, with character $-1\ \ -1\ 2$.

3. A curve of type $(k, k-1)$ on a non-singular quadric surface Q is also ACM, with postulation character

$$-1\ -1\ 0\ \dots\ 0\ 2\ 0\ \dots$$

with $\gamma(0) = \gamma(1) = -1$ and $\gamma(k) = 2$.

4. The disjoint union of two skew lines is a non-ACM curve with module $M = k$ in degree 0, and postulation character

$$-1\ -1\ 3\ -1.$$

5. The rational quartic curve (a curve of type $(1,3)$ on the non-singular quadric surface) has module $M = k$ in degree 1, and postulation character

$$-1\ -1\ 0\ 3\ -1.$$

Notice that this one dominates the one of example 4, as it should, since they are both in the liaison equivalence class of two skew lines, and the former is minimal in its class.

6. For something a bit more complicated we give in this and the next two examples curves of degree 6 and genus 3. First of all consider a curve of type $(2,4)$ on the non-singular quadric surface. It has module $M = k$ in degree 2, and character

$$-1\ -1\ 0\ 0\ 3\ -1\ 0\ \dots$$

which dominates, as it should, the character of two skew lines of example 4.

7. Let us look for an ACM curve of degree 6 and genus 3 on a cubic surface. Its character would be

$$-1\ -1\ -1\ a_0\ a_1\ a_2\ \dots$$

with all $a_i \geq 0$, and genus 3 so $\sum a_i = 3$, and degree 6, so $-3 + 3a_0 + 4a_1 + 5a_2 + \dots = 6$. The only solution is

$$-1\ -1\ -1\ 3\ 0\ \dots\ .$$

We conclude from the theorems mentioned above that there is an ACM curve with this postulation character on a cubic surface. In fact, this is the general form of a non-hyperelliptic curve of genus 3 embedded by a linear system of degree 6 into $\mathbf{P}^3$.

8. Now notice that the numerical character $-1\ -1\ -1\ 3\ 0$ of example 7 dominates the character $-1\ -1\ 3\ -1$ of two skew lines. Thus from the theorem above we conclude that there must be a curve in the liaison class of two skew lines having this postulation character. An example of such a curve is a plane quartic curve with two skew lines each meeting the quartic curve in one point.

These examples show that for fixed d, g, there may be curves in different liaison equivalence classes with the same postulation character (examples 7 and 8), and there may be curves in the same even liaison equivalence class with different postulation characters (examples 6 and 8).

5.4 Open Problems

We close with some open problems.

Problem 1 Find necessary and sufficient conditions on a numerical character $\gamma : \mathbf{Z} \to \mathbf{Z}$ for it to be the postulation character for some curve in $\mathbf{P}^3$.

If one could do this, one could find all possible postulation functions for curves in $\mathbf{P}^3$. Since one already knows the possible degree and genus for curves in $\mathbf{P}^3$, this does not seem to be too much to ask. It is easy to see that the conditions 1), 2), 3) and c) above are not sufficient. For example, the character $-1\ 1\ -1\ 1$ satisfies them all. Yet if this were the postulation character of a curve C, then $s_0 = 1$, so C would be a plane curve, hence ACM, so the character would have to be positive, which it is not. Even to determine which characters γ with $s_0 = 2$ come from curves is not a trivial task.

Problem 2 Given (d, g), give reasonable bounds on the size of the H^1-module M for curves of degree d and arithmetic genus g.

So far, the big progess has been to start with a module M, then describe its even liaison class and determine numerical invariants of those curves. If one wishes to make an exhaustive study of the Hilbert scheme $H_{d,g}$ of curves of degree d and genus g, some kind of bounds on possible modules are needed. For this problem, already the case $d = 3$, $g = 0$ is non-trivial.

5.5 References

[1] Ballico, E., Bolondi, G., and Migliore, J., The Lazarsfeld–Rao problem for liaison classes of two-codimensional subschemes of $\mathbf{P}^n$, *Amer. J. Math.* (to appear).

[2] Ellingsrud, G., Sur le schéma de Hilbert des variétés de codimension 2 dans $\mathbf{P}^e$ à cône de Cohen–Macaulay, *Ann. Sc. E.N.S. (4)* **8**, 1975, pp. 423–432.

[3] Gruson, L. and Peskine, C., Genre des courbes de l'espace projectif, *Springer LNM*, **687**, 1977, pp. 31–59

[4] Gruson, L. and Peskine, C., Genre des courbes de l'espace projectif, II, *Ann. Sc. E.N.S. (4)*, 15, 1982, pp. 401–418.

[5] Halphen, G., Mémoire sur la classification des courbes gauches algébriques, *J. Ec. Poly.*, **52**, 1882, pp. 1–200.

[6] Hartshorne, R., On the classification of algebraic space curves, in *Vector Bundles and Differential Equations*, (A. Hirschowitz, Birkhäuser, eds.), 1980, pp. 83–112.

[7] Hartshorne, R., On the classification of algebraic space curves, II, *AMS, PSPM*, **46**, 1987, pp. 145–164.

[8] Lazarsfeld, R. and Rao, A. P., Linkage of general curves of large degree, *Springer LNM*, **997**, 1983, pp. 267–289.

[9] Martin–Deschamps, M. and Perrin, D., Sur la classification des courbes gauches, *Astérisque*, (to appear).

[10] Noether, M., *Zur Grundlegung der Theorie der algebraischen Raumcurven*, Verl. König. Akad. Wiss. Berlin, 1883.

[11] Peskine, C. and Szpiro, L., Liaison des variétés algébriques, I, *Invent. Math.*, **26**, 1974, pp. 271–302.

[12] Rao, A. P., Liaison among curves in $\mathbf{P}^3$, *Invent. Math.*, **50**, 1979, pp. 205–217.

6 Plane Polynomial Curves

Avinash Sathaye
Jon Stenerson

ABSTRACT We study some properties of generator sequences of planar semigroups and give a method of construction of plane curves with one place at infinity with given generator sequences. We also discuss similar questions for polynomial curves.

6.1 Introduction

Abhyankar and Moh opened a new chapter in the theory of plane algebraic curves with their papers [AM1] [AM2].

To avoid unnecessary technicalities, assume that the ground field has characteristic zero and consider a plane curve C with one place at infinity. There is a semigroup $\Gamma(C)$ consisting of the set of orders of pole for various nonzero elements of the coordinate ring of C. The semigroup is a natural invariant of the defining equation (and hence of the embedding) of the curve in the plane.

Abhyankar and Moh showed that $\Gamma(C)$ is generated by a special set of generators (called a characteristic $\delta -$ *sequence* here) which can be computed as orders of pole of the (so-called) approximate roots of the defining equation. They also showed several important properties of the $\delta -$ *sequence*, in particular, the semigroup is generated in a special way so that given any integer a we get a " unique standard expression " $a = \sum_0^h a_i\delta_i$ and $a \in \Gamma(C)$ if and only if $a_0 \geq 0$. (See (6.3.1) for details.)

Sathaye [S1] defined a semigroup generated by such sequences to be " planar " and used the properties to show that certain curves are not embeddable in a plane (even though they are embeddable in space as complete intersections); because their semigroup is not planar. The converse that a planar semigroup (in fact, any give $\delta -$ *sequence*) comes from a plane curve was promised in [S1] but never formally published. We take this opportunity to publish it as well as some useful numerical lemmas for determining the possible generating sequences of a given semigroup. (See sections (6.2),(6.3)).

We note that the real power behind all these proofs is the "irreducibility criterion " and the concept of the "standard generation " due to Abhyankar and Moh. We especially note that in [A3] Abhyankar has given a very detailed exposition of the "irreducibilty criterion" and an exhaustive description of all plane curves with one place at infinity. We have included several references of works by Abhyankar, Moh, Richman, Russell, Sathaye and Singh related to expositions and generalizations of these concepts. We do not claim this to be exhaustive and in particular, it does not include works on problems which use the Abhyankar-Moh results.

In the last section, we present some attempts on another related question of Abhyankar. As noted above, plane curves can be constructed to represent any given δ – *sequence*. The question asks if a *polynomial curve* i.e. a curve parametrized by polynomials in one variable, can represent a given δ – *sequence*. Any simple counting of constants yields that there are not enough parameters to accomplish this. Indeed, Moh (with Sathaye) had calculated by hand that a polynomial curve cannot represent the sequence $(6, 8, 3)$. This example is not very satisfactory, since the semigroup is simply $(3, 8)\mathbb{N}$ and as such is realized by the curve with the parametrization (t^3, t^8). We rechecked the example with the help of a computer with the hope of extending to higher examples; so far without success. In particular, we propose the sequence $(6, 22, 17)$ as a possible better example, since in this case, we can argue that if it is not constructible, then the semigroup is not either! (See (4)).

Abhyankar's original question was not really just about the semigroups but rather about the control on the coefficients of the Newton-Puiseux series of the curve at infinity (since this has possible applications to the Jacobian problem). We discuss a very special case of this by analyzing the length of a string of zeros past the first characteristic term. It was interesting to learn from a recent article by Lang [11] that the question as well as the answer was already discovered by the Number Theorists (in the 60's) for totally different reasons. We present a proof and discuss some unsolved problems.

6.2 Preliminaries and Notation

6.2.1 Characteristic sequences

Following [A1] 6.4 we first define various characteristic sequences.

Let $\nu \neq 0$ be a given integer and J a subset of of integers bounded below.

We inductively define an integer $h(\nu, J) = h$ and two sequences $m(\nu, J) = (m_1, \cdots, m_h)$ and $d(\nu, J) = (d_1, \cdots, d_{h+1})$ as follows:

(1) If J is empty, then $h = 0$, $d_1 = |\nu|$ and $m(\nu, J)$ is empty.

(2) If J is nonempty, set $D = \gcd(J)$ the greatest common divisor of the set J , $d_1 = |\nu|$, $m_1 = \min J$ and $d_2 = \gcd\{m_1, d_1\}$.

If $d_2 = D$, then put $h = 1$ and stop.

(3) If $d_1, \cdots, d_{r+1}$ and $m_1, \cdots, m_r$ are defined and $D = d_{r+1}$ then put $h = r$ and stop; otherwise, define

$$m_{r+1} = \min\{p \in J \mid p \not\equiv 0 \text{ modulo } d_{r+1}\}$$

$$d_{r+2} = \gcd\{m_1, \cdots, m_{r+1}\} = \gcd\{d_{r+1}, m_{r+1}\}$$

Note that $D = d_{h+1}|d_h| \cdots |d_2|d_1$.

Some natural expressions in these numbers are also useful to define.

(1) $q_1 = m_1$ and $q_i = m_i - m_{i-1}$ for $i = 2, \cdots, h$

For convenience, we also set $q_{h+1} = m_{h+1} = \infty$.

(2) $s_i = \sum_1^i q_j d_j$ for $i = 1, \cdots, h$

(3) $r_i = s_i/d_i$ and $\delta_i = -r_i$ for $i = 1, \cdots, h$

(4) $n_i = d_i/d_{i+1}$ for $i = 1, \cdots, h$

Note that all the numbers defined here depend on ν, J and should indeed be written as $h(\nu, J), m_i(\nu, J)$, $q_i(\nu, J)$ etc. We may invoke such detailed notations if necessary.

If $u(\tau) \in k((\tau))$ is a meromorphic power series in τ, then we define, for any given integer n, the characteristic sequences of $u(\tau)$ by setting

$$J = \text{ Supp } u(\tau) = \{r \mid \text{ coefficient of } \tau^r \text{ in } u(\tau) \text{ is nonzero }\}$$

Thus $m(\nu, u(\tau)) = m(\nu, J)$ etc. [1]

Further, if $f = f(X, Y)$ is a polynomial such that

$$f(\tau^{-n}, Y) \;=\; \prod_{\omega^n=1} (Y - u(\omega\tau))$$

where n is the $Y-$degree of $f(X, Y)$ then all the power series $u(\omega\tau)$ have the same support and we simply set $m(f) = m(-n, u(\tau))$ etc. In words, we will describe this by saying that the curve defined by f has $h = h(f)$ characteristic terms $m(f) = (m_1(f), \cdots, m_h(f))$ etc. Curves defined by such polynomials are the so-called curves with one place at infinity (at least when $n \not\equiv 0$ modulo char k) and we discuss them next.

---

[1] In particular we will need $s_h(-n, u(\tau)) = \Sigma_0^h q_i d_i$ where $q_i d_i$ is obtained by setting $\nu = -n$, $J =$ Supp $u(\tau)$.

6.2.2 *Curves with one place at infinity*

First we fix some notation. Let k be the ground field, assumed algebraically closed unless otherwise stated. The characteristic char k is either assumed to be 0 or is not allowed to divide certain degrees in case it is nonzero. Let $f = f(X, Y)$ be an irreducible polynomial and let C denote the corresponding (irreducible) curve defined by the ideal (f). Let $A = k[X, Y]/(f)$ be the coordinate ring of C and let "$^{-}$" denote the corresponding residue class map. Let $K =$ qt A be the function field of C.

To avoid making repeated exceptions, we assume forever that $f \notin k[X]$. It is clear that a simple change of variables will accomplish this change and there is no loss of generality.

Recall that the curve C *has one place at infinity* if *there is a unique valuation ring* V *of* K/k *not containing* A . [2] Let v denote the valuation associated with V.

For nonzero elements h of A we define

$$\deg_v h = -v(h).$$

If $h = 0$ then we may extend this by $\deg_v 0 = -\infty$.

If the curve is a plane curve defined by say $f = f(X, Y) = 0$, then we extend this concept of degree to polynomials $h(X, Y)$ by

$$\deg_v h(X, Y) = \deg_v \bar{h}$$

where $\bar{h}$ is the image of $h(X, Y)$ modulo $f(X, Y)$.

We will drop reference to v if it is clear from the context. We will also write $deg(h, f)$ in place of $\deg_v h$ if we wish to mention f and not v.

For a curve C with one place at infinity we get a value semigroup

$$\Gamma(C) = \Gamma(A) = \{deg_v\ h \mid 0 \neq h \in A\}.$$

It is well known that to test whether the given curve has "one place at infinity" one can also check that f is irreducible as an element of $k((X^{-1}))[Y]$. (The assumption that $f \notin k[X]$ is used here!)

One of the main deductions from the Abhyankar-Moh theory is a very efficient algorithm for testing this irreducibility, without developing the full power-series factorization. Unfortunately, their method only works when the $Y-$degree of f is not divisible by char k; in particular, it works in characteristic zero.

Irreducibility Criterion of Abhyankar and Moh *Assume that* $n = \deg_Y f(X, Y) \not\equiv 0$ modulo char k *and* $f(X, Y)$ *is monic in* Y. *Then* $f(X, Y)$ *has one place at infinity iff there is a "test series"* $u(\tau) \in k((\tau))$ *such that*

$$\operatorname{ord}_\tau f(\tau^{-n}, u(\tau)) > s_h(-n, u(\tau))$$

[2] If k is not algebraically closed then we also need to assume that V is residually rational over k.

Moreover, given any series passing this test, there is a "root" $y(\tau)$ (usually called the Newton-Puiseux series of f) satisfying:

$$f(\tau^{-n}, y(\tau)) = 0$$
$$\text{ord}_\tau\ (y(\tau) - u(\tau)) > m_h(-n, u(\tau))$$

Conceptually, this says that if we have a guess $u(\tau)$ which describes the full initial piece of an expected root thru the last characteristic term, then its correctness as well the irreducibility of f can be verified simply by substituting it in f and comparing the order against numbers intrinsically computed from the guess itself! Indeed, this test can be generalized to build the test series $u(\tau)$ itself term by term and is a sharpened version of the original "Newton's algorithm" for the special case of one place. For a very detailed exposition of the irreducibility criterion see [A3]

Moreover, whenever f has a Newton-Puiseux series as described, one gets a factorization as described above, namely:

$$f(\tau^{-n}, Y) \ = \ \prod_{\omega^n=1} (Y - y(\omega\tau))$$

and we can talk about the various characteristic sequences associated with f itself as discussed in the previous section.

6.2.3 *The $g-$ sequence*

A sequence of polynomials $g_0, g_1, \cdots, g_{h+1}$ in $k[X,Y]$ is said to be a $g-$ *sequence* if the following holds:

(1) $g_0 = X, g_1, \cdots, g_{h+1}$ are monic in Y with one place at infinity.

(2) $\deg_Y\ g_{h+1}\ = n \not\equiv 0$ modulo char k. Moreover, g_{h+1} has exactly h characteristic terms. Let us write m_i for the ith characteristic term $m_i(g_{h+1})$ and similarly for all the other sequences associated with g_{h+1}. With this notation we require that $\deg_Y\ g_i = \frac{n}{d_i}$.

(3) There exist Newton-Puiseux series $y_i(\tau)$ for g_i, $1 \le i \le h+1$ such that $\text{ord}_\tau\ y_{h+1}(\tau) - y_i(\tau^{d_i}) = m_i$ for $1 \le i \le h$. We shall denote $y_{h+1}(\tau)$ simply by $y(\tau)$.

It is easy to see that every subsequence of a $g-$*sequence* is a $g-$*sequence* again. In particular, $m_j(g_i) = \frac{m_j}{d_i}$ for $j \le i \le h$. Similarly, we also get, $d_j(g_i) = \frac{d_j}{d_i}$ for $j \le i \le h$. Finally, for $j \le i \le h$ we get :

$$s_j(g_i) = \frac{s_j}{d_i^2}\ ,\ r_j(g_i) = \frac{r_j}{d_i^2} \text{ and } \delta_j(g_i) = \frac{\delta_j}{d_i^2}.$$

6.2.4 Substitution Formulas

We assume all the notation of the previous section and consider the Newton-Puiseux series

$$y(\tau) = \alpha_1\tau^{m_1} + \cdots + \alpha_i\tau^{m_i} + \cdots + \alpha_h\tau^{m_h} + \cdots + \beta\tau^{m_i+q} + \cdots$$

where we have identified the characteristic terms and one extra term past the last characteristic term. Let a test series $u(\tau)$ be defined by $u(\tau)-y(\tau) = (Z-\beta)\tau^{m_i+q} + \cdots$ higher terms . By an easy but tedious calculation we get the following results :

$$\begin{aligned} g_i(\tau^{-nd}, u(\tau)) &= \beta_i'\tau^{d(r_i)} + \cdots \text{ higher terms} && \text{for } 1 \le i \le h \\ &= \beta_i'\tau^{-d(\delta_i)} + \cdots \text{ higher terms} && \text{for } 1 \le i \le h \\ g_{h+1}(\tau^{-nd}, u(\tau)) &= (Z-\beta)\beta_{h+1}'\tau^{ds_h+q} + \cdots \text{ higher terms} \\ &= (Z-\beta)\beta_h'\tau^{-d_h\delta_h+q} + \cdots \text{ higher terms} \end{aligned}$$

Here, β_i' depends only on the coefficients $\alpha_1, \cdots, \alpha_i$ for $1 \le i \le h$ are nonzero elements of k. See [A1] 12 or [A2] 3.4.

6.2.5 A Lemma on Y−degrees

In this section we prove a useful fact about a $g-$ *sequence.*

Degrees of monomials in a $g-$ *sequence Let* $g_0 = X, g_1, \cdots, g_{h+1}$ *be a* $g-$ *sequence and assume all the notation introduced in the definition above. Let*

$$g* = \prod_0^h g_i^{p_i} \text{ where } 0 \le p_i \le n_i - 1 \text{ for } i \ge 1.$$

Then $\deg_Y g* < \deg_Y g_{h+1} = n$.

Proof. Induction on h. If $h = 0$ there is nothing to prove! Now let $h > 0$. Applying the induction hypothesis to $g_0, \cdots, g_h$, we get that

$$\deg_Y \prod_0^{h-1} g_i^{p_i} = \sum_0^{h-1} p_i \frac{n}{d_i} < \deg_Y g_h = \frac{n}{d_h}$$

Hence

$$\begin{aligned} \deg_Y g* &< \frac{n}{d_h} + p_h \deg_Y g_h \\ &\le \frac{n}{d_h} + (d_h - 1)\frac{n}{d_h} \\ &= n = \deg_Y g_{h+1} \end{aligned}$$

6.3 Semigroups of Curves with One Place at Infinity

6.3.1 Planar semigroups

A sequence of positive integers $(\delta_0, \cdots, \delta_h)$ is said to be **a characteristic** $\delta-$*sequence* if it satisfies the following three axioms:

(1) Set $d_i = \gcd\{\delta_0, \cdots, \delta_{i-1}\}$ for $1 \le i \le h+1$. Set $n_i = \frac{d_i}{d_{i+1}}$ for $1 \le i \le h$. Then $d_{h+1} = 1$ and $n_i > 1$ for all $i \ge 2$.

(2) $\delta_i n_i \in \{\delta_0, \cdots, \delta_{i-1}\}\mathbb{N}$ = the semigroup generated by $\{\delta_0, \cdots, \delta_{i-1}\}$.

(3) $\delta_i < \delta_{i-1}n_{i-1}$ for $i \ge 2$. Set $\delta_i = \delta_{i-1}n_{i-1} - q_i$, so that $q_i > 0$ for $i \ge 2$.

A pair of integers is said to be **nonprincipal** if neither of them divides the other. We say that the $\delta-$*sequence* $\{\delta_0, \cdots, \delta_h\}$ is **nonprincipal** if δ_0, δ_1 is nonprincipal.

An element of a semigroup will be called **primitive** if it is not a sum of two nonzero elements of the semigroup.

A semigroup generated by a characteristic $\delta-$*sequence* is said to be **a planar semigroup.**

The motivation for the definition comes from the Abhyankar-Moh theory which shows that the semigroup of a plane curve with one place at infinity is planar provided the defining equation is monic in Y of degree nondivisible by char k. In fact, in the next section we will show the converse of this! Let us remark that the condition on $Y-$degree is essential. Indeed in char $k = 3$, the semigroup of the rational curve parametrized by $x \mapsto t^9$ and $y \mapsto t^{12} + t^5$ can be calculated to be generated by $\{9, 12, 15, 17, 20, 23, 25, 28\}$ and we will show later in this section that this semigroup is not planar!

In this section, we deduce some purely numerical properties of a planar semigroup which help in determining the planarity of a given semigroup and in determining all possible generator characteristic $\delta-$*sequences* for it.

Now let $\Gamma = \{\delta_0, \cdots, \delta_h\}\mathbb{N}$ be a planar semigroup.

(1) Layers of a planar semigroup For a fixed i with $1 \le i \le h$, set $\delta'_j = \frac{\delta_j}{d_{i+1}}$.

The semigroup

$$\Gamma_i = \{\delta'_0, \cdots, \delta'_i\}\mathbb{N}$$

is also a planar semigroup for all i from 1 to h generated by the characteristic $\delta-$*sequence* $\{\delta'_0, \cdots, \delta'_i\}$. We may call Γ_i the ith layer of $\Gamma = \Gamma_h$.

(2) Standard Generation Every integer a has a unique expression

$$a = \sum_{0}^{h} a_i\delta_i \text{ where } 0 \leq a_i \leq n_i - 1 \text{ for } i \geq 1.$$

Moreover, $a \in \Gamma$ if and only if $a_0 \geq 0$. This kind of expression will be termed the standard expansion of a with respect to the $\delta - sequence$.

(3) The Conductor formula Set

$$c(\Gamma) = 1 - \delta_0 + \sum_{0}^{h}(n_i - 1)\delta_i$$

Then we have that $\alpha + \beta = c(\Gamma) - 1$ if and only if exactly one of α , β belongs to Γ. Consequently, $c(\Gamma)$ is the smallest element of Γ such that all integers bigger than or equal to it are in Γ.

(4) Bounds on the generator sequence If the sequence is nonprincipal, then $\max(\delta_0, \delta_1) \leq c(\Gamma) + 1$ and $\delta_i \leq c(\Gamma) - 1$ for all $2 \leq i \leq h$.

(5) Primitivity of Generators Every primitive element of Γ is one of the δ_j and conversely, every δ_j is either primitive or a multiple of some δ_r where either $r > j$ or $\delta_j = \delta_1$ and $\delta_r = \delta_0$.

(6) Prime Numbers in a Planar Semigroup There is at most one primitive prime number in Γ, unless Γ is generated by two primes. In particular, a semigroup containing two or more primitive prime numbers and not generated by them is not planar.

Proof of (1) We fix i and call the corresponding semigroup Γ'. All quantities associated with Γ' will be also denoted by primes of similar quantities of Γ. Then we note that $d'_j = \dfrac{d_j}{d_{i+1}}$ for $1 \leq j \leq i$. Consequently, $n'_j = n_j$ for $1 \leq j \leq i$ and $h' = i$.

Now all the conditions for a planar semigroup are easily checked by comparing with those of Γ.

Proof of (2) Note that the $\gcd\{\delta_0, \cdots, \delta_h\} = d_{h+1} = 1$ and hence we can write $a = \sum_0^h b_i\delta_i$ for some integers b_i. We now claim that we can assume, without loss of generality that:

$$b_i \geq 0 \text{ for } i \geq 1.$$

To see this, change b_i to $b_i + \lambda_i\delta_0$ for $i \geq 1$ and simultaneously change b_0 to $b_0 - \sum_0^h \lambda_i\delta_i$, then we can arrange that required condition by taking λ_i large enough.

Now we assume that a has been "standardized" past j , i.e.

$$a = \sum_{0}^{j} b_{i,j}\delta_i + \sum_{j+1}^{h} a_p\delta_p$$

where $0 \le a_p \le n_p - 1$ and $b_{i,j} > 0$ for $1 \le i \le j$. We now prove that the **standardization can then be improved** to $j-1$ such that:

$$a = \sum_{0}^{j-1} b_{i,(j-1)}\delta_i + \sum_{j}^{h} a_p\delta_p$$

where $0 \le a_j \le n_j - 1$ and $b_i^{j-1} \ge b_i^j$ for $0 \le i \le j-1$.

To see this, write $b_j = un_j + a_j$, with $0 \le a_j \le n_j - 1$ using the division algorithm. Also, we have $n_j\delta_j = \sum_0^{j-1} u_i\delta_i$ with $u_i \ge 0$ as guaranteed by the planarity condition. Then setting $b_{i,(j-1)} = b_{i,j} + uu_i$ for $0 \le i \le j-1$ gives the required improvement!

Continuing this improvement, we get the desired standard expression. Now we prove uniqueness. If possible, take two distinct standard expressions $a = \sum_0^h a_i\delta_i = \sum_0^h a_i'\delta_i$, then subtracting one from the other, we get

$$0 = \sum_{0}^{j}(a_i - a_i')\delta_i$$

where we assume that $(a_j - a_j')$ is the last nonzero term. Clearly, all the earlier terms are divisible by $\gcd\{\delta_0, \cdots, \delta_{j-1}\} = d_j$ and hence we have $d_j|(a_j - a_j')\delta_j$. But clearly, $d_{j+1} = \gcd\{\delta_j, d_j\}$ and hence $n_j = \dfrac{d_j}{d_{j+1}}$ divides $(a_j - a_j')$. Since we have already arranged that $0 \le a_j, a_j' \le n_j - 1$, we get $0 < |a_j - a_j'| < n_j$; a contradiction!

Now for the last remark, note that if $a_0 \ge 0$ then clearly $a \in \Gamma$. Conversely, if $a = \sum_0^h b_i\delta_i$ is any expression for an element a with $b_i \ge 0$ for all $0 \le i \le h$, then we can apply the improvement process as described above to get a standard expression $a = \sum_0^h a_i\delta_i$. Now we note that $a_0 \ge b_0$ since the coefficient of δ_0 never decreases during the improvement stages. Hence $a_0 \ge 0$ as required.

Proof of (3) Write

$$\alpha = \sum_{0}^{h} a_i\delta_i \text{ where } 0 \le a_i \le n_i - 1 \text{ for } 1 \le i \le h.$$

If $\alpha + \beta = c(\Gamma) - 1$, then we have

$$\beta = (-1 - a_0)\delta_0 + \sum_{0}^{h}(n_i - 1 - a_i)\delta_i.$$

This is evidently a standard expression. By the standardness of both the expressions, we get that

$$\begin{array}{ccc} \alpha \in \Gamma & \Longleftrightarrow & a_0 \geq 0 \\ \beta \in \Gamma & \Longleftrightarrow & (-1 - a_0) \geq 0 \end{array}$$

Clearly, exactly one of the above two conditions is always satisfied and hence exactly one of α, β belong to Γ.

For the last remark, note that $(c(\Gamma)+\nu)+(-\nu-1) = c(\Gamma)-1$ and hence, if $\nu \geq 0$, then $(-\nu - 1) \notin \Gamma$ and hence $(c(\Gamma) + \nu) \in \Gamma$ for all $\nu \geq 0$. On the other hand, since $0 \in \Gamma$ we have $c(\Gamma) - 1 \notin \Gamma$ and hence $c(\Gamma)$ is the smallest integer such that all integers bigger than or equal to it are in Γ.

Proof of (4) We note that

$$c(\Gamma) - 1 > \sum_2^h (n_i - 1)\delta_i$$

since the difference of the two sides is:

$$\begin{array}{cl} = & -\delta_0 + (n_1 - 1)\delta_1 \\ = & -\delta_0 + \frac{\delta_0\delta_1}{d_2} - \delta_1 \\ = & d_2((\frac{\delta_0}{d_2} - 1)(\frac{\delta_1}{d_2} - 1) - 1) \\ > & 0 \end{array}$$

The last step follows from the nonprincipality which guarantees that each of the ratios $\frac{\delta_0}{d_2}, \frac{\delta_1}{d_2}$ is bigger than 1.

Also, $n_i > 1$ for all $i \geq 2$ and hence we get

$$\delta_i < \frac{c(\Gamma) - 1}{n_i - 1} \leq c(\Gamma) - 1 \text{ for all } 2 \leq i \leq h \tag{1}$$

Now we may assume without loss of generality, that $\delta_0 > \delta_1$ and show that $c(\Gamma) + 1 \geq \delta_0$.

We prove this by induction on h. If $h = 0$, then $c(\Gamma) = 0$ and we are done.

For convenience, set $c_i = c(\Gamma_i)$ for $0 \leq i \leq h$.

By the induction hypothesis applied to Γ_{h-1} we get that

$$\frac{\delta_0}{d_h} \leq c_{h-1} + 1$$

Also the conductor formula gives

$$\begin{array}{rcl} c(\Gamma) = c_h & = & (c_{h-1} - 1)d_h + (n_h - 1)\delta_h + 1 \\ & = & (c_{h-1} + 1)d_h + (d_h - 1)\delta_h - 2d_h + 1 \end{array}$$

Thus we get

$$c(\Gamma)+1=(c_{h-1}+1)d_h+(d_h-1)(\delta_h-2)\geq \delta_0+(d_h-1)+(d_h-1)(\delta_h-2).$$

The result is now obvious if $\delta_h \geq 2$. Now we show that under the assumption of nonprincipality, $\delta_h = 1 \Longrightarrow h = 0$; thus completing the proof.

If $\delta_h = 1$, then $1 \in \Gamma$ and hence Γ contains all nonnegative integers. Clearly, $c(\Gamma) = 0$. Then $h \leq 1$ since otherwise we get a contradiction from (1): $\delta_i < c(\Gamma) - 1 = -1$ for $2 \leq i \leq h$.

If $h = 1$ and $c(\Gamma) = -1 + \delta_0 + (n_1 - 1)\delta_1 = 0$, then clearly we have a contradiction unless $n_1 = 1 = \delta_0$. Also, $n_1 = 1$ implies $\delta_1|\delta_0$ in contradiction to nonprincipality. Thus, $h = 0$ and we are done.

Proof of (5) It is clear that a primitive element of a semigroup belongs to any generating set and hence to $\{\delta_0, \cdots, \delta_h\}$.

Conversely, let δ_j be nonprimitive. If $n_1 = 1$, then δ_1 is not its own standard expression as in (1) but then $\delta_1 = n_1\delta_1$ is a multiple of δ_0 and we are done.

Now assume that δ_j is non-primitive and is its own standard expression (i.e. $n_j > 1$). Let

$$\begin{array}{ll} \delta_j = a + b & \\ \text{where} & 0 \neq a = \sum_0^h a_i\delta_i \in \Gamma \\ \text{and} & 0 \neq b = \sum_0^h b_i\delta_i \in \Gamma \end{array}$$

are standard expressions.

We claim that the expression $\delta_j = \sum_0^h (a_i + b_i)\delta_i$ is not standard, since otherwise, by uniqueness of standard expressions we will get $a = a_j\delta_j$, $b = b_j\delta_j$ and again by uniqueness one of a, b is zero, a contradiction!

Now we start the modification process for the expression of δ_j as described in (1). Let the situation just before the end of the modification process be

$$\delta_j = \sum_0^r (a_i + b_i + v_i)\delta_i \text{ where } a_r + b_r + v_r \neq 0 \text{ and } v_i \geq 0 \text{ for } 1 \leq i \leq r$$

Here v_i are the improvements caused by the previous steps. Now the right hand side of this expression is non-standard and hence $r > 0$. We claim that $r \geq j+1$. In fact, if r were less than j, then the right hand side will improve to a standard expression in $\{\delta_0, \cdots, \delta_{j-1}\}$, in contradiction to the uniqeuness.

The remaining improvement step must come from:

$$(*) \qquad \begin{array}{rcl} a_r + b_r + v_r & = & un_r \\ & \text{and} & \\ n_r\delta_r & = & \sum_0^{r-1} p_i\delta_i \end{array}$$

Finally, in the equation

$$\delta_j = \sum_{0}^{r-1}(a_i + b_i + v_i + up_i)\delta_i$$

all the corresponding coefficients of δ_i must match. Since all a_i, b_i, v_i, u, p_i are nonnegative, we see that

$$up_i = \begin{cases} 0 & i \neq j \\ 1 & i = j \end{cases}$$

Thus, $u = p_j = 1, p_i = 0$ for $i \neq j$ and from $(*)$ above, we get

$$n_r\delta_r = \delta_j$$

as claimed.

Proof of (6) Let a, b be two primitive prime numbers in Γ. By what we just proved, $a = \delta_i$ and $b = \delta_j$ for some i, j and we assume that $i < j$ without loss of generality. If $(i, j) = (0, 1)$ then we are done. Note that $d_{i+1} = 1$ or a and $d_{j+1} = 1$. Thus, $j = h$. If $d_{i+1} = a$, then Γ_i contains $\frac{\delta_i}{d_{i+1}} = \frac{a}{a} = 1$ and hence the semigroup generated by $\delta_0, \cdots, \delta_i = a\Gamma_i = a\mathbb{N}$ and we can change the generator sequence by simply dropping $\delta_0, \cdots, \delta_{i-1}$. Thus, we may assume that $a = \delta_0$ and $d_1 = a$. Now, if $d_2 = a$, then we get $a|\delta_1$ and again, we may drop δ_1 without loss of generality. Thus, we have $d_2 = 1$ and hence $h = 1$, so $j = 1$ and we are done!

Corollary *The semigroup* $\Gamma = \{9, 12, 15, 17, 20, 23, 25, 28\}$ *described earlier is not planar and hence degree-semigroups of plane curves with one place at infinity are not necessarily planar in positive characteristic.*

For proof, note that $17, 23$ are prime and primitive in Γ and do not generate it.

6.3.2 Characterization of Planar Semigroups

Recall that a semigroup Γ of nonnegative integers is said to be a planar semigroup if it is generated by a characteristic δ – *sequence* $(\delta_0, \cdots, \delta_h)$

As remarked earlier, the Abhyankar-Moh semigroup theorem is the motivation behind this definition, since it says:

Abhyankar-Moh Semigroup Theorem *Let f have one place at infinity. Assume that* $n = \deg_Y f(X, Y) \not\equiv 0$ modulo char k. *Then its semigroup is planar. Moreover, there is an associated g – sequence $(X = g_0, \cdots, g_h, g_{h+1} = f)$ such that the corresponding degree – sequence $(\delta_0 = deg(g_0, f), \cdots, \delta_h = deg(g_h, f))$ is a characteristic δ – sequence and generates the degree semigroup of f. Furthermore, the images modulo (f) of the "standard monomials" in $(g_0, \cdots, g_h)$ form a k–vector space basis of A, the coordinate ring of f such that distinct standard monomials have distinct* *induced degrees in the degree semigroup* Γ. *In other words* Γ *is simply the set of degrees of standard monomials.* In this section, we prove the converse of this, namely,

Theorem *Let* $(\delta_0, \cdots, \delta_h)$ *be a characteristic* $\delta - sequence$ *such that* $\delta_0 \not\equiv 0$ modulo char k *Then there exists a curve* f *with one place at infinity whose degree semigroup is generated by* $(\delta_0, \cdots, \delta_h)$. *Moreover, there is a* $g - sequence$ $(g_0, \cdots, g_h, g_{h+1})$ *such that* $f = g_{h+1}$ *and* $\delta_i = deg(g_i, f)$ *for* $0 \leq i \leq h$. *Furthermore,* $\deg_Y f = \delta_0$

Proof. Induction on h. For $h = 0$, we have $\delta_0 = 1$ and we simply take $g_0 = X$, $g_1 = Y$ to get the desired result.

Now let $h > 0$. Set $\delta_i' = \frac{\delta_i}{d_h}$ for $i = 1, \cdots, h-1$. Set the corresponding $d - sequence$ equal to $d_0' = \frac{d_0}{d_h}, \cdots, d_{h-1}' = \frac{d_{h-1}}{d_h}, d_h' = 1$. Other characteristic sequences are defined similarly. Then $(\delta_0', \cdots, \delta_{h-1}')$ clearly satisfies the induction hypothesis and we get a $g - sequence$ $(X = g_0, \cdots, g_h)$ such that $deg(g_i, g_h) = \delta_i'$ for $i = 0, \cdots, h-1$. By the definition of a $\delta - sequence$ we have:

$$\delta_h = \sum_0^{h-1} p_i \delta_i' \qquad \text{and} \qquad \delta_h = \delta_{h-1} n_{h-1} - q_h$$

where the expression for δ_h is standard (i.e. $p_0 \geq 0$ and $0 \leq p_i \leq n_i - 1$ for $1 \leq i \leq h-1$). Moreover, $q_h > 0$.

Let c and z be elements of k to be determined later. Set:

$$g_{h+1} \;=\; g_h^{d_h} + cg* \text{ where } g* \;=\; \prod_0^{h-1} g_i^{p_i}$$

Also set $u(\tau) = y_h(\tau^{d_h}) + z\tau^{m_h}$ where y_h is a "Newton-Puiseux series" of g_h, thus,

$$g_h(\tau^{-\delta_0'}, y_h(\tau)) = 0.$$

It follows that

$$g_h(\tau^{-\delta_0}, y_h(\tau^{d_h})) = 0.$$

Note that $\delta_0, \cdots, \delta_h$ is now the $\delta - sequence$ associated with the series $u(\tau)$. Hence note that $s_h(-\delta_0, u(\tau)) = -d_h\delta_h$.

Now we calculate $f(\tau^{-\delta_0}, u(\tau))$.

For convenience, set $\Phi(G(X,Y)) = G(\tau^{\delta_0}, u(\tau))$ for any polynomial $G(X,Y)$. Then by calculations in (2.4) applied to the $g-sequence$ $g_0, \cdots, g_h$ we get:

$$\Phi(g_h) = (za_h)\tau^{-d_h d_{h-1}' \delta_{h-1}' + q_h} + \cdots \text{ higher terms}$$

and

$$\Phi(g_i) = a_i \tau^{-d_h \delta_i'} + \cdots \text{ higher terms}$$

Observe that

$$-d_h d'_{h-1}\delta'_{h-1} + q_h = -d'_{h-1}\delta_{h-1} + q_h = -n_{h-1}\delta_{h-1} + q_h = -\delta_h.$$

Therefore we get:

$$\begin{aligned}\Phi(g_h^{d_h}) &= (za_h)^{d_h}\tau^{-d_h\delta_h} + \cdots \text{ higher terms}\\ \Phi(cg*) &= c(\textstyle\prod_0^{h-1} a_i^{p_i})\tau^{\sum_0^{h-1} -\delta_i p_i} + \cdots \text{ higher terms}\\ &= c(\textstyle\prod_0^{h-1} a_i^{p_i})\tau^{-d_h\delta_h} + \cdots \text{ higher terms}\end{aligned}$$

Here, the constants a_i depend only on the coefficients of the characteristic terms of $y_h(\tau)$ and hence not on z or c.

Now we choose nonzero c and z such that

$$(za_h)^{d_h} + c\prod_0^{h-1} a_i^{p_i} = 0.$$

It follows that $\operatorname{ord}_\tau \Phi(f) > s_h(-\delta_0, u(\tau)) = -d_h\delta_h$. Also, we note that $\deg_Y f = d_h \deg_Y g_h = d_h\delta'_0 = \delta_0$ since the $Y-$degree of $g*$ is smaller (See (6.2.5)). Finally, f has one place at infinity by the Irreducibility criterion. Further, the corresponding Newton-Puiseux series $y(\tau)$ of f coincides with $u(\tau)$ thru the hth characteristic term m_h and hence $\operatorname{ord}_\tau g_i(-\delta_0, y(\tau)) = \operatorname{ord}_\tau g_i(-\delta_0, u(\tau)) = -\delta_i$ and we have the complete $g-$ *sequence* $X = 0, g_1, \cdots, g_h, g_{h+1} = f$ as desired. The degree-semigroup $\Gamma(f)$ is also generated by the full $\delta-$ *sequence* $\delta_0, \cdots, \delta_h$.

6.4 Degree Semigroups of Polynomial Curves.

A polynomial curve is one with a parametrization $x = x(t)$, $y = y(t)$, where $x(t)$ and $y(t)$ are polynomials over k. Any rational curve with one place at infinity is a polynomial curve. In this section we shall assume that char $k = 0$. We discuss the following questions raised by Abhyankar:

Question 1. *Let Γ be a planar semigroup. Is Γ the semigroup of a polynomial curve?*

Question 2. *Let $(\delta_0, \ldots, \delta_{h+1})$ be a characteristic δ-sequence for Γ. Let f be a polynomial curve such that $\Gamma(f)$=Γ. Is there a g-sequence $(g_0, \ldots, g_{h+1})$ with $f = g_{h+1}$ and $\delta_i = \deg(g_i, f)$ for $0 \le i \le h$? (Note: for a polynomial curve $\deg(g_i, f) = \deg_t(g_i(x(t), y(t)))$.*

We shall see that the answer to Question 2 is “no”. We don't have an answer to Question 1 yet.

As noted in the introduction, Moh [Mo6] observed the following: If $x(t)$ and $y(t)$ are polynomials of degrees 6 and 8 respectively, and $F(X, Y)$ is a polynomial then the Abhyankar-Moh theory plus a calculation shows that $F(x, y)$ cannot have degree 3. (Since [Mo6] didn't actually display the calculation we recently rechecked it hoping that advances in either theory or computer software would make it more amenable to exposition. It is still messy and we will not present the calculation here either.) Hence there is no polynomial curve with characteristic δ-sequence (6,8,3). Of course, the semigroup generated by (6,8,3) is $(3,8)\mathbb{N}$ which is the semigroup of a polynomial curve, namely $f = X^8 - Y^3$ parametrized by $x = t^3$, $y = t^8$. Therefore the answer to Question 2 is no.

We suspect that the semigroup Γ generated by the characteristic sequence $(6, 22, 17)$ may lead to a negative answer to Question 1. First of all, this semigroup has no other characteristic δ-sequence generating it (other than (22,6,17)). To see this, observe that 6, 22, and 17 are primitive elements of Γ and so by property (5) of section 6.3.1 they occur in any characteristic δ-sequence for Γ. We now consider various possibilities for the order of 6, 22, and 17 within a characteristic δ-sequence. If either 6 or 22 appears before 17 then upon reaching 17 the sequence has gcd 1 and must end there. So in that case 17 is the last element of the sequence. The other case, where 6 and 22 follow 17, is impossible. For upon reaching either 6 or 22 the gcd drops to 1 and the sequence terminates before reaching the other. Thus we see that 17 is last number in any characteristic δ-sequence for Γ. Then consider two possibilities: the 6 comes before the 22, or the 22 comes before the 6.

In the first case the sequence looks like $(\ldots, 6, \ldots, 22, \ldots, 17)$. Upon reaching the 6 the gcd so far must either be 6,3, or 2. It cannot be 3 for then the gcd drops to 1 upon reaching the 22 and the sequence would end there. It cannot be 2 for then the gcd would not drop upon reaching 22. Therefore it is 6 and 6 is the first term of the sequence (since the sequence is nonprincipal). The next term in the sequence drops the gcd to either 3 or 2. Again it cannot be 3 or the sequence would end too soon. Therefore it is 2. It follows that the term after the 6 must be 22. The next term must drop the gcd to 1. Therefore it has to be 17. Thus we are reduced to (6,22,17).

In the second case the sequence looks like $(\ldots, 22, \ldots, 6, \ldots, 17)$. As in the last paragraph, the only possibility is that the sequence is (22,6,17).

We thus see that this semigroup has the nice property of having an essentially unique generating characteristic δ-sequence. Now we use the

Lemma. *Let C be a plane polynomial curve and let $\Gamma(C)$ have characteristic δ-sequence $(m, n, delta_2)$. Let $d = d_2 = \gcd(m, n)$, $m' = m/d$, $n' = n/d$. Then there is a corresponding g-sequence (X, Y, g_2) where g_2 is a polynomial of the form*

$$g_2(X, Y) = X^{n'} - Y^{m'} + \Sigma a_{ij} X^i Y^j.$$

The sum is over all (i, j) such that $im + jn < mn/d$. Furthermore, there exist polynomials $x(t)$ and $y(t)$ of degrees m and n such that $g_2(x(t), y(t))$

has t-degree δ_2.

Proof. That g_2 has the given form follows from [A1] section 8. The rest follows from Section 6.3 of this paper. ∎

Therefore let $x(t)$ be a polynomial of degree 6, and $y(t)$ be a polynomial of degree 22. If we can show that $\deg_t g_2(x,y) > 17$ for any polynomial $g_2(X,Y)$ of the above form, it will follow that there is no polynomial curve with degree semigroup $(6,22,17)\mathbb{N}$.

Why do we think that $\deg_t g_2(x,y)$ must be more than 17? The idea is that by careful counting we have 22 free variables and 23 equations (presumably independent). Take generic x and y of the appropriate degrees, say $x = t^{22} + a_1t^{21} + \cdots + a_{22}$, and $y = t^6 + b_1t^5 + \cdots + b_6$. Our variables are the coefficients of x and of y. There are initially 28. By an automorphism of $k[X,Y]$, $X \mapsto X + p_1Y^3 + p_2Y^2 + p_3Y + p_4$, $Y \mapsto Y + p_5$, we may remove 5 variables. By an automorphism of $k[t]$, $t \mapsto t + p_6$, we remove one more. This leaves 22 free variables.

Our equations are obtained by setting to zero the formal coefficients of t in $g_2(x,y)$. We begin with the coefficient of t^{65} and set it equal to zero and solve for one of the variables (e.g. a_1). The next equation comes from the term is t^{64}. This equation can be ignored since 64 is in the (6,22) semigroup anyway and we can remove the t^{64} term by adding to g_2 a multiple of XY^7. Continue in this manner. All terms whose t-degree are in the (6,22) semigroup may be ignored. The Abhyankar-Moh theory also tells us that no other number divisible by $2 = \gcd(6,22)$ can occur as the degree of the leading term of any $g_2(x,y)$: when you leave the (6,22) semigroup, you must leave the (6,22) ideal. Therefore, we need only solve equations coming from terms of odd degree between 65 and 19 (we want to leave the degree 17 term non-zero). In fact, there is no possible degree 19 term to worry about either, for 19 is not in the (3,11) semigroup. Thus we have a total of 23 equations!

Since these equations have no apparent dependence we suspect they may have no solution. This remains to be checked.

As a starting point for developing a general theory of the above sort of calculation we analyzed the case where we fix $F(X,Y) = X^m - Y^n$, since this is the simplest polynomial such that the degree of $F(x,y)$ is not completely predictable from the degrees of x and y. What is a lower bound on the t-degree of $F(X,Y)$? While we asked and answered this question it turned out that we were not the first.[3]

---

[3]**Note.** We learned from an article by Lang [L] that Theorems 1 and 2 have the following history: In 1965 Birch, Chowla, Hall, and Schinzel [BCHS] conjectured Theorem 1 and Davenport [D] proved it. Theorem 2 was proved by Mason [Mas] in 1984. They were interested in diophantine questions such as estimating $a^3 - b^2$ when a and b are integers. Theorem 1 says that $\deg_t (x^3 - y^2) \geq \frac{1}{2}\deg_t (x) + 1$. One easily finds polynomials x and y with integer coefficients such that equality

Theorem 1.

Let x and y be nonconstant polynomials in $k[t]$. If $x^m - y^n$ is not identically zero, then

$$\deg_t (x^m - y^n) \geq \deg_t (x)\frac{1}{n}(mn - m - n) + 1.$$

Theorem 1 is a consequence of the more general

Theorem 2. *Let a and b be nonconstant distinct polynomials of the same degree n. Let ab have k distinct roots. Then*

$$\deg_t (a - b) \geq n - k + 1.$$

Notation.

Let $X_1, \ldots, X_n$ be indeterminants. Denote the elementary symmetric polynomials in these variables by

$$e_1 = \Sigma X_i, \ldots, e_n = X_1 X_2 \cdots X_n$$

and the sum of powers symmetric polynomials by

$$p_1 = \Sigma X_i, \ldots, p_n = \Sigma X_i^n$$

Any symmetric polynomial may be expressed as as a polynomial in either the e_i or the p_i. In particular, the p_i may be expressed in terms of the e_i. These relations are the well-known Newton's formulas. For a derivation see the book by Macdonald [Mac]. The first few are:

$$\begin{aligned} p_1 &= e_1, \\ p_2 &= e_1^2 - 2e_2, \\ p_3 &= e_1^3 - 3e_1e_2 + 3e_3, \\ p_4 &= e_1^4 - 4e_1^2e_2 + 4e_1e_3 + 2e_2^2 - 4e_4 \end{aligned}$$

Let $\alpha_1, \ldots, \alpha_n$ and $\beta_1, \ldots, \beta_n$ be two sets of quantities. Let $e_i(\alpha)$ and $p_i(\alpha)$ denote the symmetric polynomials in the α_i and let $e_i(\beta)$ and $p_i(\beta)$ denote the symmetric polynomials in the β_i.

The following lemma is immediate:

---

holds. For example,

$$(t^4 + 4t)^3 - (t^6 + 6t^3 + 6)^2 = -8t^3 + 36 \tag{1}$$

It follows that there exist infinitely many integers a and b such that $|a^3 - b^2| < 9a^{3/4}$. One gets better inequalites by finding such x and y with higher degree.

Lemma. *Let $1 \le h \le n$. If $e_i(\alpha) = e_i(\beta)$ for $i = 1, \ldots, h$ then $p_i(\alpha) = p_i(\beta)$ for $i = 1, \ldots, h$.*

Proof of Theorem 2.

Factor a and b over the algebraic closure of k :

$$a = \mu(t - \alpha_1) \cdots (t - \alpha_n),$$

$$b = \nu(t - \beta_1) \cdots (t - \beta_n).$$

If $\mu \neq \nu$ then $\deg_t (a - b) = \deg_t a = \deg_t b$ and the theorem is trivially true. So assume $\mu = \nu$. Replacing a and b by $\frac{1}{\mu}a$ and $\frac{1}{\mu}b$ does not change n, k, or $\deg_t (a - b)$. Hence we may assume that a and b are monic.

Now proceed by contradiction. Suppose that $\deg_t (a - b) < n - k + 1$. Hence the first k coefficients of a and b are equal (the coefficients of $t^n, t^{n-1}, \ldots, t^{n-k+1}$). These are the symmetric polynomials in the α_i and the β_i, so we have equations

$$e_i(\alpha) = e_i(\beta) \ , \ i = 1, \ldots, k - 1.$$

So by the lemma

$$p_i(\alpha) = p_i(\beta) \ , \ i = 1, \ldots, k - 1. \tag{2}$$

Call the α_i and β_i collectively γ_i. In other words, define

$$\gamma_i = \alpha_i, \text{ for } i = 1, \ldots, n,$$

$$\gamma_{n+i} = \beta_i, \text{ for } i = 1, \ldots, n.$$

Among the γ there are k distinct numbers by assumption. Renumber so that these are $\gamma_1, \ldots, \gamma_k$. Collect terms so that the equations (2) take the form

$$u_1\gamma_1^i + \cdots + u_n\gamma_k^i = 0 \ , \ i = 1, \ldots, k - 1.$$

where u_i is the multiplicity of γ_i as a root of a minus the multiplicity of γ_i as a root of b. The u_i are not all zero, since this would imply $a = b$, and we have assumed a and b distinct. Also observe that $\Sigma u_i = 0$ since this sum is the number of roots of a minus the number of roots of b and these have the same degree. Putting this in matrix form we have

$$\begin{pmatrix} 1 & \cdots & 1 \\ \gamma_1 & \cdots & \gamma_k \\ \vdots & & \vdots \\ \gamma_1^{k-1} & \cdots & \gamma_k^{k-1} \end{pmatrix} \begin{pmatrix} u_1 \\ \vdots \\ u_k \end{pmatrix} = 0.$$

Since the u_i are not all zero, the Vandermonde matrix in this equation must be singular. On the other hand, the γ_i are distinct so that the matrix must have full rank. This is a contradiction. ∎

Proof of Theorem 1.

If $\deg_t(x^m) \neq \deg_t(y^n)$ then $\deg_t(x^m - y^n)$ is the maximum of the two degrees and the inequality stated in the theorem holds easily in this case. So assume that $\deg_t(x^m) = \deg_t(y^n)$. Then we have for some s, $\deg_t(x) = sn/d$, and $\deg_t(y) = sm/d$ where $d = \gcd(m,n)$. Apply theorem 2 to the case where $a = x^m$, $b = y^n$. We get

$$\begin{aligned}\deg_t(x^m - y^n) &\geq \frac{smn}{d} - \frac{sn}{d} - \frac{sm}{d} + 1\\ &= \frac{s}{d}(mn - m - n) + 1\\ &= \frac{\deg_t(x)}{n}(mn - m - n) + 1.\end{aligned}$$

∎

As noted before, one can ask whether it is possible to find x and y of arbitrarily large degree such that equality holds in the above inequality. This may not be possible if one of m , n divides the other. For example, $\deg_t(x^4 - y^2) = \deg_t(x^2 - y) + \deg_t(x^2 + y) \geq 2\deg_t(x)$ (assuming $x^4 - y^2 \neq 0$). This is strictly greater than $\deg_t(x) + 1$ given by the formula in this case.

Conjecture 3. *Let m, n be nonprincipal (i.e. neither divides the other). For any positive integer s, there exist polynomials x and y, with complex coefficients and degrees sn, and sm respectively such that*

$$\begin{aligned}\deg_t(x^m - y^n) &= \deg_t(x)\frac{1}{n}(mn - m - n) + 1\\ &= s(mn - m - n) + 1\\ &= \deg_t(x^m) - \deg_t(x) - \deg_t(y) + 1\\ &= \deg_t(y^n) - \deg_t(x) - \deg_t(y) + 1\end{aligned}$$

Write $m' = sm$, $n' = sn$, $h = m' + n'$. Proving the conjecture reduces to first finding a solution $(\gamma_1, \ldots, \gamma_h)$ to the system

$$\begin{aligned}m(\gamma_1 + \cdots + \gamma_{n'}) &= n(\gamma_{n'+1} + \cdots \gamma_h),\\ &\vdots\\ m(\gamma_1^{h-2} + \cdots \gamma_{n'}^{h-2}) &= n(\gamma_{n'+1}^{h-2} + \cdots \gamma_h^{h-2}).\end{aligned} \tag{3}$$

Furthermore, the solution must be nondegenerate in the sense that the next equation in the sequence,

$$m(\gamma_1^{h-1} + \cdots \gamma_{n'}^{h-1}) = n(\gamma_{n'+1}^{h-1} + \cdots \gamma_h^{h-1})$$

must not be satisfied, or Theorem 1 would imply that $x^m = y^n$.

We also note that $m(1 + \cdots n' \text{ terms }) + n((1 + \cdots m' \text{ terms }) = 0$. Together with this, put the system (3) in matrix form:

$$\begin{pmatrix} 1 & \dots & 1 \\ \gamma_1 & \dots & \gamma_h \\ \gamma_1 & \dots & \gamma_h \\ \vdots & & \vdots \\ \gamma_1^{h-2} & \dots & \gamma_h^{h-2} \end{pmatrix} \begin{pmatrix} m \\ \vdots \\ m \\ -n \\ \vdots \\ -n \end{pmatrix} = 0.$$

Consider the more general question: For what sequences $(u_1, \dots, u_h) \neq 0$ is there a nondegenerate solution $(\gamma_1, \dots, \gamma_h)$ to the system

$$\begin{pmatrix} \gamma_1 & \dots & \gamma_h \\ \vdots & & \vdots \\ \gamma_1^{h-2} & \dots & \gamma_h^{h-2} \end{pmatrix} \begin{pmatrix} u_1 \\ \vdots \\ u_h \end{pmatrix} = 0? \qquad (4)$$

Lemma. *The system (4) has infinitely many non-zero (but perhaps degenerate) solutions.*

Proof. Think of the equations as $h-2$ hypersurfaces in $\mathbb{P}^{h-1}$. By Bezout's Theorem, the locus of solutions has dimension 1 or more. ■

Lemma. *Suppose that the sum of all the u_i is zero, but that no proper subset of the u_i sums to zero. Then the system has a nondegenerate solution.*

Proof. Let $\gamma_1, \dots, \gamma_h$ be a non-zero solution. Suppose it is degenerate. Then we have:

$$\begin{pmatrix} 1 & \dots & 1 \\ \gamma_1 & \dots & \gamma_h \\ \vdots & & \vdots \\ \gamma_1^{h-2} & \dots & \gamma_h^{h-2} \\ \gamma_1^{h-1} & \dots & \gamma_h^{h-1} \end{pmatrix} \begin{pmatrix} u_1 \\ \vdots \\ u_h \end{pmatrix} = 0.$$

Hence, two of the γ must be equal. Without loss of generality say $\gamma_{h-1} = \gamma_h$. We may therefore rewrite the system, absorbing all references to γ_h into γ_{h-1}, and letting u_{h-1} become $u_{h-1} + u_h$, which is not zero by hypothesis. Finally, remove the last equation. We have:

$$\begin{pmatrix} 1 & \dots & 1 \\ \gamma_1 & \dots & \gamma_{h-1} \\ \vdots & & \vdots \\ \gamma_1^{h-2} & \dots & \gamma_{h-1}^{h-2} \end{pmatrix} \begin{pmatrix} u_1 \\ \vdots \\ u_{h-1} \end{pmatrix} = 0$$

Therefore two of the remaining γ must be equal. Continuing this process we see that all of the γ are equal. But this corresponds to just one point of $\mathbb{P}^{n-2}$, namely $(1, 1, \dots, 1)$, and we have seen that there must be other solutions which are then nondegenerate.■

Corollary. *Let m and n be relatively prime. Then there exist polynomials x and y of degrees n and m respectively such that* $\deg_t(x^m - y^n) = \deg_t(x)\frac{1}{n}(mn - m - n) + 1$

Proof. In this case we have $u_1 = \cdots = u_m = n$ and $u_{m+1} = \cdots = u_{m+n} = -m$. Since m and n are relatively prime, no proper subset of the u_i sums to zero. ∎

Question.

The hypothesis of the theorem doesn't hold if we let x and y have degrees sn and sm for s other than 1. Consider the example (1) where x has degree 4 and y has degree 6. Then $(u_1, \ldots, u_{10}) = (3, 3, 3, 3, -2, -2, -2, -2, -2, -2)$ and there are clearly proper subsets adding to zero (3 + 3 - 2 - 2 - 2). On the other hand $\deg_t(x^3 - y^2) = 4$, the minimum possible. This shows that the hypothesis of the theorem is not necessary.

Is there a condition similar to ours which is necessary and sufficient?

6.5 References

[A1] Abhyankar, S. S. *"Expansion Techniques in Algebraic Geometry"*, Tata Institute of Fundamental Research, 1977.

[A2] Abhyankar, S. S. *On the semigroup of a meromorphic curve,* Proc. Int. Symp. on Algebraic Geometry, Kyoto 1977, 249-414.

[A3] Abhyankar, S., S. *Irreducibility criterion for germs of analytic functions of Two complex Variables.* Advances in Mathematics, 74(2), (1989), 190-257.

[AM1] Abhyankar, S.S., and Moh, T.T., *Newton-Puiseux expansion and Tschirnhausen transformation I,* J. Reine Angew. Math. 260 (1973), 47-83.

[AM2] Abhyankar, S.S., and Moh, T.T., *Newton-Puiseux expansion and Tschirnhausen transformation II,* J. Reine Angew. Math. 261 (1973), 29-54.

[AM3] Abhyankar, S.S., and Moh, T.T., *Embeddings of the line in the plane,* J. Reine Angew. Math. 276 (1975),148-166.

[AS] Abhyankar, S.S., and Singh,B., *Embeddings of certain curves in the affine plane,* Amer. J. Math. 100(1978), 99-175.

[BCHS] Birch,B., Chowla,S., Hall, M., and Schinzel, A., *On the difference* $x^3 - y^2$, K. Norske Vid. Selsk. Forrh. (Trondheim), 38(1965), 65-69.

[D] Davenport, H., *On* $f^3(t) - g^2(t)$, K. Norske Vid. Selsk. Forrh.(Trondheim), 38(1965), 86-87.

[L] Lang,S., *Old and new conjectured diophantine inequalities,* Bull. Amer. Math. Soc. (N.S.) 23(1) (1990),37-75.

[Mac] Macdonald, I.G. "Symmetric Functions and Hall Polynomials." Oxford University Press, New York, 1979.

[Mas] Mason, R.C., *Equations over function fields,* Springe Lecture Notes 1068 (1984)149-157, in Number Theory, proceedings of the Noordwijkerhout, 1983.

[Mo1] Moh,T.T., *On characteristic paris of algebroid plane curves for characteristic p,* Bull. Inst. Math. Acad. Sinica, 1(1973), 75-91.

[Mo2] Moh,T.T., *On approximate roots of a polynomial,* J. Reine Angew. Math. 278(1975), 301-306.

[Mo3] Moh,T.T., *On then concept of approximate roots for algebra,* J. Alg. 65(1980),301-306.

[Mo4] Moh, T.T., *On two fundamental theorems for the concept of approximate roots,* J. Math. Soc. Japan 34(1982), 637-652.

[Mo5] Moh,T.T. *On analytic irreducibility at* ∞, Proc. Amer. Math. Soc. 44(1982), 22-24.

[Mo6] Moh, T.T., *On the jacobian conjecture and the configuration of roots,* J. Reine Angew. Math. 340(1983), 331-340.

[Ri] Richman, D. R., *On the computation of mainimal polynomials,* J. of Algebra, 103(1986), 1-17.

[Ru] Russell, P. *Hamburger-Noether expansions and approximate roots of polynomials,* manuscripta math. 31(1980), 25-95.

[S1] Sathaye, A., *On planar curves,* Amer. J. Math. 99(5)(1977),1105-1135.

[S2] Sathaye,A., *Generalized Newton-Puiseux expansion and Abhyankar-Moh semigroup theorem,* Invent. math. 74(1983), 149-157.

Part III

Algebraic Surfaces

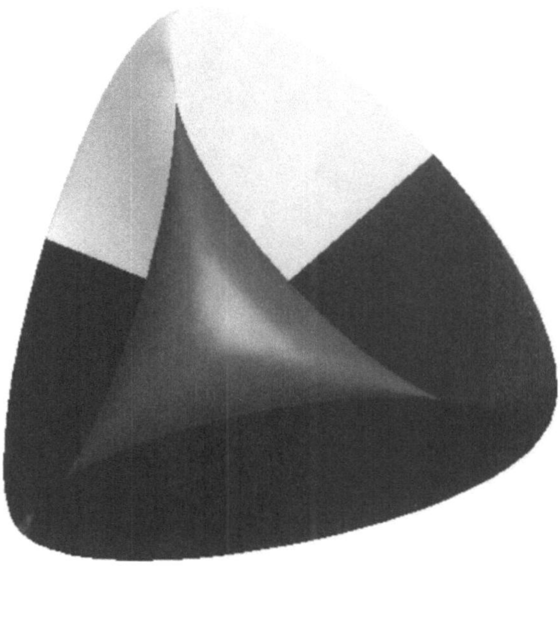

7
A Sharp Castelnuovo Bound for the Normalization of Certain Projective Surfaces

Nadia Chiarli

7.1 Introduction

Let $\mathbf{P}$ be the projective r-space ($r \geq 3$) over an algebraically closed field k. Consider a reduced, irreducible, complete, non-degenerate surface $X \subseteq \mathbf{P}$ of degree d, and let $\bar{X}$ be its normalization: we shall assume $\bar{X}$ to be smooth (which is the case, e.g., when X is the generic projection of a given smooth projective surface $\bar{X}$). Let $\Sigma^{(n)}$ be the linear system of curves cut out on $\bar{X}$ by the hypersurfaces of $\mathbf{P}$ of degree n; in other words $\Sigma^{(n)}$ corresponds to the image of the canonical map

$$\rho_n : H^0(\mathbf{P}, O_{\mathbf{P}}(n)) \to H^0(\bar{X}, O_{\bar{X}}(nD))$$

(where D is the pull-back, via the normalization morphism $\nu : \bar{X} \to X$, of the generic hyperplane section X' of$X$).

When $X = \bar{X}$, by Serre's vanishing theorem it follows that $\Sigma^{(n)}$ is complete for $n \gg 0$, and Lazarsfeld proved that actually $\Sigma^{(n)}$ is complete if $n \geq d - r + 2$ ([11]).

On the contrary, when X is singular $\Sigma^{(n)}$ is certainly non-complete for $n \gg 0$, that is it has a non-zero deficiency $\delta^{(n)} = \dim \operatorname{coker} \rho_n$. (See remark before Theorem 1, and Theorem 1 itself).

The aim of this paper is:

- to show that, for $n \gg 0$, the deficiency $\delta^{(n)}$ is a linear polynomial $p(n)$ in n, and write it down explicitely (Th. 1)

- to find, under suitable assumptions on X, a sharp Castelnuovo bound N, that is a non-negative integer N such that for $n \geq N$ we have $\delta^{(n)} = p(n)$ (Th. 6)

As a corollary we get that $N = d - 3$ is the sharp Casteluovo bound in the case when $X \subseteq \mathbf{P}^3$ is the generic projection of a smooth projective $\bar{X}$.

Notation

I_X: the ideal sheaf of X

D: the pull-back, via ν, of a generic hyperplane section X' of X

J: the conductor of O_X in $\nu_* \, O_{\bar{X}}$

Γ: the subscheme of X corresponding to J

Δ: the pull-back, via ν, of Γ

$p_a(Y)$: the arithmetic genus of Y

$K_{\bar{X}}$: the canonical divisor on $\bar{X}$

$\varepsilon^{(n)}$: the non-negative integer $h^0(\bar{X}, O_{\bar{X}}(nD)) - h^0(X, O_X(n)) \quad \forall\, n \geq 0$

Remark

1. $H^i(\bar{X}, O_{\bar{X}}(nD) \cong H^i(X, (\nu_* O_{\bar{X}})(n)) \quad \forall i$ This follows from the projection formula, using the fact that ν is affine.
2. $\varepsilon^{(n)} \leq \delta^{(n)} \quad \forall\, n \geq 0$, with equality if and only if $h^1(\mathbf{P}, I_X(n)) = 0$; moreover $\varepsilon^{(n)} = \delta^{(n)} = \chi(G(n))$ for $n \gg 0$, where $G = \text{coker}\ (O_X \to \nu_* O_{\bar{X}})$.

The first assertion follows form the exact sequences

$$
\begin{array}{ccccccc}
 & & & & 0 & & \\
 & & & & \downarrow & & \\
0 \to I_X & \to & O_{\mathbf{P}} & \to & O_X & \to 0 \\
 & & & & \downarrow & & \\
 & & & & \nu_* O_{\bar{X}} & & \\
 & & & & \downarrow & & \\
 & & & & G & & \\
 & & & & \downarrow & & \\
 & & & & 0 & &
\end{array}
$$

by the definitions of $\varepsilon^{(n)}$ and $\delta^{(n)}$, while the second one is a consequence of Serre's vanishing theorem.

The next theorem shows that $\chi(G(n)$ is a linear polynomial in n: this proves as a by-product that $\Sigma^{(n)}$ is non-complete for $n \gg 0$ (a classically well known fact, altough not explicitely written in the literature).

Theorem 1. *Set* $p(n) := \chi(G(n))$ *for all* $n \geq 0$. *Then:* $p(n) = [p_a(X') - p_a(D)]n + p_a(\bar{X}) - p_a(X)$

Proof. From the exact sequence

$$0 \to O_X \to \nu_* O_{\bar{X}} \to G \to 0$$

it follows $\chi(G(n)) = \chi(O_{\bar{X}}(nD)) - \chi(O_X(n))$ where, by the Riemann-Roch theorem on $\bar{X}$

$$\chi(O_{\bar{X}}(nD)) = \frac{1}{2} nD \cdot (nD - K_{\bar{X}}) + 1 + p_a(\bar{X})$$

(see [8], 1.6, p. 362). Moreover

$$\chi(O_X(n)) = \mathrm{P}_X(n) = \frac{1}{2} dn^2 + [\frac{1}{2} d - p_a(X') + 1]n + p_a(X) + 1$$

where $\mathrm{P}_X(n)$ denotes the Hilbert polynomial fo X (see [8], 1.2, p. 366). By an easy computation, using the adjunction formula, we get the conclusion.

Corollary 2. *Assume* X *subcanonical* (e.g. $r = 3$). *Then:*

i) $p_a(X') - p_a(D) = (1/2) D \cdot \Delta$

ii) $p(n) = (1/2) D \cdot \Delta + p_a(\bar{X}) - p_a(X)$

Proof. Let X be subcanonical, i.e. $\omega_X = O_X(a)$. Then $K_{\bar{X}} = aD - \Delta$ and, since

$$\omega_{X'} = (\omega_X \otimes_{O_X} O_X(X')) \otimes_{O_{X'}}$$

we get $\omega_{X'} = O_{X'}(a+1)$. In other words X' is $(a+1)$-subcanonical. Moreover we have $2p_a(X') - 2 = (a+1)d$ and, by the adjunction formula:

$$2p_a(D) - 2(K_{\bar{X}} + D) \cdot D = (a+1)d - D \cdot \Delta$$

Now i) follows by an easy computation and ii) follows from i) and Theorem 1.

Proposition 3. *We have, for any fixed* n*:*

i) $p(n) \leq \varepsilon^{(n)}$ *if* $h^1(X, O_X(n)) = 0$

ii) $p(n) = \varepsilon^{(n)}$ *if and only if* $h^1(X, O_X(n)) - h^1(\bar{X}, O_{\bar{X}}(nD)) = h^2(X, O_X(n)) - h^2(\bar{X}, O_{\bar{X}}(nD))$

Proof. Consider the exact sequence $0 \to O_X \to \nu_* O_{\bar{X}} \to G \to 0$, twist by n and take cohomology. Since the support of G is 1-dimensional we have $h^2(X, G(n)) = 0$, whence $h^2(X, O_X(n)) \geq h^2(\bar{X}, O_{\bar{X}}(nD))$ for all $n \geq 0$. Now both i) and ii) follow directly from the definition of $p(n)$ and $\varepsilon^{(n)}$.

From now on we assume that X is arithmetically Cohen-Macaulay and a-subcanonical, e.g. a complete intersection. Hence Γ is a locally Cohen-Macaulay curve (see [8]): this allows us to apply a Lemma by Mumford (see [13], Lemma C, p.122) in order to get the following:

Lemma 4. *Assume* Γ *reduced and let* I_Γ *be the ideal sheaf of* $O_{\mathbf{P}}$ *corresponding to* Γ. *Then* $h^1(\mathbf{P}, I_\Gamma(m)) = 0$ *for* $m < 0$.

Proof. By descending induction on m. The case $m = -1$ follows immediately from the exact sequence $0 \to I_\Gamma(-1) \to O_{\mathbf{P}}(-1) \to O_\Gamma(-1) \to 0$, using Mumford's Lemma.

Now, let Γ' be the generic hyperplane section of Γ and let $I_{\Gamma'}$ be the ideal sheaf of $O_{\mathbf{P}}$ corresponding to Γ'. Our claim follows from the exact sequence

$$0 \to I_\Gamma(m-1) \to I_\Gamma(m) \to I_{\Gamma'}(m) \to 0$$

by the induction hypothesis, using the fact that

$$h^0(\mathbf{P}, I_{\Gamma'}(m)) = 0, \text{ when } m < 0 .$$

Remark. The assumption on X is set in order to have $h^1(\mathbf{P}, I_X(n)) = 0$ for all $n \geq 0$. From this indeed it follows that $\delta^{(n)} = \varepsilon^{(n)}$ for all $n \geq 0$, which allows us to look for the non-negative integer N such that $\varepsilon^n = p(n)$ (for $n \geq N$), rather than $\delta^{(n)} = p(n)$ (for $n \geq N$), which is easier.

Proposition 5. *We have:*

i) $h^i(\bar{X}, O_{\bar{X}}(nD) = h^{2-i}(X, J(a-n)$ *for* $i = 0, 1, 2$

ii) *if* Γ *is reduced and* $n \geq a+1$, *then* $h^1(\bar{X}, O_{\bar{X}}(nD) = h^1(X, J(a-n)) = 0$

Proof. We have $K_{\bar{X}} = aD - \Delta$ whence, by duality

$$h^i(\bar{X}, O_{\bar{X}}(nD)) = h^{2-i}(\bar{X}, O_{\bar{X}}(K_{\bar{X}} - nD)) =$$
$$h^{2-i}\{\bar{X}, O_{\bar{X}}[(a-n)D - \Delta]\} \qquad (i = 0, 1, 2)$$

On the other hand

$$\nu_*\{O_{\bar{X}}[(a-n)D - \Delta]\} =$$
$$\nu_*[O_{\bar{X}}(a-n)D \otimes_{O_{\bar{X}}} O_{\bar{X}}(-\Delta)] = O_{\bar{X}}(a-n) \otimes_{O_X} J =$$
$$= J(a-n), \text{ by the projection formula.}$$

Now, since ν is an affine morphism, we have:

$$h^{2-i}\{\bar{X}, O_{\bar{X}}[(a-n)D - \Delta]\} = h^{2-i}(X, J(a-n)), \text{ whence}$$
$$h^i(\bar{X}, O_{\bar{X}}(nD)) = h^{2-i}(X, J(a-n)),$$

and this proves i). From i) and Lemma 4 we have ii).

Theorem 6. *Assume* Γ *reduced. Then the following are equivalent, for any fixed integer* n:

i) $\delta^{(n)} = p(n)$

ii) $h^1(X, J(a-n)) = 0$ *and* $h^0(X, O_X(a-n)) = h^0(X, J(a-n))$

iii) $h^1(X, J(a-n)) = h^0(X, O_X(a-n)) = 0$

iv) $n \geq a+1$

Proof. From our assumption on X and Γ we have:

1. $\delta^{(n)} = \varepsilon^{(n)}$ for all $n \geq 0$.
2. $h^1(\bar{X}, O_{\bar{X}}(nD)) = h^1(X, J(a-n))$ and $h^2(\bar{X}, O_{\bar{X}}(nD)) = h^0(X, J(a-n))$ (see proposition 5, i))
3. By duality on X we have $h^2(X, O_X(n)) = h^0(X, O_X(a-n))$ for all n

Hence the equivalence between i) and ii) is just proposition 3.

ii) $\Rightarrow$ iv) From the exact sequence

$$0 \to J \to O_X \to O_\Gamma \to 0$$

twisting by $a-n$ and taking cohomology we get (as $h^1(X, J(a-n)) = 0$, by assumption)

$$0 \to H^0(X, J(a-n)) \to H^0(X, O(a-n)) \to H^0(\Gamma, O_\Gamma(a-n)) \to 0$$

whence $h^0(\Gamma, O_\Gamma(a-n)) = 0$, which implies $a-n<0$ and we are done.

iv) $\Rightarrow$ iii) We have $h^1(X, J(a-n)) = 0$, by proposition 5 ii), and $h^0(X, O_X(a-n)) = 0$, by Mumford's Lemma applied to X.

iii) $\Rightarrow$ ii) Follows easily by using the exact sequence

$$0 \to J \to O_X \to O_\Gamma \to 0$$

Corollary 7. *Let X be the complete intersection of $r-2$ hypersurfaces of* $\mathbf{P}$ *of degrees e_i $(i : 1, 2, ..., r-2)$, and set $e = \Sigma_i e_i$. Assume Γ reduced. Then $\delta^{(n)} = p(n)$ if and only if $n \geq e - r$. In particular if $r = 3$, i.e. if X is a surface in the projective 3-space, then $\delta^{(n)} = p(n)$ if and only if $n \geq d-3$.*

Remark. Corollary 7 shows that $d-3$ is e.g. the sharp Castelnuovo bound for the surfaces of $\mathbf{P}^3$ that are the generic projection of a smooth surface embedded in a higher dimensional projective space.

Remark. The assumption of Γ being reduced is actually used in our main Theorem 6 in order to prove iv) $\Rightarrow$ iii), and in particular in order to have $h^1(X, J(m)) = 0$ for $m < 0$, which could perhaps be true more generally.

Acknowledgment

Part of this research was done during a stay at the Math. and Stat. Dept. of Queen's University at Kingston (Ont., Canada), that the author wishes to thank for warm hospitality and support: a preliminary version of this paper is being published on the Queen's Papers in Pure and Applied Mathematics.

The author wishes also to thank Ciro Ciliberto for encouragement and fruitful discussions.

7.2 References

[1] Hartshorne, R., *Algebraic Geometry.* G.M.T. (52) Springer-Verlag, 1977.

[2] Lazarsfeld, R., A sharp Castelnuovo bound for smooth surfaces, *Duke Math. J.* **55**, (2) 1987.

[3] Mumford, D., Lectures on curves on an algebraic surface, *Ann. of Math. Studies*, **59**, Princeton Univ. Press, 1966.

[4] Szpiro, L., *Lectures on equations defining space curves.* Tata Institute of Fund. Res., Bombay, 1979.

8
Abhyankar's Work on Desingularization

S. B. Mulay

8.1 Introduction

It is not our aim in this article to present an exhaustive account of Abhyankar's work related to singularities. Such an account can not be given in a short article. We shall present only a modest exposition of some of the ideas that play a central role in his resolution of singularities. Even this would be lengthy and necessarily very technical if we were to discuss the desingularization of Arithmetic Surfaces, or that of solids, or the expansions related to Canonical Desingularization. Therefore, we shall restrict ourselves mainly to the case of surfaces (over an algebraically closed ground field).

8.2 Algebraic Curves

For the understanding of the problem of desingularization it is imperative to begin with the case of plane curves. Let C be a plane curve. The desingularization problem asks whether C can be transformed into a non-singular curve (not necessarily planar) by a finite sequence of quadratic transformations. Obviously it is sufficient to show that the multiplicity of a singular point can be reduced by a finite sequence of such transformations. To study the effect of a quadratic transformation Abhyankar introduces the numerical character e (called 'magnitude'). It is a measure of the infinitely near equimultiple points. The model is provided by the singularity of $z^d + x^{de} = 0$, where $e \geq 1$, at the origin. To see this, let P be a d-fold point of the curve C. Assume that we have chosen Weierstrass coordinates i.e. P is the origin with respect to the coordinates and $f := z^d + f_1(x)z^{d-1} + ... + f_d(x) = 0$ is a local equation of our curve at P. The coefficients $f_i(x)$ can be expanded as power series in x.

Define $e(z,x) = min\{(1/i)ord(f_i(x))|1 \leq i \leq d\}$ and let

$$e = sup\{e(z,x)|\ (z,x) \text{ Weierstrass}\}.$$

Now, since P is d-fold $e \geq 1$. Also $e = \infty$ if and only if f is a $d-th$

power of some power series. In particular if P is a nonsingular point of C then $e = \infty$.

If f is analytically irreducible and P is singular then the pair (d, de) is the so called 'first characteristic pair' of C at P. The above definition of e extends the notion of the first characteristic pair to the case of a general (possibly reducible) power series.

If d is not divisible by the characteristic of the field then the absence of the term z^{d-1} in the expression of f is a necessary and sufficient condition on Weierstrass coordinates (z, x) to have the property $e(z, x) = e$. Thus, from any Weierstrass coordinates (z, x), by applying the Tschirnhausen transformation $z \to z - (1/d)f_1(x)$, we may at once obtain such a coordinate system.

When we make the quadratic transformation $x = x' \quad z = x'z'$ centered at P ('blow up the point P') we get $f = x'^d f'$. The proper transform of C is given by the equation $f' = 0$. The line $x' = 0$ is the 'exceptional line'. Every point P' that comes out of P has coordinates $z' = \alpha,\ x' = 0$ At such a point $z' - \alpha,\ x'$ are Weierstrass coordinates and we have the pair (d', e') corresponding to f'. Now we state the

Reduction Theorem for Curves:

If $e < \infty$ then either $d' < d$ or $(d', e') = (d, e-1)$.

Assuming C to be free of multiple components we can ensure $e < \infty$ at each singular point of C. Then, it follows from the theorem that by at most e quadratic transformations P will be transformed into points of lesser multiplicity.

Obviously, C can be desingularized by repeated application of the Reduction Theorem. In fact, the following stronger version of desingularization is obtained by proving an analog of the Reduction Theorem for certain pairs of curves.

Monomialisation Theorem for Curves:

By a finite sequence of quadratic transformations the plane can be transformed to a surface V such that the proper transform of C is a nonsingular curve on V and the total transform of C has only normal crossings i.e. at every point Q of V there exist local coordinates z, x with $f = u(z, x)z^a x^b$; $u(0, 0) \neq 0$.

Philosophically, the proof can be thought of as a method of approximating curve-singularities by general cusps. Abhyankar has extended this philosophy to higher dimensions in his monograph 'Weighted Expansions for Canonical Desingularization'. The above argument for curves can be redressed using weighted expansions; this has the advantage of not needing to invoke the Weierstrass Preparation Theorem.

8.3 Algebraic Surfaces

Next, we turn to the case of a surface S in 3-space; we shall tacitly assume S to be free of multiple components. The desingularization problem asks whether S can be transformed into a nonsingular surface (not necessarily spatial) by a finite sequence of quadratic and monoidal transformations. Although it is possible to treat the characteristic $p > 0$ and the characteristic 0 case in a unified manner, desingularization is easier in the later case. Hence, we shall consider the characteristic 0 case before proceeding to discuss the peculiarities of the positive characteristic. As before we consider a d-fold point P, where $d > 1$, of our surface S and analyse the singularity locally. Here the model is provided by the singularity of $z^d + u(x,y)x^a y^b = 0$, where $u(0,0) \neq 0$ and $a + b \geq d$, at the origin. Let $e(P) = [a/d] + [b/d]$ where [ ] denotes the integral part. If $e(P) = 0$ i.e. $a, b < d$ then P is an isolated d-fold point. Otherwise either $z = 0, x = 0$ or $z = 0, y = 0$ is a d-fold curve of S through P. Let $\bar{a}, \bar{b}$ be defined by $a \equiv \bar{a} \bmod d \ , b \equiv \bar{b} \bmod d$. Abhyankar calls P a 'Good point' of the surface S provided $\bar{a} + \bar{b} < d$.

Note that if P is good then it can not be an isolated d-fold point. Say, the curve $C: \ z = 0, x = 0$ is a d-fold curve of S through P. Make the monodial transformation $x = x', y = y', z = z'x'$ centered at C ('blow up the curve C'). Then the proper transform of S is given by $z'^d + u(x',y')x'^{a-d}y'^b = 0$. Every point P' that comes out of P has coordinates $z' = \alpha$, $x' = 0$ $y' = 0$. Letting $w = z' - \alpha$, at P' the local equation of our new surface is $w^d + d\alpha w^{d-1} + \cdots + u(x',y')x'^{a-d}y'^b = 0$. Since the characteristic of the ground field is assumed to be 0 we have $d \neq 0$ and further if α also is nonzero then P' is at most $(d-1)$-fold for the proper transform of S. Now assume that $\alpha = 0$. Then we have $e(P') = e(P) - 1$. At this point it is important to observe two things.

1. If $e(P') \leq 0$ then P' can not be a d-fold point of the proper transform, and

2. If P' is a d-fold point of the proper transform then it is also a Good point of the proper transform.

Therefore, we can iterate the process till P gets transformed into points of lesser multiplicity. To summarize, the singularity at P behaves like a curve-singularity in the sense that it takes the same effort to resolve it as it takes to resolve a curve-singularity.

In general, let $f = 0$ be a local equation of S at P. By choosing a Weierstrass coordinate system we may assume that P is the origin with respect to the coordinates and $f = z^d + \cdots + f_d(x,y)$. Since we are in the characteristic zero case we can also effect the Tschirnhausen transformation $z \to z - (1/d)f_1(x,y)$ if necessary, and thereby assume $f_1(x,y) = 0$. One immediate advantage of having $f_1(x,y) = 0$ is that then every d-fold curve through P has to lie in the plane $z = 0$. The coefficients $f_i(x, y)$ are power series in x, y. Let

$$g = \gcd\{f_i(x,y)^{1/i} | 2 \le i \le d\}.$$

Definition : P is 'pre-Good' for S if for some choice of coordinates x, y, z as above we have $g = x^a y^b$; $(a, b) \neq (0, 0)$

Geometrically, a pre-Good d-fold point of S is

1. not an isolated d-fold point of S,
2. not a singular point of any d-fold curve of S,
3. at most two d-fold curves of S pass through P and if there are two such, they meet nontangentially at P.

By *ord* we mean the (x, y)-order. Let $e(P) = ord(g) := a + b$ and

$$E(P) = min\{ \frac{d}{m} ord(\frac{f_m(x,y)}{g^m}) | \; 2 \le m \le d\}.$$

Definition : P is a 'Good' point of S if it is a pre-Good point of S with $E(P) < d$.

Geometrically, P is 'Good' for S provided for any finite sequence

$$P = P_1 \leftarrow \cdots\cdots \leftarrow P_n \quad (n \ge 1)$$

of monoidal transforms centered at nonsingular d-fold curves, the following holds : If P_n is d-fold for the proper transform S_n of S then P_n is a pre-Good point of S_n.

Here are some examples of surfaces having various types of singularities at the origin.

$$\begin{aligned} z^2 - x^2 y & \quad (\text{Good}), \\ z^2 - xy & \quad (\text{non} - \text{pre} - \text{Good}), \\ z^2 - x^3 y & \quad (\text{pre} - \text{good but non} - \text{Good}). \end{aligned}$$

Assume that P is a pre-Good point and $C : \; z = 0, \; x = 0$ is a d-fold curve of S through P. As before, we blow up the curve C and study the proper transform of S at a point P' which has come out of P. Absence of the term z^{d-1} in f together with the 0 characteristic imply, in exactly the same manner as in the model example, that if P' is given by $z' = \alpha, \; x' = 0, \; y' = 0$ with $\alpha \neq 0$ then P' can not be a d-fold point of the proper transform. Hence we only need to consider the case $\alpha = 0$. Then, we get $e(P') =$

$e(P) - 1$. P' is certainly pre-Good for the proper transform if $e(P') > 0$. Otherwise P' is either not a d-fold point of the proper transform or it is an isolated d-fold point of the proper transform. For a Good point P the second possibility does not exist since $E(P) < d$; in fact, it is easy to see that if P' is d-fold then it is a Good point of the proper transform. Thus Goodness is equimultiply-stable under such monoidal transformations. Observe that after at most $e(P)$ monoidal transformations a Good point P will be transformed into points of lesser multiplicity.

Next, we consider the case of a 'Bad' i.e. non-Good point P. Make the quadratic transformation $x = x', y = x'y', z = z'x'$ centered at P and let P' be a point of the proper transform of S with coordinates $z' = \alpha, y' = \beta, x' = 0$. Then, once again we employ the facts $f_1(x,y) = 0$ and $d \neq 0$ (in the ground field) to conclude that P' can not be a d-fold point of the proper transform unless $\alpha = 0$. So we focus our attention at the points in the plane $z' = 0$. In terms of equations, in this case we get $f = x'^{df}$ where $f' = z'^{d} + f_2 z'^{d} - 2 + \cdots + f'_d = 0$ gives the proper transform. Each coefficient f'_m is simply a product of some (nonnegative) power of x' with the proper transform of $f_m(x,y)$. The coefficients of f give the array $f_2(x,y) = 0, \cdots, f_d(x,y) = 0$ of plane curves and our quadratic transformation, in essence, induces the quadratic transformation $x = x', y = y'x'$ on each of these curves. For our model surface, $z'^{d} + x^{a+b-d}y'^{bv}(x',y') = 0$ is the proper transform. Since P is Bad i.e. $d \leq \bar{a} + \bar{b}$ it follows that $\overline{a+b-d} + \bar{b} < \bar{a} + \bar{b}$. Obviously P' is 'closer' to being Good in comparison to P. In general, we can use the Monomialization and the Principalization theorems for plane curves to ensure that after finitely many quadratic transformations the resulting surface behaves exactly like our model surface. Thus, by a finite sequence of quadratic transformations P gets transformed into points that are either Good d-fold points or have multiplicity less than d. It is important to note that Good-ness is not equimultiply-stable under quadratic transformations. We can only say that along an infinite equimultiple sequence of quadratic transforms, Good-ness is achieved infinitely often.

Let d be the maximum of the multiplicities of points of S and let $B(d)$ be the set of all non-Good d-fold points of S. Bad points are of two kinds; points that are not pre-Good and those pre-Good points which get transformed into isolated equimultiple points by monoidal transformations centered at equimultiple curves. ¿From the geometric definition of pre-Good points it at once follows that there are finitely many non-pre-Good d-fold points. On the other hand, localizing at the generic points of d-fold curves and using the desingularization theorem for plane curves one can conclude that there are only a finite number of bad d-fold points of the second kind. Thus, $B(d)$ is a finite set.

By putting together the finiteness properties seen in the above two paragraphs and the discussion on Good points, we get

Reduction Theorem for Surfaces : There exists a composition of quadratic transformations centered at d-fold points which transforms S into such a surface S^* that all the d-fold points of S^* are Good. Further, if $d \geq 2$ then by a finite sequence of monoidal transformations centered at d-fold curves, S^* is transformed into a surface S_1 devoid of d-fold points.

S can be desingularized by iterating this entire process finitely many times. The 'Monomialisation Theorem for Surfaces' is obtained by proving an analog of the Reduction Theorem for certain triples of surfaces.

We note that in our discussion of the Reduction Theorem we only used the property of d being not divisible by the characteristic of the ground-field. In other words, the Reduction Theorem is applicable whenever the maximum multilicity is not divisible by the characteristic.

Let's turn to the char $p > 0$ case. Consider our model surface $z^d + u(x,y)x^a y^b = 0$. Suppose $d = p = a, b = 0$ and $u(x,y) = 1 + v(x,y)$ with $ord(v(x,y)) \geq p$. Making the monoidal transformation $x = x', y = y', z = z'x'$ we see that at $P' : z' = -1, y' = 0, x' = 0$ the proper transform of our surface has a local equation $w^p + v(x', y') = 0$ where $w = z' + 1$. So the situation could have become worse in some sense !

In view of the above phenomenon, it is necessary to amend the notion of a Good point especially when d is a multiple of p. If $d = p^n$ and the local equation of our model surface looks like

$$z^{p^n} + u(x,y)(x^r y^s)^{p^n} = 0 \quad \text{where} \quad 0 < ord(u(x,y)) < p^n$$

then the origin should be called a Good point. Because, in this case, after finitely many monoidal transformations there is a reduction in the multiplicity.

In general we have $f = z^d + f_1(x,y)z^{d-1} + \cdots + f_d(x,y) = 0$ as a local equation of S at P. Let $q = p^n$ be the p-share of d i.e. $d = mp^n$ where m is prime to p. If $E(P) < d$ as well as $ord(g^i) < ord(f_i(x,y))$ for $1 \leq i \leq q$ then P is a Good point of S.

The essential difficulty for nonzero characteristic arises in dealing with the Bad d-fold points of the surface. Even in the case of $z^q + F(x,y)$ (where $q = p^n$), Good-ness can not be achieved by applying the ready-made Monomialisation Theorem for Curves. Abhyankar does an intricate analysis of the effect of a quadratic transformation on $F(x,y)$ in [4] to establish an 'enhanced' version of the Monomialisation. It can be thought of as Monomialisation modulo q-th powers. One of the novel features of his algorithm is his [ ] function. To define it, we have to express $F(x,y) = x^a y^b G(x,y)$ where $G(0,y) \neq 0 \neq G(x,0)$ and assume that $F(x,y)$ is free of the q-th powers. Let $c = ord(G(0,y))$. Clearly $ord(G(x,y)) \leq c$. Define

$$[q; b, c] = \begin{cases} c & \text{if } b = 0 \bmod q and c > q/p \\ c + (q/p) & \text{if} b \neq 0 \bmod q \text{ and } c \geq q/p \\ 0 & \text{otherwise.} \end{cases}$$

Note that when $c = 0$ the situation is exactly characteristic 0 type; also, it is almost equally stable when $[q; b, c] = 0$. The key lies in the observation that unlike the *ord*( ), the [ ]-function is non-increasing, and more importantly nonconstant, along any sequence of Bad equimultiple quadratic transforms of P.

In general $f_q(x, y)$ is the right candidate to play the role similar to that of $F(x, y)$. But now, there is an additional complication at an equimultiple quadratic transform P' of P which is not in the plane $z' = 0$. Say P' has coordinates $z' = \alpha \neq 0, y' = 0 = x'$. Upon substituting $z = x'(w+\alpha), y = x'y', x = x'$ in f and factoring out x'^d we get the proper transform $f' = w^d + h_1(x', y')w^{d-1} + \cdots + h_d(x', y')$. Here, $h_{i+1}(x', y')$ is a combination of the transforms of $f_1(x, y), \cdots, f_i(x, y)$. Thus, the effect of the transformation on the coefficients is rather complex. Nevertheless, if the inequalities $(1/i)ord(f_i(x, y)) \geq (1/q)ord(f_q(x, y))$ hold for $1 \leq i \leq q$ then the translation by α has limited impact on the low order terms in $f_q(x, y)$. Abhyankar shows that eventually these inequalities will hold in a stable sense i.e. after a finite sequence of transformations the $(x, y) - ord$ valuation does not split in the algebraic extension given by $z^q + f_1(x, y)z^{q-1} + \cdots + f_q(x, y) = 0$ and as a consequence we get the inequalities. Putting these results together, in [6], he proves the non-existence of an infinite sequence of Bad equimultiple quadratic transforms.

Although the language of Good and Bad points is not used in Abhyankar's articles [4], [6] etc., the philosophy is very much there and can hardly escape detection (at least retrospectively). It is interesting to note that Abhyankar's resolution-algorithm for surfaces is characteristic-free in the sense that it works regardless of the characteristic provided we are in the equicharacteristic case. Most of the results as present in [4], are valid even in the non-equicharacteristic case. Abhyankar has used them to desingularize Arithmetic Surfaces (see [7], [8]), or more generally Excellent Surfaces (see [3]).

8.4 References

[1] Abhyankar, S.S.: Resolution of Singularities of Embedded Algebraic Surfaces,Academic Press,1966.

[2] Abhyankar, S.S.: Good points of a Hypersurface, Advances in Mathematics,Vol.68,No.2,April 1988.

[3] Abhyankar, S.S.: Resolution of Singularities of Algebraic Surfaces, Proceedings of the Bombay Colloquium on Algebraic Geometry 1968.

[4] Abhyankar, S.S.: Nonsplitting of Valuations in Extensions of Two Dimensional Regular Local Domains, Math. Annalen 170, 87-144, 1967.

[5] Abhyankar, S.S.: Uniformization in a p-cyclic extension of a two dimensional regular local domain of residue field characteristic p, Wis-

senschaftliche Abhandlungen des Landes Nordrhein-Westfallen, Band 33, 1966.

[6] Abhyankar, S.S.: An algorithm on polynomials in one indeterminate with coefficients in a two dimensional regular local domain, Annali di Mathematica pura ed applicata (IV), Vol. LXXI, pp.25-60, 1966.

[7] Abhyankar, S.S.: Uniformization of Jungian Local Domains, Math. Annalen 159, 1-43, 1965.

[8] Abhyankar, S.S.: Resolution of Singularities of Arithmetical Surfaces, Arithmetical Algebraic Geometry, Harper & Row Inc. 1966.

[9] Abhyankar, S.S.: Weighted Expansions for Canonical Desingularization, Lecture Notes in Mathematics, 910, Springer-Verlag, 1982.

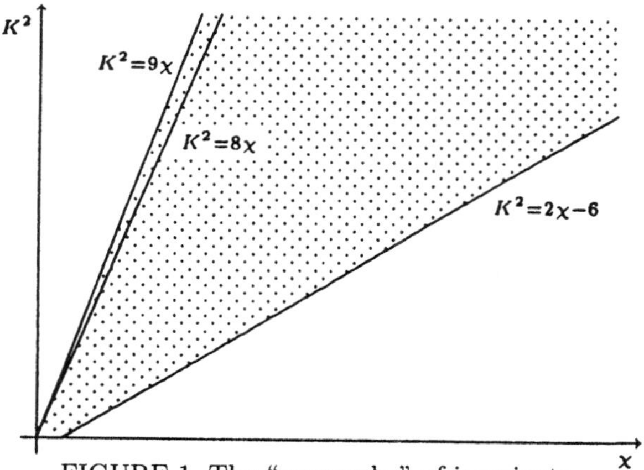

FIGURE 1. The "geography" of invariants.

or even connected. As Catanese has shown in [4], for each number k, there exists a pair of invariants (K^2, χ) such that the moduli space has at least k connected components. All of these components have different dimensions, but nevertheless they all parametrize surfaces of the same homeomorphism class.

Because of these difficulties, the structure of the moduli space of surfaces with given invariants is known only for very few pairs (K^2, χ), mostly for those on or close to the Noether line $K^2 = 2\chi - 6$, which were studied in great detail by Horikawa [10] relax. Most of these surfaces admit a fibration by curves of genus two over the projective line, and therefore a rational morphism of degree two onto a rational ruled surface. This morphism makes a birationally equivalent model of the given surface into a (ramified) twofold covering of a ruled surface; this point of view turns out to be very useful for the study of their deformations.

In this paper, which is an extended version of my talk at the Purdue conference, I want to report on my recent work [20] on moduli of surfaces of general type with a fibration by curves of genus two over a base curve of arbitrary genus g. For such surfaces $K^2 \leq 8\chi$, and in fact, as Persson [16] and Xiao [23] have shown, for every such pair of invariants, there does exist a surfaces of this type. Of course, there will usually be other surfaces with the same invariants; so my results won't give a complete picture of the moduli space of all surfaces with invariants K^2 and χ, but a recent results of Beauville (Appendix to [8]), Catanese [8], and Siu [21] implies that the surfaces in question form connected components of their own in the moduli space, so one gets at least results on some components of the moduli space of surfaces of general type with invariants farther away from the Noether line. Furthermore, if $K^2 < 3\chi$, every surface with an infinite fundamental group admits a fibration by hyperelliptic curves ([10], [18]), and if even

9
Moduli Spaces for Special Surfaces of General Type

Wolfgang K. Seiler

9.1 Introduction

Classification of algebraic varieties consists of two parts: First find a set of discrete invariants like dimensions, genera, ... to describe the basic topological properties of the variety, and then try to make the set of all varieties with given discrete invariants into an algebraic variety, the so called *moduli space.* The prototypical example is classification of Riemann surfaces: The genus is sufficient to fix the topological (and even differentiable) structure, and the set of all Riemann surfaces of a given genus can be made into an irreducible quasi-projective variety; its dimension is $3g - 3$ for $g \geq 2$.

For complex algebraic surfaces the situation is not so simple: Instead of the genus, several topological invariants must be considered, and except for special classes of surfaces like ruled surfaces, K3-surfaces, abelian surfaces or elliptic surfaces, only very little is known about the structure of the set of all surfaces with given invariants. In this paper, I shall exclude these special surfaces and only consider surfaces of general type, that is, surfaces for which a sufficiently high multicanonical linear system defines a birational morphism into some projective space. Also, all surfaces are assumed to be projective, nonsingular and minimal.

There are natural choices for a set of discrete invariants, for example the Chern numbers of the surface, or, equivalently, the self-intersection number K^2 of the canonical divisor and the Euler-Poincaré-characteristic of the sheaf of holomorphic functions. As Gieseker [10] has shown, the set of all surfaces with a fixed pair of invariants has the structure of a quasi-projective variety, but so far it is not even clear yet for which invariants this variety is non-empty. It is known that the condition

$$2\chi - 6 \leq K^2 \leq 9\chi$$

must be satisfied if surfaces of general type with invariants (K^2, χ) are to exist; the first inequality goes back to M. Noether, the second to Miyaoka and Yau. Also, $K^2 > 0$ and $\chi > 0$ for every surface of general type; hence the invariants of surfaces of general type lie in the shaded region of Figure 1. The moduli space, however, even if it is nonempty, need not be irreducible

$K^2 < 8\chi/3$, the fibres have genus two. For $K^2 < 3\chi - 10$, Persson shows in [17] that every surface of general type admits a fibration by hyperelliptic curves.

I shall impose a second condition and only study surfaces for which the degree two map onto the ruled surface is regular and — up to the contraction of rational (-2)-curves into rational double points — is a twofold (ramified) covering. This restriction is mainly technical: If χ is big enough with respect to K^2, the surfaces excluded by this condition have singular models which are degenerations of the surfaces studied here, and a more or less canonical extension of the methods used here should lead to results on the components of the moduli space parametrizing them. This investigation should also lead to new examples of invariants (K^2, χ) such that the moduli space of surfaces with these invariants has more than a given number of connected components. Also, it might be possible to extend my results to hyperelliptic fibrations of higher genus, so I hope that the techniques presented here will lead to a better understanding of quite a few further components of moduli spaces for surfaces of general type well before Abhyankar's seventieth birthday.

9.2 Deformations of Surfaces

The invariants of the surfaces studied here lie on the lines

$$K^2 = 2\chi + 6g - 6$$

parallel to the Noether line (see Figure 2). In the picture, the rays for which I can get a complete picture are drawn as thick lines, the parts where I only get partial information as thin lines.

The first problem in dealing with deformations of these surfaces is to understand the deformations of the underlying ruled surfaces. The geometry of ruled surfaces has been studied for more than a hundred years and can be found, for example, in textbooks like ([9], Chap. V, §2), but surprisingly deformations of ruled surfaces so far have only been investigated in the case of a rational [15] or elliptic [22] base curve; so I have to begin with deformations of ruled surfaces over a curve of genus greater or equal to two.

Let $p : W \to C$ be such a surface. We can write $W = \mathbf{P}(\mathcal{E})$ as the $\mathbf{P}$: 1-bundle associated to a vector bundle $\mathcal{E}$ of rank two on the base curve C, and $\mathcal{E}$ can be normalized in such a way that $\mathcal{E}$ has a nontrivial global section, whereas $H^0(C, \mathcal{E} \otimes \mathcal{L}) = 0$ for every line bundle $\mathcal{L}$ of negative degree. The degree of the determinant of a normalized vector bundle $\mathcal{E}$ is called $-e$ and is a very fundamental invariant of the surface. Most of the problems one encounters when deforming ruled surfaces are due to the fact that e is not invariant under deformations. Geometrically, $-e$ is the smallest self-intersection number of a section of W over C; as Nagata has shown, it is at

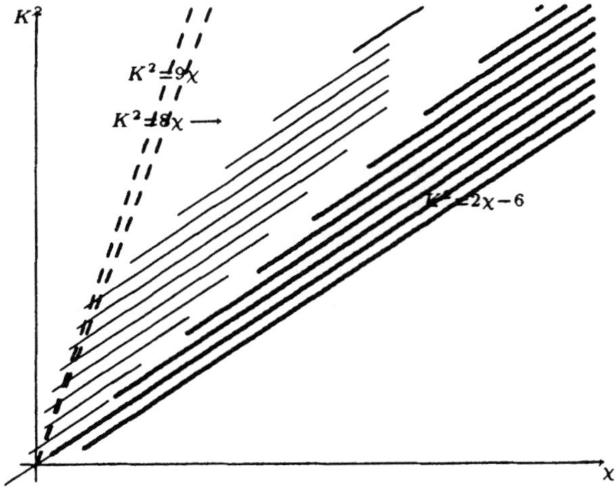

FIGURE 2. Invariants of the surfaces studied here.

most equal to the genus g of the base curve. If C_0 is any such section, its numerical equivalence class and that of a fibre f generate the Néron-Severi group of W, hence every divisor D on W is numerically equivalent to one of the form $aC_0 + bf$; the pair (a, b) is called the *bidegree* of D.

Let ΘW be the tangent sheaf of W. One easily shows that $H^2(W, \Theta W)$ vanishes, hence, by the theorem of Kodaira-Nirenberg-Spencer, there exists a family $\mathcal{W} \to \mathcal{C} \to S$ of ruled surfaces, called the *versal deformation* of W, such that every deformation of W is induced — at least infinitesimally — by this family. The base manifold S has the same dimension as $H^1(W, \Theta W)$, so this dimension must be computed next.

The exact sequence

$$0 \to \Theta W/C \to \Theta_W \to p^*\Theta C \to 0$$

shows that

$$h^1(W, \Theta W) = h^1(W, \Theta_{W/C}) + h^1(W, p^*\Theta_C) = h^1(W, \Theta_{W/C}) + 3g - 3$$

and by Riemann-Roch

$$h^1(W, \Theta_{W/C}) = h^0(W, \Theta_{W/C}) + 3g - 3.$$

This reduces the problem to computing the dimension of the linear system of $\Theta_{W/C}$. For this one has to distinguish several cases, depending on the sign of e and decomposability of $\mathcal{E}$; the final result is

Theorem 1. *Let $W \cong \mathbf{P}(\mathcal{E})$ be a ruled surface, $\mathcal{L} = \det\mathcal{E}$, and $e = -\deg\mathcal{L}$. In the case of a rational base curve,*

$$h^1(W, \Theta_W) = \begin{cases} 0 & \text{for } W \cong \mathbf{P}^1 \times \mathbf{P}^1 \\ e-1 & \text{otherwise} \end{cases}$$

for an elliptic base curve E,

$$h^1(W, \Theta_W) = \begin{cases} 4 & \text{for } W \cong E \times \mathbf{P}^1 \\ 2 & \text{for all other cases with} e = 0 \\ e+2 & \text{for indecomposable } \mathcal{E} \\ e+1 & \text{otherwise,} \end{cases}$$

and for a base curve C of genus g geq2,

$$h^1(W, \Theta_W) = \begin{cases} 6g-3 & \text{for } W \cong C \times \mathbf{P}^1 \\ 6g-6+h^0(C, \mathcal{L}^{-1)} & \text{for indecomposable } E \\ 6g-5+h^0(C, \mathcal{L}^{-1}) & \text{for decomposable } \mathcal{E} \neq 0_C^w \end{cases}$$

The next problem is to decide which surfaces lie in the versal deformation of a given ruled surface W. Because of the invariance of plurigenera under smooth deformations [11], it is clear that every such surface must be ruled again, and the genus of the base curve, which is half the second Betti number of the ruled surface, must remain invariant. The number e, however, can jump: In the case of a rational base curve, two ruled surfaces can be deformed into each other, *iff* their e-invariants differ by an even number ([15], Chap. 1, Theorem 4.2); for ruled surfaces over an elliptic curve, the same holds by the results of [22], and in fact, the same is true in general: It is quite clear that surfaces over base curves of the same genus g can be deformed into each other if both have the same value of e, because the moduli space of curves of genus g is irreducible, and the Jacobian of any curve is irreducible of dimension g, so that the claim is true for surfaces $\mathbf{P}(\mathcal{E})$ with decomposable $\mathcal{E} = \mathcal{O}_C \oplus \mathcal{L}$, and every surface $\mathbf{P}(\mathcal{E})$ with an indecomposable vector bundle $\mathcal{E}$ can be deformed into the surface $\mathbf{P}(\mathcal{O}_C \oplus \det\mathcal{E})$ belonging to a decomposable vector bundle. Deformations changing e can be constructed using Nagata's elementary operations, i.e. by blowing up one point on a ruled surface and contracting the strict transform of the fibre containing that point. This procedure either lowers or raises the number e by one; letting the points vary and assuming e to be positive, one gets families containing surfaces with invariant $e+1$ and those with invariant $e-1$. For negative e, the vector bundle $\mathcal{E}$ must be stable because of the definition of a normalized bundle, so one can use the moduli space of stable vector bundles (see e.g. [19]) and estimate the dimensions of the subsets belonging to bundles with a fixed e-invariant. It is rather straightforward to see that ruled surfaces with different parity of e cannot be deformed into each other, so everything put together leads to

Theorem 2. *A generic surface in the versal deformation of a ruled surface*

W over a base curve of genus g has invariant

$$e' = \begin{cases} -g & \text{if } e \equiv g \bmod 2 \\ -(g-1) & \text{if } e \not\equiv g \bmod 2. \end{cases}$$

For $g \geq 2$ or $g = 1$ and e even, it belongs to a stable vector bundle $\mathcal{E}$, in all other cases it belongs to a decomposable vector bundle $\mathcal{E}$.

Now that deformations of ruled surfaces are known, the next problem is to investigate the connection between deformations of a ruled surface W and deformations of a twofold covering X of W. I shall consider three invariants of X:

— the genus g of the base curve

— the invariant e of the ruled surface

— the integer n for which the ramification divisor B of X over W has bidegree $(6, n + 3e)$.

It is clear that B has degree six on all fibres of W, because the corresponding fibre of X is a curve of genus two which is a twofold covering of $\mathbf{P}^1$, ramified in six points. The reason for writing the second component of the bidegree of B in the form $n + 3e$ is that, with this notation, n is invariant under deformations.

Of course, the numbers g, e and n are neither arbitrary, nor are they independent of each other. The genus g must be nonnegative, and Fujita's result [7] on the positivity of the relative canonical sheaf $K_{X/C}$ implies — after a short calculation with Leray's spectral sequence — that n is always nonnegative, too. To get bounds on e, let C_0 be a section of W with minimal self-intersection number, i.e. $C_0^2 = -e$; then

$$BC_0 = (6C_0 + (n+3e)f)\,C_0 = -6e + n + 3e = n - 3e$$

therefore, $e \leq n/3$ if C_0 is not a component of B. If C_0 [11] is a component of B, it has to be a simple component, because the ramification divisor of a twofold covering is reduced, and therefore $(B - C_0)C_0 = n - 2e$ must be nonnegative. Together with Nagata's bound, we get the inequalities

$$-g \leq e \leq \frac{n}{2}$$

and $e > n/3$ forces C_0 to be a component of the ramification divisor. In the sequel, I shall always assume that X is a surface of general type; for $g \geq 2$ this is no restriction, but for $g = 1$ the case $n = 0$ has to be excluded, and for $g = 0$ all of the cases $n \leq 4$.

The vector space of first order deformations of X is $H^1(X, \Theta_X)$; for the computation of its dimension, let (g, e, n) denote the invariants of X, and suppose that X is a surface of general type. By the theorem of Hirzebruch-Riemann-Roch,

$$\chi(X, \Theta_X) = 6g - 6 - 6n$$

since $H^0(X, \Theta_X)$ vanishes for a surface of general type by [14], this implies that

$$h^1(X, \Theta_X) - h^2(X, \Theta_X) = 6n + 6 - 6g.$$

Generalizing Horikawa's methods from [10], I can show

Theorem 3. *Assume that* either $n > 6g - 6$ *and* $n > 6$ *for* $g = 0$, *and that the ramification* divisor B *of* X *over* W *is nonsingular, or that* $n + 3e$ *is bigger than* $10g - 10$, *and* $n - e > 4$ *for* $g = 0$, *and that* B *is the sum of an irreducible nonsingular curve and the minimal section of the ruled surface such that the two components meet transversally. In both cases, the vector space of all first order deformation has dimension* $7n$, *and all first order deformations can be extended to arbitrary order.*

In order to generalize this result to arbitrary ramification divisors, one would have to work with resolutions of arbitrary rational double points instead of A_1 double points only, which would complicate matters quite a lot. There is, however, another approach: The surface X is uniquely determined by the triple $(W, \mathcal{F}, B)$, where W is the ruled surface, B the ramification divisor, and $\mathcal{F}$ is a line bundle such that $\mathcal{F}^2 \cong \mathcal{O}_W(B)$. If $n > 0$ and $n > 6$ for $g = 0$, Xiao's results in ([23], §3 and §6), imply that, vice versa, $(W, \mathcal{F}, B)$ is uniquely determined by X. Using the result of Beauville, Catanese and Siu mentioned above, and also Brieskorn's resolution theorem ([2], [1]), one finds that instead of the deformations of X one can just as well study deformations of the triple $(W, \mathcal{F}, B)$.

We already know the versal deformation $\mathcal{W} \to \mathcal{C} \to S$ of a ruled surface W, and because of the representability of the Picard functor, the versal deformation of the pair $(W, \mathcal{F})$ is simply the Poincaré bundle over the relative Picard scheme of $\mathcal{W}$ over S. So the main problem is to extend the divisor B or, equivalently, a section $b \in H^0(W, \mathcal{O}_W(B))$ defining B, to a deformation of the pair $(W, \mathcal{F})$. According to general deformation theory, the obstruction against such an extension lies in $H^1(W, \mathcal{O}_{\mathcal{W}}(\mathcal{B}))$, hence there is certainly no problem if this whole group vanishes. In that case, every deformation of the pair $(W, \mathcal{F})$ can be extended to a deformation of the triple $(W, \mathcal{F}, B)$. We already know from Theorem 2 that the general point of the versal deformation of W has e-invariant $-g$ or $-(g-1)$, depending on the parity of n; therefore, if every deformation can be extended to a deformation of the triple, also the general triple in the versal deformation of $(W, \mathcal{F}, B)$ has the same e-invariant. Filling in all the details and excluding the more difficult cases where n is small, I get

Theorem 4. *Let* (g, e, n) *be integers such that* g *is nonnegative,* $n > 2g - 2$ *and* $n > 6$ *for* $g = 0$, *and* $e < \frac{1}{3}(n + 2 - 2g)$. *Then all triples* $(W, \mathcal{F}, B)$

with invariants (g, e, n) *lie in the same irreducible component of the moduli space. The general triple in this component corresponds to a triple whose ruled surface* W *has the minimal possible value* $-g$ *or* $-(g-1)$ *as its e-invariant.*

This gives one component of the moduli space; the rest of paper will deal with the question under which conditions a surface with bigger e still lies in this same component, and whether or not there are additional components parametrizing surfaces with a fibration of the type considered here. The main technical problem is that special divisors on ruled surfaces belonging to an indecomposable vector bundle are much more difficult to handle than those on surface belonging to a decomposable vector bundle, because for the latter ones everything can be translated into a problem about special divisors on curves.

9.3 Moduli Space of Surfaces

Let us begin with surfaces X whose ruled surface belongs to a decomposable vector bundle $\mathcal{O}_C \oplus \mathcal{L}$. Since we can always write $\mathcal{F}$ as $\mathcal{F} = \mathcal{O}_W(3C_0) \otimes p^* \mathcal{M}$,

$$H^i\left((W, \mathcal{O}_W(B)\right) \cong \bigoplus_{i=0}^{6} H^i\left((C, \mathcal{M}^2 \otimes \mathcal{L}^i\right)$$

for these surfaces. The triple $(W, \mathcal{F}, b)$, where $b \in H^0(W, \mathcal{F}^2)$ is an equation for B, is thus given by a 10-tuple

$$(C, \mathcal{L}, \mathcal{M}, b_0, b_1, b_2, b_3, b_4, b_5, b_6)$$

where b_i is the projection from $b \in H^0\left((W, \mathcal{O}_W(B)\right)$ to $H^0(C, \mathcal{M}^2 \otimes \mathcal{L}^i)$. Here b_6 vanishes, *iff* the minimal section C_0 is a component of B. Since C_0 cannot occur with higher multiplicity, b_5 and b_6 cannot vanish simultaneously.

Using the deformation theory of linear system on a curve, which for lack of space I cannot discuss here in detail, I can prove

Theorem 5. *If* $n > 3g - 3$ *and* $n > \frac{3}{2}g$, *every triple* $(W, \mathcal{F}, B)$ *belonging to a 10-tuple with* $b_6 \neq 0$ *can be deformed into a triple whose ruled surface* W *has the minimal possible e-invariant.*

So we are left with two remaining cases:

— the case that the ruled surface W belong to an indecomposable vector bundle, and

— the case that the ramification divisor B contains the zero section.

For both cases, a result of Burns and Wahl ([3], Theorem 2.14; see also [12], Theorem 4.1) is very useful: If one deforms a surface with rational double points, the vector space of first order deformations of the surface maps surjectively onto the sum of the vector spaces of first order deformations of the rational double points. Since the general point of the versal deformation of a rational double point corresponds to a nonsingular point, this means that generic surfaces in the versal deformation of the given surface are nonsingular — provided that every first order deformation of the given surface is unobstructed. Again, the only problem in proving unobstructedness is the ramification divisor B; using deformation theory of linear systems on the ruled surface W, one can show that a first order deformation of B can be extended to any order if $B^2 = 12n$ is big enough:

Theorem 6. *If $n > 6g$ or $n > 8g - 8$, every generic surface in the versal deformation has a nonsingular ramification divisor.*

In almost all cases, this nonsingular ramification divisor must even be an irreducible curve: A nonsingular curve is irreducible *iff* if is connected; assuming that $B = B' + B''$ with $B'B'' = 0$, the conditions imposed on the bidegree of a curve on a ruled surface ([9], Prop. 2.20 and 2.21) show that one of the components must be the minimal section C_0, and $n = 2e$. Since every surface with $e > n/3$ must contain C_0 as a component of its ramification divisor, this rules out generic surfaces with $n/3 < e < n/2$. In fact, I can even show

Theorem 7. *If $n > 6g$ or $n > 8g - 8$, and $e < n/2$, every generic surface in the versal deformation has the minimal e-invariant*

$$e' = \begin{cases} -g & \text{if } e \equiv g \bmod 2 \\ -(g-1) & \text{if } e \not\equiv g \bmod 2. \end{cases}$$

So we are left with the case $e = n/2$. Here the ramification divisor is disconnected, so these surfaces clearly cannot have deformations with a connected ramification divisor; hence, in the moduli space, they must form components of their own. e having the same parity as n, the equation $n = 2e$ implies that n is divisible by four. I assume that n is bigger than $4g-4$; then e is bigger than $2g - 2$, and this implies that the vector bundle $\mathcal{E}$ defining the ruled surface can be decomposed into a sum $\mathcal{E} = \mathcal{O}_C \oplus \mathcal{L}$ of line bundles. Therefore, deformations of these surfaces can be studied via deformations of 10-tuples, as above, but now of course with the additional condition $b_6 \equiv 0$. Since b_5 and b_6 cannot vanish simultaneously, $b_5 \in H^0(C, \mathcal{M}^2 \otimes \mathcal{L}^5)$ is a nonzero section of the line bundle $\mathcal{M}^2 \otimes \mathcal{L}^5$, forcing this bundle to be trivial because it has degree zero. Hence $\mathcal{M}$ is determined uniquely up to 2-torsion by $\mathcal{L}$; using the irreducibility of the moduli space of curves [6] and deformation theory of linear systems, one gets

Theorem 8. *If $n > 4g - 4$, all surfaces with $e = n/2$ are parametrized by*

the same component of the moduli space; the dimension of this component is $7n$.

Finally, one still has to show that all the components of the moduli space mentioned above are nonempty. Construction of surfaces with given invariants is easy for $g = 0$, where the degrees of all curves on the ruled surface are know (e.g. [9], Chap. V, Cor. 2.18); these results can be extended to special surfaces with $g > 0$ by constructing suitable coverings of the $g = 0$ surfaces.

Putting everything together and using the geometric invariants

$$\chi(\mathcal{O}_X) = g - 1 + n \qquad \text{and} \qquad K^2 = 8g - 8 + 2n$$

instead of n and g, the results of this paper can be summarized in

Theorem 9. *Let K^2 and χ be integers such that*

$$G = \frac{K^2 - 2\chi}{6} + 1 \geq 0$$

is an integer, and $\chi > 3g - 3$, resp. $\chi > 5$ for $g = 0$; if $g = 1$ and χ is odd, also assume that $\chi > 2$, and if $g = 2$ and χ is even, assume that $\chi > 5$.

a) *Then the moduli space of surfaces of general type with invariants χ and K^2 has an irreducible component Z_1 of dimension*

$$\dim Z_1 = 7(\chi + 1 - g)$$

parametrizing surfaces with a fibration by curves of genus two over a curve of genus g such that the canonical model of the surface is a twofold covering of a ruled surface. Generically, the surface itself is a double covering of a ruled surface. The ruled surface has an invariant e with

$$-g \leq e < \frac{1}{2}(\chi + 1 - g) \qquad \text{and} \qquad e \equiv \chi + 1 - g \bmod 2.$$

All even values of e satisfying these conditions really occur; odd values occur, if they are nonnegative, or the minimal possible value, or if $\chi > 3g$.

b) *If K^2 is divisible by eight, there is a second component Z_2 with the same dimension*

$$\dim Z_2 = 7(\chi + 1 - g)$$

parametrizing surfaces with a fibration by curves of genus two over a curve of genus g such that the canonical model of the surface is a twofold covering of a ruled surface. Generically, the surface itself is a double covering of a ruled surface. The ruled surface has invariant $e = \frac{1}{2}(\chi + 1 - g)$.

c) *If in addition* $\chi > 5g - 5$, *then* Z_2 *is a connected component of the moduli space and parametrizes all surfaces with* $e = \frac{1}{2}(\chi + 1 - g)$. *Also,* Z_2 *is generically nonsingular then.*

d) *If even* $\chi > 7g - 1$ *or* $\chi > 9g - 9$, *then* Z_1 *is a generically nonsingular connected component too. All surfaces of the type consider here are parametrized by points of either* Z_1 *or* Z_2, *with* Z_2 *occurring* iff K^2 *is divisible by eight.*

9.4 References

[1] Artin, M., Algebraic Construction of Brieskorn's Resolution, *J. Algebra*, **29**, 1974, pp. 330–348.

[2] Brieskorn, E., Die Auflösung der Rationalen Singularitäten Holomorpher Abbildungen, *Math. Ann.*, **178**, 1968, pp. 255–270.

[3] Burns, D.M., J.M. Wahl, Local Contributions to Global Deformations of Surfaces, *Inv. Math.*, **26**, 1974, pp. 67–88.

[4] Catanese, F., On the Moduli Spaces of Surfaces of General Type, *J. Diff. Geom.*, **19**, 1984, pp. 483–515.

[5] Catanese, F., *Moduli and Classification of Irregular Kaehler Manifolds and Algebraic Varieties With Albanese General Type Fibrations*, preprint 1990.

[6] Deligne, P., D. Mumford, The Irreducibility of the Space of Curves of a Given Genus, *Publ. Math. IHES*, **36**, (1969), pp. 75–110.

[7] Fujita, T., On Kähler Fiber Spaces Over Curves, *J. Math. Soc.*, **30**, 1978, pp. 779–794.

[8] Gieseker, D., Global Moduli for Surfaces of General Type, *Inv. Math.*, **43**, 1977, pp. 233–282.

[9] Hartshorne, R., *Algebraic Geometry*, Springer, 1977.

[10] Horikawa, E., Algebraic Surfaces of General Type With Small c_1 : 2, I: *Ann. Math.*, **104**, 1976, pp. 357–387, II: *Inv. Math.*, **37**, 1976, pp. 121–155, III: *Inv. Math.*, **47**, 1978, pp. 209–248, IV: *Inv. Math.*, **50**, 1979, pp. 103–128, V: *J. Fac. Sci. Univ. Tokyo, Sect. A. Math.* **283**, 1981, pp. 745–755.

[11] Iitaka, S., Deformations of Compact Complex Surfaces II, *J. Math. Soc. Japan*, **22**, 1970, pp. 247–261.

[12] Kas, A., Ordinary Double Points and Obstructed Surfaces, *Topology*, **16**, 1977, pp. 51–64.

[13] Maruyama, M., *On Classification of Ruled Surfaces*, Tokyo, 1970.

[14] Matsumura, H., On Algebraic Groups of Birational Transformations, *Atti Acad. Naz. Lincei, Rend. Cl. Sci. Fis. Mat. Natur.*, **8**, 1963, pp. 151–155.

[15] Morrow,J., K. Kodaira, *Complex Manifolds*, New York, 1971.

[16] Persson, U., On Chern Invariants of Surfaces of General Type, *Comp. Math.*, **43**, 1981, pp. 3–58.

[17] Persson, U., An Introduction to the Geography of Surfaces of General Type, *Proc. Symp. Pure Math.*, **46**, 1987, pp. 195–218.

[18] Reid, M., π_1 for Surfaces With Small c_1 : 2, in *Algebraic Geometry Copenhagen* Lecture Notes in Mathematics, **732**, 1979, pp. 534–544.

[19] Seshadri, C.S., Moduli of π-Vector Bundles Over an Algebraic Curve, in *Questions on Algebraic Varieties*, C.I.M.E. Lecture Notes, **51**, Firenze 1970, pp. 139–260.

[20] Seiler, W.K., Modulräume von Flächen Allgemeinen Typs Mit Einer Faserung Durch Kurven vom Geschlecht zwei, submitted as *Habilitationsschrift*, Dept. of Mathematics, Universität Mannheim.

[21] Siu, Y.T., Strong Rigidity for Kaehler Manifolds and the Construction of Bounded Holomorphic Functions, in (R. Howe, ed.), *Discrete Groups in Analysis* Birkhäuser, 1987, pp. 124–151.

[22] Suwa, T., On Ruled Surfaces of Genus 1, *J. Math. Soc. Japan*, **21**, 1969, pp. 291–311.

[23] Xiao Gang, *Surfaces Fibrées en Courbes de Genre Deux*, Lecture Notes in Mathematics **1137**, Springer 1985.

Part IV

Analytic Functions

10

A Stationary Phase Formula for p-ADIC Integrals and its Applications

Jun-ichi Igusa

10.1 Introduction

The classical stationary phase formula is for the oscillator integral:

$$I(\lambda) = \int_{I\!R}^{n} \Phi(x) \ \exp(i\lambda f(x)) \ dx.$$

If K is a locally compact field, there are three closely related functions such that the above $I(\lambda)$ becomes one for $K = I\!R$. Another of the three functions is the zeta function:

$$Z(s) = \int_{K^n} \Phi(x) \mid f(x) \mid_K^s \ dx, \quad Re(s) > 0,$$

In which $f(x)$ is a K-analytic function, $\Phi(x)$ is a Schwartz-Bruhat function and dx is a Haar measure on K^n. In this paper we shall only consider the non-archimedean case. We denote by O_K the maximal compact subring of K with πO_K as its maximal ideal, take the characteristic function of O_K^n as $\Phi(x)$ and a convergent power series on O_K^n as $f(x)$, and normalize the absolute value on K and the Haar measure on K^n in the usual way. by a simple adjustment, we may assume that the coefficients of $f(x)$ are in O_K but not all in πO_K. If then we apply the reduction mod π to $f(x)$, i.e., the canonical homomorphism $O_K \to O_K / \pi O_K = I\!F_q$ to the coefficients of $f(x)$, we get a nonzero polynomial $\overline{f}(x)$. We denote by $\overline{S}$ the set of $I\!F_q$-rational singular points of the hypersurface defined by $\overline{f}(x) = 0$ and take a lifting S of $\overline{S}$ to O_K^n.

Furthermore, we denote by N the number of $I\!F_q$-rational points of $\overline{f}(x) = 0$. Then we have the following formula:

$$Z(s) = (1 - q^{-n}N) + (N - \text{card}(s))(1 - q^{-1})q^{(s+n)} \,/\, (1 - q^{-(s+1)})$$
$$+ \sum_{\xi \in S} q^{-n} \cdot \int_{O_K^n} \mid f(\xi + \pi x) \mid_K^s \ dx.$$

This is the stationary phase formula in the title.

If now $e(\xi)$ denotes the smallest order in π of the coefficients of $f(\xi+\pi x)$, we can apply the above formula to each

$$f_\xi(x) = \pi^{-e(\xi)} f(\xi + \pi x)$$

and repeat the same procedure indefinitely. We observe that if $f(x)$ is a polynomial, then $f(x)$, $f_\xi(x)\ldots$, all have the same degree, hence $\overline{f}(x)$, $\overline{f}_\xi(x)\ldots$, belong to a finite set, and that $Z(s)$ can be expanded into a convergent power series in $t = q^{-s}$. Furthermore, the coefficients of this power series depend only on the above finite set. A closer examination of this situation might lead to a new proof of the rationality of $Z(s)$ in t, which is still open if $\mathrm{char}(K) > 0$. At any rate we shall systematically apply the stationary phase formula to the computation of various p-adic integrals and determine $Z(s)$ for the discriminant $f(x) = f(x_1, x_2)$ of the binary cubic form $\det(ux_1 + vx_2)$ in u, v, where x_1, x_2 are symmetric matrices of degree 3. This is the reduced case of one of the 9 types among 29 types of K-split irreducible regular prehomogeneous vector spaces for which $Z(s)$ has not been computed.

10.2 p-adic Stationary Phase Formula

We express an element $f(x)$ of $K[[x_1, \ldots, x_n]]$ as

$$f(x) = \sum_{i_1,\ldots,i_n} c_{i_1,\ldots,i_n}\, x_1^{i_1} \ldots x_n^{i_n} = \sum_i c_i x^i$$

with $c_i = c_{i_1\ldots i_n}$ in K for every $i = (i_1, \ldots, i_n)$ in $\mathbb{N}^n$; then $f(x)$ is convergent on O_K^n if and only if c_i tends to O as $\mid i \mid = i_1 + \ldots + i_n$ tends to ∞. Therefore the smallest order e in π of all c_i is finite and by replacing $f(x)$ by $\pi^{-e} f(x)$ we may assume that $e = 0$. If $\overline{c}_i$ denotes the image of c_i in $\mathbb{F}_q = O_K \,/\, \pi O_K$, then

$$\overline{f}(x) = \sum_i \overline{c}_i\, x^i$$

becomes an element of $\mathbb{F}_q[x_1, \ldots, x_n]$ different from 0. If $c_0 = 0$ and c_i is in $\pi^{|i|-1} O_K$ for every i, then $f(x)$ is called a special restricted power series, abbreviated as SRP, in $x_1, \ldots, x_n$.

We take a subset $\overline{E}$ of $\mathbb{F}_q^n$ and denote by E its preimage in O_K^n under $O_K^n \to \mathbb{F}_q^n$. We fix a lifting R of $\mathbb{F}_q^n$ in O_K^n and denote by S the subset of R which is mapped bijectively to the set of singular points of $\overline{f}(x) = 0$ in $\overline{E}$. Also we define ν, σ as

$$\begin{cases} \nu = q^{-n} & \{\text{the number of } \overline{\xi} \text{ in } \overline{E} \text{ for which } \overline{f}(\overline{\xi}) \neq 0\} \\ \sigma = q^{-n} & \{\text{the number of smooth points of } \overline{f}(x) = 0 \text{ in } \overline{E}\} \end{cases}$$

and put $t = q^{-s}$ for $Re(s). > 0$; then our *stationary phase formula*, abbreviated as SPF, can be stated as follows:

$$\int_E \mid f(x) \mid_K^s \ dx = \nu + \sigma \cdot (1 - q^{-1})t \,/\, (1 - q^{-1}t)$$
$$+ \sum_{\xi \in S} q^{-n} \cdot \int_{O_K^n} \mid f(\xi + \pi x) \mid_K^n \ dx.$$

The proof is as follows:

By definition we have

$$\int_E \mid f(x) \mid_K^s \ dx = \sum_{\overline{\xi} \in \overline{E}} q^{-n} \cdot \int_{O_K^n} \mid f(\xi + \overline{\pi} x) \mid_K^s \ dx,$$

in which ξ is the element of R which is mapped to $\overline{\xi}$. If $\overline{\xi}$ is restricted by $\overline{f}(\overline{\xi}) \neq 0$, we get the first part and if $\overline{\xi}$ runs over $\overline{S}$, we obviously get the third part. If $\overline{\xi}$ is a smooth point of $\overline{f}(x) = 0$ in $\overline{E}$, then $\overline{f}(\overline{\xi}) = 0$ and $(\partial \overline{f} \,/\, \partial x_i)(\overline{\xi}) \neq 0$ for some i, $1 \leq i \leq n$. If we put

$$\psi_j(x) = \begin{cases} \pi^{-1}\{f(\xi + \pi x) - f(\xi)\} & j = i \\ x_j & j \neq i, \end{cases}$$

then $\psi_1(x), \ldots, \psi_n(x)$ are SRP's in $x_1, \ldots, x_n$ and

$$\partial(\psi_1, \ldots, \psi_n) \,/\, \partial(x_1, \ldots, x_n)(0) = (\partial f \,/\, \partial x_i)(\xi) \not\equiv 0 \bmod \pi.$$

Therefore, $y = \psi(x)$, i.e., $y_j = \psi_j(x)$ for $1 \leq j \leq n$, gives a measure-preserving K-bianalytic map of O_K^n to itself; cf. ([1], p. 213). Since $c = \pi^{-1} f(\xi)$ is in O_K, therefore, we get

$$\begin{aligned} \textstyle\int_{O_K^n} \mid f(\xi + \pi x) \mid_K^s \ dx &= \textstyle\int_{O_K^n} \mid \pi(y_i + c) \mid_K^s \ dy \\ &= t \cdot \textstyle\int_{O_K} \mid y_i \mid_K^s \ dy_i \\ &= t \cdot (1 - q^{-1}) \,/\, (1 - q^{-1}t). \end{aligned}$$

Since the number of such $\overline{\xi}$ is $q^n \cdot \sigma$, we get the second part.

If a, b are in $\mathbb{N}$, then for our later purpose we put

$$(a, b) = 1 - q^{-a_t b}, \qquad (a, b)_+ = 1 + q^{-a_t b}$$

and further $(a) = (a, 0), (a)_+ = (a, 0)_+$. We shall apply SPF repeatedly taking O_K^n as E and denoting S, ν, σ by $S(\overline{f})$, $\nu(\overline{f})$, $\sigma(\overline{f})$. We see by an induction on n that the number of $\mathbb{F}_q$-rational points of $\overline{f}(x) = 0$ is at most equal to $\deg(\overline{f}) q^{n-1}$. Therefore we always have

$$\mathrm{card}(S(\overline{f})) \leq \deg(\overline{f}) \ q^{n-1}.$$

We define $f_{\xi_1, \ldots, \xi_k}(x)$ for ξ_k in $S(\overline{f}_{\xi_1, \ldots, \xi_{k-1}})$ by an induction on $k \geq 0$ as

$$f_{\xi_1,\ldots,\xi_k}(x) = \begin{cases} f(x) & k=0 \\ \pi^{-e(\xi_1,\ldots,\xi_k)} f_{\xi_1,\ldots,\xi_{k-1}}(\xi_k+\pi x) & k\geq 1, \end{cases}$$

in which $e(\xi_1,\ldots,\xi_k)$ denotes the smallest order in π of the coefficients of $f_{\xi_1,\ldots,\xi_{k-1}}(\xi_k+\pi x)$. We observe that $\overline{f}_{\xi_1,\ldots,\xi_k}(x)$ depends only on $\overline{\xi}_1,\ldots,\overline{\xi}_k$.

If now the degrees of all such $\overline{f}_{\xi_1,\ldots,\xi_k}(x)$ are bounded by d, e.g., if $f(x)$ itself is a polynomial of degree d, then we can expand $Z(s)$ as

$$Z(s) = \sum_{k\geq 0} q^{-kn}\left\{\sum_{\xi_1,\ldots,\xi_k} \nu(\overline{f}_{\xi_1,\ldots,\xi_k})t^{E(\xi_1,\ldots,\xi_k)}\right\}$$

$$+(1)t\,/\,(1,1)\cdot\sum_{k\geq 0} q^{-kn}\left\{\sum_{\xi_1,\ldots,\xi_k} \sigma(\overline{f}_{\xi_1,\ldots,\xi_k})t^{E(\xi_1,\ldots,\xi_k)}\right\},$$

in which $E(\xi_1,\ldots,\xi_k) = e(\xi_1)+\ldots+e(\xi_1,\ldots,\xi_k)$. We observe that the two power series are convergent for $\mid t\mid< \min\{1, d^{-1}q\}$, which is equal to 1 for all $q\geq d$.

As an example of the above procedure, we start from $f_0(x) = x_1^2+x_2^3$. If we put $f_1(x)=x_1^2+\pi x_2^3$, $f_2(x)=\pi x_1^2+x_2^3$, $f_3(x)=x_1^2+\pi^2x_2^3$ and denote the corresponding $z(s)$ for $f(x)=f_i(x)$ by $Z_i(s)$, then we successively get

$$\begin{aligned} Z_0(s) &= (1)+(1)^2q^{-1}t\,/\,(1,1)+q^{-2}t^2\cdot Z_1(s), \\ Z_i(s) &= (1)+q^{-1}t\cdot z_{i+1}(s) \qquad \text{for } i=1,2, \\ Z_3(s) &= (1)+q^{-1}t^2\cdot Z_0(s). \end{aligned}$$

Therefore all $Z_i(s)$ have been computed and in particular

$$\int_{O_K^2} \mid x_1^2+\pi x_2^3\mid_K^s\ dx = (1)(1-q^{-3}t^3+q^{-3}t^4-q^{-5}t^5)\,/\,(1,1)(5,6). \quad (F_0)$$

1-bis. We shall give further examples of $Z(s)$ for some $f(x)$ in $O_K[x_1,\ldots,x_n]$. In general if a desingularization of the hypersurface $f(x)=0$ in Aff^n can easily be constructed, e.g., if $n=2$, then it can be used for the computation of $Z(s)$. The length of the computation depends on the complexity of the set of K-rational singular points of $f(x)=0$. In some cases, e.g., if $f(x)=\det(x)$ for x in M_d, this set has such a regularity that $Z(s)$ can easily be computed by induction. Furthermore if $f(x)$ has a large group of similarities, then what we have called a "key lemma" and its supplement, cf. ([1], pp. 212–213), can be effectively applied. For example if $D(x)$ denotes the discriminant of a binary cubic form

$$C_x(u,v) = x_1u^3+x_2u^2v+x_3uv^2+x_4v^3,$$

i.e., if we put

$$D(x) = x_2^2x_3^2 - 4x_1x_3^3 - 4x_2^3x_4 + 18x_1x_2x_3x_4 - 27x_1^2x_4^2,$$

then the key lemma with supplement and (F_0) immediately imply

$$\int_{O_K^4} |\, D(x)\,|_K^s \, dx = (1)(2)(5,5)\,/\,(1,1)(2,2)(5,6), \qquad (F_1)$$

This under the assumption that q is relatively prime to 6; cf. [2]. In the same way and also under the same assumption, we get

$$\begin{aligned}&\textstyle\int_{O_K^3} |\, D(1,x_1,x_2,x_3)\,|_K^s \, dx \\ &= (1)(1-q^{-2}t+q^{-2}t^2-q^{-5}t^5)\,/\,(1,1)(5,6).\end{aligned} \qquad (F_2)$$

Although the computations become lengthy, we can equally use a desingularization or SPF to get (F_1) and (F_2). For example, if we apply SPF to $f(x) = D(1,x_2,x_2,x_3)$, we get

$$\int_{O_K^3} |\, f(x)\,|_K^s \, dx = (1) + (1)^2q^{-1}t\,/\,(1,1) + \sum_c q^{-3}\cdot \int_{O_K^3} |\, f_c(x)\,|_K^s \, dx,$$

in which $f_c(x) = D(1, 3c+\pi x_1, 3c^2+\pi x_2, c^3+\pi x_3)$ and c runs over a lifting of $\mathbb{F}_q$ to O_K. We apply SPF three times starting from $f_c(x)$; then we get

$$\int_{O_K^3} |\, f(x)\,|_K^s \, dx = (1)(3,3)t^2\,/\,(1,1) + q^{-3}t^6 \cdot \int_{O_K^3} |\, f(x)\,|_K^s \, dx.$$

In this way we not only get (F_2) above, but also

$$\begin{aligned}&\textstyle\int_{O_K^3} |\, D(1,3c+\pi x_1, 3c^2+\pi x_2, c^3+\pi x_3)\,|_K^s \, dx \\ &= (1)t^2(1-q^{-3}t^3+q^{-3}t^4-q^{-5}t^5)\,/\,(1,1)(5,6),\end{aligned} \qquad (F_3)$$

which we shall use later.

10.3 Orbital Structure over $\mathbb{F}_q$

We start the computation of $Z(s)$ for the discriminant $f(x)$ of $\det(ux_1 + vx_2)$, where x_1, x_2 are in the space Sym_3 of symmetric matrices of degree 3; we observe that the degree of $f(x)$ is 12. We first use our key lemma to shorten the computation and as its preliminary we determine the orbital structure of the x-space under the group of similarities of $f(x)$ over $\mathbb{F}_q$.

We denote a binary cubic form by $C_y(u,v) = y_1u^3 + y_2u^2v + y_3uv^2 + y_4v^3$ and define a representation ρ_0 of GL_2 in Aff^4 as

$$C_{\rho_0(g_0)y}(u,v) = C_y((u,v)g_0);$$

then we have $D(\rho_0(g_0)y) = \det(g_0)^6 D(y)$ for every g_0 in GL_2 and y in Aff^4. We put $X = (\mathrm{Sym}_3)^2$ and define a representation ρ of $\tilde{G} = GL_3 \times GL_2$ ion X as

$$\rho(g)x = \rho(g_1, g_2)(x_1, x_2) = (g_1 x_1 {}^t g_1, g_1 x_2 {}^t g_1)^{{}^t g_2}.$$

If we introduce a morphism $\phi : X \to Aff^4$ as $C_{\phi(x)}(u, v) = \det(ux_1 + vx_2)$, then we have

$$f(x) = D(\phi(x)), \qquad \phi(\rho(g)x) = \det(g_1)^2 \rho_0(g_2)\phi(x),$$

hence $f(\rho(g)x) = \det(g_1)^8 \det(g_2)^6 f(x)$, for every $g = (g_1, g_2)$ in $\tilde{G}$ and $x = (x_1, x_2)$ in X. We observe that $G_m = GL_1$ is mapped isomorphically to $\mathrm{Ker}(\rho)$ as $\tau \mapsto (\tau 1_3, \tau^{-2} 1_2)$; we put $G = Im(\rho)$. We denote by X' (reps. Y) the subset of X consisting of all x with x_1, x_2 linearly independent (resp. $f(x) \neq 0$); then X' contains Y and they are both G-stable.

We shall assume that q is odd and determine the $G(\mathbb{F}_q)$-orbital structure of $(X' - Y)(\mathbb{F}_q)$. By definition we have

$$\mathrm{card}(X'(\mathbb{F}_q)) = (q^6 - 1)(q^6 - q) = q^{12}(5)(6).$$

As for $\mathrm{card}(Y(\mathbb{F}_q))$, if μ_m denotes the group of mth roots of unity, then SL_3 is transitive on $\phi^{-1}(\phi(x))$ for every x in Y with $(\mu_2)^2$ as its fixer. Since $\mathrm{card}(\phi(Y)(\mathbb{F}_q)) = q^4(1)(2)$, therefore, by using Galois cohomology we get

$$\mathrm{card}(Y(\mathbb{F}_q)) = q^4(1)(2) \cdot \mathrm{card}(SL_3(\mathbb{F}_q)) = q^{12}(1)(2)^2(3).$$

We also have $\rho(\tilde{G}(\mathbb{F}_1)) = G(\mathbb{F}_q)$. Therefore for every ξ in $X(\mathbb{F}_q)$ if $\tilde{G}_\xi$ denotes the fixer of ξ in $\tilde{G}$, then we have

$$\mathrm{card}(G(\mathbb{F}_q)\xi) = \mathrm{card}(\tilde{G}(\mathbb{F}_q)) \,/\, \mathrm{card}\ (\tilde{G}_\xi(\mathbb{F}_q)),$$

in which $\mathrm{card}(\tilde{G}(\mathbb{F}_q)) = q^{13}(1)^2(2)^2(3)$. We shall denote by $[m]$ the mth power map and choose λ from $\mathbb{F}_q^x - (\mathbb{F}_q^x)^{[2]}$. In the following theorem R is the functor $R_{\mathbb{F}_{q^2}/\mathbb{F}_q}$, cf. [5], and U^d is a d-dimensional connected unimpotent group:

Theorem 1. *There exist ten $G(\mathbb{F}_q)$-orbits in $(X' - Y)(\mathbb{F}_q)$; a table of an orbital representative ξ, $\tilde{G}_\xi$ and card $(G(\mathbb{F}_q)\xi)$ is as follows:*

$\mathrm{codim}(G\xi) = 1$, two $G(\mathbb{F}_q)$-orbits:

$$\xi = (e_{12} + e_{21}, ie_{11} + e_{33}), \quad i = 1,\ \lambda,\ \tilde{G}_\xi \,/\, (G_m)^2 = \mu_2,$$
$$\mathrm{card}(G(\mathbb{F}_q)\xi) = (1/2)q^{11}(2)^2(3).$$

$\mathrm{codim}(G\xi) = 2$, three $G(\mathbb{F}_q)$-orbits:

$$\xi = (e_{11}, e_{23} + e_{32}), \tilde{G}_\xi \,/\, (G_m)^3 = \mu_2,$$
$$\mathrm{card}(G(\mathbb{F}_q)\xi) = (1/2)q^{10}(1)_+(2)(3);$$
$$\xi = (e_{11}, e_{22} - \lambda e_{33}), \tilde{G}_\xi \,/\, (G_m \times R(G_m)) = \mu_2,$$
$$\mathrm{card}(G(\mathbb{F}_q)\xi) = (1/2)q^{10}(1)(2)(3);$$
$$\xi = (e_{12} + e_{21}, e_{13} + e_{22} + e_{31}), \tilde{G}_\xi = (G_m)^2U^1,$$
$$\mathrm{card}(G(\mathbb{F}_q)\xi) = q^{10}(2)^2(3).$$

$\mathrm{codim}(G\xi) = 3$, one $G(\mathbb{F}_q)$-orbit:

$$\xi = (e_{11}, e_{13} + e_{22} + e_{31}), \tilde{G}_\xi / U^2 = (G_m)^2,$$

$\mathrm{card}(G(\mathbb{F}_q)\xi) = q^9(2)^2(3)$.
$\mathrm{codim}(G\xi) = 4$, three $G(\mathbb{F}_q)$-orbits:

$$\xi = (e_{11}, e_{22}), \tilde{G}_\xi \,/\, (G_m)^3U^2 = \mu_2,$$
$$\mathrm{card}(G(\mathbb{F}_q)\xi) = (1/2)q^8(1)_+(2)(3);$$
$$\xi = (e_{12} + e_{21}, \lambda e_{11} + e_{22}), \tilde{G}_\xi \,/\, (G_m \times R(G_m))U^2 = \mu_2,$$
$$\mathrm{card}(G(\mathbb{F}_q)\xi) = (1/2)q^8(1)(2)(3);$$
$$\xi = (e_{12} + e_{21}, e_{13} + e_{31}), \tilde{G}_\xi = G_m \times GL_2,$$
$$\mathrm{card}(G(\mathbb{F}_q)\xi) = q^8(2)(3).$$

$\mathrm{codim}(G\xi) = 5$, one $G(\mathbb{F}_q)$-orbit:

$$\xi = (e_{12}, e_{11} + e_{21}), \tilde{G}_\xi = (G_m)^3U^3,$$
$$\mathrm{card}(G(\mathbb{F}_q)\xi) = q^7(1)_+(2)(3).$$

This theorem can easily be proved by actually decomposing $(X' - Y)(\mathbb{F}_q)$ into $G(\mathbb{F}_q)$-orbits. However if we just want its verification, then we can proceed as follows: We show that for each ξ its fixer has the structure given in the theorem. Since they are different, the ten $G(\mathbb{F}_q)$-orbits are distinct. Furthermore

$$\sum_\xi \mathrm{card}(G(\mathbb{F}_q)\xi) = q^{11}(2)(3)(1 + 2q^{-1} + q^{-3} + q^{-5}).$$

We have only to observe, finally, that this is equal to $\mathrm{card}((X' - Y)(\mathbb{F}_q))$.

10.4 Ten Partial Integrals and $Z(s)$

If we denote by i_X the morphism $X = (\mathrm{Sym}_3)^2 \to I(x) = \wedge^2\mathrm{Sym}_3$, then $f(x)$ can be written as $f_0(i_X(x))$ with a homogeneous polynomial function f_0 on $I(X)$ of degree 6. Furthermore if we put $I(X)' = i_X(X')$ and $U = I(X)'(O_K)$, then by Lemma 8 in ([1], pp. 220-221) we get

$$Z(s) = \int_{X(O_K)} |\, f(x)\,|_K^s \, dx = (2)\,/\,(5,6)(6,6) \cdot \int_U |\, f_0(i)\,|_K^s \, di,$$

$$\int_{X'(O_K)} |\, f(x)\,|_K^s \, dx = (2) \cdot \int_U |\, f_0(i)\,|_K^s \, di.$$

In fact we have only to take $|\, f_0(x)\,|_K^s$ and $|\, f_0\,|_K^s$-times the characteristic function of U and "ϕ" in that lemma. Therefore we get

$$(5,6)(6,6)Z(s) = \int_{X'(O_K)} |\, f(x)\,|_K^s \, dx.$$

We define ξ in $X(O_K)$ by the same expression as in Theorem 1 with the understanding that the new λ is in $O_K^x - (O_K^x)^{[2]}$. If we denote the set of ten such ξ in the order in Theorem 1 by

$$R = \{\xi_1, \xi_1', \xi_2, \xi_2'', \xi_3, \xi_4, \xi_4', \xi_4'', \xi_5\},$$

then we get

$$\int_{X'(O_K)} |\, f(x)\,|_K^s \, dx = \mathrm{card}(Y(\mathbb{F}_q))q^{-12} + \sum_{\xi \in R} \mathrm{card}(G(\mathbb{F}_q)\bar{\xi})q^{-12} I_\xi,$$

in which

$$I_\xi = q^{12} \cdot \int_{\xi + \pi X(O_K)} |\, f(x)\,|_K^s \, dx.$$

Since we already know $\mathrm{card}(Y(\mathbb{F}_q))$ and $\mathrm{card}(G(\mathbb{F}_q)\bar{\xi})$, we have only to compute I_ξ for all ξ in R. The proof of the following theorem will be given later:

Theorem 2. *Suppose that q is relatively prime to 6 and put $I_\xi = I_1$, $I_1', \ldots, I_5$ respectively for $\xi = \xi_1, \xi_1', \ldots, \xi_5$; then a table for them is as follows:*

$$\begin{aligned}
I_1 &= I_1' = (1)t\,/\,(1,1)\\
I_2 &= (1)^2t^2\,/\,(1,1)^2,\\
I_2' &= (2)t^2\,/\,(2,2),\\
I_2'' &= (1)t^2(1-q^{-3}t^3+q^{-3}t^4-q^{-5}t^5)\,/\,(1,1)(5,6),\\
I_3 &= (1)t^3\{1-2q^{-2}t^2+(1)q^{-2}t^3+q^{-3}t^4+q^{-5}t^5+(-2q^{-5}+q^{-6})t^6\\
&\quad -2q^{-6}t^7+2q^{-7}t^8+(q^{-6}-q^{-7}+uq^{-8})t^9+(-3q^{-8}+q^{-9})t^{10}\\
&\quad +(q^{-8}-2q^{-9}-q^{-10})t^{11}+(2q^{-10}-q^{-11})t^{12}+(-q^{-10}+3q^{-11})t^{13}\\
&\quad -(1)q^{-11}t^{14}-(1)_+q^{-12}t^{15}+(2q^{-13}-q^{-14})t^{16}+(-q^{-13}+2q^{-14})t^{17}\\
&\quad -q^{-16}t^{18}\}\,/\,(1,1)(2,2)(5,6)(9,12).\\
I_4 &= (1)^2t^4\{1+q^{-1}t-q^{-2}t^2-q^{-3}t^3+q^{-2}t^4-3q^{-4}t^5+2(1)q^{-4}t^6\\
&\quad +2q^{-5}t^7-(1+q^{-1}-2q^{-2})q^{-5}t^8-2(1)q^{-7}t^9-2q^{-8}t^{10}+2q^{-9}t^{11}\\
&\quad +(1-2q^{-1}+2q^{-2})q^{-8}t^{12}-(3-q^{-1}+q^{-2})q^{-10}t^{13}-q^{-11}t^{14}\\
&\quad +3q^{-12}t^{15}-(1-2q^{-1})q^{-12}t^{16}-(1-q^{-1}-q^{-2})q^{-13}t^{17}-2q^{-15}t^{18}\\
&\quad +2(1)q^{-15}t^{19}+2q^{-16}t^{20}-(1)_+q^{-17}t^{21}\}\,/\,(1,1)(2,2)(3,4)(5,6)(9,12).\\
I_4' &= (2)t^4\{1+(1)q^{-2}t^4+(1)q^{-3}t^5+(2-3q^{-1})q^{-4}t^6+2(1)q^{-5}t^7\\
&\quad +2(1)q^{-6}t^8+(1)q^{-7}t^9+(1)q^{-8}t^{10}\}\,/\,(5,6)(6,8).\\
I_4'' &= (1)(2)(5,5)t^4\,/\,(1,1)(2,2)(5,6),\\
I_5 &= (1)^2t^5\{1+q^{-1}t-q^{-2}t^2+(1-2q^{-1})q^{-2}t^3+(1-2q^{-1})q^{-3}t^4\\
&\quad +(1+2q^{-1}-q^{-2})q^{-3}t^5-2(1)q^{-5}t^6+(2-5q^{-1})q^{-5}t^7+(1)^2q^{-6}t^8\\
&\quad +(1)_+q^{-7}t^9-3(1)q^{8}t^{10}+(1-4q^{-1}+2q^{-2})q^{-8}t^{11}-(1)q^{-8}t{-}10t^{12}\\
&\quad -(2-q^{-1})q^{-10}t^{13}-(1-3q^{-1}+q^{-2})q^{-11}t^{14}-(1-2q^{-1})q^{-12}t^{15}\\
&\quad -(1-q^{-1}-q^{-2})q^{-12}t^{16}-(2-q^{-1}+q^{-2})q^{-13}t^{17}+(1)q^{-15}t^{18}\\
&\quad -(1-2q^{-1}-2q^{-2})q^{-14}t^{19}+(1)q^{-16}t^{20}-(2)_+q^{-16}t^{21}+(2)q^{-17}t^{22}\\
&\quad +q^{-18}t^{23}+(q-1^{-1}-q^{-2})q^{-19}t^{24}\}\,/\,(1,1)(2,2)(5,6)(6,8)(9,12).
\end{aligned}$$

We put all our results together and we get the following theorem:

Theorem 3. *If $f(x) = f(x_1, x_2)$ is the discriminant of the binary cubic form $\det(ux_1 + vx_2)$ in u, v for x_1, x_2 in Sym_3 and K is a locally compact nonarchimedean field with $q = \mathrm{card}(O_K\,/\,\pi O_K)$ relatively prime to 6, then for $Re(s) > 0$ we have*

$$\int_{\mathrm{Sym}_3(O_k)^2} |\, f(x)\,|_K^s\, dx$$
$$= (1)(2)^2(3)\,/\,(1,1)(2,2)(5,6)(6,8)(9,12)\cdot F(q^{-1}, q^{-s}),$$

in which $(a,b) = 1 - q^{-(a+bs)}$, $(a) = 1 - q^{-a}$ and

$$\begin{aligned}
F(x,y) &= 1+x^2y^2+x^3y^3+x^4y^4-2x^5y^5+2x^5y^6+(1-x)x^6y^7\\
&+ x^7y^8-x^8y^9+(1-x)x^8y^{10}-2x^{10}y^{11}+2x^{10}y^{12}-x^{11}y^{13}\\
&- x^{12}y^{14}-x^{13}y^{15}-x^{15}y^{17}.
\end{aligned}$$

We might mention that $F(x, y)$ is absolutely irreducible. We recall that the b-function of the above $f(x)$ is

$$b(s) = (s+1)^4\{(s+\frac{3}{4})(s+\frac{7}{6})(s+\frac{5}{6})(s+\frac{5}{4})\}^2;$$

cf. [4].

10.5 Computation of I_1, I_1', I_2, I_2', I_2'', I_4''

We shall compute the ten partial integrals from easy integrals to difficult ones. We change our notation as follows:

$$x = (x_1, x_2) \longmapsto (\sum_{i\le j} x_{ij}e_{ij}, \sum_{i\le j} y_{ij}e_{ij}).$$

Also we denote by T_ξ the tangent space of $G\xi$ of ξ. We shall use the fact that if a K-subpace N of X and its reduction mod π are supplements to T_ξ and its reduction mod π, respectively, then

$$I_\xi = \int_{N(O_K)} |\, f(\xi + \pi x^*)\,|_K^s \, dx^*;$$

cf. ([1], p. 213).

If ξ is either ξ_1 or ξ_1', then T_ξ is defined by $y_{22} = 0$, hence we can take $x^* = (0, y_{22}e_{22})$. Then $\phi(\xi + x^*) = (0, -1, 0, iy_{22})$, where i is either 1 or λ, hence $f(\xi + x^*) = 4iy_{22}$. Therefore if I is either I_1 or I_1', then

$$I = \int_{O_K} |\, 4i\pi y_{22}\,|_K^s \, dy_{22} = t \cdot \int_{O_K} |\, x\,|_K^s \, dx = (1)t \,/\, (1,1).$$

If $\xi = \xi_2$, then T_ξ is defined by $x_{22} = x_{33} = 0$, hence we can take $x^* = (x_{22}e_{22} + x_{33}e_{33}, 0)$. Then $\phi(\xi + x^*) = (x_{22}x_{33}, 0, -1, 0)$, hence $f(\xi + x^*) = 4x_{22}x_{33}$. Therefore

$$I_2 = t^2 \cdot \int_{O_{K^2}} |\, x_1x_2\,|_K^s \, dx = (1)^2t^2 \,/\, (1,1)^2.$$

If $\xi = \xi_2'$, then T_ξ is defined by $\lambda x_{22} + x_{33} = x_{23} = 0$, hence we can take $x^* = (x_{23}e_{23} + x_{33}e_{33}, 0)$. Then $\phi(\xi + x^*) = (-x_{23}^2, x_{33}, -\lambda, 0)$, hence $f(\xi + x^*) = \lambda^2(x_{33}^2 - 4\lambda x_{23}^2)$. Therefore

$$I_2' = t^2 \cdot \int_{O_{K^2}} |\, x_1^2 - \lambda x_2^2\,|_K^s \, dx = (2)t^2 \,/\, (2,2).$$

If $\xi = \xi_2''$, then T_ξ is defined by $2x_{23} - y_{33} = x_{33} = 0$, hence we can take $X^* = (x_{23}e_{23} + x_{33}e_{33}, 0)$ as before. Then $\phi(\xi + x^*) = (-x_{33}, 2x_{23}, 0, -1)$, hence $f(\xi + x^*) = 32x_{23}^3 - 27x_{33}^2$. Therefore by (F_0) we get

$$\begin{aligned} I_2'' &= t^2 \cdot \textstyle\int_{O_{K^2}} |\, x_1^2 + \pi x_2^3\,|_K^s \, dx \\ &= (1)t^2(1 - q^{-3}t^3 + q^{-3}t^4 - q^{-5}t^5) \,/\, (1,1)(5,6). \end{aligned}$$

If $\xi = \xi_4''$, then T_ξ is defined by $x_{22} - 2y_{23} = 2x_{23} - y_{33} = x_{33} = y_{22} = 0$, hence we can take $x^* = (x_{22}e_{22} + x_{23}e_{23} + x_{33}e_{33}, y_{22}e_{22})$. Then $\phi(\xi + x^*) = (-x_{33}, 2x_{23}, -x_{22}, -y_{22})$, hence by (F_1) we get

$$I_4'' = t^4 \cdot \int_{O_{K^4}} | D(x) |_K^x \, dx = (1)(2)(5,5)t^4 \,/\, (1,1)(2,2)(5,6).$$

10.6 Computation of I_3

If $\xi = \xi_3$, then T_ξ is defined by $2(x_{13} - x_{22}) - y_{33} = x_{23} = x_{33} = 0$, hence we can take $x^* = (x_{23}e_{23} + x_{33}e_{33}, y_{33}e_{33})$. Then $f(\xi + x^*) = (-x_{23}^2, x_{33}, y_{33}, -1)$, hence $f(\xi + x^*) = D(1, -y_{33}, -x_{33}, x_{23}^2)$. Therefore

$$I_3 = \int_{O_{K^3}} | f_1(x) |_K^s \, dx,$$

in which $f_i(x) = D(1, \pi^i x_1, \pi^i x_2, \pi^{2i} x_3^2)$ for $i \geq 0$. We apply SPF twice to the above integral and we get

$$I_3 = (1)t^3 + (1)q^{-1}t^4 + q^{-2}t^6 \cdot \int_{O_{K^3}} | D(1, x_1, x_2, \pi x_3^2) |_K^s \, dx;$$

we shall show in §8 that this integral is equal to

$$(1)^2(1,1)_+ \,/\, (1,1) + (1)(1,1)_+ q^{-2}t^2 + q^{-4}t^6 \cdot \int_{O_K^2} | f_0(x) |_K^s \, dx.$$

We observe that the set of $\mathbb{F}_1$-rational singular points of $\overline{f}_0(x) = 0$ is the disjoint union of $\mathbb{F}_q \times \{(0,0)\}$ and $\{(3c^2, 3c^4, c^3);\ c \in \mathbb{F}_q^x\}$. Therefore by SPF we get

$$\int_{O_{K^3}} | f_0(x) |_K^s \, dx$$

$$= (1) + (1)^3 q^{-1} t \,/\, (1,1) + q^{-2}(1)F_3 + q^{-2}t^2 \cdot \int_{O_{K^3}} | g_1(x) |_K^s \, dx,$$

in which F_3 is an in §1-bis and $g_i(x) = \pi^{-2i} f_0(x_1, \pi^i x_2, \pi^i x_3)$ for $i \geq 0$. Since $g_1(\pi x_1, x_2, x_3) = \pi^{-2} f_1(x)$, we can write

$$\int_{O_{K^3}} | g_1(x) |_K^s \, dx = q^{-1}t^{-2} I_3 + \int_{O_K^x \times O_K^2} | g_1(x) |_K^s \, dx.$$

Since $g_1(x) \equiv x_1^2(x_2^2 - 4x_1x_3^2) \bmod \pi$ for $i \geq 1$, by SPF we get

$$\int_{O_{K^2}} | g_i(x) |_K^s \, dx_2 dx_3 = (1)(1 - \chi(x_1))q^{-1} + (1)^2(1 + x(x_1))q^{-1}t \,/\, (1,1)$$
$$+ q^{-2}t^2 \cdot \int_{O_{K^2}} | g_{i+1}(x) |_K^s \, dx_2 dx_3,$$

in which χ is the unique nontrivial character of $O_K^\times / (O_K^\times)^{[2]}$. This implies

$$\int_{O_{K^2}} | g_i(x) |_K^s \, dx_2 dx_3 = (1)\{(2.1) - (0,1)q^{-1}\chi(x_1)\} / (1,1)(2,2)$$

for every $i \geq 1$, hence

$$\infty_{O_K^\times \times O_K^2} | g_i(x) |_K^s \, dx = (1)^2(2,1) / (1,1)(2,2).$$

By putting these together we get the expression for I_3 in Theorem 2. We have also shown that

$$\begin{aligned} \textstyle\int_{O_{K^3}} | D(1, x_1, x_2, x_3^2) |_K^s \, dx = (1)\{1 - (2 - q^{-1})q^{-2}t - q^{-3}t^2 \\ + q^{-4}t^3\} / (1,1)(2,2) + (1)q^{-2}F_3 + q^{-3}I_3, \end{aligned} \qquad (F_4)$$

which we shall use in §9.

10.7 Computation of I_4'

If $\xi = \xi_4'$, then T_ξ is defined by $x_{13} - y_{23} = \lambda x_{23} - y_{13} = x_{33} = y_{33} = 0$, hence we can take $x^* = (x_{33}e_{33}, y_{13}e_{13} + y_{23}e_{23} + y_{33}e_{33})$. Then $\phi(\xi + x^*) = (-x_{33}, -y_{33}, \lambda x_{33} + 2y_{13}y_{23}, \lambda y_{33} - y_{13}^2 - \lambda y_{23}^2)$, hence

$$I_4' = t^4 \cdot \int_{O_K^4} | f_1(x) |_K^s \, dx,$$

in which

$$f_i(x) = D(x_1, x_2, -\lambda x_1 + 2\pi^i x_3 x_4, -\lambda x_2 + \pi^i(x_3^2 + \lambda x_4^2))$$

for $i \geq 0$. Since $D(x_1, x_2, -\lambda x_1 - \lambda x_2) = 4\lambda(\lambda x_1^2 - x_2^2)^2$, if we put

$$J_i = \int_{O_K^2 \times A} | f_i(x) |_K^s \, dx, \qquad A = O_K^2 - \pi O_K^2,$$

then $J_i = (2)^2 + q^{-2}t^4 J_{i-1}$ for every $i \geq 1$, hence

$$J_i = (2)^2(2i, 4i) / (2,4) + (q^{-1}t^2)^{2i} J_0,$$

and hence

$$\begin{aligned} t^{-4}I_4' &= \textstyle\sum_{i \geq 0} q^{-2i} J_{2i+1} \\ &= (2)^2 / (2,4) \cdot \{1 / (2) - q^{-2}t^4 / (6,8)\} + q^{-2}t^4/(6,8) \cdot J_0. \end{aligned}$$

We shall apply SPF to J_0 after the following remarks:

If $\theta = \lambda^{1/2}$, then $L = K(\theta)$ is the unique unramified quadratic extension of K and $0_L = O_K + O_k\theta$; if $y = x_3 + x_4\theta$, then y is in 0_L^x if and only if (x_3, x_4) is in A; if $N(y) = x_3^2 - \lambda x_4^2$, then N maps 0_L^x surjectively to 0_K^x and it gives rise to the following isomorphism:

$$O_L^x \,/\, (O_L^x)^{[2]}(= \mathbb{F}_{q^2}^x \,/\, (\mathbb{F}_{q^2}^x)^{[2]}) \cong O_K^x \,/\, (O_K^x)^{[2]}(= \mathbb{F}_q^x \,/\, (\mathbb{F}_q^x)^{[2]}).$$

If we denote by B the subset of O_K^2 consisting of all (x_3, x_4) such that $x_3 + \theta x_4$ is in $(O_L^x)^{[2]}$, i.e., $x_3^2 - \lambda x_4^2$ is in $(O_K^x)^{[2]}$, then B is contained in A and

$$\int_A \psi(x_3^2 + \lambda x_4^2, 2x_3x_4)dx = 2 \cdot \int_B \psi(x_3, x_4)dx$$

for any continuous function ψ on A. In fact if we put $\Psi(x_3 + \theta x_4) = \psi(x_3, x_4)$, then this becomes equivalent to

$$\int_{O_L^x} \Phi(y^2)dy = 2 \cdot \int_{(C_L^x)^{[2]}} \Psi(y)dy,$$

which is obvious. Therefore we can write

$$J_0 = 2 \cdot \int_{O_K^2 \times B} |\, g(x) \,|_K^s \, dx,$$

in which $g(x) = D(x_1, x_2, -\lambda x_1 + x_4, -\lambda x_2 + x_3)$. We shall apply SPF to "$f(x)$" $= g(x)$ and "E" $= O_K^2 \times B$.

We shall first assume that $q \equiv 1 \bmod 4$ and later come back to the case where $q \equiv -1 \bmod 4$. We can easily verify that $a = (a_1, a_2, a_3, a_4)$ in $\overline{E}$ is a singular point of $\overline{g}(x) = 0$ if and only if either $a_1 = a_2 = a_4 = 0$, a_3 in $\mathbb{F}_q^x$ or $a_1 = b$, $a_2 = 3bc$, $a_3 = b(c^3 + 3\overline{\lambda}c)$, $a_4 = b(3c^2 + \overline{\lambda})$ for some b, c in $\mathbb{F}_q^x$ such that (a_3, a_4) is in $\overline{B}$. We observe that (a_3, a_4) is in $\overline{B}$ if and only if $c^2 - \overline{\lambda}$ is in $(\mathbb{F}_q^x)^{[2]}$, i.e., $c = (1/2)(\overline{\lambda}u + u^{-1})$ for some u in $\mathbb{F}_q^x$, which is unique up to $u \longmapsto (\overline{\lambda}u)^{-1} \neq u$. The numbers of the two kinds of singular points of $\overline{g}(x) = 0$ in $\overline{E}$ are $q - 1$ and $(1/2)(q-1)^2$, respectively, hence $\mathrm{card}(S) = (1/2)(q^2 - 1)$. As for points of $\overline{g}(x) = 0$ in $\overline{E}$, there are three kinds: $a_1 = a_2 = 0$ and (a_3, a_4) in $\overline{B}$; $a_1 = 0$, a_2 in $\mathbb{F}_q^x$ and $a_3 = \lambda a_2 + a_4^2 \,/\, 4a_2$ for a_4 in $\mathbb{F}_q$; $a_1 = b$ in $\mathbb{F}_q$, $a_2 = b(c + 2d)$, $a_3 = b(cd^2 + \lambda c + s\lambda d)$, $a_4 = b(2cd + d^2 + \lambda)$ for some c, d in $\mathbb{F}_q$ such that $c^2 - \overline{\lambda}$ is in $(\mathbb{F}_q^x)^{[2]}$. In particular the number of all such points is $(1/2)(q+1)(q^2 - 1)$. Therefore by SPF we get

$$J_0 = (2)(1-q^{-1}-q^{-2}) + (1)(2)q^{-1}t \,/\, (1,1) + 2q^{-4} \cdot \sum_{a \in S} \int_{O_K^4} |\, g(a + \pi x) \,|_K^s \, dx.$$

Since $g(\pi x_1, \pi x_2, x_3, \pi x_4) = D(\pi x_1, \pi x_2, \pi(x_4 - \lambda x_1), x_3 - \pi\lambda x_2)$, the partial sum over the first kind of singular points of $\overline{g}(x) = 0$ in $\overline{E}$ gives $q(1)F_3$. Similarly the partial sum over the second kind of singular points becomes

$$q \cdot \sum_c \int_{O_K^{\times} \times O_k^3} | D(x_1, 3c + \pi x_2, 3c^2 + \pi x_3, c^3 + \pi x_4) |_K^s \, dx,$$

in which $\overline{c}^2 - \overline{\lambda}$ is in $(\mathbb{F}_q^{\times})^{[2]}$, hence it gives $(1) \cdot (q-1) / 2 \cdot F_3$. Therefore we get

$$2q^{-4} \cdot \sum_{a \in S} \int_{O_K^4} | g(x) |_K^s \, dx = (2) q^{-2} F_3.$$

by putting these together we get the expression for I_4' in Theorem 2.

We remark that we get the same expression for J_0, hence for I_4', also in the case where $q \equiv -1 \bmod 4$. In fact the difference may come only from the classification of singular points of $\overline{g}(x) = 0$ in $\overline{E}$. There are more singular points of the first kind, i.e., those defined by $a_2 = a_3 = 0$ and $a_4 = \lambda a_1$ for a_1 in $\mathbb{F}_q^{\times}$. On the other hand in the parameterization $c = (1/2)(\overline{\lambda} u + u^{-1})$ for the second kind of singular points we have to assume that $\overline{\lambda} u^2 + 1 \neq 0$. Therefore we have to add $q(1)F_3$ to the first partial sum while we have to subtract $q(1)F_3$ from the second partial sum.

10.8 Computation of I_4

If $\xi = \xi_4$, then T_ξ is defined by $x_{23} = x_{33} = y_{13} = y_{33} = 0$, hence we can take $x^* = (x_{23}e_{23} + x_{33}e_{33}, y_{13}e_{13} + y_{33}e_{33})$. Then $\phi(\xi + x^*) = (-x_{23}^2, x_{33}, y_{33} - y_{13}^2)$, hence

$$I_4 = T^4 \cdot J_{1,1}^*,$$

in which

$$\begin{aligned} f_{i,j}^*(x) &= D(\pi^i x_1^2, x_2, x_3, \pi^j x_4^2), f_{i,j}(x) = f_{i,j}^*(x_1, x_2, x_3), \\ J_{i,j}^* &= \textstyle\int_{O_K^4} | f_{i,j}^*(x) |_K^s \, dx, \quad J_{i,j}(x) = \int_{O_K^3} | f_{i,j}(x) |_K^s \, dx, \\ J_{i,j}' &= \textstyle\int_{O_K^2 \times O_K^{\times}} | f_{i,j}(x) |_K^s \, dx, j_i = j_{i,0} \end{aligned}$$

for all $i, j \geq 0$. We shall first show that J_0, J_1, $J_2, \ldots,$ are related as follows:

$$J_1 = (1)^2 (1,1)_+ / (1,1) + (1)(1,1)_+ q^{-2} t^2 + q^{-4} t^6 J_0,$$

$$J_2 = (1)^2 \{1 + (1) q^{-1} t^2 - q^{-3} t^3\} / (1,1)(2,2) + (1)(1,1)_+ q^{-2} t^3 + q^{-4} t^6 J_1,$$

$$J_1 = (1)^2 (2,3) / (1,1)(1,2) + (1) q^{-2} t^3 + q^{-3} t^6 J_{i-3}$$

$$+(1)^2(0,1)\,/\,(1,1)(1,2)\cdot\begin{cases} q^{-(i+1)/2}t^i & i=3,5,\dots \\ (1)\,/\,(2,2)\cdot q^{-i/2-1}t^{i+1} & i=4,6,\dots. \end{cases}$$

Since $f_{i,0}(x) \equiv x_2^2(x_3^2 - 4x_2) \bmod \pi$ for $i \geq 1$, we split O_K^2 in $O_K^3 = O_K \times O_K^2$ into

$$\begin{aligned} A_1 &= O_K \times \pi O_K, \\ A_2 &= \pi O_K \times O_K^x, \\ A_3 &= (O_K^x)^2 - A_4, \\ A_4 &= \{(x_2, x_3) \in (O_K^x)^2; 4x_2 \equiv x_3^2 \bmod \pi\} \end{aligned}$$

We apply SPF, if necessary, to the corresponding partial integrals, say M_1, M_2, M_3, M_4, J_i. We clearly have $M_3 = (1)(1-2q^{-1})$. If we put

$$g(x) = \pi^{-1}D(\pi^i x_1^2, (x_3\,/\,2)^2 + \pi x_2, x_3, 1),$$

then the coefficients of $g(x)$ are in O_K and $\partial g\,/\,\partial x_2 \equiv -x_3^4\,/\,4 \bmod \pi$, hence

$$M_4 = q^{-1}t\cdot\int_{O_K^2\times O_K^x} |\, g(x)\,|_K^s\, dx = (1)^2 q^{-1}t\,/\,(1,1).$$

As for M_1, M_2, by using SPF we get the following relations:

$$M_1 = (1)q^{-1} + \begin{cases} (1)q^{-1}t^2 + (1)q^{-3}t^3 + q^{-4}t^6 J_0 & i=1 \\ (1)q^{-2}t^3 + (1)q^{-3}t^4 + q^{-4}t^6 J_1 & i=2 \\ (1)q^{-2}t^3 + q^{-3}t^6 J_{i-3} & i\geq 3, \end{cases}$$

$$M_2 = \begin{cases} (1)^2 q^{-1}t + q^{-2}t^2 J'_{1,1} & i=1 \\ q^{-1}t^2 J'_{i-2,1} & i \geq 2. \end{cases}$$

Therefore we have only to show that

$$J'_{i,1} = (1)^2\,/\,(1,2) + (1)^2(0,1)\,/\,(1,1)(1,2)\cdot\begin{cases} q^{-(i+1)/2}t^i \\ (1)\,/\,(2,2)\cdot q^{-i/2-1}t^{i+1} \end{cases}$$

respectively for $i = 1,3,\dots$ and $i = 0,2,\dots$.

In order to prove the above formulas for $J'_{i,1}$, we show by an induction on j and by SPF that

$$J'_{i,1} = (1)^2\{1 + q^{-1}t^2 + \dots + (q^{-1}t^2)^{j-1}\} + (q^{-1}t^2)^j J'_{i-2j,j+1} \qquad (*)$$

for $0 \leq j \leq i/2$. If $i = 2k-1$ for some $k \geq 1$, we put $j = k-1$ in (*); then we get

$$J'_{i,1} = (1)^2(k-1, 2k-2)\,/\,(1,2) + (q^{-1}t^2)^{k-1}J'_{1,k}.$$

On the other hand by SPF we have

$$J'_{1,k} = (1)^2 + (1)^2 q^{-1}t + q^{-2}t^2 J'_{1,k+1}.$$

hence $J'_{1,k} = (1)^2 / (1,1)$. By putting these together we get the first formula for $J'_{i,1}$. If $i = 2(k-1)$ for some $k \geq 1$, we put $j = k-1$ in (*); then we get

$$J'_{i,1} = (1)^2(k-1, 2k-2) / (1,2) + (q^{-1}t^2)^{k-1} J'_{0,k}$$

We apply SPF to $J'_{0,k}$ and we get

$$J'_{0,k} = (1)^2 + (1)^3 q^{-1}t / (1,1) + q^{-2}t^2 J'_{0,k+1},$$

hence $J'_{0,k} = (1)^2(2,1) / (1,1)(2,2)$. Therefore we get the second formula for $J'_{i,1}$.

Before we proceed further we observe that since the relation between J_1 and J_0 has been established, the results of §7 are valid, hence $J_0 = F_4$, J_1, $J_2 \ldots$ are all known. We further observe that

$$\int_{O_K^2 \times O_K^\times \times O_K} | f^*_{i,1}(x) |_K^s \, dx = J'_{i,1}.$$

In fact the integral on the LHS can be computed exactly in the same way as $J'_{i,1}$.

If now $i \geq 1$, then by splitting O_K^4 as $x_3 \not\equiv 0$ and $x_3 \equiv 0$ and π and by SPF we get

$$J^*_{i,1} = J'_{i,1} + (1)^2 q^{-1}t + (1)t^4 J_{i-1} + q^{-2}t^2 J^*_{i+1,1},$$

hence

$$J^*_{1,1} = (1)^2 q^{-1}t / (2,2) + \sum_{i\geq 0} (q^{-1}t)^{2i} J'_{i+1,1} + (1)q^{-2}t^4 \cdot \sum_{i \geq 0} (q^{-1}t)^{2i} J_i.$$

If we incorporate the above expressions for $J'_{i,1}$ and J_i, we get

$$\sum_{i\geq 0} (q^{-1}t)^{2i} J'_{i+1,1} = (1)^2(1 - q^{-3}t^3 + q^{-3}t^4 - q^{-5}t^5) / (1,1)(2,2)(5,6)$$

and

$$\sum_{i\geq 0} (q^{-1}t)^{2i} J_i = S_0 + S_1 + S_2,$$

in which $(9,12)(q^{-1}t)^{-2i} S_i = J_i$

$$+(q^{-1}t)^6 / (6,6) \cdot \{(1)^2(2,3) / (1,1)(1,2) + (1)q^{-2}t^3\}$$

$$+(1)^2(0,1) / (1,1)(1,2)(15,18) \begin{cases} q^{-8}t^9\{1 + (1)q^8t^{10} / (2,2)\} & i = 0 \\ q^{-9}t^{11}\{q^{-7}t^8 + (1) / (2,2)\} & i = 1 \\ q^{-9}t^{11}\{1 + (1)q^{-8}t^{10} / (2,2)\} & i = 2. \end{cases}$$

By putting all these together we get the expression for I_4 in Theorem 2.

10.9 Computation of I_5

If $\xi = \xi_5$, then T_ξ is defined by $x_{13} - y_{23} = x_{22} = x_{23} = x_{33} = y_{33} = 0$, hence we can take $x^* = (x_{13}e_{13} + x_{22}e_{22} + x_{23}e_{23} + x_{33}e_{33}, y_{33}e_{33})$. Then $\phi(\xi + x^*) = (x_{22}x_{33} - x_{23}^2 - x_{13}^2 x_{22}, x_{22}y_{33} + 2x_{13}x_{23}, -x_{33}, -y_{33})$, hence

$$I_5 = t^4 \cdot \int_{O_K^5} \mid D(x_1, x_2, \pi(-x_1x_5 + 2x_3x_4), \pi(-x_2x_5 + x_3^2 + \pi x_4^2 x_5)) \mid_K^s dx.$$

In general if

$$\begin{aligned} f(x) &= \pi^{j_1} x_2^2 y_3^2 - 4\pi^{j_2} x_1 y_3^3 - 4\pi^{j_3} x_2^3 y_4 + 18\pi^{j_4} x_1 x_2 y_3 y_4 - 27\pi^{j_5} x_1^2 y_4^2, \\ y_3 &= -\pi^{k_1} x_1 x_5 + 2\pi^{k_2} x_3 x_4, y_4 = -\pi^{k_3} x_2 x_5 + \pi^{k_4} x_3^2 + \pi^{k_5} x_4^2 x_5, \end{aligned}$$

and if E is any compact open subset of O_K^5, then we shall write

$$\int_E \mid f(x) \mid_K^s dx = \int_E (j_1, j_2, j_3, j_4, j_5; k_1, k_2, k_3, k_4, k_5).$$

If we put

$$I = \int_{O_K^5} (1,2,0,1,1;0,0,0,0,1), \quad J = \int_{O_K^5} (1,2,0,1,1;0,0,0,1,0),$$

then $I_5 = t^5 I$; we shall obtain two linear equations for I, J involving some known integrals and the following:

$$\begin{aligned} &\textstyle\int_{O_K^4 \times O_K^\times} (0,0,0,0,0;\ 0,0,0,0,0) \\ &= q^2 t^4 \{(1)1 \,/\, 2 \cdot t^{-4}(I_4 + I_4') - (1)^2(2,1) \,/\, (1,1)\} \end{aligned} \qquad (F_5)$$

$$\begin{aligned} &\textstyle\int_{O_K^4 \times O_K^\times} (2,1,2,1,0;\ 0,0,1,1,0) \\ &= (1)^2\{(1)+q^{-2}t-q^{-2}t^2+(1)q^{-2}t^3\} \,/\, (1,1)(3,4)+(1)^2 q^{-1}t^2 \,/\, (3,4) \cdot F_2, \end{aligned} \qquad (F_6)$$

$$\int_{O_K \times O_K^\times \times O_K^3} (1,2,0,1,1;\ 0,0,0,1,0) = (1)^3 \,/\, (1,1)^2, \qquad (F_7)$$

$$\int_{O_K^\times \times O_K^3 \times O_K^\times} (0,0,0,0,0;\ 0,1,0,1,1) = (1)^3 \,/\, (1,1). \qquad (F_8)$$

We shall postpone their verification and obtain linear equations for I and J.

We apply SPF twice to I and we get

$$\begin{aligned} I = \ & (1)^2 \,/\, (1,1) \cdot (1 + q^{-1}t - q^{-2}t^2) + (1)q^{-2}t^3 F_2 \\ & + q^{-2}t^2 \cdot \textstyle\int_{O_K^5} (1,0,2,1,1;\ 0,1,0,1,0). \end{aligned}$$

We apply SPF to the above integral and split one of the integrals into two as $x_5 \not\equiv 0$ and $x_5 \equiv 0 \bmod \pi$; then it becomes

$$(1)^2 + q^{-1}t^3F_5 + q^{-1}t^2 \cdot \int_{O_K^4 \times O_K^\times}(1,2,0,1,1;\ \ 0,0,0,1,0) \\ +q^{-2}t^3 \cdot \int_{O_K^5}(0,0,0,0,0;\ \ 1,0,1,0,1).$$

We apply SPF to the first integral; then it becomes $(1)^3 \,/\, (1,1) + q^{-1}tF_6$. We split the second integral into two as $x_3 \not\equiv 0$ and $x_3 \equiv 0 \bmod \pi$; then it becomes $(1)F_2 + q^{-1}tJ$. By putting these together we get

$$I - q^{-5}t^6J = (1)^2\{1 + q^{-1}t - q^{-3}t^3 + (1)q^{-3}t^4\} \,/\, (1,1) \\ +(1)(2,2)_+q^{-2}t^3F_2 + q^{-3}t^5F_5 + q^{-4}t^5F_6.$$

We split the integral J into $M_1 + M_2$, where M_1 is over x_1, $x_2 \equiv 0 \bmod \pi$ and M_2 is over the remaining part. After replacing x_1, x_2 by πx_1, πx_2, we split the integral M_1 into two as $x_4 \not\equiv 0$ and $x_4 \equiv 0 \bmod \pi$; then we get

$$M_1 = (1)q^{-2}t^3F_1 + q^{-3}t^4I.$$

If we split the integral M_2 into two as $x_2 \not\equiv 0$ and $x_2 \equiv 0$, $x_1 \not\equiv 0 \bmod \pi$, then we get

$$M_2 = F_7 + q^{-1}t \cdot \int_{O_K^\times \times O_K^4}(2,1,2,1,0;\ \ 0,0,1,1,0).$$

We split the above integral into three as x_4, $x_5 \not\equiv 0 \bmod \pi$, $x_5 \not\equiv 0$, $x_4 \equiv 0 \bmod \pi$ and $x_5 \equiv 0 \bmod \pi$; then it becomes

$$(1)^3 + (1)^2q^{-1}t + q^{-1}t \cdot \int_{O_K^\times \times O_K^4}(1,0,2,1,1,;\ \ 1,0,1,0,0).$$

We apply SPF twice to the above integral; then it becomes

$$(1)^3{+}(1)^2q^{-1}t{+}(1)^3q^{-1}t{+}(1)^2q^{-2}t^3F_2{+}q^{-2}t^3{\cdot}\int_{O_K^\times \times O_K^4}(0,0,0,0,0;\ \ 0,1,0,1,1).$$

We split the above integral into two as $x_5 \not\equiv 0$ and $x_5 \equiv 0 \bmod \pi$, and apply SPF twice to the second integral; then it becomes

$$F_8{+}(1)^3q^{-1}t \,/\, (1,1){+}(1)^2q^{-2}t^2{+}q^{-3}t^3{\cdot}\{\int_{O_K^5}(1,0,2,1,1;0,1,0,1,0){-}q^{-1}t^2J\}.$$

We have seen that the above integral is equal to

$$(1)^2\{1{-}q^{-1}t{+}(1)q^{-1}t^2\} \,/\, (1,1){+}(1)q^{-2}t^3F_2{+}q^{-1}t^3F_5{+}q^{-2}t^3F_6{+}q^{-3}t^4J.$$

Therefore we get

$$(1+(2,2)q^{-8}t^{10})J - q^{-3}t^4 I = (1)^2 q^{-1}t\{(1)+q^{-1}t-(2-q^{-1})q^{-3}t^3 \\ +(1)q^{-4}t^5+q^{-5}t^6-q^{-7}t^8+(1)q^{-7}t^9\}(1,1)+(1)q^{-2}t^3F_1+(1)((1) \\ +q^{-5}t^6)q^{-4}t^5F_2+q^{-8}t^{11}F_5+q^{-9}t^{11}F_6+F_7+q^{-4}t^5F_8.$$

If we solve the above two linear equations for I, J with known coefficients, we get the expression for $I_5 = t^5 I$ in Theorem 2.

8-bis. We apply SPF to the integral F_6; then it becomes

$$(1)^3+(1)^2q^{-1}t+q^{-1}t^2\cdot\int_A(0,0,0,0,0;\ \ 1,0,1,1,0),$$

in which $A = O_K^4 \times O_K^x$. We split this integral into two as $x_4 \not\equiv 0$ and $x_4 \equiv 0$ mod π, and apply SPF twice to the second integral; then it becomes

$$(1)^3q^{-1}t(1+q^{-1}t-q^{-2}t^2)\,/\,(1,1)+(1)^2(3,4)_+F_2 \\ +q^{-3}t^3\cdot\textstyle\int_A(1,0,2,1,1;\ \ 0,1,0,1,0).$$

We apply SPF twice to this integral; then it becomes $(1)^2+(1)^3q^{-1}t^2\,/\,(1,1)+$ $q^{-2}t^3F_6$. By putting these together we get the expression for f_6.

We apply SPF to the integral F_7; then it becomes

$$(1)^3(1,1)_+\,/\,(1,1)+q^{-2}t^2\cdot\int_A |\, f_0(x)\,|_K^s\, dx,$$

in which $A = O_K^3 \times O_K^x \times O_K$ and $f_0(x) \equiv 4x_4^3(x_1x_3 - x_2x_4)(x_3^2 - x_4^2x_5)$ mod π. We keep on applying SPF and we get

$$\int_A |\, f_i(x)\,|_K^s\, dx = (1)^3(1,1)_+\,/\,(1,1)+q^{-2}t^2\cdot\int_A |\, f_{i+1}(x)\,|_K^s\, dx,$$

in which $f_i(x) \equiv 4x_2x_4^3x_5$ mod π for every $i \geq 1$, hence $F_7 = (1)^3\,/\,(1,1)^2$.

As for F_8, by restricting x_5 to $\lambda(O_K^x)^{[2]}$ or to $(O_K^x)^{[2]}$ and by applying SPF to the second integral we get $(1)^2\,/\,2$ or

$$\tfrac{1}{2}\cdot\textstyle\int_A |\, D(x_1,x_2,-x_1x_5^2+2\pi x_3x_4,-x_2x_5^2+\pi(x_3^2+x_4^2x_5))\,|_K^s\, dx \\ =\tfrac{1}{2}\cdot(1)^2(1-2q^{-1})+q^{-1}t\cdot\textstyle\int_A |\, f_0(x)\,|_K^s\ dx,$$

in which $A = O_K^x \times O_K^3 \times O_K^x$ and $f_0(x) \equiv 2^5x_1^3x_5^3(x_3 - x_4x_5)^2$ mod π. We keep on applying SPF; we have

$$\int_A |\, f_0(x)\,|_K^s\, dx = (1)^3+q^{-1}t\cdot\int_A |\, f_1(x)\,|_I^s\, dx,$$

in which $f_1(x) \equiv 2^4x_1^2x_5^4(x_2 - x_4^2)^2$ mod π, and

$$\int_A |\, f_i(x)\,|_K^s\, dx = (1)^3+q^{-1}t\cdot\int_A |\, f_{i+1}(x)\,|_K^s\, dx$$

with $f_{2j}(x) \equiv 2^5 x_1^3 x_3^2 x_5^3$ and $f_{2j+1}(x) \equiv 2^4 x_1^2 x_2^2 x_5^4 \bmod \pi$ for every $j \geq 1$. Therefore the LHS is equal to $(1)^3 / (1,1)$, hence $F_8 = (1)^3 / (1,1)$.

Finally if we put

$$J_\pm = \int_{O_K^4} | D(x_1, x_2, -ux_1 + 2x_3x_4, -ux_2 + x_3^2 + ux_4^2) |_K^s \, dx$$

for $u = 1$, λ, then clearly we have

$$F_5 = (1) / 2 \cdot (J_+ - J_-), t^{-4} I_4' = (2) + q^{-2} t^4 J_-.$$

On the other hand if we split 0_K^2 as $x_1^2 \not\equiv x_2^2$, $x_1 \equiv x_2 \equiv 0 \bmod \pi$ and the remaining part, then we get

$$\begin{aligned} &\textstyle\int_{O_K^4} | D(x_1, x_2, -x_1 + 2\pi x_3 x_4, -x_2 + \pi(x_3^2 + x_4^2)) |_K^s \, dx \\ &= (1)^2 + q^{-2} t^4 J_+ + 2q^{-1} t \cdot \textstyle\int_A | f_0(x) |_K^s \, dx, \end{aligned}$$

in which $A = O_K^x \times O_K^3$ and $f_0(x) \equiv 2^5 x_1^3 (x_3 - x_4)^2 \bmod \pi$. This integral can be computed in the same way as before. We have

$$\int_A | f_i(x) |_K^s \, dx = (1)^2 + q^{-1} t \cdot \int_A | f_{i+1}(x) |_K^s \, dx,$$

in which $f_1(x) \equiv 2^4 x_1^2 (x_2 - x_4^2)^2$, $f_{2j}(x) \equiv 2^5 x_1^3 x_3^2$ and $f_{2j+1}(x) \equiv 2^4 x_1^2 x_2^2 \bmod \pi$ for every $j \geq 1$. Therefore we get

$$t^{-4} I_4 = (1)^2 + q^{-2} t^4 J_+ + 2q^{-1} t (1)^2 / (1,1).$$

The expression for F_5 follows from these.

10.10 References

[1] J. Igusa, On the arithmetic of a singular invariant, *Amer. J. Math.*, **110**, 1988, pp. 197–233.

[2] J. Ingusa, A problem on certain P-adic zeta functions, to appear in a volume dedicated to I.I. Piatetski-Shapiro.

[3] T. Kimura, The b-functions and holonomy diagrams of irreducible regular prehomogeneous vector spaces, *Nagoya Math. J.*, **85**, 1982, pp. 1–80.

[4] M. Sato and T. Kimura, A classification of irreducible prehomogeneous vector spaces and their relative invariants, *Nagoya Math. J.*, **65**, 1977, pp. 1–155.

[5] A. Seil, Adeles and Algebraic Groups, Institute for Advanced Study, Princeton, 1962; *Progress in Math.*, **23**, Birkhäuser, 1982.

Part V

Groups and Coverings

11

The $\mathbf{Q}$-admissibility of $2A_6$ and $2A_7$

Walter Feit

11.1 Introduction

Let L be a finite extension of the field K. Then L is K-adequate if L is a maximal subfield of a division algebra with center K.

A finite group G is K – *admissible* if $G = \mathrm{Gal}(L/K)$ for some Galois extension L of K which is K-adequate.

These concepts were first introduced by Schacher [5], who also gave a characterization of K-adequate extensions of number fields. This in particular implies that if G is $\mathbf{Q}$-adequate then every Sylow group of G is meta-cyclic. The converse to this is still open. For instance it is not known whether $PSL_2(p)$ is $\mathbf{Q}$-adequate for any prime $p \geq 7$.

The only alternating groups A_n with meta-cyclic Sylow groups are those for $n \leq 7$. In these cases the double covers $2A_n$ also have meta-cyclic Sylow groups. The groups A_5 and $2A_5$ were handled by Sonn [6]. Actually much more is now known in this case, see [3]. The $\mathbf{Q}$-admissibility of A_6, A_7 and $2A_7$ was handled in an earlier paper, [J. Number Theory 26 (1987), 210-226]. Unfortunately there is an error in this paper which invalidates the results there. A few years ago I was able to show that A_6 is $\mathbf{Q}$-adequate. See [2], Section 5. This paper contains the following Theorem which subsumes these earlier results.

Theorem A *A_n and $2A_n$ are $\mathbf{Q}$-admissible for $n = 6$ or 7.*

The proof consists of constructing an explicit polynomial (see (3.23)) and applying several recent results (see Section 12.4), especially a Theorem of Mestre [3] as well as Serre's Theorem [7]. Independently of this Fein and Schacher [1] have used similar methods to show that A_6 and A_7 are $\mathbf{Q}$-admissible. In fact they have shown that there exist infinitely many $\mathbf{Q}$-adequate fields L with $\mathrm{Gal}(L/\mathbf{Q}) = A_n$ for $n = 6$ or 7. By looking more closely at the Hilbert irreducibility theorem applied to $F_T(x)$ (see (4.1)) it should be possible to prove an analogous result for $2A_n$ for $n = 6$ or 7.

11.2 Notation

The notation in this paper is standard but we list some of it here to avoid confusion.

If p is a prime and $a \in \mathbf{Z}, a \neq 0$ then $\nu_p(a)$ is the exact power of p which divides a. If $a, b \in \mathbf{Z}, ab \neq 0$ then $\nu_p(a/b) = \nu_p(a) - \nu_p(b)$.

Let $f(x)$ be a monic polynomial in $F[x]$ with distinct roots, where F is a field. Define $Tr_f(\alpha)$ to be the trace of α in $F[x]/(f(x))$. Then $Tr_f(\alpha^2)$ is a nondegenerate quadratic form over $\mathbf{Q}$. Let $w_p(f)$ be the Witt invariant of this quadratic form over $\mathbf{Q}_p$ for $p = \infty$ or p a prime.

Let $D(f)$ denote the discriminant of the polynomial f.

If $a, b \in \mathbf{Q}^\times$ write $a \sim b$ if $a = bc^2$ for some $c \in \mathbf{Q}$.

11.3 Some Polynomials

In this section some polynomials are constructed which will be needed below.

For $u, t \in \mathbf{Q}^\times$ define

$$f(x : u, t) = x^4 - tx^3 - utx^2 + ut \in \mathbf{Q}[x]. \tag{3.1}$$

Lemma 3.2 *Choose u, t so that $f(x : u, t)$ has distinct roots. Then*

$$w_p(f) = (X, -2tY)_p(t, 2Y)_p \tag{3.3}$$

for all p, where

$$X(t, u) = X = -8u + 2tu^2 - 9t, \tag{3.4}$$

$$Y(t, u) = Y = 16t^2u^3 + 4t^3u^2 - 128tu^2 - 144t^2u - 27t^3 + 256u, \tag{3.5}$$

and

$$D(f) \sim tY. \tag{3.6}$$

Proof Let $y = \frac{1}{x}, T = \frac{1}{t}, U = \frac{1}{u}$. Then

$$h(y) = \frac{f(x : u, t)}{x^4ut} = y^4 - y^2 - Uy + TU. \tag{3.7}$$

Thus $w_p(f) = w_p(h)$ for all p and $D(h) \sim D(f)$.

Let s_j denote the sum of the jth powers of the roots of $h(y)$. By Newton's formulas

$$(s_{i+j-2}) = \begin{pmatrix} 4 & 0 & 2 \\ 0 & 2 & 3U \\ 2 & 3U & -4TU + 2 \end{pmatrix}. \tag{3.8}$$

for $1 \leq i, j \leq 3$. The formula for the discriminant of a quartic yields that

$$\begin{aligned} D(h) &= 16TU + 4U^2 - 128T^2U^2 - 144U^3 - 27U^4 + 256T^3U^3 \\ &= \tfrac{1}{t^3u^4} Y \sim tY. \end{aligned} \tag{3.9}$$

This proves (3.6).

The principal minors of (3.8) are a_1, a_2, a_3 with $a_1 \sim 1, a_2 \sim 2$ and

$$a_3 \sim -8TU + 2 - 9U^2 \sim tX. \tag{3.10}$$

Thus Tr_f is equivalent to the diagonal form $\{1, 2, 2a_3, a_3 D(f)\}$ which by (3.9) and (3.10) is equivalent to the diagonal form $\{1, 2, 2tX, XY\}$. Let $w(f) = w_p(f)$. Then

$$\begin{aligned} w(f) &= (2, 2tY)(2tX, XY) = (2, t)(2, Y)(2t, XY)(X, -Y) \\ &= (2, t)(2, Y)(2, XY)(t, XY)(X, -Y) \\ &= (X, -2tY)(t, 2Y) \end{aligned}$$

as required. □

For $a, b, m \in \mathbf{Q}^\times$ define

$$g(x : a, b, m) = x^3 - amx + bm. \tag{3.12}$$

Lemma 3.13 *Suppose that $g(x : a, b, m)$ has distinct roots. Then*

$$w_p(g) = (2, D(g))_p \tag{3.14}$$

for all p and

$$D(g) \sim 4a^3m - 27b^2. \tag{3.15}$$

Proof The formula for the discriminant of a cubic yields (3.15). For $\epsilon = 1$, condition (ii) of Serre's Theorem [3] Theorem 3.2 is always satisfied. Thus (3.14) follows from condition (i). □

Define

$$P(x : u, t, a, b, m) = f(x : u, t)g(x : a, b, m). \tag{3.16}$$

Lemma 3.17 *$D(P) \sim D(f)D(g)$ and*

$$w_p(P) = w_p(f)w_p(g)(D(f), D(g))_p$$

for every p.

Proof This follows directly from the definitions. □

Let $Y_0 = Y(161, 5)$ be defined by (3.5). Then by direct computation

$$Y_0 = 337,314,473 \quad \text{is aprime.} \tag{3.18}$$

Define

$$f(x) = f(x : 161, 5) = x^4 - 161x^3 - 805x^2 + 805 \tag{3.19}$$

$$g(x) = g(x : 9, 18, m_0) = x^3 - 9m_0x + 18m_0, \tag{3.20}$$

where

$$m_0 = 161Y_0 + 3 = 54,307,630,156. \tag{3.21}$$

The prime factorization fo m_0 is

$$m_0 = 2^2.13^2.19.37.114,277. \tag{3.22}$$

Let

$$P(x) = f(x)g(x). \tag{3.23}$$

Lemma 3.24 *$D(P) \sim 1$ and $w_p(P) = 1$ for all p.*

Proof By (3.6) and (3.15)

$$D(g) \sim m_0 - 3 \sim 161Y_0 \sim D(f). \tag{3.25}$$

Thus $D(P) \sim 1$. By (3.18) $(-1, Y_0)_p = 1$. Hence (3.25) implies that

$$(D(f), D(g))_p = (161Y_0, 161Y_0)_p = (-1, 161Y_0)_p = (-1, 161)_p. \tag{3.26}$$

By (3.4) $X_0 = X(161, 5) = 3^8$. By (3.3), (3.14), (3.26) and Lemma (3.17)

$$w_p(P) = (161, 2Y_0)_p(2, 161Y_0)_p(-1, 161)_p. \tag{3.27}$$

Thus $w_\infty(P) = 1$. Since $Y_0 \equiv -8 (\text{mod } 161)$, (3.27) implies that

$$w_p(P) = (161, 8)_p = (\frac{2}{p}) = 1,$$

for $p = 7$ or 23. If $p = Y_0$ then by (3.18) and (3.27)

$$w_p(P) = (\frac{2.7.23}{p}) = (\frac{p}{7})(\frac{p}{23}) = (-1)^2 = 1.$$

Hence $w_p(P) = 1$ for $p = \infty$ or an odd prime and so $w_p(P) = 1$ for all p by the product formula.□

Lemma 3.28 *Let $c = 7^2.23^2.13^3 19^2$. Let $s \in \mathbf{Z}$ and let $Q_0(x)$ be a polynomial of degree at most 6 in $\mathbf{Z}[x]$. Let $h(x) = P(x) - csQ_0(x)$.*

(i) If $p = 13$ or 19 then $h(x)$ has a root in $\mathbf{Q}_p$ and the Galois group of $h(x)$ over $\mathbf{Q}_p$ contains a noncyclic group of order 9.

(ii) If $p = 7$ or 23 then $h(x)$ has a root in $\mathbf{Q}_p$ and the Galois group of $h(x)$ over $\mathbf{Q}_p$ contains a dihedral group of order 8.

Proof (i) By (3.22)

$$h(x) \equiv x^3 f(x) (\text{mod } p).$$

Direct computation shows that

$$\begin{aligned} f(x) &\equiv (x-6)(x^3 - 3x^2 - 6x + 2)(\text{mod } 19) \\ f(x) &\equiv (x+5)(x^3 + 3x^2 - x + 5)(\text{mod } 13). \end{aligned}$$

Thus by Hensel's Lemma $h(x)$ has a root in $\mathbf{Q}_p$. Furthermore a splitting field L of $f(x)$ contains an unramified subfield of degree 3. Thus it suffices to show that 3 divides the ramification index of L.

Let $h(x) = \sum_0^7 a_i x^i$. By (3.19), (3.20) and (3.22) we see that

$$a_0 \equiv a_1 \equiv a_2 \equiv 0 (\text{mod } p)$$

and $\nu_p(a_3) = 0$. Furthermore $\nu_{13}(a_0) = 2$ and $\nu_{19}(a_0) = 1$. Thus by Newton's polygons we see that 3 divides the ramification index of L.

(ii) By (3.21) $m_0 \equiv 3 (\text{mod } p)$. Thus by (3.19) and (3.20)

$$h(x) \equiv x^4 (x^3 - 27x + 54) \quad (\text{mod } p).$$

Hence

$$\begin{aligned} h(x) &\equiv x^4 (x+6)(x-3)^2 \quad (\text{mod } 23) \\ &\text{and} \\ h(x) &\equiv x^4 (x-1)(x-3)^2 \quad (\text{mod } 23) \end{aligned}$$

Thus by Hensel's Lemma $h(x)$ has a root in $\mathbf{Q}_p$. Furthermore

$$h(x) = h_3(x) h_4(x)$$

where $h_4(x) = \sum_0^4 b_i x^i$ with $b_4 = 1$,

$$b_0 \equiv b_1 \equiv b_2 \equiv b_3 \equiv 0 \quad (\text{mod } p)$$

and $\nu_p(b_0) = 1$. Thus by Newton's polygons we see that 4 divides the ramification index of L. Let $H = \text{Gal}(L/\mathbf{Q}_p)$. Then $H \subseteq \Sigma_4$ and H is meta-cyclic since L is tamely ramified. Furthermore H contains an element of order 4. Thus $|H| = 4$ or 8. Since $p \equiv 3 \pmod 4$ there is no ramified cyclic extension of $\mathbf{Q}_p$ of order 4. Hence $|H| = 8$ and so H is a Sylow 2-group of Σ_4. Thus H is dihedral. □

11.4 The Proof of Theorem A

In this section we will need a variety of results. These are all listed in [3] and so it will be convenient to refer to them there.

Let $P(x)$ be defined by (3.23). Let $Q(x)$ be the polynomial defined by Mestre [3] Theorem 4.1. Let c be as in Lemma 3.28 and let

$$F_T(x) = P(x) - cTQ(x) \in \mathbf{Q}[T][x]. \tag{4.1}$$

Let $G = \mathrm{Gal}(L/\mathbf{Q}(T))$ where L is a splitting field of $F_T(x)$. By [3] Theorem 4.1 $G \subseteq A_7$.

Lemma 4.2 *$F_T(x)$ is irreducible in $\mathbf{Q}[T][x]$ and $G \simeq A_7$.*

Proof By Lemma 3.28 $72 \mid |G|$. Thus either $G = A_7$ or $G \simeq A_6$. Hence if the result is false $F_T(x)$ has a root in $Q[T]$ and so $P(x)$ has a root in $\mathbf{Q}$. However this contradicts Lemma 3.28. □

By the Hilbert irreducibility theorem there exist an integer t (actually infinitely many integers t) such that $F_t(x) = P(x) - ctQ(x)$ is irreducible in $\mathbf{Q}$. By Lemma 3.28 and Schacher's criterion, [3] Theorem 2.1, the splitting field L_0 of $F_t(x)$ is $\mathbf{Q}$-adequate. By Mestre's result [3] Theorem 4.1, $w_p(F_t(x)) = 1$ for all p. Hence by Serre's Theorem, [3] 3.2 there exists a field M with $L_0 \subseteq M$ and $\mathrm{Gal}(M/Q_0) \simeq 2A_7$. By [3] Corollary 2.3 $2A_7$ is $\mathbf{Q}$-admissible.

By [3] Lemma 6.6 A_6 and $2A_6$ are also $\mathbf{Q}$-admissible.

11.5 References

[1] B. Fein and M. Schacher, $\mathbf{Q}$-admissibility questions for alternating groups, *J. Alg.*

[2] B. Fein and M. Schacher, Minimality questions for simple algebras, *Ring Theory 1989, Israel Math. Soc. Conf. Proc.* 1989, pp. 280–292.

[3] P. Feit and W. Feit, The K-admissibility of $SL(2,5)$, *Geometriae Dedicata*, **36**, 1990, pp. 1–13.

[4] J.-F. Mestre, Extensions reguliéres de $\mathbf{Q}(T)$ de groupe de Galois $\tilde{A}_n$, *J. Algebra*, **13**, 1990, pp. 483–496.

[5] M. Schacher, Subfields of division rings, *J. Alg.*, **9**, 1968, pp. 451–477.

[6] J. Sonn, $SL(2,5)$ and Frobenius Galois groups over $\mathbf{Q}$, *Canad. J. Math.*, **32**, 1980, pp. 281–293.

[7] J.-P. Serre, L'invariant de Witt de la forme $Tr(x^2)$, *Comment. Math. Helv.*, **59**, 1984, pp. 651–676.

12

Groups Which Cannot be Realized as Fundamental Groups of the Complements to Hypersurfaces in C^N

Anatoly S. Libgober

12.1 Introduction

Finding restrictions imposed on a group by the fact that it can appear as a fundamental group of a smooth algebraic variety is an important problem particularly attributed to J.P. Serre. It has rather different aspects in characteristic p and zero and here we will address exclusively the latter case. Most restrictions described in the literature seem to rely on Hodge theory or some clever use of it (cf. [2]). A prototype of such restrictions is evenness of $rk(\pi_1 / \pi_1' \otimes \mathbf{Q})$ where π_1' is the commutator subgroup of the fundamental group π_1. In the case of open non-singular varieties one can apply mixed Hodge theory. This was done by J. Morgan who obtained restrictions on the nilpotent quotients of the fundamental groups [13]. Here I shall describe a different (but also by no means complete) type of restriction on the fundamental groups of open varieties which are complements to hypersurfaces in $\mathbf{C}^n$. It is implicitly contained in previous work on Alexander polynomial of plane curves [4]. For example many knot groups cannot occur as fundamental groups of the complement to an algebraic curve. This gives automatically the same restrictions on the fundamental groups of complements to arbitrary hypersurfaces in $\mathbf{C}^n$ as follows from the well-known argument using Zariski Lefschetz type theorem: For a generic plane H relative to given hypersurface V in $\mathbf{C}^n$ the natural map $\pi_1(H - H \cap V,\ p_0) \to \pi_1(\mathbf{C}^n - V, p_0)(p_0 \in H)$ is an isomorphism, i.e., possible fundamental groups of the complement to hypersurfaces in $\mathbf{C}^n$ are precisely the fundamental groups of the complements to plane curves. Therefore from now on I shall work with plane curves only.

12.2 Alexander Polynomials of Plane Curves

So let C be such a curve having, say, μ components. Let μ denote the number of irreducible components of such a curve C and d be its degree. Then according to van Kampen's theorem [10] $\pi_1(\mathbf{C}^2 - C)$ is generated by d generators belonging to a generic line L which are images of the standard generators of $\pi_1(L - L \cap C)$ (the latter is a free group on d generators). Here the standard generator is a loop which is formed by a path leading from a base point to a point near $C \cap L$, then going once around the nearby poaint of $C \cap L$ and then returning back to the base point traversing the old path in the opposite direction. Generators corresponding to two points of $C \cap L$ belonging to the same irreducible component of C are conjugate in $\pi_1(\mathbf{C}^2 - C)$. The conjugating loop is composed of the path leading from the base point to the vicinity of the first point, a path leading from the vacinity of the first point to the vicinity of the second point which is a pushout into $\mathbf{C}^2 - C$ of a path in the locus of non-singular points of C connecting the first and second points and then returning to the base point along the path comprising the standard generator corresponding to the second point.

In particular this implies that $H_1(\mathbf{C}^2 - C, \mathbf{Z})$ is generated by at most μ generators. On the other hand, linking coefficients with each irreducible component of C provide surjection of this group on $\mathbf{Z}^\mu$. Hence $H_1(\mathbf{C}^2 - C, \mathbf{Z}) = \mathbf{Z}^\mu$ and consequently $\pi_1(\mathbf{C}^2 - \mathbf{C})$ is a normal closure of its μ elements where $\mu = rk\pi_1 / \pi_1'$. This simple (topological) condition prohibits some groups (like free products of $\mathbf{Z}$ with perfect groups) from being fundamental groups of the complement to plane curves.

To state an algebro-geometric restriction, let us consider the homomorphism $\phi : H_1(\mathbf{C}^2 - C) \to \mathbf{Z}$ given by the total linking number with C and the infinite cyclic cover determined by ϕ. Denote it $(\widetilde{\mathbf{C}^2 - C})_\phi$. Then $H_1(\widetilde{\mathbf{C}^2 - \mathbf{C}})_\phi, \mathbf{Q}$ is a module over the group ring of $\mathbf{Z}$ over $\mathbf{Q}$, i.e., over $![t, t^{-1}]$, which as it turns out is a torsion module over this ring (cf. [4]). Hence it is isomorphic to $\bigoplus \mathbf{Q}[t, t^{-1}] / (\lambda_i)$ for some Laurent polynomials λ_i determined up to a unit of $\mathbf{Q}[t, t^{-1}]$. The order $\Delta = \Pi\lambda_i$ of $H_1(\widetilde{\mathbf{C}^2 - \mathbf{C}})_\phi, \mathbf{Q}$ is called the Alexander polynomial of curve C. The special property which is satisfied by the fundamental groups of the complement to plane curves is contained in the following theorem:

Theorem. *The Alexander polynomial Δ of a curve is cyclotomic. In fact Δ divides the product of the characteristic polynomials of monodromies of all singular points of the union of C and the line in infinity.*

In the case when curve C is transversal to the line in infinity one can give a very simple proof of the cyclotomic property of Δ (in this restricted case Kohno's algebraic description of the Alexander polynomial [10] or Randell's interpretation [8] of it as the characteristic polynomial of the monodromy operator of the 2-dimensional singularity defined by the equation of the curve can be used). In fact one has the following:

Proposition

If curve C is transversal to the line in infinity then the automorphism of $H_1(\widetilde{(\mathbf{C}^2 - C)}_\phi, \mathbf{Q})$ induced by the deck transformation of the infinite cyclic cover has a finite order dividing the order of the curve C.

To see this, let us put $\pi_{1,a} = \pi_1(\mathbf{C}^2 - C)$ and $\pi_1 = \pi_1(\mathbf{CP}^2 - \overline{C})$ where $\overline{C}$ is the projective completion of C in $\mathbf{CP}^2$. Then one has the following diagram:

$$\begin{array}{ccccccccc} 0 & \to & \pi_1^\phi & \to & \pi_1 & \to & \mathbf{Z}/d\mathbf{Z} & \to & 0 \\ & & \uparrow & & \uparrow & & \uparrow & & \\ 0 & \to & \pi_{1,a}^\phi & \to & \pi_{1,a} & \to & \mathbf{Z} & \to & 0 \end{array}$$

The surjections on the right ends of each row are homomorphisms of the fundamental group on H_1 for $\mathbf{CP}^2 - \overline{C}$ and $\mathbf{C}^2$ respectively. The groups on the left side are the kernels of these surjections. The middle vertical arrow is induced by inclusion and is a surjection because any loop in $\mathbf{CP}^2 - \overline{C}$ can be moved to miss the line in infinity. π_1^ϕ and $\pi_{1,a}^\phi$ are the fundamental groups of the cyclic coverings having as the Galois groups $\mathbf{Z}/d\mathbf{Z}$ and $\mathbf{Z}$ respectively. The action of the Galois group on the homology of each of these cyclic coverings coincides with the action induced by each of these two sequences on the abelianization of the left terms. We will need the following (cf. [11], p. 509):

Lemma *If C is transversal to the line in infinity then* $\mathrm{Ker}(\pi_{1,a} \to \pi_1) = \mathbf{Z}$ *and belongs to the center of* $\pi_{1,a}$.

We shall postpone the proof for a moment. This lemma implies that the left arrow in the diagram above is an isomorphism because the kernel of the middle map injects by the map ϕ into $\mathbf{Z}$. Hence $H_1(\widetilde{(\mathbf{C}^2 - C)}_\phi, \mathbf{Z} \simeq \pi_{1,a}^\phi)/(\pi)1, a^\phi)' \simeq \pi_1^\phi/(\pi_1^\phi)' \simeq H_1(\widetilde{\mathbf{CP}^2 - C_\phi}, \mathbf{Z})$. Moreover the action of the generator of $\mathbf{Z}$ in the diagram above on abelianization $\pi_{1,a}^\phi$ coincides with the action of $\mathbf{Z}/d\mathbf{Z}$ on the abelianization of π_1/π_1' under the above isomorphism and our claim about the order of the action of the automorphism induced by deck transformation follows.

As far as the proof of the lemma is concerned notice that the fact that the kernel of $\pi_1(\mathbf{C}^2 - C) \to \pi_1(\mathbf{CP}^2 - C)$ is a normal closure of a loop γ in the vicinity of the line in infinity L_∞ which surrounds this line follows from the comparison of van Kampen's presentations of the fundamental groups of the complement in $\mathbf{CP}^2$ to $C \cup L_\infty$ and C which use as generators the standard generators of $\pi_1(D - D \cap C \cap L_\infty)$ for a generic line D. To see that γ is in the center of $\pi_1(\mathbf{C}^2 - C)$, select a line L_t which is small perturbation of L_∞ and write the van Kampen presentation for $\pi_1(\mathbf{C}^2 - C, L_t \cap D)$ usiong generators in $\pi_1(L_t - \cap L_t(C \cup L_\infty))$. Let D_∞ be a small disk in D about $D \cap L_\infty$, parameterizing lines in the pencil of lines containing L_∞ and L_t.

Then the union of affine parts (i.e., in $\mathbf{C}^2$) of these lines minus $C \cup L_\infty$ is the trivial fibration over $D_\infty - D_\infty \cap L_\infty$ with fibre $\mathbf{C}$ minus d points because this fibration extends over D_∞, (as follows from our assumption that C is transversal to L_∞). The generator of π_1 (D_∞ which can be identified with v hence commutes with all generators of $(\mathbf{C}^2 - C)$ and the lemma follows.

Note that if the line in infinity is not transversal to the curve then the relationship between $\pi_1(\mathbf{C}^2 - C)$ and $\pi_1(\mathbf{CP}^2 - C)$ is more subtle. For example if C is given in $\mathbf{C}^2$ by equation $x^2 = y^3$ then $\pi_1(\mathbf{C}^2 - c) = \{a, b, /aba = bab\}$ and $\pi_1(\mathbf{CP}^2 - C) = \mathbf{Z}/3\mathbf{Z}$ as follows for example from Abhyankar's calculations [1].

A series of calculations of the fundamental groups of the complements is discussed in [5]. In particular the class of possible fundamental groups includes braid groups (Zariski, Moishezon) and groups of torus knots of type (p, q) (M. Oka). The cyclotomic property of Alexander polynomials of algebraic curves discussed above prohibits, however, many knot or link groups to be fundamental groups of an algebraic curve. For example, the fundamental group of figure eight knot (Alexander polynomial is t^2-3t+1) cannot occur as a fundamental group of a complement to a plane algebraic curve. The lower central series of this group at the same time stabilizes in the second term. Indeed if Γ_i is the lower central series of π_1 (i.e., $\Gamma_1 = \pi_1$ and $\Gamma_{i+1} = [\pi_1, \Gamma_i]$ then $Hom(\Gamma_2/\Gamma_3, \mathbf{Z})$ can be identified with the kernel of the cup product map: $\Lambda^2 H^1 \to H^2$ (cf. [8]) which is trivial for a space with $H^1 = \mathbf{Z}$. Hence the methods of [13] does not outrule these groups as the fundamental groups of an open algebraic variety.

Finally note that the theory of Alexander polynomials of plane curves has a high dimensional generalization in which the first homology of the infinite cyclic cover are replaced by the first non vanishing homotopy group of the infinite cyclic cover which imposes a non-trivial restriction on the possible homotopy types of the complements to hypersurfaces in $\mathbf{C}^n$ (cf. [6]).

12.3 References

[1] S. Abhyankar, Tame coverings and the fundamental groups of algebraic varieties (Par VI) *Amer. J. of Math.*, **81**, 1959, pp. 46–94.

[2] F.E.A. Johnson and E.G. Rees, On the fundamental group for a complex algebraic manifold, *Bull. London Math. Soc.*, **19**, 1987, pp. 463–466.

[3] T. Kohno, An algebraic computation of the Alexander polynomial of plane algebraic curve, *Proc. Japan Acad. Ser. A. Math.*, **59**, 1983, pp. 94–97.

[4] A. Libgober, Alexander polynomials of plane algebraic curves and cyclic multiple planes, *Duke Math. J.*, **49**, 1982, pp. 833–851.

[5] A. Libgober, Alexander invariants of plane algebraic curves, *Proc. Symp. Pure Math.*, **40**, *Amer. Math. Soc.*, Providence, RI, Part 2, pp. 135–144.

[6] A. Libgober, Homotyopy groups of the complements to algebraic hypersurfaces, *Bull. A.M.S.*, 1985.

[7] J. Morgan, The algebraic topology of smooth algebraic varieties, *Inst. Hautes Etudes Sci. Publ. Math.* **48**, 1978, pp. 137–204.

[8] R. Randell, Minor fibres and the Alexander polynomials of plane algebraic curves, *Proc. Symp. Pure Math.*, **40**, Providence, RI, Part 2, pp. 415–420.

[9] D. Sullivan, On the intersection form of compact 3-manifold, *Topology*, **14**, pp. 275–277.

[10] van Kampen, On the fundamental group of an algebraic curve, Amer. J. Math., **55**, 1933.

[11] O. Zariski, On the irregularity of cyclic multiple Planes, *Ann. of Math.*, **32**, 1981, pp. 309–318.

13

Unramified Coverings of the Affine Line in Positive Characteristic

Madhav V. Nori

13.1 Introduction

Let p be a prime number. We shall show that there is a finite unramified covering C of the affine line $\mathbf{A}^1$, defined over the field of p elements, so that $\mathbf{A}^1$ is the quotient of C by the fixed–point free action of $SL_n(\mathbf{F}_{p^k})$.

More generally, $SL_n(\mathbf{F}_{p^k})$ can be replaced by (the universal central extension of) any finite simple group of Lie type in characteristic p. These results were obtained independently by J-P. Serre, and for the case of $SL_n(\mathbf{F}_{p^k})$ was done also by Burt Totaro (as yet unpublished).

13.2 Unramified Coverings

The idea is as follows. Let A be a connected algebraic group defined over $\mathbf{F}_p$ and $\phi : A \to A$ be a homomorphism so that there are natural numbers m and n satisfying

$$\phi^m = \pi^n$$

where $\pi : A \to A$ is the Frobenius morphism. Let

$$G = \{a \in A(\overline{\mathbf{F}}_p) \mid \phi(a) = a\}.$$

The quotient of the variety A by the right translations of G is obtained by the Lang morphism

$$L : A \to A \quad \text{given by} \quad L(a) = a\phi(a)^{-1}.$$

Putting $m = 1$, we get $G = A(\mathbf{F}_{p^n})$ and putting $m = 2$ and n odd (and suitable A) the groups of Ree and Suzuki are obtained in G. In fact the Lang morphism above is a minor modification of the original one designed to include the Ree–Suzuki groups (see the remark at the end of the paper).

Now, if $f : \mathbf{A}^1 \to A$ is a morphism (defined over $\mathbf{F}_p$) and C is the fibre-product of f and L,

$$\begin{array}{ccccc} & C & \longrightarrow & A & \\ L^* & \downarrow & & \downarrow & L \\ & \mathbf{A}^1 & \xrightarrow{f} & A & \end{array}$$

we obtain a fixed-point free action of G on C with $C/G = \mathbf{A}^1$.

We wish to choose f so that C is absolutely irreducible. We first observe that

LEMMA 1: *If for every $\sigma \in G$, there is a closed point x of $\mathbf{A}^1$, that is a maximal ideal of $\mathbf{F}_p[T]$, whose Frobenius is conjugate to σ, then C is irreducible.*

Proof: Let D be an irreducible component of C and let H be its stabilizer. Now H contains the decomposition groups of closed points of D lying above various closed points x of $\mathbf{A}^1$, and therefore every $\sigma \in G$ is conjugate to an element of H. Thus

$$G = \bigcup_{g \in G} gHg^{-1}$$

and therefore $H = G$. Thus $C = D$ and the lemma is proved.

LEMMA 2: *Let r be the exponent of the commutative group $G/[G,G]$. Assume that for every $\sigma \in G$ there is a closed point x of $\mathbf{A}^1$ so that (a) the Frobenius at x is conjugate to σ and (b) the residue field at x contains $\mathbf{F}_{p^r}$. Then C is absolutely irreducible.*

Proof: $\mathbf{A}^1_{\mathbf{F}_{p^r}}$ is the quotient of $C_{\mathbf{F}_{p^r}}$ by the action of G, and by assumption, there are closed points x' of $\mathbf{A}^1_{\mathbf{F}_{p^r}}$ with Frobenius conjugate to arbitrary $\sigma \in G$, and thus by the argument in Lemma 1, we see that $C_{\mathbf{F}_{p^r}}$ is irreducible. To prove that C is absolutely irreducible, we need to know that the function field $k(C)$ of C is linearly disjoint from $\overline{\mathbf{F}}_p$, or equivalently, that if

$$K = k(C) \cdot \overline{\mathbf{F}}_p,$$

$$Gal(K/\mathbf{F}_p(T)) \hookrightarrow G \times \widehat{\mathbf{Z}}$$

is an isomorphism. The projections to G and $\widehat{\mathbf{Z}}$ are both surjective (because C is irreducible) and denoting the kernal of these projections by P and Q respectively, we see that

$$\widehat{\mathbf{Z}}/P \cong Gal(K/\mathbf{F}_p(T))/Q \times P \cong G/Q$$

and so $P = s\widehat{\mathbf{Z}}$ with s dividing r. The irreducibility of $C_{\mathbf{F}_{p^r}}$ shows that $P \to \widehat{\mathbf{Z}}/r\widehat{\mathbf{Z}}$ is surjective. Thus $P = \widehat{\mathbf{Z}}$ and the lemma is proved.

LEMMA 3: *For every finite field F containing $\mathbf{F}_p$ and for every $\sigma \in G$, there is an $\alpha \in A(F)$ so that the Frobenius at α in the Lang covering $L : A \to A$ is a conjugate of σ.*

Proof: By a theorem of S. Lang, $H^1(F, A) = 0$. Galois cohomology applied to the exact sequence

$$1 \to G \to A \to A/G \to 1$$

therefore yields a surjection

$$(A/G)(F) \to H^1(F, G)$$

and this is simply a re-statement of the assertion in the lemma.

In view of the above lemmas, we have an $f : \mathbf{A}^1 \to A$ so that C is absolutely irreducible from the lemma below.

LEMMA 4: *Given closed points x_i of $\mathbf{A}^1$ with residue fields F_i and elements $a_i \in A(F_i)$ for $1 \le i \le e$, there is a morphism $f : \mathbf{A}^1 \to A$ so that*

$$f(x_i) = a_i \text{ for } 1 \le i \le e$$

For the above lemma we need the hypothesis (H) on A.

(H) There are closed subgroups U and V of A so that

(a) $U(F)$ and $V(F)$ generate $A(F)$ for all finite fields F containing $\mathbf{F}_p$,

(b) as varieties both U and V are isomorphic to affine space.

For example, if $A = SL_n$, we may take U (resp. V) to be the groups of strictly upper (resp. lower) triangular matrices in LS_n.

The hypothesis (H) is satisfied by any connected semi-simply connected algebraic group, and for any extension of such a group by a connected unipotent algebraic subgroup (see J. Tits "Abstract and Algebraic Simple groups", Annals of Math. 1964, vol. 80, 313-329, for the semi-simple case).

Proof of Lemma 4 (given (H)): It is clear that if $\alpha \in U(F_i)$, there is a morphism

$$g : \mathbf{A}^1 \to U$$

so that $g(x_i) = \alpha$, and $g(x_j) = 1$ if $j \neq i$. Doing the same for V, and taking products of such morphisms, we obtain $f_i : \mathbf{A}^1 \to A$ so that

$$f_i(x_i) = a_i \text{ and } f_i(x_j) = 1 \text{ if } j \neq i.$$

Putting $F(t) = f_1(t) f_2(t) \cdots f_e(t)$ the lemma follows.

Summarizing the above, we obtain

THEOREM: *Let A be an algebraic group defined over $\mathbf{F}_p$ that satisfies (H). Let $\phi : A \to A$ be a homomorphism so that $\phi^m = \pi^n$ for some natural numbers m and n (and $\pi : A \to A$ is the Frobenius morphism). Let G be the fixed points of ϕ in $A(\overline{\mathbf{F}}_p)$. Then there is an absolutely irreducible unramified Galois covering of the affine line, defined over $\mathbf{F}_p$, with Galois group equal to G.*

13.3 A remark on the Lang morphism

We need to check that $L : A \to A$ identifies A with the quotient A/G. Following the exposition of A. Borel we define an action ρ of A on A by

$$\rho(a)b = ab\phi(a)^{-1}$$

Once it is checked that the orbits of this action are open, the connectedness of A implies that this action is actually transitive. The stabilizer of the identity element 1 of A is precisely G, and therefore

$$a \mapsto \rho(a)1 = L(a)$$

gives the quotient morphism of A by the right–action of G.

Note that $h(a) = a \cdot \phi(a) \cdot \phi^2(a) \cdots \phi^{m-1}(a)$ defines a morphism and $h(\rho(a)b) = a \cdot h(b)(\pi^n(a))^{-1}$. For a fixed $b \in A(\overline{\mathbf{F}}_p)$,

$$a \mapsto ah(b)(\pi^n(a))^{-1}$$

is an étale morphism because π^n induces zero on tangent–spaces, and therefore

$$a \mapsto \rho(a)b$$

is also an étale morphism. Thus the orbit of b is open, and this completes the proof.

A beautiful exposition of both the theorems of Serge Lang on connected algebraic groups defined over finite fields used in this note is found in the book of A. Borel "Linear Algebraic Groups", 1969, published by Benjamin.

Part VI

Young Tableaux

14
Abhyankar's Work on Young Tableaux and Some Recent Developments

Sudhir R. Ghorpade

14.1 Introduction

This paper is meant to be a survey of Abhyankar's work on Young tableaux and some of the subsequent developments motivated by it. Leaving the technical details to the remaining sections, we now attempt to give an overview by narrating some background, a little bit of history, and some stories in the two paragraphs below. In keeping with the spirit of Abhyankar's work on Young tableaux, much of this paper is of an elementary nature except perhaps for a few words thrown here and there.

The initial motivation for Abhyankar's work on Young tableaux came from a discussion with V. Deodhar in 1982 following which he investigated the singularities of Schubert varieties in flag manifolds, and showed that for a class of such varieties, the tangent cone at the so called distinguished point is given by an ideal of the type $I_p(L)$, the ideal generated by $p \times p$ minors in L where L is a ladder shaped subset of a rectangular matrix with indeterminate entries. This led him to study such determinantal ideals in general. The combinatorial notion of Young tableaux was invented by Alfred Young in 1902 [Y] for its applications to invariant theory and the representation theory of the symmetric group. About 70 years later, Doubilet-Rota-Stein [DRS] obtained the remarkable Straightening Law (or Standard Basis Theorem) using a two-sided variant of Young tableaux, called (Young) bitableaux. This result, which we shall soon describe in detail, was clearly seen to be of great importance for the study of determinantal varieties and numerous applications in this direction were soon given by DeConcini-Eisenbud-Procesi [DEP]. Learning of these developments, Abhyankar began the project of "understanding" it all in his own inimitable manner. By 1983, he had a new proof of the Straightening Law using enumerative techniques. This was a remarkable display of tour de force, and all the combinatorial results were independently proved only from the basic principles. The multitude of enumerative formulas Abhyankar obtained, and the implied insight appears to be incredible. At any rate, that's how it seemed to me when I would travel to Bombay University to attend the series of lectures by Professor Abhyankar on this topic while I was doing my M.Sc. at I.I.T. Bombay. I might add that what attracted me to begin attending these lectures was the little note in the seminar notice saying that only High School Algebra background would suffice to understand these lectures! Now apart from being of an interest in itself, the enumeration has its advantages. In short, counting counts. Indeed, it turns out that Abhyankar's counting yields explicit formulas for the Hilbert functions of a class of determinantal ideals which includes the ideal generated by $p \times p$ minors of a matrix with indeterminate entries. The Hilbert functions are in fact polynomials for *all* parameter values in the case of these ideals. Primality of these ideals, and in fact, of a more general class of ideals, which include the ladder determinantal ideals mentioned earlier, can be deduced from the Standard Basis Theorem (=SBT). Now to prove the SBT itself, the enumeration helps in reducing the work at least by half. The other half is proved using the so called Universal Determinantal Identity which generalises the quadratic relations satisfied by the Plucker coordinates in Grassmann variety. Following the counting of standard bitableaux and its applications to enumertive proof of SBT and explicit computations of Hilbert functions of determinantal ideals, Abhyankar observed that the a certain set of the so called indexed monomials has the same cardinality as the set of standard bitableaux under consideration. This result makes it possible to give an explicit basis for the homogeneous coordinate ring of the ladder determinantal varieties indicated earlier, and also to prove their irreducibility.

When Abhyankar wrote all these results, it became a paper of about 500 pages, and so it came out as a book [A3]. One of the main aims of this paper is to give a short and motivated account of several of Abhyankar's ideas on Young tableaux, thereby hoping to make [A3] more accessible. This paper differs from [A2], which gives a summary of [A3] but no proofs, in the respect that most sections, i.e., except 15.5 and 15.7, contain fairly self- contained proofs. In Sections 15.2 to 15.6 below, we describe the work of Abhyankar, and in the last section, we discuss some problems suggested by this work, and their status.

14.2 Preliminaries plus Preview

In this section we set some notation, define Young tableaux and formally state the Standard Basis Theorem (=SBT) and give a preview of the succeeding material by way of motivating Abhyankar's strategy for an enumerative proof of the SBT.

Given any two integers A and B, by $[A, B]$ we denote the set of integers between A and B, i.e., $[A, B] = \{C \in \mathbb{Z} : A \leq C \leq B\}$. Let q be a positive integer. Given a positive integer p, a q-sided strictly increasing sequence of positive integers such that the length of each side is p, i.e., a map $a : [1,q] \times [1,p] \to \mathbb{Z}$ such that $1 \leq a(k,1) < \cdots < a(k,p)$ for $k = 1, \ldots, q$, is called a *multivector* of *width* q and *length* p; if $q = 1$, then a is called a *univector*, whereas if $q = 2$, then a is called a *bivector*. By a *(Young) tableau* of width q we mean a finite sequence $T = (T[1], T[2], \ldots, T[d])$ of multivectors of width q (and varying length); here the nonnegative integer d is called the *depth* of the tableau T, and it may be denoted by $\operatorname{dep}(T)$ (empty tableau being the only one of depth 0). Tableaux of width 1 and 2 are called *unitableaux* and *bitableaux* respectively, whereas those of higher width are generally referred to as *multitableaux*. Thus a bitableau T of depth d may be visualised as an array of the following type:

$$\begin{array}{ccc|ccc}
T[1](1,p_1) > & \cdots & > T[1](1,1) & T[1](2,1) < & \cdots & < T[1](2,p_1) \\
T[2](1,p_2) > & \cdots & > T[2](1,1) & T[2](2,1) < & \cdots & < T[2](2,p_2) \\
 & & \vdots & \vdots & & \\
T[d](1,p_d) > & \cdots & > T[d](1,1) & T[d](2,1) < & \cdots & < T[d](2,p_d)
\end{array}$$

Here we have let $p_1, p_2, \ldots, p_d$ denote the lengths of $T[1], T[2], \ldots, T[d]$ respectively. If we erase one of the sides in the above array, then we "see" a unitableau. To "see" a multitableau of width greater than 2, one can imagine pages of a book opened up so that any two pages are at some distance from each other; the book evidently has q pages. The sum of row lengths of a tableau, which is also the number of elements on (any) one of its sides, is called its *area*; thus for a tableau T as above, the area equals $p_1 + p_2 + \cdots + p_d$. Given q positive integers $m(1), \ldots, m(q)$, we say that a tableau T is *bounded* by $m = (m(1), \ldots, m(q))$ if the entries on the k-th side are no larger than $m(k)$, i.e., for each $k \in [1,q]$ we have $T[e](k,i) \leq m(k)$ for all permissible values of e and i. One more definition. A tableau T is said to be *standard* if the row lengths decrease and there is an increase along each column, i.e., in the above notation we have $p_1 \geq p_2 \geq \ldots \geq p_d$ and for each $e \in [1, d-1]$ we have $T[e](k,i) \leq T[e+1](k,i)$ for $1 \leq k \leq q$ and $1 \leq i \leq p_{e+1}$. The condition of standardness may be briefly written as $T[1] \leq T[2] \leq \cdots \leq T[d]$ where for two multivectors $a[1]$ and $a[2]$ of width q and lengths p_1 and p_2 respectively, we set $a[1] \leq a[2]$ iff $p_1 \geq p_2$ and $a[1](k,i) \leq a[2](k,i)$ for all $k \in [1,q]$ and $i \in [1,p_2]$; note that the latter defines a partial order on the set of all multivectors of width q. The set of all standard tableaux of width q which are bounded by m will be denoted by $\operatorname{stab}(q,m)$, and given a nonnegative integer V, the subset of $\operatorname{stab}(q,m)$ consisting of standard tableaux of area V will be denoted by $\operatorname{stab}(q,m,V)$.

Now that we know what Young tableaux are, one might ask what are they good for. We cite two possible reasons and of these, the second one leads us to motivate the contents of some of the succeeding sections.

The irreducible representations of the symmetric group S_n of degree n, being equinumerous with the conjugacy classes in S_n, are in one-to-one correspondence with the partitions of n. Given a partition $\lambda = (p_1, p_2, \ldots, p_d)$ of n [so that $p_1 \geq p_2 \geq \cdots \geq p_d \geq 1$ and $p_1 + p_2 + \cdots + p_d = n$], the corresponding irreducible S_n–module is denoted by S^λ and is called the *Specht module* corresponding to λ. It turns out that the degree of the irreducible representation corresponding to λ, i.e., $\dim S^\lambda$, is exactly equal to the number of standard unitableaux of depth d and row lengths $p_1, p_2, \ldots, p_d$ in which every integer between 1 and n occurs exactly once (or equivalently, they are bounded by n and are "strictly standard" in the sense that they are standard and there is strict increase along the columns as well. Actually this is what A. Young meant by standard). Furthermore, the character table of S_n can also be constructed using these tableaux. Indeed, we more or less know everything about representations of S_n—thanks to tableaux. For details concerning these results, see G.D. James's monograph [J], or the forthcoming book of B. Sagan [S]. More about the connection of Young tableaux with Representation theory, Invariant theory, and Combinatorics, can be found in [Ku] and [St].

To see another domain of applicability, let X be an $m(1) \times m(2)$ matrix whose entries X_{ij}'s are independent indeterminates over some field K, and let $K[X]$ denote the ring of polynomials in these $m(1)m(2)$ variables with coefficients in K. Let p be a positive integer such that $p \leq \min\{m(1), m(2)\}$. Note that a $p \times p$ minor of X is uniquely determined once we prescribe the row and column indices, and thus by a bivector bounded by $m = (m(1), m(2))$ and of length p. A bitableau T bounded by m can be made to correspond to the product of the corresponding minors, or in other words, to a *monomial in the minors of* X; it is a homogeneous polynomial whose degree equals the area of T. Now we can state the Standard Basis Theorem or the Straightening Law of Doubilet-Rota-Stein. It says that *the monomials in the minors of* X *corresponding to standard bitableaux bounded by* $m = (m(1), m(2))$ *form a* K*-vector space basis of* $K[X]$.

The above basis is quite different from the usual basis (of monomials in X_{ij}'s) and is more useful when one wants to study determinantal ideals and such. The idea of Abhyankar's enumerative proof of SBT is as follows. Fix a nonnegative integer V and consider the Vth homogeneous component $K[X]_V$ of $K[X]$. It suffices to show that the monomials in the minors of X corresponding to standard bitableaux bounded by m and of area V form a K-vector space basis of $K[X]_V$ (for any given V). But now the advantage is that $K[X]_V$ is a finite dimensional vector space and we also know its dimension; indeed

$$\begin{aligned}\dim_K K[X]_V &= \text{no. of monomials of degree } V \text{ in } m(1)m(2) \text{ variables} \\ &= \binom{V + m(1)m(2) - 1}{m(1)m(2) - 1}.\end{aligned}$$

So all we need to do is to show that the cardinality of the set $\text{stab}(2, m, V)$ of all standard bitableaux bounded by m and of area V equals $\dim_K K[X]_V$, and to prove that the corresponding monomials in the minors of X generate $K[X]_V$ or that they are linearly independent. Abhyankar proves the generation part by using the Laplace expansion for suitable determinants; these ideas are essentially similar to those of Hodge in his proof of the Basis Theorem for Grassmann varieties [Ho], and also to those in [DRS] or [DEP]. The problem of finding $|\text{stab}(2, m, V)|$ appears formidable and to tackle this Abhyankar generalizes the set by throwing in some more parameters p and a. By giving particular values to these new parameters we get back the set $\text{stab}(2, m, V)$. In greater detail, we fix a positive integer p and a bivector a bounded by m and of length p, and consider the set $\text{stab}(2, m, p, a, V)$ which, by definition, consists of all standard bitableaux T bounded by m, dominated by a, and of area V, where a bitableau T is said to be *dominated* by the bivector a if the tableau obtained by placing a on the top of T is also standard, i.e., $a \leq T[e]$ for all $e \in [1, \text{dep}(T)]$. More generally, we can consider a to be a multivector of width q and define the set $\text{stab}(q, m, p, a, V)$ in a completely analogous manner. The thought of "monomials in minors" corresponding to multitableaux is rather intriguing, but we shall come to that in a later section. Right now we just regard multitableaux as combinatorial objects waiting to be counted. As to how Abhyankar enumerates this set, we see that in the next section.

14.3 Enumeration of Standard Tableaux

Let us fix positive integers q, p, a q–tuple $m = (m(1), \ldots, m(q))$ of positive integers, a multivector a of width q and length p such that a is bounded by m, and a nonnegative integer V. We wish to count the set

$$\text{stab}(q, m, p, a, V) = \{T \in \text{stab}(q, m) : a \leq T \text{ and the area of } T \text{ is } V\}$$

where we have written $a \leq T$ to briefly indicate that T is dominated by a, i.e., $a \leq T[e]$ for all $e \in [1, \text{dep}(T)]$; note that when $V = 0$, this set consists only of the empty tableau.

The first step is to decompose this set into disjoint subsets which are easier to handle. To this effect, we define the *shape* of a tableau $T \in \text{stab}(q, m, p, a, V)$ to be the p–tuple $v = (v(1), \ldots, v(p))$ where

$$\begin{aligned} v(i) &= \text{the number of rows of length } i \\ &= |\{e \in [1, \text{dep}(T)] : \text{ the length of } T[e] \text{ is } i\}|; \end{aligned}$$

we clearly have $v(1) + 2v(2) + \ldots + pv(p) = V$. Given a p-tuple $v = (v(1), \ldots, v(p))$ of nonnegative integers, we define

$$\text{stas}(q, m, p, a, v) = \{T \in \text{stab}(q, m, p, a, V) : \text{the shape of } T \text{ is } v\}.$$

Obviously

$$|\text{stab}(q, m, p, a, V)| = \sum_{v} |\text{stas}(q, m, p, a, v)|$$

where the sum is taken over all p–tuples $v = (v(1), \ldots, v(p))$ of nonnegative integers such that $v(1) + 2v(2) + \ldots + pv(p) = V$. The next reduction is to note that we clearly have

$$|\text{stas}(q, m, p, a, v)| = \prod_{k=1}^{q} |\text{stas}(1, m(k), p, a^{(k)}, v)|$$

where for $k \in [1, q]$, $a^{(k)}$ denotes the univector of length p obtained from a by restricting to its kth side so that $a^{(k)}(1, i) = a(k, i)$ for $1 \le i \le p$; evidently $a^{(k)}$ is bounded by $m(k)$. So it suffices to consider unitableaux.

Now let us fix a positive integer μ [to serve as a "bound" for unitableaux], and a p-tuple $v = (v(1), \ldots, v(p))$ of nonnegative integers. Let α, β denote univectors bounded by μ and of length p. If $\beta \ge \alpha$, then any $T \in \text{stas}(1, \mu, p, \beta, v)$ can be preaugmented by β to give a unitableaux $T^* \in \text{stas}(1, \mu, p, \alpha, v^*)$, where the "new shape" $v^* = (v^*(1), \ldots, v^*(p))$ is given by

$$v^*(i) = \begin{cases} v(i) & \text{if } 1 \le i \le p-1 \\ v(p)+1 & \text{if } i = p \end{cases}$$

and the new unitableaux T^* is given by $T^*[1] = \beta$, $T^*[e] = T[e-1]$ for $e = 2, \ldots, \text{dep}(T^*) = \text{dep}(T) + 1$. Conversely, any T^* in $\text{stas}(1, \mu, p, \alpha, v^*)$ can be obtained in this way [note that v^* is characterized by the property that $v^*(p) \ne 0$]. It follows that we have the recurrence relation

$$|\text{stas}(1, \mu, p, \alpha, v^*)| = \sum_{\beta \ge \alpha} |\text{stas}(1, \mu, p, \beta, v)|. \tag{14.1}$$

Furthermore, we clearly have that

$$\text{if } p \ne 1 \text{ and } v(p) = 0, \text{ then } |\text{stas}(1, \mu, p, \alpha, v)| = |\text{stas}(1, \mu, p-1, \tilde{\alpha}, \hat{v})| \tag{14.2}$$

where $\tilde{\alpha}$ is a univector of length $p-1$ and $\hat{v}$ is a $(p-1)$–tuple, obtained by restricting α and v respectively, i.e., $\tilde{\alpha}(1, i) = \alpha(1, i)$ and $\hat{v}(i) = v(i)$ for $1 \le i \le p-1$. Lastly, it is clear that

$$\text{if } p = 1 \text{ and } v(p) = 0, \text{ then } |\text{stas}(1, \mu, p, \alpha, v)| = 1. \tag{14.3}$$

Now we seek a function of the parameters (μ, p, α, v) satisfying the above relations. After some experimentation, Abhyankar arrived at the function ν defined as follows.

$$\nu(\mu, p, \alpha, v) = \det\left(\begin{bmatrix} \mu - \alpha(1, j) \\ v(i) + v(i+1) + \ldots + v(p) + j - i \end{bmatrix}\right)$$

where the entries of the above $p \times p$ determinant are the so called *twisted binomial coefficients* which are defined as follows.

$$\begin{bmatrix} U \\ A \end{bmatrix} = \binom{U+A}{A} = \begin{cases} \frac{(U+1)\cdots(U+A)}{A!} & \text{if } A \ge 0 \\ 0 & \text{if } A < 0 \end{cases}$$

where A is an integer and U can be any element in an overring of $\mathbb{Z}$, in particular, it can be an indeterminate over $\mathbb{Q}$ in which case the binomial coefficient is a univariate polynomial with rational coefficients. These twisted binomial coefficients are convenient for our purposes and will often appear in this paper. At this point we would also like to remark on the convention that unless mentioned otherwise, all matrices will be of order p, and whenever we write $\det(a_{ij})$, it may be tacitly assumed that the corresponding matrix is of order p so that the indices i and j vary between 1 and p.

To verify (15.1), let us write $\bar{\alpha}_j = \mu - \alpha(1,j)$, $\bar{\beta}_j = \mu - \beta(1,j)$, and $\bar{v}(j) = v(j) + v(j+1) + \cdots + v(p)$ for $1 \le j \le p$. It is readily seen that

$$\sum_{\beta \ge \alpha} \nu(\mu, p, \beta, v) = \sum_{0 \le \bar{\beta}_p \le \bar{\alpha}_p} \; \sum_{\bar{\beta}_p < \bar{\beta}_{p-1} \le \bar{\alpha}_{p-1}} \cdots \sum_{\bar{\beta}_2 < \bar{\beta}_1 \le \bar{\alpha}_1} \det\left(\begin{bmatrix} \bar{\beta}_j \\ \bar{v}(i) + j - i \end{bmatrix}\right).$$

In the first sum, only the first column of the above determinant varies, and so by using the multilinearity of determinants, we can eliminate the first sum. Now we need the following

General Progression Lemma (3.1). *Given integers e, x, y with $x < y$, we have*

$$\sum_{x < s \le y} \begin{bmatrix} s \\ e \end{bmatrix} = \begin{bmatrix} y \\ e+1 \end{bmatrix} - \begin{bmatrix} x \\ e+1 \end{bmatrix}.$$

This lemma can be proved by induction on $y - x$ using the Basic Binomial Identity

$$\begin{bmatrix} U \\ A \end{bmatrix} = \begin{bmatrix} U-1 \\ A \end{bmatrix} + \begin{bmatrix} U \\ A-1 \end{bmatrix} \tag{14.4}$$

whose proof is straightforward. The General Progression Lemma can be applied to each entry in the resulting determinant, which yields the difference of two determinants of which the first one has $\bar{\beta}_1$ replaced by $\bar{\alpha}_1$ and $\bar{v}(1)$ by $\bar{v}(1) + 1$, and the second one vanishes because two of its columns are identical. Continuing in this manner, we see that

$$\sum_{\beta \ge \alpha} \nu(\mu, p, \beta, v) = \det\left(\begin{bmatrix} \bar{\alpha}_j \\ \bar{v}(i) + 1 + j - i \end{bmatrix}\right) = \nu(\mu, p, \alpha, v^*)$$

which shows that the function ν satisfies a relation of the form (15.1). It is easy to see that ν also satisfies relations of the form (15.2) and (15.3). Hence by (double) induction on p and $v(p)$, it follows that

$$|\mathrm{stab}(q, m, p, a, V)| = F^{(8)}(q, m, p, a, V)$$

where we define

$$F^{(8)}(q, m, p, a, V) = \sum_{v} \prod_{k=1}^{q} \det\left(\begin{bmatrix} m(k) - a(k,j) \\ v(i) + v(i+1) + \cdots + v(p) + j - i \end{bmatrix}\right)$$

where the sum is taken over all p–tuples $v = (v(1), \ldots, v(p))$ of nonnegative integers such that $v(1) + 2v(2) + \cdots + pv(p) = V$.

Now we shall make transformations to obtain different versions of the above formula. One of the motivations in doing so is to be able to deduce that by giving particular values to the parameters q, p and a, we have $|\text{stab}(2, m, V)| = \dim_K K[X]_V$. To this end, we begin by noting that the transformation

$$u(i) = v(i) + v(i+1) + \cdots + v(p) + p - i, \quad i = 1, 2, \ldots, p$$

maps the set of all p-tuples $v = (v(1), \ldots, v(p))$ of nonnegative integers with $v(1) + 2v(2) + \cdots + pv(p) = V$ bijectively onto the set of all p–tuples $u = (u(1), \ldots, u(p))$ of nonnegative integers with $u(1) > u(2) > \cdots > u(p)$ and $u(1) + \cdots + u(p) = V + \frac{p(p-1)}{2}$. Hence if we let

$$F^{(7)}(q, m, p, a, V) = \sum_u H^{(7)}(q, m, p, a, u)$$

where the sum is taken over all p–tuples u with the above properties, and where for any p–tuple $u = (u(1), \ldots, u(p))$ of integers we have put

$$H^{(7)}(q, m, p, a, u) = \prod_{k=1}^{q} \det \left(\begin{bmatrix} m(k) - a(k, j) \\ u(i) + j - p \end{bmatrix} \right),$$

then we have: $F^{(8)}(q, m, p, a, V) = F^{(7)}(q, m, p, a, V)$.

We will now make further transformations with *the assumption that q is even.* First of all, we will try to get rid of the clumsy restriction $u(1) > u(2) > \cdots > u(p) \geq 0$ on the parameter u in the above summation. Let us begin by introducing some

NOTATION. Let $\mathbb{Z}(p)$ denote the set of all integer p–tuples, i.e.,

$$\mathbb{Z}(p) = \{v = (v(1), \ldots, v(p)) : v(i) \in \mathbb{Z} \text{ for } 1 \leq i \leq p\}$$

and for any $U \in \mathbb{Z}$, let

$$\mathbb{Z}(p, U) = \{v \in \mathbb{Z}(p) : v(1) + \cdots + v(p) = U\}.$$

Let $u = (u(1), \ldots, u(p))$ be any element of $\mathbb{Z}(p)$. Observe that we clearly have the following.

(i) If $u(p) < 0$ then $u(p) + j - p < 0$ for all $j \in [1, p]$, and therefore in this case we have $H^{(7)}(q, m, p, a, u) = 0$.

(ii) If $u(i) = u(i')$ for some $i \neq i'$ in $[1, p]$, then $H^{(7)}(q, m, p, a, u) = 0$.

(iii) Let $\tau \in S_p$ be any permutation of $[1, p]$ and let $u[\tau] \in \mathbb{Z}(p)$ denote the permuted p–tuple, i.e., $u[\tau] = (u(\tau(1)), \ldots, u(\tau(p)))$. Then

$$H^{(7)}(q, m, p, a, u[\tau]) = [\text{sgn}(\tau)]^q H^{(7)}(q, m, p, a, u) = H^{(7)}(q, m, p, a, u).$$

Now given any p–tuple $u = (u(1), \dots, u(p))$ of distinct integers, there exists a unique permutation $\tau \in S_p$ such that $u(\tau(1)) > u(\tau(2)) > \dots > u(\tau(p))$. Hence, in view of the observations (i), (ii), and (iii), we see that

$$F^{(7)}(q,m,p,a,V) = \frac{1}{p!} \sum_{u \in \mathbb{Z}(p, V + \frac{p(p-1)}{2})} H^{(7)}(q,m,p,a,u)$$

where the above summation is *essentially finite*, i.e., all except finitely many summands are zero. The advantage of letting the parameters $u(1), \dots, u(p)$ take all integer values is that it facilitates further transformations. For example, the equations

$$v(i) = u(i) + i - p, \quad i = 1, 2, \dots, p$$

determine a bijective transformation of $\mathbb{Z}(p, V + \frac{p(p-1)}{2})$ onto $\mathbb{Z}(p, V)$. Thus if we let

$$F^{(6)}(q,m,p,a,V) = \frac{1}{p!} \sum_{v \in \mathbb{Z}(p,V)} \prod_{k=1}^{q} G^{(6k)}(q,m,p,a,v)$$

where for any $v \in \mathbb{Z}(p)$ and $k \in [1, q]$ we have put

$$G^{(6k)}(q,m,p,a,v) = \det\left(\begin{bmatrix} m(k) - a(k,j) \\ v(i) + j - i \end{bmatrix}\right),$$

then we have: $F^{(7)}(q,m,p,a,V) = F^{(6)}(q,m,p,a,V)$.

Remark: It may be interesting to note that if q is odd, and $p > 1$ then, in view of (i), (ii), (iii) above, we have that $F^{(7)}(q,m,p,a,V) = F^{(6)}(q,m,p,a,V) = 0$. For a direct proof of this identity, see [Gh1].

Our next step is to get rid of the factor $\frac{1}{p!}$ in $F^{(6)}$ by showing that one of the determinants, say $G^{(6k)}$, can be pulled out of the product and be replaced by $p!$ times the product of entries on its principal diagonal. We start with the following elementary observations.

(a) If (a_{ij}) is a $p \times p$ matrix and $\tau \in S_p$ is a permutation, then $\det(a_{ij}) = \operatorname{sgn}(\tau) \det\left(a_{i\tau(j)}\right)$.

(b) Let $f : \mathbb{Z}(p) \to \mathbb{Q}$ be any map such that $\{v \in \mathbb{Z}(p) : f(v) \neq 0\}$ is finite, and let $\sigma \in S_p$ be any permutation. Then

$$\sum_{v \in \mathbb{Z}(p,V)} f(v(1), \dots, v(p)) = \sum_{v \in \mathbb{Z}(p,V)} f(v(\sigma(1)), \dots, v(\sigma(p))).$$

(c) Let f be as in **(b)**, and let any $U \in \mathbb{Z}$ and $u \in \mathbb{Z}(p, U)$ be given. Then

$$\sum_{v \in \mathbb{Z}(p,V)} f(v(1)+u(1), \dots, v(p)+u(p)) = \sum_{v \in \mathbb{Z}(p,V+U)} f(v(1), \dots, v(p)).$$

Now let us fix some $k \in [1,q]$. Given any $v \in \mathbb{Z}(p,V)$, let us expand the kth determinant in $F^{(6)}$ to get

$$G^{(6k)}(q,m,p,a,v) = \sum_{\tau \in S_p} \operatorname{sgn}(\tau) \prod_{j=1}^{p} \begin{bmatrix} m(k) - a(k,j) \\ v(\tau(j)) + j - \tau(j) \end{bmatrix};$$

also let us take the transpose of the remaining matrices and use **(a)** to write

$$\prod_{\substack{1 \le n \le q \\ n \ne k}} G^{(6n)}(q,m,p,a,v) = [\operatorname{sgn}(\tau)]^{q-1} \prod_{\substack{1 \le n \le q \\ n \ne k}} \det \left(\begin{bmatrix} m(n) - a(n,i) \\ v(\tau(j)) + i - \tau(j) \end{bmatrix} \right).$$

Now substitute the expressions on the right in the corresponding summand in $F^{(6)}$, interchange the summations, and observe that $[\operatorname{sgn}(\tau)]^q = 1$. This shows that $p!F^{(6)}(q,m,p,a,V)$ is equal to

$$\sum_{\tau \in S_p} \sum_{v \in \mathbb{Z}(p,V)} \left(\prod_{j=1}^{p} \begin{bmatrix} m(k) - a(k,j) \\ v(\tau(j)) + j - \tau(j) \end{bmatrix} \right) \prod_{\substack{1 \le n \le q \\ n \ne k}} \det \left(\begin{bmatrix} m(n) - a(n,i) \\ v(\tau(j)) + i - \tau(j) \end{bmatrix} \right)$$

$$= \sum_{\tau \in S_p} \sum_{v \in \mathbb{Z}(p,V)} \left(\prod_{j=1}^{p} \begin{bmatrix} m(k) - a(k,j) \\ v(j) + j - \tau(j) \end{bmatrix} \right) \prod_{\substack{1 \le n \le q \\ n \ne k}} \det \left(\begin{bmatrix} m(n) - a(n,i) \\ v(j) + i - \tau(j) \end{bmatrix} \right)$$

where the last equality follows from **(b)**. Finally, we use **(c)** with $u(j) = j - \tau(j)$ to conclude that the above expression is equal to $p!F^{(5k)}(q,m,p,a,V)$, where we have put

$$F^{(5k)}(q,m,p,a,V) = \sum_{v \in \mathbb{Z}(p,V)} \left(\prod_{i=1}^{p} \begin{bmatrix} m(k) - a(k,i) \\ v(i) \end{bmatrix} \right) \prod_{\substack{1 \le n \le q \\ n \ne k}} G^{(6n)}(q,m,p,a,v).$$

It follows that for any $k \in [1,q]$ we have: $F^{(6)}(q,m,p,a,V) = F^{(5k)}(q,m,p,a,V)$. Let us summarise some of what we have proved so far.

THEOREM (3.2). *Let the notation be as above. We have*

$$|\operatorname{stab}(q,m,p,a,V)| = F^{(8)}(q,m,p,a,V) = F^{(7)}(q,m,p,a,V).$$

Moreover, if q is even then we have

$$|\operatorname{stab}(q,m,p,a,V)| = F^{(6)}(q,m,p,a,V) = F^{(5k)}(q,m,p,a,V)$$

where k can be any element of $[1,q]$.

Even with all these formulas, it is not clear that by giving particular values to q, p, and a, we get the desired equality between $|\operatorname{stab}(2,m,V)|$

and $\dim_K K[X]_V$. So we must work harder! Henceforth in this section we forget about multitableaux and restrict ourselves to the case when $q = 2$. Let us fix some $k \in [1, 2]$ and let $k' = 3 - k$ so that k' is 2 or 1 according as k is 1 or 2. To simplify the manipulations we shall use the following notation.

$$r(i) = m(k) - a(k,i) \quad \text{and} \quad s(i) = m(k') - a(k',i), \quad \text{for } i = 1, 2, \ldots, p;$$
$$R = \sum_{i=1}^{p} r(i) \quad \text{and} \quad S = \sum_{i=1}^{p} s(i).$$

With this notation, $F^{(5k)}(q, m, p, a, V)$, in our case of $q = 2$, can be written as

$$\sum_{v \in \mathbb{Z}(p,V)} \left(\prod_{i=1}^{p} \begin{bmatrix} r(i) \\ v(i) \end{bmatrix} \right) \det \left(\begin{bmatrix} s(j) \\ v(i) + j - i \end{bmatrix} \right).$$

Let us denote this expression by $F^{(40)}(r, s, V)$. We shall need the following lemmas on (twisted) binomial coefficients.

LEMMA (3.3). *Given any integers U and A, we have*

(i) $\binom{U}{A} = (-1)^A \begin{bmatrix} -U-1 \\ A \end{bmatrix}$ *and* $\begin{bmatrix} U \\ A \end{bmatrix} = (-1)^A \binom{-U-1}{A}$.

(ii) $\binom{U}{A} = \binom{U}{U-A}$ *iff either $U \geq 0$ or $U < A < 0$; and*

$\begin{bmatrix} U \\ A \end{bmatrix} = \begin{bmatrix} A \\ U \end{bmatrix}$ *iff either $U + A \geq 0$ or $U < 0$ and $A < 0$.*

Proof: Obvious.

LEMMA (3.4). *Given any integers U, A, and B, we have*

$$\begin{bmatrix} U \\ A \end{bmatrix} \begin{bmatrix} U+A \\ B \end{bmatrix} = \begin{bmatrix} U \\ A+B \end{bmatrix} \begin{bmatrix} B \\ A \end{bmatrix}.$$

Proof: Straightforward.

LEMMA (3.5). *Given any positive integer p', integers U, E, and any p'-tuple $e \in \mathbb{Z}(p', E)$, we have*

(i) $$\sum_{u \in \mathbb{Z}(p',U)} \prod_{i=1}^{p'} \binom{e(i)}{u(i)} = \binom{E}{U}$$

(ii) $$\sum_{u \in \mathbb{Z}(p',U)} \prod_{i=1}^{p'} \begin{bmatrix} e(i) \\ u(i) \end{bmatrix} = \begin{bmatrix} E + p' - 1 \\ U \end{bmatrix}.$$

where both the sums above are essentially finite.

Proof: Let Y be an indeterminate over $\mathbb{Q}$. Consider the identities

$$\prod_{i=1}^{p'} (1+Y)^{e(i)} = (1+Y)^E \quad \text{and} \quad \prod_{i=1}^{p'} (1+Y)^{-e(i)-1} = (1+Y)^{-E-p'}.$$

Use Binomial Theorem and equate the coefficients of Y^U on both sides to deduce (i) and (ii); note that the latter follows in view of (i) of (3.3). Essential finiteness is obvious.

LEMMA (3.6). *Given any integers T, P, U, U' such that $U+U'\geq 0$ and $T-U'\geq 0$, we have*

$$\begin{bmatrix}T\\P\end{bmatrix}\begin{bmatrix}U\\P+U'\end{bmatrix}=\sum_{d^*\in\mathbb{Z}}(-1)^{d^*}\begin{bmatrix}T+U-d^*\\P\end{bmatrix}\binom{T-U'}{d^*}\begin{bmatrix}U-d^*\\T\end{bmatrix}$$

where the summation on the right is essentially finite.

Proof: Without loss of generality, we may and do assume that $P\geq 0$ and $T\geq 0$. Now by (ii) of (3.3), we see that

$$\begin{aligned}
&\text{LHS}\\
&=\begin{bmatrix}P\\T\end{bmatrix}\begin{bmatrix}P+T+U'-T\\U\end{bmatrix}\\
&=\begin{bmatrix}P\\T\end{bmatrix}\sum_{d^*\in\mathbb{Z}}\begin{bmatrix}P+T\\U-d^*\end{bmatrix}\begin{bmatrix}U'-T-1\\d^*\end{bmatrix} && \text{by (3.5)}\\
&=\sum_{d^*\in\mathbb{Z}}(-1)^{d^*}\binom{T-U'}{d^*}\begin{bmatrix}P\\T\end{bmatrix}\begin{bmatrix}P+T\\U-d^*\end{bmatrix} && \text{by (i) of (3.3)}\\
&=\sum_{d^*\in\mathbb{Z}}(-1)^{d^*}\binom{T-U'}{d^*}\begin{bmatrix}P\\T+U-d^*\end{bmatrix}\begin{bmatrix}U-d^*\\T\end{bmatrix} && \text{by (3.4).}
\end{aligned}$$

Now, since $T-U'\geq 0$, for the above summand to be nonzero, we must have $T-U'\geq d^*\geq 0$; also $P\geq 0$ and hence $P+T-U'-d^*\geq 0$. The desired equality now follows from (ii) of (3.3).

Naturally, we wish to apply the last lemma to the relevant product in $F^{(40)}(r,s,V)$. Before we do that, let us make the following elementary observations. Of these, the first observation is an easy consequence of the fact that each side of the bivector a is a strictly increasing sequence, and the second one is quite obvious.

(1) For any i,j in $[1,p]$, we have: $r(i)+i-j\geq 0$ and $s(j)+j-i\geq 0$.

(2) Suppose for any i,j in $[1,p]$, we are given a map $A_{ij}:\mathbb{Z}\to\mathbb{Q}$ such that $\{d^*\in\mathbb{Z}:A_{ij}(d^*)\neq 0\}$ is finite. Then

$$\prod_{j=1}^{p}\sum_{d^*\in\mathbb{Z}}A_{ij}(d^*)=\sum_{d\in\mathbb{Z}(p)}\prod_{j=1}^{p}A_{ij}(d(j))=\sum_{D\in\mathbb{Z}}\sum_{d\in\mathbb{Z}(p,D)}\prod_{j=1}^{p}A_{ij}(d(j)).$$

We are now ready to transform

$$F^{(40)}(r,s,V)=\sum_{v\in\mathbb{Z}(p,V)}\left(\prod_{i=1}^{p}\begin{bmatrix}r(i)\\v(i)\end{bmatrix}\right)\det\left(\begin{bmatrix}s(j)\\v(i)+j-i\end{bmatrix}\right).$$

For a moment, let us fix some $v \in \mathbb{Z}(p, V)$, and just look at the corresponding summand. First we note that this summand can obviously be written as

$$\det\left(\begin{bmatrix} r(i) \\ v(i) \end{bmatrix}\begin{bmatrix} s(j) \\ v(i)+j-i \end{bmatrix}\right) = \sum_{\sigma\in S_p} \operatorname{sgn}(\sigma) \prod_{j=1}^{p} \begin{bmatrix} r(\sigma(j)) \\ v(\sigma(j)) \end{bmatrix}\begin{bmatrix} s(j) \\ v(\sigma(j))+j-\sigma(j) \end{bmatrix}$$

and now for a given $\sigma \in S_p$, we can apply (3.6) to write the above product as

$$\prod_{j=1}^{p} \sum_{d^*\in\mathbb{Z}} (-1)^{d^*} \begin{bmatrix} r(\sigma(j))+s(j)-d^* \\ v(\sigma(j)) \end{bmatrix}\begin{bmatrix} s(j)-d^* \\ r(\sigma(j)) \end{bmatrix}\binom{r(\sigma(j))+\sigma(j)-j}{d^*}$$

and by observation **(2)**, the above expression can be written as

$$\sum_{D\in\mathbb{Z}} (-1)^D \sum_{d\in\mathbb{Z}(p,D)} \prod_{j=1}^{p} \begin{bmatrix} r(\sigma(j))+s(j)-d(j) \\ v(\sigma(j)) \end{bmatrix}\begin{bmatrix} s(j)-d(j) \\ r(\sigma(j)) \end{bmatrix}$$
$$\binom{r(\sigma(j))+\sigma(j)-j}{d(j)}.$$

Notice that the parameter v figures in only one term. So let us sum the above expression as v varies over $\mathbb{Z}(p, V)$. Then, in view of (ii) of (3.5), we get

$$\sum_{D\in\mathbb{Z}} (-1)^D \sum_{d\in\mathbb{Z}(p,D)} \begin{bmatrix} R+S-D+p-1 \\ V \end{bmatrix}\left(\prod_{j=1}^{p} \begin{bmatrix} s(j)-d(j) \\ r(\sigma(j)) \end{bmatrix}\binom{r(\sigma(j))+\sigma(j)-j}{d(j)}\right).$$

Now it's time to bring back $\operatorname{sgn}(\sigma)$ and sum over $\sigma \in S_p$; we can move this summation sign for σ right before the last product in the above expression, and then collate the terms to form determinants. It follows that $F^{(40)}(r, s, V)$ is equal to

$$\sum_{D\in\mathbb{Z}} (-1)^D \sum_{d\in\mathbb{Z}(p,D)} \begin{bmatrix} R+S-D+p-1 \\ V \end{bmatrix} \det\left(\begin{bmatrix} s(j)-d(j) \\ r(i) \end{bmatrix}\right.$$
$$\left.\binom{r(i)+i-j}{d(j)}\right).$$

We are pretty much done now. For any $D \in \mathbb{Z}$, let us denote the sum

$$\sum_{d\in\mathbb{Z}(p,D)} \det\left(\begin{bmatrix} s(i)-d(i) \\ r(j) \end{bmatrix}\binom{r(j)+j-i}{d(i)}\right)$$

by $F_D^{(30)}(r,s)$ or alternately by $F_D^{(3k)}(m,p,a)$ [recall that $r(i) = m(k) - a(k,i)$ and $s(i) = m(k') - a(k',i)$]. Given any $d \in \mathbb{Z}(p,D)$, in order that the corresponding summand in $F_D^{(30)}(r,s)$ be nonzero, we must have

$$\prod_{i=1}^{p} \binom{r(\sigma(i)) + \sigma(i) - i}{d(i)} \neq 0 \quad \text{for some } \sigma \in S_p$$

and hence, in view of the observation **(1)**, we must have $0 \le D \le R$. Thus

$$\{D \in \mathbb{Z} : F_D^{(30)}(r,s) \neq 0\} \subseteq [0,R]. \tag{14.5}$$

Now we define

$$F^{(30)}(r,s,V) = \sum_{D\in\mathbb{Z}} (-1)^D F_D^{(30)}(r,s) \begin{bmatrix} V \\ R+S+p-1-D \end{bmatrix}$$

and remark that this expression may alternately be denoted by $F^{(3k)}$ (m,p,a,V). In view of (15.5) and (ii) of (3.3), we can conclude that

$$F^{(40)}(r,s,V) = F^{(30)}(r,s,V)$$

or in other words,

$$F^{(5k)}(2,m,p,a,V) = F^{(3k)}(m,p,a,V).$$

Note that $F^{(3k)}$ can be regarded as a polynomial in the parameter V. This is quite significant as we shall soon realise. First, we show that the much sought after equality $|\mathrm{stab}(2,m,V)| = \dim_K K[X]_V$ can now be deduced. To this end, let us suppose that we are in the following
Minimal Case: $p = \min\{m(1), m(2)\}$ and $a(k,i) = i = a(k',i)$ for all $i \in [1,p]$.
In this case, it is clear that $a \le T$ for any $T \in \mathrm{stab}(2,m)$, and hence $\mathrm{stab}(2,m,p,a,V)$ is the same as $\mathrm{stab}(2,m,V)$. To compute $|\mathrm{stab}(2,m,V)|$ using the above formula, we may and do assume that $k \in [1,2]$ is chosen such that $p = \min\{m(1), m(2)\} = m(k')$. For the sake of brevity, let us denote $m(k)$ by p'. Given any $D \in \mathbb{Z}$ and $d \in \mathbb{Z}(p,D)$, the corresponding summand in $F_D^{(3k)}(m,p,a)$ is now equal to

$$\left\{\prod_{i=1}^{p} \binom{p'-i}{d(i)}\right\} \det\left(\begin{bmatrix} p-i-d(i) \\ p'-j \end{bmatrix}\right).$$

For the above product to be nonzero, we must have $0 \le d(i) \le p' - i$ for all $i \in [1,p]$; let us assume that this is the case. Now consider the $p \times p$ matrix whose $(i,j)^{\text{th}}$ entry is

$$\begin{bmatrix} p-i-d(i) \\ p'-j \end{bmatrix}.$$

We successively make the column transformations $C_1 - C_2$, $C_2 - C_3$, $\ldots$, $C_{p-1} - C_p$ [where C_i stands for the ith column], and use the Basic Binomial Identity (15.4). Then for $j \leq p-1$, the $(i,j)^{\text{th}}$ entry of the resulting matrix becomes

$$\begin{bmatrix} p-i-1-d(i) \\ p'-j \end{bmatrix}.$$

Next, we make the column transformations $C_1 - C_2$, $C_2 - C_3$, $\ldots$, $C_{p-2} - C_{p-1}$ and use (15.4), following which we make the column transformations $C_1 - C_2$, $C_2 - C_3$, $\ldots$, $C_{p-3} - C_{p-2}$ and use (15.4), and so on until we reach a stage when we just make $C_1 - C_2$ and use (15.4). Note that the jth column is transformed $p - j$ times each time reducing by one the top entry in the corresponding binomial coefficient. Thus the $(i,j)^{\text{th}}$ entry in the resulting matrix equals

$$\begin{bmatrix} p-i-(p-j)-d(i) \\ p'-j \end{bmatrix} = \begin{bmatrix} j-i-d(i) \\ p'-j \end{bmatrix} = \binom{p'-i-d(i)}{p'-j}.$$

Since $0 \leq d(i) \leq p' - i$, it follows that $0 \leq p' - i - d(i) < p' - j$ whenever $1 \leq j < i \leq p$, and hence if $j < i$, then the binomial coefficient above must be zero. Thus the resulting matrix is triangular, and therefore

$$\det\left(\begin{bmatrix} p-i-d(i) \\ p'-j \end{bmatrix}\right) = \prod_{i=1}^{p} \binom{p'-i-d(i)}{p'-i}.$$

Moreover, if $d(i) > 0$ for some $i \in [1,p]$, then $0 \leq p' - i - d(i) < p' - i$, and hence the above product is zero; on the other hand, if $d(i) = 0$ for all $i \in [1,p]$, then the above product is clearly equal to 1. It follows that in this minimal case, for any $D \in \mathbb{Z}$, we have

$$F_D^{(3k)}(m,p,a) = \begin{cases} 1 & \text{if } D = 0 \\ 0 & \text{if } D \neq 0 \end{cases}$$

and therefore

$$F^{(3k)}(m,p,a,V) = \begin{bmatrix} V \\ R+S+p-1 \end{bmatrix}.$$

Finally we note that we now have $p = m(k')$, $a(k,i) = i = a(k',i)$, and so

$$\begin{aligned} R+S+p-1 &= (p-1) + \sum_{i=1}^{p} [m(k) - i] + [m(k') - i)] \\ &= p - 1 + pm(k) + pm(k') - p(p+1) \\ &= m(k')m(k) + m(k')^2 - m(k')^2 - 1 \\ &= m(1)m(2) - 1. \end{aligned}$$

Thus we have proved the following

THEOREM (3.7). *Given any* $k \in [1,2]$ *we have*
(i)$|\text{stab}(2,m,p,a,V)| = F^{(3k)}(m,p,a,V)$.
(ii)$|\text{stab}(2,m,V)| = \begin{bmatrix} V \\ m(1)m(2)-1 \end{bmatrix} = \dim_K K[X]_V$.

Remark: In Abhyankar's original proof (cf. [A1], [A3]), the above assertion (ii) is deduced from the previous assertion by first converting $F^{(3k)}$ into another formula, called $F^{(2k)}$, by making several transformations and using a Binomial Lemma similar to (3.6), and then manipulating the determinants for the above particular values of p and a. The proof we have given here is somewhat shorter. The crucial thing, as we indicated earlier, is the fact that $F^{(3k)}$ is a "polynomial formula" in V for $|\text{stab}(2,m,p,a,V)|$.

14.4 Universal Determinantal Identity

An essential ingredient of the Straightening Law, or the SBT, is the fact that any monomial in minors can be "straightened out" so as to be expressible as a sum of standard monomials in minors, i.e., monomials in minors corresponding to standard tableaux. This can be done using a certain universal determinantal identity satisfied by the minors of a matrix, which we shall now describe.

Let $m = (m(1), m(2))$ be a pair of positive integers and $X = (X_{ij})$ be an $m(1) \times m(2)$ matrix whose entries are independent indeterminates over a field K, and let $K[X]$ be the corresponding ring of polynomials. If a is a bivector of length p and bounded by m, then it corresponds to a $p \times p$ minor of X, which we denote by $\text{mor}(X,a)$; thus

$$\text{mor}(X,a) = \det\left(X_{a(1,i)a(2,j)}\right)_{1\le i,j\le p}$$

Given a bitableau $T = (T[1], T[2], \ldots, T[d])$ bounded by m, we have the corresponding monomial in minors of X which we denote by $\text{mom}(X,T)$; thus

$$\text{mom}(X,T) = \prod_{e=1}^{d} \text{mor}(X,T[e]).$$

Note that $\text{mom}(X,T)$ is a homogeneous polynomial in $K[X]$ and its degree is equal to the area of T. The so called Universal Determinantal Identity implies that for any bitableau T bounded by m, we have

$$\text{mom}(X,T) = \sum_{S\in\text{stab}(2,m)} \text{fin}(m,T,S)\,\text{mom}(X,S) \qquad (14.6)$$

where the sum is taken over all standard bitableaux S bounded by m, and the coefficient $\text{fin}(m,T,S)$ is an integer such that $\text{fin}(m,T,S) = 0$ for all except finitely many such S, and

$$\text{fin}(m,T,S) \neq 0 \;\Rightarrow\; \text{area of } S = \text{ area of } T, \text{ and } S[1] \le T[1] \qquad (14.7)$$

where the last inequality holds provided $\operatorname{dep}(T) \neq 0 \neq \operatorname{dep}(S)$. Note that the identity (15.6) clearly implies that

$$\{\operatorname{mom}(X, S) : S \in \operatorname{stab}(2, m)\} \text{ generates } K[X] \tag{14.8}$$

and, in view of (15.7), it also shows that for any nonnegative integer V,

$$\{\operatorname{mom}(X, S) : S \in \operatorname{stab}(2, m, V)\} \text{ generates } K[X]_V. \tag{14.9}$$

To see (15.6), we note that if a bitableau $T = (T[1], \ldots, T[d])$ of depth d is not standard, then there exists $e \in [1, d-1]$ such that $T[e] \not\leq T[e+1]$. Thus it suffices to consider two bivectors a and b bounded by m such that $a \not\leq b$, and express the product of the corresponding pair of minors as a sum of "better products" of pairs of minors. Now these bivectors may have different length, but we can reduce to the case of maximal size minors by means of the following lemma whose proof is left as an exercise. Let us first assume, without loss of generality, that $m(1) \leq m(2)$.

LEMMA (4.1). *Given any bivector a bounded by m and of length p, let $\alpha_1, \alpha_2, \ldots, \alpha_{m(1)-p}$ be integers such that $\alpha_1 < \alpha_2 < \cdots < \alpha_{m(1)-p}$ and*

$$[1, m(1)] \setminus \{a(1,1), \ldots, a(1,p)\} = \{\alpha_1, \ldots, \alpha_{m(1)-p}\}.$$

Let $\beta_1, \beta_2, \ldots, \beta_{m(1)}$ be defined by

$$\beta_j = \begin{cases} a(2,j) & \text{if } 1 \leq j \leq p \\ m(1) + m(2) + 1 - \alpha_{m(1)+1-j} & \text{if } p+1 \leq j \leq m(1). \end{cases}$$

Then $\beta = (\beta_1, \beta_2, \ldots, \beta_{m(1)})$ is a univector bounded by $m(1) + m(2)$ and of length $m(1)$. Moreover, the above correspondence $a \mapsto \beta$ is an order-preserving bijection of the set of all bivectors bounded by m onto the set of all univectors bounded by $m(1) + m(2)$ and of length $m(1)$. Furthermore, if we let $\overline{X} = (X \mid I^*)$ be the $m(1) \times (m(1) + m(2))$ matrix obtained by adjoining the $m(1) \times m(1)$ antiidentity matrix $I^* = (\delta_{i\, m(1)+1-j})$ [where δ is the usual Kronecker delta] to X, then $\operatorname{mor}(X, a) = \pm \det[\beta]$, where $[\beta]$ denotes the $m(1) \times m(1)$ submatrix of $\overline{X}$ formed by the columns whose indices are $\beta_1, \ldots, \beta_{m(1)}$.

To obtain the desired identity for $m(1) \times m(1)$ minors of $\overline{X}$ [or for that matter, any $m(1) \times (m(1) + m(2))$ matrix], we first note that univectors bounded by $m(1)+m(2)$ and of length $m(1)$ can alternatively be thought of as subsets of $[1, m(1)+m(2)]$ of cardinality $m(1)$, and we may tacitly do so when it is convenient. Given a subset T^* of $[1, m(1) + m(2)]$ of cardinality $m(1)$, or in other words, a univector T^* bounded by $m(1) + m(2)$ and of length $m(1)$, we let $[T^*]$ denote the $m(1) \times m(1)$ submatrix of $\overline{X}$ corresponding to T^*. Now let T_1^* and T_2^* be univectors bounded by $m(1) + m(2)$ and of length $m(1)$ such that $T_1^* \not\leq T_2^*$. Then there exists a unique $v \in [1, m(1)]$ such that

$$T_1^*(1, i) \leq T_2^*(1, i) \quad \text{for } 1 \leq i < v \quad \text{and } T_1^*(1, v) > T_2^*(1, v). \tag{14.10}$$

Let us take

$$B_1^* = \{T_1^*(1,i) : v \le i \le m(1)\} \text{ and } B_2^* = \{T_2^*(1,i) : 1 \le i \le v\}$$

and consider the $2m(1) \times 2m(1)$ matrix

$$y = \begin{array}{|c|c|} \hline [T_1^*] & [\widehat{B}_2^*] \\ \hline [B_1^*] & [\widehat{T}_2^*] \\ \hline \end{array}$$

where $[\widehat{B}_1^*]$ (resp: $[\widehat{B}_2^*]$) denotes the $m(1) \times m(1)$ submatrix of $[T_1^*]$ (resp: $[T_2^*]$) obtained by taking the columns corresponding to B_1^* (resp: B_2^*) and filling the remaining columns by zeros. Thus, for example, the first v columns of $[\widehat{B}_2^*]$ are the same as those of $[T_2^*]$, whereas the last $m(1)-v+1$ coulmns of $[\widehat{B}_1^*]$ are the same as those of $[T_1^*]$. Now we can expand y by Laplace developement considering the two blocks formed by the first $m(1)$ rows and by the last $m(1)$ rows. This gives

$$\det y = \sum_{\substack{A_k^* \subseteq B_k^* \\ |A_1^*|=|A_2^*|}} \pm \det\left([(T_1^* \setminus A_1^*) \cup A_2^*]\right) \det\left([(T_2^* \setminus A_2^*) \cup A_1^*\right) \tag{14.11}$$

where the sum is taken over all pairs $A^* = (A_1^*, A_2^*)$ with $A_1^* \subseteq B_1^*$, $A_2^* \subseteq B_2^*$, and $|A_1^*| = |A_2^*|$. On the other hand, if in y we subtract the lower block from the upper block [i.e., $(m(1)+i)^{\text{th}}$ row from i^{th} row, for $1 \le i \le m(1)$], and expand the resulting matrix by Laplace expansion as before, then in view of the fact that

$$|B_1^*| + |B_2^*| > m(1) \tag{14.12}$$

we see that in each term, one of the submatrices has a column of zero, and hence

$$\det y = 0. \tag{14.13}$$

Combining this with (15.11), and pulling out the term corresponding to $A_1^* = \emptyset = A_2^*$, we get

$$\det([T_1^*])\det([T_2^*]) = \sum_{\substack{A_k^* \subseteq B_k^* \\ |A_1^*|=|A_2^*|\neq 0}} \pm \det\left([S_1^*(T_1^*, A_1^*)]\right) \det\left([S_2^*(T_2^*, A_2^*)]\right) \tag{14.14}$$

where the sum is taken over all nonempty subsets A_1^*, A_2^* of B_1^*, B_2^* respectively such that $|A_1^*| = |A_2^*|$, and for any such pair (A_1^*, A_2^*), we have let $S_1^*(T_1^*, A_1^*)$ be the univector corresponding to $(T_1^* \setminus A_1^*) \cup A_2^*$ and $S_2^*(T_2^*, A_2^*)$ be the univector corresponding to $(T_2^* \setminus A_2^*) \cup A_1^*$. By the choice of B_1^* and

B_2^*, and the strict inequality in (15.10), it follows that for any (A_1^*, A_2^*) as above, we have

$$S_1^*(T_1^*, A_1^*) < T_1^*. \tag{14.15}$$

Thus, in view of (15.15), we can deduce (15.6) and (15.7) by repeated applications of (15.14); this is facilitated by restricting to bitableaux of a fixed area V which enables us to work only in the finite set $\text{stab}(2, m, V)$. Consequently, we get (15.8) and (15.9). In [A3], the condition (15.13) is generalised so as to obtain the identity (14) with weaker assumptions; an identity like (15.14), which clearly holds for any matrix of appropriate size, may be referred to as a universal determinantal identity. For details concerning this section, see [A3]. As remarked earlier, the ideas outlined here are fairly classical and go back at least to Hodge [Ho]; for proofs of (15.8) along similar lines, see [DEP], [DKR], [Se].

Finally, we use the results of the previous section to prove the

STANDARD BASIS THEOREM (4.2). *Given any nonnegative integer V, $\{\text{mom}(X, S) : S \in \text{stab}(2, m, V)\}$ forms a K-vector space basis of $K[X]_V$. Moreover, $\{\text{mom}(X, S) : S \in \text{stab}(2, m)\}$ forms a K-vector space basis of $K[X]$.*

Proof: The first assertion follows from (15.8) and (ii) of (3.7), and the second assertion follows from the first one.

For another proof, also based on enumeration, of the above theorem, see [AG].

14.5 Indexed Monomials

In this section, we describe a certain set of monomials in $X = (X_{ij})_{\substack{1 \le i \le m(1) \\ 1 \le j \le m(2)}}$ and indicate how this set can be enumerated. It turns out that this set of monomials has the same cardinality as that of $\text{stab}(2, m, p, a, V)$. Here, as usual, $m = (m(1), m(2))$ is a pair of positive integers, p is a positive integer, a is a bivector bounded by m and of length p, and V is a nonnegative integer. We proceed to give a brief sketch of the main ideas of Abhyankar concerning the enumeration of these so called indexed monomials. For details, see Sections 5 and 8 of [A3]. Some applications of the results in this section, which may be regarded as a motivation to obtain these results, are indicated in the next section.

Let us begin by noting that a monomial in the $m(1)m(2)$ variables X_{ij}'s can be thought of as a map from the rectangle $[1, m(1)] \times [1, m(2)]$ into the set $\mathbb{N}$ of nonnegative integers. Indeed, if $t : [1, m(1)] \times [1, m(2)] \to \mathbb{N}$ is any map, then

$$X^t \stackrel{\text{def}}{=} \prod_y X_{y(1)\, y(2)}^{t(y)},$$

where the product is taken over all pairs $y = (y(1), y(2)) \in [1, m(1)] \times$

$[1, m(2)]$, is a monomial in the X_{ij}'s. Conversely, a monomial in X_{ij}'s obviously corresponds to such a map. In this terminology, we have the notions of *absolute value* abs(t) and *support* supp(t) of a monomial t, which are defined as follows.

$$\text{abs}(t) = \deg_{(} X^t) = \sum_{y \in [1,m(1)] \times [1,m(2)]} t(y)$$

and

$$\text{supp}(t) = \{y \in [1, m(1)] \times [1, m(2)] : t(y) \neq 0\}.$$

Let mon$(2, m)$ denote the set of all maps $t : [1, m(1)] \times [1, m(2)] \to \mathbb{N}$, and given any nonnegative integer V, let mon$(2, m, V)$ be the set of all elements $t \in$ mon$(2, m)$ such that abs$(t) = V$. Now we have the key notion of *index* of a subset of $[1, m(1)] \times [1, m(2)]$; the index of a monomial will be the index of its support, and roughly speaking, it is equal to the largest integer j such that the principal diagonal of a $j \times j$ minor of X divides this monomial. More precisely, for any subset Y of $[1, m(1)] \times [1, m(2)]$, we define

$$\text{ind}(Y) = \max\{j : \exists\, y_1, \ldots, y_j \in Y \text{ such that } y_i(k) < y_{i+1}(k)\ \forall\ k \in [1, 2] \text{ and } i \in [1, j-1]\}$$

and for any monomial $t \in$ mon$(2, m)$, we define

$$\text{ind}(t) = \text{ind}(\text{supp}(t)).$$

Given a bivector a bounded by m and of length p, we can draw vertical and horizontal lines in the rectangle $[1, m(1)] \times [1, m(2)]$ emanating from $a(k, i)$, where k can be 1 or 2, and i varies between 1 and p, and consider the monomials satisfying an index condition when restricted to the region above [or to the left of] $a(k, i)$. More precisely, given any $k \in [1, 2], i \in [1, p]$, and $t \in$ mon$(2, m)$, we let

$$t_{ki} = \{y \in \text{supp}(t) : y(k) < a(k, i)\}$$

and we define

$$\text{mon}(2, m, p, a) = \{t \in \text{mon}(2, m) : \text{ind}(t) \leq p \text{ and } \text{ind}(t_{ki}) < i\ \forall k \in [1, 2], \text{ and } i \in [1, p]\}$$

and for any $V \in \mathbb{N}$, we define

$$\text{mon}(2, m, p, a, V) = \text{mon}(2, m, p, a) \cap \text{mon}(2, m, V).$$

The elements of this set may be referred to as *indexed monomials.* We wish to find the cardinality of this set. The basic strategy is similar to that employed in finding $|\mathrm{stab}(q,m,p,a,V)|$, namely, to first find recurrence relations satisfied by the cardinality, and then discover an explicit formula satisfying the same recurrence relations. To obtain a recurrence relation satisfied by $|\mathrm{mon}(2,m,p,a,V)|$, we enlarge the rectangle $[1,m(1)]\times[1,m(2)]$ on one of its sides, say the kth side, and look at $[1,m^*(1)]\times[1,m^*(2)]$ where $m^*(k) = m(k)+1$ and $m^*(k') = m(k')$ [k' equals 2 or 1 according as k is 1 or 2, i.e., $k' = 3-k$]. Now a monomial on the larger rectangle can be split into a monomial on the original rectangle and a monomial defined on the remaining line. The latter can be counted easily. Keeping track of conditions on index and such, it is not hard to see that we get the recurrence relations described below.

Basically, we have 2 recurrence relations, of which the first one tells how to express $|\mathrm{mon}(2,m^*,p,a^*,V)|$ in terms of $|\mathrm{mon}(2,m,p,a,V)|$ when the vector a^* has length p and is such that $[m^*(k)-a^*(k,p)] = [m(k)-a(k,p)]+1$, and the second one tells us how to express $|\mathrm{mon}(2,m^*,p+1,a^*,V)|$ in terms of $|\mathrm{mon}(2,m,p,a,V)|$ when the vector a^* has length $p+1$ and is such that $[m^*(k)-a^*(k,p+1)] = 0$. To describe both the relations in one stroke, we can consider a bivector b of length $p+1$, and, for any subset u of $[1,p]$, let

$$M[p,b,k,u] = \{a : \begin{array}{l} a(k,i) = b(k,i)\ \forall i \in [1,p] \text{ and} \\ b(k',j) = a(k',j)\ \forall j \in [1,p]\setminus u, \text{ and} \\ b(k',i) < a(k',i) < b(k',i+1)\ \forall\ i \in u\}; \end{array}$$

note that the elements of $M[p,b,k,u]$ are bivectors of length p. For any $U \in [0,p]$, let

$$M(p,b,k,U) = \bigcup_{\substack{u\subseteq[1,p] \\ |u|=U}} M[p,b,k,u].$$

Recall that $m^* = (m^*(1),m^*(2))$ is given by $m^*(k) = m(k)+1$ and $m^*(k') = m(k')$, and let a positive integer p^* and a bivector a^* bounded by m^* and of length p^* be given. Consider the following conditions on p^* and b.

$$\left\{ \begin{array}{l} p^* = p,\ a^*(k,p^*) \neq m^*(k),\ b(\hat{k},i) = a^*(\hat{k},i), \text{ and} \\ b(\hat{k},p+1) = m(\hat{k})+1 \text{ for all } \hat{k} \in [1,2] \text{ and } i \in [1,p] \end{array} \right. \tag{14.16}$$

or

$$p^* = p+1,\ a^*(k,p^*) = m^*(k), \text{ and } b = a^*. \tag{14.17}$$

Finally, let

$$C^* = p + m(k') - b(k',p+1) + \sum_{i=1}^{p} [a(k',i) - b(k',i)].$$

Now the said recurrence relations can be stated as follows.

LEMMA (5.1). *Let the notation be as above. Assume that either (15.16) holds or (15.17) holds. Then*

$$|\mathrm{mon}(2, m^*, p^*, a^*, V)| =$$

$$\sum_{U=0}^{p} \sum_{a \in M(p,b,k,U)} \sum_{\substack{v_1, v_2 \in \mathbb{N} \\ v_1+v_2=V}} |\mathrm{mon}(2, m, p, a, v_1)| \begin{bmatrix} C^* \\ v_2 - U \end{bmatrix}.$$

Let us also note that we can easily compute $|\mathrm{mon}(2, m, p, a, V)|$ in the initial case, i.e.,

$$\begin{cases} \text{if the bivector } a \text{ has length } p = 1, \text{ and if} \\ m(k) - a(k,p) = 0, \text{ then for every } V \in \mathbb{N}, \\ |\mathrm{mon}(2, m, p, a, V)| = \begin{bmatrix} m(k') - a(k', 1) \\ V \end{bmatrix}. \end{cases}$$

By making double induction on p and $m(k) - a(k,p)$, we see that the cardinality of $\mathrm{mon}(2, m, p, a, V)$ is determined by the above recurrence relations and initial condition. In 1984, Abhyankar obtained a formula satisfying the same properties; this formula is denoted by $F^{(1k)}(m, p, a, V)$ and it can be defined as follows.

$$F^{(1k)}(m, p, a, V) = \sum_{D \in \mathbb{Z}} (-1)^D F_D^{(1k)}(m, p, a) \begin{bmatrix} V \\ C(m, p, a) - D \end{bmatrix}$$

where $C(m, p, a) = (p - 1) + C_1 + C_2$ with $C_k = \sum_{i=1}^{p} [m(k) - a(k, i)]$, and for any $D \in \mathbb{Z}$,

$$F_D^{(1k)}(m, p, a) = \sum_{E \in \mathbb{Z}} (-1)^{C_k - E} \binom{E}{D + E - C_k} H_E^{(1k)}$$

where for any $E \in \mathbb{Z}$,

$$H_E^{(1k)} = \sum_{e \in \mathbb{Z}(p,E)} \det \left(\begin{bmatrix} m(k) - a(k, j) + j - i \\ m(k) - a(k, j) - e(i) \end{bmatrix} \right).$$

It may be noted that all the summations above are essentially finite. Thus we have the following

THEOREM (5.2). *For any $k \in [1, 2]$, we have $|\mathrm{mon}(2, m, p, a, V)| = F^{(1k)}(m, p, a, V)$.*

Now we already have enumerative formulas for $|\mathrm{stab}(2, m, p, a, V)|$, and notable among those is $F^{(3k)}(m, p, a, V)$ which is a polynomial in V and which has an expression similar to $F^{(1k)}(m, p, a, V)$. Transforming the summations and using some properties of binomial coefficients, in a manner somewhat similar to that in Section 15.3, Abhyankar proved that for every $D \in \mathbb{Z}$, we have

$$F_D^{(3k)}(m,p,a) = F_D^{(1k)}(m,p,a).$$

Combining this with (3.7) and (5.2), we get the following

THEOREM (5.3). *The sets* $\mathrm{stab}(2,m,p,a,V)$ *and* $\mathrm{mon}(2,m,p,a,V)$ *have the same cardinality. Moreover, the common cardinality is given by the equivalent formulas* $F^{(3k)}(m,p,a,V)$ *or* $F^{(1k)}(m,p,a,V)$.

Note that from (5.3), we can easily deduce that $|\mathrm{stab}(2,m,V)| = \dim_K K[X]_V$; indeed, $\mathrm{mon}(2,m,V)$ simply consists of all monomials of degree V in $m(1)m(2)$ indeterminates.

14.6 Determinantal Ideals and their Hilbert Functions

Let K be a field and let $X = (X_{ij})$ be an $m(1) \times m(2)$ matrix whose entries are independent indeterminates over K, and let $K[X]$ denote the corresponding ring of polynomials. Given a positive integer $p \le \min\{m(1), m(2)\}$, let I_p denote the ideal in $K[X]$ generated by all $p \times p$ minors of X. Also, given any bivector a bounded by the pair $m = (m(1), m(2))$ of positive integers and of length p, let $I(p,a)$ denote the ideal in $K[X]$ generated by all $(p+1) \times (p+1)$ minors of X, all $i \times i$ minors of the submatrix $(X_{uv})_{\substack{1\le u\ lea(1,i)-1 \\ 1\le v\le m(2)}}$, and all $j \times j$ minors of $(X_{uv})_{\substack{1\le u\le m(1) \\ 1\le v\le a(2,j)-1}}$, where i and j range over $[1,p]$. Equivalently, $I(p,a)$ is the ideal in $K[X]$ generated by $\mathrm{mor}(X,b)$ with b ranging over the set of all bivectors bounded by m such that $a \not\le b$ (verify!). Note that in the case when $a(1,i) = i = a(2,i)$ for all $i \in [1,p]$, we have $I(p,a) = I_{p+1}$. The study of the ideals I_p is classical, and the primality of such ideals was first proved by E. Pascal in 1888 [P] in the form of Second Fundamental Theorem of Invariant Theory. For more recent proofs, see: [M], [HE], [DEP], [A1]. It may be noted that ideals similar to $I(p,a)$ have also been studied in [HE], [Mus2]. In this section, we shall describe explicit K–vector space bases for the ideal $I(p,a)$ and its residue class ring $K[X]/I(p,a)$, using the ideas of straightening. $I(p,a)$ is clearly a homogeneous ideal, and these bases induce bases for the homogeneous components of $I(p,a)$ and $K[X]/I(p,a)$. In fact, the set $\{\mathrm{mom}(X,S) : S \in \mathrm{stab}(2,m,p,a,V)\}$ is a K–basis for $K[X]_V/I(p,a)_V$. Consequently, $F^{(3k)}(m,p,a,V)$ gives the Hilbert function of $I(p,a)$. Thus the ideal $I(p,a)$ is *Hilbertian* in the sense that its Hilbert function equals its Hilbert polynomial for *all* values of the parameter V. As another consequence of the Straightening Law, we can deduce that $I(p,a)$ is, in fact, a homogeneous prime ideal and the corresponding projective variety is rational, i.e., the quotient field of $K[X]/I(p,a)$ is a pure transcendental extension of K. The indexed monomials can be used to give another basis for

$I(p,a)$. This basis has the advantage that it can be adapted to give bases for a more general class of ideals such as the ladder determinantal ideal $I_p(L)$ discussed in the Introduction; moreover, we can also use it to prove the primality of these ideals.

To describe a basis of $K[X]/I(p,a)$, let us define

$$\text{stab}(2,m,p,a) = \{T \in \text{stab}(2,m) : a \leq T[e] \text{ for all } e \in [1, \text{dep}(T)]\}.$$

THEOREM (6.1). *Given m,p,a as above, we have the following.*

(i) $\{\text{mom}(X,S) : S \in \text{stab}(2,m) \setminus \text{stab}(2,m,p,a)\}$ is a K–basis of $I(p,a)$, whereas $\{\text{mom}(X,S) : S \in \text{stab}(2,m,p,a)\}$ is a K–basis of $K[X]/I(p,a)$.

(ii) For every $V \in \mathbb{N}$, $\{\text{mom}(X,S) : S \in \text{stab}(2,m,V) \setminus \text{stab}(2,m,p,a,V)\}$ is a K–basis of $I(p,a)_V$, whereas $\{\text{mom}(X,S) : S \in \text{stab}(2,m,p,a,V)\}$ is a K–basis of $K[X]_V/I(p,a)_V$.

(iii) For every $V \in \mathbb{N}$, $\dim_K K[X]_V/I(p,a)_V = |\text{stab}(2,m,p,a,V)|$.

Proof: Since $I(p,a)$ is generated by $\{\text{mor}(X,b) : a \not\leq b\}$, given any polynomial $B \in I(p,a)$, we can find bivectors $b_1, \ldots, b_n$ with $a \not\leq b_i$ for each $i \in [1,n]$, elements $r_1, \ldots, r_n$ in K, and (usual) monomials $M_1, \ldots, M_n$ in the X_{ij}'s such that

$$B = \sum_{i=1}^{n} r_i \, \text{mor}(X, b_i) M_i.$$

For each $i \in [1,n]$, the monomial M_i can be regarded as a monomial in minors of X corresponding to a bitableau U_i such that every row of U_i is of length 1. Now if we let T_i denote the bitableau obtained by augmenting U_i with b_i, i.e, $T_i[1] = b_i$ and $T_i[e] = U_i[e-1]$ for $e = 2, 3, \ldots, \text{dep}(T_i) = \text{dep}(U_i) + 1$, then we have $\text{mor}(X, b_i) M_i = \text{mom}(X, T_i)$. By (15.6) and (15.7) of Section 15.4, we see that each $\text{mom}(X, T_i)$ can be written as an integral linear combination of $\text{mom}(X, S_{ij})$ for some $S_{ij} \in \text{stab}(2,m)$ with $S_{ij}[1] \leq T_i[1] = b_i$. Hence we can find $S_1, \ldots, S_N$ in $\text{stab}(2,m)$ and elements $t_1, \ldots, t_N$ in K such that $a \not\leq S_i[1]$ (so that $S_i \in \text{stab}(2,m) \setminus \text{stab}(2,m,p,a)$), and write

$$B = \sum_{i=1}^{N} t_i \, \text{mom}(X, S_i). \tag{14.18}$$

Moreover, given any $S \in \text{stab}(2,m) \setminus \text{stab}(2,m,p,a)$, we have $\text{dep}(S) \neq 0$ and $a \not\leq S[1]$. Hence one of the following statements must hold.

1. $\exists i \in [1,p]$ such that $S[1](1,\ell) \geq a(1,\ell)$ for $1 \leq \ell < i$, and $S[1](1,i) < a(1,i)$.

2. $\exists j \in [1,p]$ such that $S[1](2,\ell) \geq a(2,\ell)$ for $1 \leq \ell < j$, and $S[1](2,j) < a(2,j)$.

3. the length of $S[1]$ is $\geq p+1$.

In each of these cases, it is easily seen that $\mathrm{mor}(X, S[1])$ can be written as a sum of multiples of the minors which are amongst generators of $I(p,a)$; therefore $\mathrm{mor}(X, S[1]) \in I(p,a)$. Thus $\{\mathrm{mom}(X,S) : S \in \mathrm{stab}(2,m) \setminus \mathrm{stab}(2,m,p,a)\}$ is a subset of $I(p,a)$. Note that this set is linearly independent being a subset of the basis $\{\mathrm{mom}(X,S) : S \in \mathrm{stab}(2,m)\}$ of $K[X]$. This proves (i). Assertions (ii) and (iii) follow immediately from (i).

COROLLARY (6.2). *For any $k \in [1,2]$, $F^{(3k)}(m,p,a,V)$ is the Hilbert function as well as the Hilbert polynomial of $I(p,a)$. In particular, if $\mathcal{V}(p,a)$ denotes the projective variety corresponding to $I(p,a)$, then*

$$\dim \mathcal{V}(p,a) = \deg_V F^{(3k)}(m,p,a,V) = (p-1) + \sum_{\hat{k}=1}^{2}\sum_{i=1}^{p}[m(\hat{k}) - a(\hat{k},i)]$$

and the degree of $\mathcal{V}(p,a)$ equals $F_0^{(3k)}(m,p,a)$.

Proof: Follows from (6.1) and a well–known theorem of Hilbert (see, for example, pp. 235-236 of [ZS]).

Now we discuss the primality of the ideal $I(p,a)$. A key step in Abhyankar's proof is the following

LEMMA (6.3). *The $p\times p$ minor of X corresponding to a is not a zero divisor in $K[X]/I(p,a)$, i.e.,*

$$B \in K[X],\ \mathrm{mor}(X,a)B \in I(p,a) \Rightarrow B \in I(p,a).$$

Proof: By SBT, we can write $B = B_0 + B_1$ with

$$B_0 = \sum_{i=1}^{n_0} r_i \,\mathrm{mom}(X,T_i) \quad \text{and} \quad B_1 = \sum_{j=1}^{n_1} t_j \,\mathrm{mom}(X,S_j)$$

where $T_i \in \mathrm{stab}(2,m) \setminus \mathrm{stab}(2,m,p,a)$, $S_j \in \mathrm{stab}(2,m,p,a)$, r_i, s_j are elements of K for $1 \le i \le n_0$, $1 \le j \le n_1$. Assume that $\mathrm{mor}(X,a)B \in I(p,a)$. By (6.1), we see that $B_0 \in I(p,a)$, and hence $\mathrm{mor}(X,a)B_1 \in I(p,a)$. But we clearly have

$$\mathrm{mor}(X,a)B_1 = \sum_{j=1}^{n_1} t_j \,\mathrm{mom}(X,S_j^*)$$

where for each $j \in [1,n_1]$, S_j^* is the bitableau obtained by augmenting a with S_j. Since $a \le S_j$, $S_j^* \in \mathrm{stab}(2,m,p,a)$. In view of (6.1), it follows that we must have $t_j = 0$ for each $j \in [1,n_1]$. Thus $B_1 = 0$, and hence $B = B_0 \in I(p,a)$.

Once we have this lemma at our disposal, proving the primality is basically a clever application of Elementary Linear Algebra. For the sake of simplicity, we just consider the ideal I_{p+1} and outline a proof that it is prime. Proof in the general case, i.e., for $I(p,a)$, is based on analogous ideas and later we shall indicate how one proceeds in this case. Just as before, [A3] is the reference for greater details. Now we introduce some obvious notation.

NOTATION. Given any subset Y of $[1, m(1)] \times [1, m(2)]$, by $K[Y]$ we denote the ring of polynomials in the indeterminates X_{ij} with (i, j) ranging over Y, and by $K(Y)$ we denote the quotient field of $K[Y]$. By $E[p]$ we will denote the bivector of length p given by $E[p](1, i) = i = E[p](2, i)$ for all $i \in [1, p]$. Note that $I(p, E[p]) = I_{p+1}$.

To show that I_{p+1} is prime, we define a subset $y[p+1]$ of $[1, m(1)] \times [1, m(2)]$, and a homomorphism $f[p+1] : K[X] \to K(y[p+1])$, and prove that I_{p+1} equals the kernel of $f[p+1]$. The set $y[p+1]$ is obtained by cutting out the subrectangle $[p+1, m(1)] \times [p+1, m(2)]$, i.e.,

$$y[p+1] = \{(i, j) \in [1, m(1)] \times [1, m(2)] : \text{either } i \leq p \text{ or } j \leq p\}.$$

If $(i, j) \notin y[p+1]$, then we consider the $(p+1) \times (p+1)$ submatrix M_{ij} of X formed by the rows $1, 2, \ldots, p, i$ and the columns $1, 2, \ldots, p, j$. Clearly, we can find a unique element $\widetilde{A}[X, p+1, i, j] \in K[y[p+1]]$ such that

$$\det M_{ij} = \operatorname{mor}(X, E[p]) X_{ij} - \widetilde{A}[X, p+1, i, j]. \tag{14.19}$$

We define $f[p+1]$ to be the unique K-homomorphism of $K[X]$ into $K(y[p+1])$ such that

$$f[p+1](X_{ij}) = \begin{cases} X_{ij} & \text{if } (i, j) \in y[p+1] \\ \frac{\widetilde{A}[X, p+1, i, j]}{\operatorname{mor}(X, E[p])} & \text{if } (i, j) \notin y[p+1]. \end{cases}$$

Let us write $x_{ij} = f[p+1](X_{ij})$. To show that $I_{p+1} \subseteq \ker f[p+1]$, it suffices to show that the rank of the $m(1) \times m(2)$ matrix $x = (x_{ij})$ is equal to p. Given any $i \in [p+1, m(1)]$, consider the $(p+1) \times m(2)$ submatrix, say $\widehat{M}$, formed by the rows $1, 2, \ldots, p, i$ of x. Since $\operatorname{mor}(x, E[p]) = \operatorname{mor}(X, E[p]) \neq 0$, the first p rows of x are linearly independent, and so $\operatorname{rank}(\widehat{M}) \geq p$. On the other hand, for any $j \in [p+1, m(2)]$, the determinant of the $(p+1) \times (p+1)$ submatrix of $\widehat{M}$ formed by its columns $1, 2, \ldots, p, j$, is zero, and hence the jth column of $\widehat{M}$ is linearly dependent on the first p columns of $\widehat{M}$, and therefore $\operatorname{rank}(\widehat{M}) \leq p$. This shows that the ith row of x is linearly dependent on the first p rows of x, for any $i \in [p+1, m(1)]$. It follows that $\operatorname{rank}(x) = p$ and $I_{p+1} \subseteq \ker f[p+1]$. To prove the reverse inclusion, let us take any polynomial $B(X)$ in $K[X]$. In view of (15.19), we see that upon multiplying $B(X)$ by a large enough power of $\operatorname{mor}(X, E[p])$, we have

$$\operatorname{mor}(X, E[p])^e B(X) = B^{**}(X) + \sum_{(i,j) \notin y[p+1]} B^*_{ij}(X) \det M_{ij}$$

for some $B^{**}(X) \in K[y[p+1]]$, polynomials $B^*_{ij}(X) \in K[X]$, and a non-negative integer e. Suppose $B(X) \in \ker f[p+1]$. Since $\det M_{ij} \in I_{p+1}$ and

$I_{p+1} \subseteq \ker f[p+1]$, it follows that $B^{**}(X) \in \ker f[p+1]$. But $f[p+1]$ is identity on $K[y[p+1]]$, so we must have $B^{**}(X) = 0$. Hence $\text{mor}(X, E[p])^e B(X) \in I_{p+1}$, and so, in view of (6.3), we conclude that $B(X) \in I_{p+1}$. This proves that I_{p+1} is prime.

In the general case, the ideal $I(p, a)$ is proved to be the kernel of a homomorphism $f(p, a) : K[X] \to K(y[p, a])$. The set $y[p, a]$ is the union of all "hooks" corresponding to the nodes $(a(1, i), a(2, i))$ as i ranges over $[1, p]$; more precisely,

$$y[p, a] = \bigcup_{i=1}^{p} (\{a(1, i)\} \times [a(2, i), m(2)]) \cup ([a(1, i), m(1)] \times \{a(2, i)\}).$$

Note that $y[p, E[p]] = y[p + 1]$. The map $f(p, a)$ is defined to be identity on $y[p, a]$, and zero above the first hook, and in the region below the last hook or between any two hooks it is defined by expanding a minor of an appropriate size as described below. If $(i, j) \in [1, m(1)] \times [1, m(2)]$ is such that $i \geq a(1, 1)$ and $j \geq a(2, 1)$ and $(i, j) \notin y[p, a]$, then we can find a unique integer $u \in [1, p]$ which is largest with the property that $a(1, u) < i$ and $a(2, u) < j$. Now consider the $(u+1) \times (u+1)$ submatrix $M_{ij}(u)$ formed by the rows $a(1, 1), \ldots, a(1, u), i$ and the columns $a(2, 1), \ldots, a(2, u), j$ of X. Let $E[p, a, u]$ denote the bivector of length u obtained by restricting a to $[1, 2] \times [1, u]$. We can find a unique polynomial $\widetilde{A}(X, p, a, i, j) \in K[y[p, a]]$ such that

$$\det M_{ij}(u) = \text{mor}(X, E[p, a, u]) X_{ij} - \widetilde{A}(X, p, a, i, j).$$

Now we define $f(p, a)$ to be the unique K-homomorphism of $K[X]$ into $K(y[p, a])$ such that

$$f(p, a)(X_{ij}) = \begin{cases} 0 & \text{if } i < a(1, 1) \text{ or } j < a(2, 1) \\ X_{ij} & \text{if } (i, j) \in y[p, a] \\ \frac{\widetilde{A}(X, p, a, i, j)}{\text{mor}(X, E[p, a, u])} & \text{if } (i, j) \notin y[p, a],\ i \geq a(1, 1), \\ & \quad \text{and } j \geq a(2, 1). \end{cases}$$

The inclusion $I(p, a) \subseteq \ker f(p, a)$ is proved in an analogous manner, except that we now have more cases to consider. The proof of the desired equality can then be completed using (6.3) and by inducting on p. For details, see [A3].

From the definition of $f(p, a)$ it is clear that it induces a K-isomorphism of the quotient field of $K[X]/I(p, a)$ onto $K(y[p, a])$, and thus we can deduce the following

THEOREM (6.4). *The quotient field of $K[X]/I(p, a)$ is a pure transcendental extension of K, and its transcendence degree is equal to $|y[p, a]|$. In other words, the irreducible projective variety $\mathcal{V}(p, a)$ defined by $I(p, a)$, is rational and its dimension is equal to $|y[p, a]| - 1$.*

Remark: Note that by the definition of $y[p,a]$, we clearly have

$$|y[p,a]| = \sum_{i=1}^{p}[m(1)-a(1,i)+m(2)-a(2,i)+1] = p+\sum_{\hat{k}=1}^{2}\sum_{i=1}^{p}[m(\hat{k})-a(\hat{k},i)]$$

and thus the dimension of $\mathcal{V}(p,a)$ given by Theorem (6.4) is in agreement with Corollary (6.2), as is to be expected.

Now we proceed to show how indexed monomials can be used to give another set of bases for the homogeneous components of $I(p,a)$ and $K[X]/I(p,a)$. The ideal $I(p,a)$ being prime, cannot, in general, contain the usual monomials. So we must adjust them suitably. Let us explain how one can do this.

Given any bivector b bounded by m, we associate to it a monomial, denoted by $d(b)$, which represents the product of the indeterminates on the principal diagonal of the submatrix of X corresponding to b; thus

$$X^{d(b)} = \prod_i X_{b(1,i)\,b(2,i)}$$

where i varies from 1 to the length of b. We also allow the bivector to have length 0, in which case we set $X^{d(b)} = 1$, i.e., $d(b)(y) = 0$ for all $y \in [1,m(1)] \times [1,m(2)]$. Now let us fix some $V \in \mathbb{N}$. To a monomial $t \in \text{mon}(2,m,V)$, we shall associate a bivector $b = g(t)$, and this will have the property that $X^{d(b)}$ divides X^t. The basis of indexed monomials for $K[X]_V$, which induces bases for $I(p,a)_V$ and $K[X]_V/I(p,a)_V$, is given by $\{w(t) : t \in \text{mon}(2,m,V)\}$ where

$$w(t) = \frac{X^t}{X^{d(g(t))}}\text{mor}(X, g(t)).$$

In order to define the associated bivector $b = g(t)$, we first order the elements of the rectangle lexicographically, i.e., for any $y = (y(1), y(2))$, $\tilde{y} = (\tilde{y}(1), \tilde{y}(2))$ in $[1,m(1)] \times [1,m(2)]$, we define

$$y < \tilde{y} \iff \text{either } y(1) < \tilde{y}(1) \text{ or } y(1) = \tilde{y}(1) \text{ and } y(2) < \tilde{y}(2).$$

And we put the reverse lexicographic order on monomials, i.e., for any t, t^* in $\text{mon}(2,m,V)$, we define

$$t < t^* \iff \exists y \text{ such that } t(y) > t^*(y) \text{ and } t(\tilde{y}) = t^*(\tilde{y})\ \forall \tilde{y} < y;$$

here y and $\tilde{y}$ denote elements of $[1,m(1)] \times [1,m(2)]$. Note that this gives a total order on $\text{mon}(2,m,V)$, and that $\text{mon}(2,m,V)$ is a finite set, so in particular, we may speak of the least element in any nonempty subset of $\text{mon}(2,m,V)$. The associated bivector $b = g(t)$ can now be defined by considering the four cases below.

Case 0: $\text{ind}(t) \le p$, $\text{ind}(t_{1i}) < i$, and $\text{ind}(t_{2j}) < j$ for all $i, j \in [1,p]$.

In this case we take $b = g(t)$ to be the empty bivector of length 0 so that $X^{d(b)} = 1$.

Case 1: $\mathrm{ind}(t_{1i}) \geq i$ for some $i \in [1,p]$.
In this case we let u be the smallest such i and $b = g(t)$ be the unique bivector of length u such that its associated monomial $d(b)$ is the least among $d(b^*)$ where b^* ranges over the nonempty set of all bivectors bounded by m and of length u such that $b^*(1,u) < a(1,u)$ and $X^{d(b^*)}$ divides X^t.

Case 2: $\mathrm{ind}(t_{1i}) < i$ for all $i \in [1,p]$ and $\mathrm{ind}(t_{2j}) \geq j$ for some $j \in [1,p]$.
In this case we let u be the smallest such j and $b = g(t)$ be defined exactly as in Case 1 except that the condition $b^*(1,u) < a(1,u)$ is replaced by $b^*(2,u) < a(2,u)$.

Case 3: $\mathrm{ind}(t_{1i}) < i$, $\mathrm{ind}(t_{2j}) < j$ for all $i,j \in [1,p]$ and $\mathrm{ind}(t) > p$.
In this case $b = g(t)$ is defined to be the unique bivector of length $p+1$ such that $d(b)$ is the least among all $d(b^*)$ where b^* ranges over the nonempty set of all bivectors bounded by m and of length $p+1$ such that $X^{d(b^*)}$ divides X^t.

We now show that $\{w(t) : t \in \mathrm{mon}(2,m,V)\}$ is a linearly independent subset of $K[X]$. To this end, we first note that for any bivector b bounded by m, the associated monomial $d = d(b)$ is the least among all the monomials in the expansion of the minor of X corresponding to b; in other words, we can write

$$\mathrm{mor}(X,b) = X^d + \sum_{r>d} \epsilon_r X^r \quad \text{with } \epsilon_r \in \{0,1,-1\}.$$

Consequently, for any $t \in \mathrm{mon}(2,m,V)$, we can write

$$w(t) = X^t + \sum_{s>t} \epsilon_s X^s \quad \text{with } \epsilon_s \in \{0,1,-1\}. \tag{14.20}$$

Using this we can deduce the following

LEMMA (6.5). *$\{w(t) : t \in \mathrm{mon}(2,m,V)\}$ is a K-linearly independent subset of $K[X]$.*

Proof: Suppose a nontrivial linear combination $\sum \alpha_t w(t)$ were zero, then upon letting $t^* = \min\{t : \alpha_t \neq 0\}$, by (15.20) we see that

$$\alpha_{t^*} X^{t^*} + \sum_{s^*>t^*} \beta_{s^*} X^{s^*} = 0 \quad \text{for some } \beta_{s^*} \in K,$$

and hence $\alpha_{t^*} = 0$, which is a contradiction.

Note that if $t \in \mathrm{mon}(2,m,V)$ is such that $t \notin \mathrm{mon}(2,m,p,a)$, then $\mathrm{mor}(X,g(t)) \in I(p,a)$, and hence $w(t) \in I(p,a)$. In fact, this conclusion holds even when we restrict to subsets of the rectangle, which are "saturated" in the sense that if a principal diagonal of a minor is there, then so is the minor. More precisely, we make the following

Definition (6.6). A subset Y of $[1, m(1)] \times [1, m(2)]$ is said to be *saturated* if for any bivector b bounded by m, we have

$$(b(1,i), b(2,i)) \in Y \ \forall i \Rightarrow (b(1,i), b(2,j)) \in Y \ \forall i,j$$

where i and j vary from 1 to the length of b.

Given any subset Y of $[1, m(1)] \times [1, m(2)]$, we let

$$\begin{aligned}
\text{mon}(2,m;Y) &= \{t \in \text{mon}(2,m) : \text{supp}(t) \subseteq Y\} \\
\text{mon}(2,m,V;Y) &= \text{mon}(2,m,V) \cap \text{mon}(2,m;Y) \\
\text{mon}(2,m,p,a;Y) &= \text{mon}(2,m,p,a) \cap \text{mon}(2,m;Y) \\
\text{mon}(2,m,p,a,V;Y) &= \text{mon}(2,m,p,a,V) \cap \text{mon}(2,m;Y)
\end{aligned}$$

and we let $I^*(p,a,Y)$ denote the ideal in $K[Y]$ generated by $\{\text{mor}(X,b) : b \not\leq a\} \cap K[Y]$.

Theorem (6.7). *Let the notation be as above. We have the following*

(i) $\{w(t) : t \in \text{mon}(2,m,V)\}$ *is a K–basis of* $K[X]_V$.

(ii) $\{w(t) : t \in \text{mon}(2,m,V) \backslash \text{mon}(2,m,p,a,V)\}$ *is a K–basis of* $I(p,a)_V$, *whereas* $\{w(t) : t \in \text{mon}(2,m,p,a,V)\}$ *is a K–basis of* $K[X]_V/I(p,a)_V$.

(iii) For any saturated subset Y of $[1, m(1)] \times [1, m(2)]$, $\{w(t) : t \in \text{mon}(2,m,V;Y)\}$ *is a K–basis of* $K[Y]_V$, *and moreover,* $\{w(t) : t \in \text{mon}(2,m,V;Y) \backslash \text{mon}(2,m,p,a,V;Y)\}$ *is a K–basis of* $I^*(p,a,Y)_V$, *whereas* $\{w(t) : t \in \text{mon}(2,m,p,a,V;Y)\}$ *is a K–basis of* $K[X]_V/I^*(p,a,Y)_V$.

Proof: Clearly $|\text{mon}(2,m,V)| = \dim_K K[X]_V$, and hence (i) follows from (6.5). Next, by (5.2) and (6.1), we see $|\text{mon}(2,m,V) \setminus \text{mon}(2,m,p,a,V)| = \dim_K I(p,a)_V$, and hence (6.5) implies the first part of (ii), whereas the second part of (ii) follows from (i) and the first part of (ii). Finally, since Y is saturated, for any $t \in \text{mon}(2,m,V;Y)$, $w(t) \in K[Y]_V$, and moreover if $t \notin \text{mon}(2,m,p,a,V;Y)$, then $w(t) \in I(p,a)_V$. Also we obviously have $|\text{mon}(2,m,V;Y)| = \dim_K K[Y]_V$. Hence (iii) can be deduced from (i) and (ii).

Corollary (6.8). *For any saturated subset Y of* $[1, m(1)] \times [1, m(2)]$, *we have* $I^*(p,a,Y) = I(p,a) \cap K[Y]$, *and hence the ideal* $I^*(p,a,Y)$ *is prime.*

Remark: Note that a ladder shaped subset L of $[1, m(1)] \times [1, m(2)]$ is clearly saturated, and as a special case of (6.8), it follows that the ideal $I_p(L)$ discussed in the Introduction is prime. Primality in this special case is also proved in [N].

14.7 Some Recent developments

In this section, we shall briefly discuss some of the questions suggested by the work of Abhyankar that was described in the previous sections, and indicate the progress that has taken place in the last five years. Most of the problems mentioned here were posed in [A3].

14.7.1 Enumeration of Standard Multitableaux

Here one has the combinatorial problem of finding a nice polynomial formula for $|\text{stab}(q, m, p, a, V)|$ where q can be any positive integer. Recall that in Section 15.3, we obtained formulas $F^{(8)}$ and $F^{(7)}$ for this, and the formulas $F^{(6)}$ and $F^{(5k)}$ in the case of even q. A polynomial formulas was, however, obtained only when $q = 2$, and we have already seen how crucial that was. An answer to this problem in the case of even q, was given in 1986 by Ghorpade (cf. [Gh1]). It turns out that the higher dimensional determinants introduced by Cayley in 1843 [C] appear naturally in such a formula. For definitions and basic properties of higher dimensional determinants, see the last chapter of [MM] or Section 4 of [Gh1]. One of the key steps in obtaining this formula is to prove the so called Multiproduct Lemmas which say that if $A_1, A_2, \ldots, A_q$ are any integers and V is an indeterminate over $\mathbb{Q}$, then

$$\binom{V}{A_1}\binom{V}{A_2}\cdots\binom{V}{A_q} = \sum_{G=0}^{S_q} \phi_G(A_1, \ldots, A_q)\binom{V}{S_q - G}$$

$$\begin{bmatrix} V \\ A_1 \end{bmatrix}\begin{bmatrix} V \\ A_2 \end{bmatrix}\cdots\begin{bmatrix} V \\ A_q \end{bmatrix} = \sum_{G=0}^{S_q} (-1)^G \phi_G(A_1, \ldots, A_q)\begin{bmatrix} V \\ S_q - G \end{bmatrix}$$

where for any $k \in [1, q]$, $S_k = A_1 + \cdots + A_k$, and for any $G \in \mathbb{Z}$, ϕ_G is the symmetric function of $A_1, \ldots, A_q$ defined by

$$\phi_G(A_1, A_2, \ldots, A_q) = \sum_{\substack{e_1, \ldots, e_q \in \mathbb{Z} \\ e_q = G}} \prod_{k=1}^{q} \binom{S_k - A_k}{A_k}\binom{A_k}{e_k - e_{k-1}}$$

with the convention that $e_0 = 0$. For details, see [Gh1]. It may be interesting to know whether a polynomial formula for $|\text{stab}(q, m, p, a, V)|$ can be obtained in the case of odd q.

14.7.2 Straightening Law in Higher Dimensions

Now that we have the notion of higher dimensional determinant, to a multivector b bounded by $m = (m(1), \ldots, m(q))$ and of width q, we can correspond a "multiminor" of an $m(1) \times m(2) \times \cdots \times m(q)$ matrix $X = (X_{i_1 i_2 \ldots i_q})$ whose entries are indeterminates over a field K. A multitableau gives a monomial in multiminors, and we may ask whether the ones corresponding to standard multitableaux give a basis of $K[X]$. In [AG] it is proved that the "standard monomials in multiminors" are always linearly independent. However, as a consequence of the formula for $|\text{stab}(q, m, p, a, V)|$, it can be seen that they do not, in general, form a basis, i.e., they are too few to generate $K[X]$. For details, see [AG] and [Gh1]. It may be interesting to find a basis for $K[X]$ which naturally extends this set. It may also be interesting to investigate the primality of the ideal in $K[X]$ generated by all $p \times p \times \cdots \times p$ minors of the multimatrix X.

14.7.3 Bijection between Standard Bitableaux and Indexed Monomials

From (5.8), we know that the sets $\text{stab}(2, m, p, a, V)$ and $\text{mon}(2, m, p, a, V)$ have the same cardinality, and so we may ask whether one can construct an explicit bijection between them. Using the correspondence given by G. Robinson, C. Schensted, and D. Knuth, Abhyankar and Kulkarni have obtained such a bijection. As a corollary, we get a bijective proof of the identity $|\text{stab}(2, m, V)| = \dim_K K[X]_V$, which was proved in (3.7) using enumeration. For details, see [AK2]. A variation of this bijection which, roughly speaking, corresponds to using columns instead of rows, is described in [AK3].

14.7.4 Standard Multitableaux and Monomials

Let X be a q–dimensional matrix as in 7.2. Analogous to $\text{mon}(2, m, p, a, V)$, we can define the set $\text{mon}(q, m, p, a, V)$ of monomials on a hypercube, i.e., in $K[X]$. From 7.2, we already know that the two sets $\text{stab}(q, m, p, a, V)$ and $\text{mon}(q, m, p, a, V)$ cannot, in general, have the same cardinality (since $\text{mon}(2, m, V)$ generates $K[X]_V$). Using the Robinson–Schensted–Knuth correspondence, or rather the Abhyankar–Kulkarni version of it, it is not difficult to see that for $q > 1$, we can get an injection of $\text{stab}(q, m, p, a, V)$ into $\text{mon}(q, m, p, a, V)$ which is a surjection only if $q = 2$ (and in some pathological cases). One way to do this could be to note that the bijection in 7.3 is valid even when the entries come from finite totally ordered sets M_1 and M_2 instead of $[1, m(1)]$ and $[1, m(2)]$; now multitableaux become bitableaux with entries on one of its sides coming from a set of $(q-1)$–tuples, ordered suitably. And $\text{stab}(q, m, p, a, V)$ corresponds to a set of monomials on the hypercube, which is, in general, smaller than $\text{mon}(q, m, p, a, V)$ when $q > 2$. For other proofs of this, see [AJ1] and [AJ2]. It may be interesting to find an enumerative formula for $|\text{mon}(q, m, p, a, V)|$ for any q.

14.7.5 Ladder Determinantal Varieties

We have already discussed these in the Introduction and in the Remark following (6.7). A problem concerning these is to find the Hilbert function of the ideal $I_p(L)$ of all $p \times p$ minors in a ladder shaped subset L of the rectangle $[1, m(1)] \times [1, m(2)]$, or, more generally, of the ideal $I^*(p, a, Y)$ for any saturated subset Y of the rectangle. Note that when $Y = [1, m(1)] \times [1, m(2)]$, $F^{(3k)}$ gives the Hilbert function of $I^*(p, a, Y) = I(p, a)$. For a ladder L, an explicit formula for the Hilbert function of $I_p(L)$, in the case of $p=2$, was obtained by Kulkarni in 1985 [K]. In 1987, Abhyankar and Kulkarni [AK1] showed that the ideal $I^*(p,a,Y)$ is Hilbertian, i.e., its Hilbert function is a polynomial for all nonnegative integers. However, there remained the problem of finding an explicit formula for the Hilbert function of $I_p(L)$ for any p. This has recently been solved, and for details we refer to [Gh2]. For the connection of determinantal and ladder determinantal ideals with Schubert varieties and such, see [G], [Mu], [Mus1], [Mus2], [Se].

14.7.6 Coefficients of the Hilbert Function of $I(p,a)$

Recall that for any $k \in [1,2]$,

$$F^{(3k)}(m,p,a,V) = \sum_{D\in\mathbb{Z}} (-1)^D F_D^{(3k)}(m,p,a)\begin{bmatrix} V \\ C(m,p,a)-D\end{bmatrix}$$

gives the Hilbert function as well as the Hilbert polynomial of $I(p,a)$. The coefficients $F_D^{(3k)}(m,p,a)$ have the interesting property that they are all nonnegative and the positive ones among them appear consecutively, i.e., $\{D\in\mathbb{Z} : F_D^{(3k)}(m,p,a)\neq 0\}$ is an interval of the type $[0,D^*]$, for some nonnegative integer D^*. In particular, $F_0^{(3k)}(m,p,a)>0$, which follows indirectly from (6.2) or from the fact that $F_D^{(3k)} = F_D^{(1k)}$ and that the latter satisfy recurrence relations as in (5.1) [these recurrence relations also show that $F_D^{(3k)}$ is always nonnegative]. Direct combinatorial proofs of the nonnegativity of $F_D^{(3k)}$ and the positivity of $F_0^{(3k)}$ were given by Udpikar; he also proved that $\{D\in\mathbb{Z} : F_D^{(3k)}(m,p,a)\neq 0\}$ is an interval of the form $[0,D^*]$. For details, see [U].

14.8 References

[A1] S. S. Abhyankar, *Combinatoire des tableaux de Young, varietes determinantielles et calcul de fonctions de Hilbert,* Rend. Sem. Mat. Univers. Politecn. Torino, 42 (1984), 65–88.

[A2] S. S. Abhyankar, *Determinantal loci and enumerative combinatorics of Young tableaux,* Algebraic Geometry and Commutative Algebra in Honor of Masayoshi Nagata, Volume 1, Kinokuniya Company Ltd., Tokyo (1987), 1–26.

[A3] S. S. Abhyankar, *Enumerative Combinatorics of Young Tableaux,* Marcel Dekker, New York, 1988.

[AG] S. S. Abhyankar and S. R. Ghorpade, *Young tableaux and linear independence of standard monomials in multiminors of a multimatrix,* to appear in Discrete Math.

[AJ1] S. S. Abhyankar and S. B. Joshi, *Generalised rodeletive correspondence between multitableaux and multimonomials,* preprint.

[AJ2] S. S. Abhyankar and S. B. Joshi, *Generalised codeletion and standard multitableaux,* Canadian Math. Soc. Conference Proc., Volume 10 (1989), 1–24.

[AK1] S. S. Abhyankar and D. M. Kulkarni, *On hilbertian ideals,* Linear Algebra and its Appl., 116 (1989), 53–79

[AK2] S. S. Abhyankar and D. M. Kulkarni, *Bijection between indexed monomials and standard bitableaux,* Discrete Math., 79 (1989/90) 1–48.

[AK3] S. S. Abhyankar and D. M. Kulkarni, *Coinsertion and standard bitableaux,* preprint.

[C] A. Cayley, *On the theory of determinants,* Trans. Camb. Phil. Soc., 8, (1843), 75–88.

[DEP] C. DeConcini, D. Eisenbud, and C. Procesi, *Young diagrams and determinantal varieties,* Inv. Math., 56 (1980), 129–165.

[DKR] J. Desarmenien, J. P. S. Kung, and G. C. Rota, *Invariant theory, Young bitableaux, and combinatorics,* Adv. in Math., 27 (1978), 63–92.

[DRS] P. Doubilet, G. C. Rota, and J. Stein, *Foundations of Combinatorics IX: Combinatorial Methods in Invariant Theory,* Studies in Appl. Math., 53 (1974), 185–216.

[G] A. Galigo, *Computations of some Hilbert functions related with Schubert calculus,* Algebraic Geometry, Sitges (Barcelona), Lecture Notes in Math. No. 1124, Springer–Verlag, New York, 1983.

[Gh1] S. R. Ghorpade, *Young multitableaux and higher dimensional determinants,* to appear in Adv. in Math.

[Gh2] S. R. Ghorpade, *Hilbert functions of ladder determinantal varieties,* forthcoming.

[Ho] W. V. D. Hodge, *Some enumerative results in the theory of forms,* Proc. Camb. Phil. Soc., 39 (1943), 22–30.

[HE] M. Hochster and J. A. Eagon, *Cohen–Macaulay rings, Invariant theory, and the generic perfection of determinantal loci,* Amer. J. Math., 93 (1971), 1020–1058.

[J] G. D. James, *The Representation Theory of Symmetric Groups,* Lecture Notes in Math. No. 682, Springer–Verlag, New York, 1977.

[K] D. M. Kulkarni, *Thesis*, Purdue Univ., 1985.

[Ku] J. P. S. Kung, *Young Tableaux in Combinatorics, Invariant Theory, and Algebra*, Academic Press, New York, 1982.

[M] K. R. Mount, *A remark on determinantal loci,* J. London Math. Soc., 42 (1967), 595–598.

[MM] T. Muir and W. Metzler, *A Treatise on the Theory of Determinants,* Longman's Green and Co., London, 1933.

[Mu] S. B. Mulay, *Determinantal loci and the flag variety,* Adv. in Math., 74 (1989), 1–30.

[Mus1] C. Musili, *Some properties of Schubert varieties,* J. Indian Math. Soc., 36 (1974), 131–145.

[Mus2] C. Musili, *Applications of standard monomial theory,* preprint.

[N] H. Narasimhan, *Irreducibility of ladder determinantal varieties,* J. of Algebra, 102 (1986), 162–185.

[P] E. Pascal, *Sopra le relacioni che possono sussistere identicanemte fra formacioni simboliche del tipo invariantivo nella teoria generale delle forme algebriche,* Mem. Accad. Lincei Roma, Series V, 4a (1888), 375–387.

[S] B. Sagan, *The Symmetric Group: Representations, Combinatorial Algorithms, and Symmetric Functions*, to appear in Wadsworth and Brooks/Cole, Monterey.

[Se] C. S. Seshadri, *Introduction to Standard Monomial Theory,* Brandeis Lecture Notes 4, 1985.

[St] D. Stanton (ed.), *Invariant theory and tableaux*, Springer–Verlag, New York, 1990.

[U] S. G. Udpikar, *Thesis,* Punjab Univ., 1987.

[Y] A. Young, *The Collected Papers of Alfred Young,* Univ. of Toronto Press, 1977.

[ZS] O. Zariski and P. Samuel, *Commutative Algebra,* Vol. II, Van Nostrand, New York, 1960.

15 Abhyankar's Recursive Formula Regarding Standard Bi-Tableau

Shrinivas G. Udpikar

ABSTRACT Let $X = (X_{ij})$ be an $m(1)$ by $m(2)$ matrix whose entries X_{ij}, $1 \leq i \leq m(1)$, $1 \leq j \leq m(2)$; are indeterminants over a field K. Let $K[X]$ be the polynomial ring in these $m(1)m(2)$ variables over K.
In [1], Abhyankar enumerates standard Young bi-tableau with certain conditions and deduces that standard monomials in minors of X form a base for the vector space $K[X]$ over K, a well known result which has been proved by DeConcini-Eisenbud Procesi using straightening formula [2]. The polynomial expression enumerating these bi-tableau involves certain integer valued functions $F_D^{(Lk)}(m,p,a)$, which characterize the Hilbert polynomial of certain determinantal ideals. Using Abhyankar's recursive formula developed in [1], we prove certain properties of these integer valued functions.

15.1 Notation and definitions

We assume all notations and conventions of [1]. Let Z, N, N^* and Q denote the set of all integers, nonnegative integer, positive integers and rational numbers respectively.

For any two integers A and B, we let

$$[A, B] = \{D \in Z : A \leq D \leq B\}.$$

For any $p \in N$, we denote by $Z(p), N(p)$ and $N^*(p)$, the set of all maps from $[1,p]$ to Z, N and N^*, we let

$$Z(p, D) = \{d \in Z(p) : d(1) + d(2) + \cdots + d(p) = D\} \text{ and}$$

$$N(p, D) = \{d \in N(p) : d(1) + d(2) + \cdots + d(p) = D\}.$$

For any $p \in N$ and $A \in Z$, we let $J(p) =$ all subsets of $[1,p]$ and $J(p, A) =$ all subsets of $[1,p]$ with cardinality A.

Finally for any $p \in N$ we let

$$\text{rec}\,(2,p) = \{k, i) : k \in [1,2] \text{ and } i \in [1,p]\}.$$

Let us recall some definitions from [1].

For $p \in N$ by a multivector of width 2 and length p (or a bi-vector of length p) we mean a map $a:\ \mathrm{rec}\,(2,p) \to N^*$, which to every $k \in [1,2]$ and $i \in [1,p]$ associates $a(k,i) \in N^*$ such that

$$a(k,1) < a(k,2) < \cdots < a(k,p) \text{ for } k \in [1,2].$$

By a multivector of width 2, we mean a multivector of width 2 and length p for some $p \in N$. In this case we put $\mathrm{len}\,(a) = p$.

For any $p \in N$, $m \in N^*(2)$ and a bi-vector a of length p, we say that a is bounded by m and denote this by $a \leq m$, to mean that

$$a(k,i) \leq m(k) \text{ for } k \in [1,2] \text{ and } i \in [1,p].$$

Given any two bi-vectors a and a' of lengths p and p' respectively, we define $a \leq a'$ to mean that $p \geq p'$ and $a(k,i) \leq a(k',i)$ for $k \in [1,2]$ and $i \in [1,p']$.

We note that this defines a partial order on the set of all bi-vectors.

For any $p \in N$, $m \in N^*(2)$, we put

$\mathrm{vec}\,(2)$ = the set of all bi-vectors,

$\mathrm{vec}\,[2,p]$ = the set of all bi-vectors of length p,

$\mathrm{vec}\,(2,m,p) = \{a \in \mathrm{vec}\,(2):\ a \leq m\}$.

Given any $d \in N$, by a bi-tableau of depth d, we mean a mapping T of $[1,d]$ into $\mathrm{vec}\,(2)$, which to every $j \in [1,d]$, associates $T(j) \in \mathrm{vec}\,(2)$. The depth of the bi-tableau T is denoted by $\mathrm{dep}\,(T)$.

A bi-tableau T of depth d is said to be standard if the length of $T[e]$ is positive for each $e \in [1,d]$ and

$$T[1] \leq T[2] \leq \cdots \leq T[d].$$

Given any $m \in N^*(2)$, $a \in \mathrm{vec}\,(2)$ and a bi-tableau T of depth d, we say that

(i) T is bounded by m if $T[e] \leq m$ for $e = 1,2,...,d$,

(ii) T is predominated by a if $a \leq T[e]$ for $e = 1,2,...,d$.

The Area of a bi-tableau T of depth d is denoted by $\mathrm{are}(T)$ and is defined as

$$\mathrm{are}\,(T) = \begin{cases} \sum_{e=1}^{d} \mathrm{len}\,(T[e]) & \text{if } d \neq 0 \\ 0 & \text{if } d = 0. \end{cases}$$

By $\mathrm{stab}\,(2,m,p,a,V)$ we denote the set of all standard bi-tableau of area V, which are bounded by m and predominated by bi-vector a of length p. By $\mathrm{Stab}\,(2,m,p,a,V)$ we denote the cardinality of the set $\mathrm{stab}\,(2,m,p,a,V)$.

Let V be an element in an overring of Q and r be any integer.

$\binom{V}{r}$ denotes binomial coefficient defined as below.

$$\binom{V}{r} = \begin{cases} \dfrac{V(V-1)\cdots(V-r+1)}{r!} & \text{if } r \geq 0 \\ 0 & \text{if } r < 0 \end{cases}$$

We define the twisted binomial coefficient $\left[\begin{smallmatrix}V\\r\end{smallmatrix}\right]$, by putting

$$\begin{bmatrix}V\\r\end{bmatrix} = \begin{cases} \begin{pmatrix}V+r\\r\end{pmatrix} & \text{if } r \geq 0 \\ 0 & \text{if } r < 0 \end{cases}$$

We note that if $V,\ r \in Z$ be such that $r > V \geq 0$ then $\binom{V}{r} = 0$. Also note that $\binom{V}{r} \neq 0$ if and only if either $V \geq r \geq 0$ or $V < 0 \leq r$.

15.2 Integer Valued Functions $\mathbf{F_D^{(Lk)}(m,p,a)}$

In [1], Abhyankar defines several integer valued functions $F_D^{(Lk)}(m,p,a)$ where $L \in \{1,1^*,2,2^*,3,3^*\}$, $k \in [1,2]$, $D \in Z$, $m \in N^*(2)$, $p \in N^*$ and $a \in \text{vec}\,(2,m,p)$ and proves their equality. For the purposes of this paper we define below $F_D^{(3k)}(m,p,a)$.

Let $m \in N^*(2)$, $p \in N^*$, $a \in \text{vec}\,(2,m,p)$ and $k \in [1,2]$. Let

$$k' = \begin{cases} 2 & \text{if } k = 1 \\ 1 & \text{if } k = 2. \end{cases}$$

We put

$$\begin{aligned} C[m,p,a,k] &= \sum_{i=1}^{p}(m(k) - a(k,i)) \\ C[m,p,a,k'] &= \sum_{i=1}^{p}(m(k') - a(k',i)) \\ C[m,p,a] &= C[m,p,a,k] + C[m,p,a,k'] + p - 1 \end{aligned}$$

and

$$C[m,p,a] = \min\{C[m,p,a,k],\ C[m,p,a,k']\}.$$

For $i,j \in [1,p]$ and $d \in Z(p)$, let $G^{(3k)}(m,p,a,d)$ denote the $p \times p$ matrix whose $(i,j)^{\text{th}}$ entry is

$$G_{ij}^{(3k)}(m,p,a,d) = \begin{bmatrix} m(k') - a(k',i) - d(i) \\ m(k) - a(k,j) \end{bmatrix} \begin{pmatrix} m(k) - a(k,j) + j - i \\ d(i) \end{pmatrix}$$

Then for every $D \in Z$ we define

$$F_D^{(3k)}(m,p,a) = \sum_{d \in Z(p,D)} \det\ G^{(3k)}(m,p,a,d).$$

In view of the usual properties of binomial coefficients and determinants, it is clear that the above sum may be replaced by sum over $d \in N(p,D)$ and is essentially finite.

It may also be noted that $F_D^{(3k)}(m,p,a)$ can be expressed as an alternative notation as follows.

Fix $k \in [1,2]$ and put $m(k) - a(k,i) = r(i) = r_i$ and $m(k') - a(k',i) = s(i) = s_i$ for $i = 1,2,...,p$, $C[m,p,a,k] = r_1 + r_2 + \cdots + r_p = R$ and $C[m,p,a,k'] = s_1 + s_2 + \cdots + s_p = S$.

Then for strictly decreasing sequences r and s of integers given by $r: r_1 > r_2 > \cdots > r_p \geq 0$ and $s: s_1 > s_2 > \cdots > s_p \geq 0$, we have

$$F_D^{(3k)}(m,p,a) = \sum_{d \in Z(p,D)} \det \left\{ \begin{bmatrix} s_i - d_i \\ \\ r_j \end{bmatrix} \begin{pmatrix} r_j + j - i \\ \\ d_i \end{pmatrix} \right\}$$

where we have put $d(i) = d_i$ for $i = 1,2,...,p$. We rename this as $F_D^{(30)}(r,s,p)$. In what follows we shall write for every $D \in Z$, $F_D^{(3k)}(m,p,a) = F_D(m,p,a)$ and $F_D^{(30)}(r,s,p) = F_D(r,s,p)$.

We now state the following known results to be used later.

Theorem (2.1). *Let there be given any $m \in N^*(2)$, $p \in N^*$ and $a \in$ vec$(2,m,p)$. Then we have*

i) $F_0(m,p,a)$ is a positive integer,

ii) For all $D \in Z$, $F_D(m,p,a)$ is a nonnegative integer and

iii) $\{D \in Z : F_D(m,p,a) \neq 0\} \subseteq [0, C[(m,p,a)]$

where $C(m,p,a) = \min\{C[m,p,a,k],\ C[m,p,a,k']\}$.

Proof. Refer to theorem (9.13), lemma (9.15) and theorem (9.17) of [1].

Theorem (2.2) *In the above notation, for every $V \in N$, we have*

$$\text{Stab}\,(2,m,p,a,V) = F(m,p,a,V)$$

$$\stackrel{\text{def}}{=} \sum_{D \in Z} (-1)^D F_D(m,p,a) \begin{bmatrix} V \\ c[m,p,a] - D \end{bmatrix}$$

Proof. Refer to theorem (9.7) of [1].

We note that $F(m,p,a,V)$ defined in the theorem (2.2) above, is a polynomial in V with rational coefficients and is of degree $C[m,p,a]$. It happens to be the Hilbert Function as well as the Hilbert Polynomial of a certain determinantal ideal in the polynomial ring $K[X]$, where K is any field and $X = (X_{ij})$ is an $m(1) \times m(2)$ matrix whose elements X_{ij}, $1 \leq i \leq m(1)$, $1 \leq j \leq m(2)$ are indeterminants over the field K. (Refer to theorem (20.14) of [1]).

In [1], Abhyankar raised the following problems.

Given any $m \in N^*(2)$, $p \in N^*$ and $a \in \text{vec}\,(2, m, p)$, does there exist a nonnegative integer $C^*(m, p, a)$ such that

i) $\{D \in Z : F_D(m, p, a) \neq 0\} = [0, C^*(m, p, a)]$?

ii) If so, then is it possible that

$$C^*(m, p, a) = c(m, p, a) = \min\{C[m, p, a, 1],\ C[m, p, a, 2]\}?$$

(Refer to remark (9.18) of [1]).

Both these problems are settled in an affirmative manner in [3].

15.3 Abhyankar's Recursive formula

We follow the usual convention that the sum over an empty family is zero. Before we list some further notation, let us give a few definitions.

Let there be given any $m \in N^*(2)$, $p \in N^*$, $b \in \text{vec}\,[2, p+1]$, $u \in J(p)$, $U \in [0, p]$ and $k \in [1, 2]$. Further let

$$k' = \begin{cases} 2 & \text{if } k = 1, \\ 1 & \text{if } k = 2. \end{cases}$$

We define the sets $M[p, b, k, u]$ nd $M(p, b, k, U)$ as follows:

$$M[p, b, k, u] = \left\{ \begin{array}{l} a \in \text{vec}\,[2, p] :\ a(k, i^*) = b(k, i^*),\ \forall i^* \in [1, p], \\ a(k', i) \in [b(k', i) + 1, b(k', i+1) - 1],\ \ \forall i \in u \\ \text{and } a(k', j) = b(k', j),\ \ \forall j \in [1, p] \setminus u \end{array} \right\}$$

and $M(p, b, k, U) = \displaystyle\bigcup_{u \in J(p,U)} M[p, b, k, u]$.

Now we set the following notation. Given any $m \in N^*(2)$ be such that

(1) $m^*(k') = m(k')$ and $m^*(k) = m(k) + 1$ holds. Further let $p^* \in N^*$ and $a^* \in \text{vec}(2, m^*, p^*)$ be such that either

([(1')] $p^* = p$ and $m^*(k) - a^*(k, p^*) \neq 0$ and $a^*(\hat{k}, i) = b(\hat{k}, i)$ for all $\hat{k} \in [1, 2]$ holds or

(1*) $p^* = p + 1$, $m^*(k) - a^*(k, p^*) = 0$ and $a^* = b$ holds.

Finally for every $m \in N^*(2)$, $p \in N^*$, $a \in \text{vec}[2, p]$, $U \in [0, p]$ and $D \in Z$, we set:

$$S_U(m, p, a; D) = \sum_{I \in [0, U]} \binom{U}{I} F_{D-I}(m, p, a).$$

We now state a useful recursive formula in the following:

Theorem (3.1)

Let there be given any $m \in N^*(2)$, $p \in N^*$, $k \in [1,2]$, $b \in \text{vec}\,[2,p+1]$ and $m^* \in N(2)$, $p^* \in N^*$, $a^* \in \text{vec}\,(2,m^*,p^*)$ such that (1) holds and either (1′) or (1) holds. Then for every $D \in Z$, we have*

$$F_D(m^*,p^*,a^*) = \sum_{I\in[0,p]} \sum_{U\in[0,p]} \sum_{a\in M(p,b,k,U)} \binom{U}{I} F_{D-I}(m,p,a)$$

Proof: Refer to theorem (9.8*), section 9 of [1].

In the above recursive formula, by interchanged the summations over I and U and by noting the fact that $\binom{U}{I} = 0$ if $I > U \geq 0$, we see that

$$F_D(m^*,p^*,a^*) = \sum_{U\in[0,p]} \sum_{a\in M(p,b,k,U)} S_U(m,p,a;D)$$

In what follows, we shall prove some results regarding integer valued functions $F_D(m,p,a)$, using recursive formula stated above. First we prove the following theorem.

Theorem (3.2): *Let $k \in [1,2]$, $p \in N^*$ and $b \in \text{vec}\,[2,p+1]$ be given. Then we have the following.*

I) $M(p,b,k,0) = M[p,b,k,\emptyset]$ *is nonempty set consisting of the unique element* $\tilde{b} \in \text{vec}\,[2,p]$ *obtained from* b *by putting* $\tilde{b}(\hat{k},i)$ *for all* $\hat{k} \in [1,2]$ *and* $i \in [1,p]$.

II) *If* $M(p,b,k,1) = \emptyset$ *then* $\{b(k',j)\}$, $j = 1,2,...,p$; *is a sequence of consecutive integers i.e;* $b(k',j+1) = b(k',j)+1$, *for all* $j \in [1,p-1]$ *and consequently* $M(p,b,k,U) = \emptyset$ *for every* $U \geq 1$.

Proof: (I) Since $\emptyset$ is the only subset of $[1,p]$ of cardinality zero, we have $M(p,b,k,0) = m[p,b,k,\emptyset]$. The rest follows from the definition of $m[p,b,k,\emptyset]$.

(II) Let $u_1 = \{i\}$ where i is some arbitrarily chosen element of $[1,p-1]$. Then by definition of $M[p,b,k,u_1]$ we have $M[p,b,k,u_1] = \emptyset \Leftrightarrow [b(k',i)+1, b(k',i+1)-1] = \emptyset \Leftrightarrow b(k',i+1) = b(k',i)+1$, so that $\{b(k',j)\}$, $j = 1,2,...,p$, is an increasing sequence of consecutive integers.

Finally if $U > 1$ and $u \in J(p)$ is any set of cardinality U, then there is an $i \in u$ such that $i \in [1,p-1]$. Now if $a \in M(p,b,k,U)$, then we must have $a(k',i) \in [b(k',j)\}$, $j = 1,2,...,p$ is a sequence of consecutive integers. Hence $M(p,b,k,u) = \emptyset$ for every $U > 1$ if $M(p,b,k,1) = \emptyset$. Hence the theorem.

15.4 Some Results

We begin with the following theorem.

Theorem (4.1): *Given any $p \in N^*$, there exist integers $r_1 > r_2 > \cdots > r_p \geq 0$ and $s_1 > s_2 > \cdots > s_p \geq 0$ such that $F_D^{(30)}(r,s,p) = F_D(r,s,p) > 0$ for every $D \in [0.R]$ where $R = \min\{R, S\}$.*

Proof: We have

$$F_D(r,s,p) = \sum_{d \in N(p,D)} \det \left\{ \begin{bmatrix} s_i - d_i \\ r_j \end{bmatrix} \begin{pmatrix} r_j + j + i \\ d_i \end{pmatrix} \right\}$$

Take $r_1, r_2, ..., r_p$ consecutive integers, so that for every $j \in [1,p]$, we have $r_j + j - i = r_1 + 1 - i = r_i$ for $i = 1, 2, ..., p$.

With this,

$$\text{(i)} \qquad F_D(r,s,p) = \sum_{d \in N(p,D)} \binom{r_1}{d_1}\binom{r_2}{d_2} \cdots \binom{r_p}{d_p} \det \left\{ \begin{bmatrix} s_i - d_i \\ r_j \end{bmatrix} \right\}$$

Now choose $s_1, s_2, ..., s_p$ such that $(s_i - s_{i+1}) > R$ for $i = 1, 2, ..., p-1$ and $s_p > R$. Clearly for any $d \in N(p,D)$ with $D \leq R$, we have $0 \leq d_i \leq D \leq R$ for $i = 1, 2, ..., p$. Hence

$$(s_i - d_i) - (s_{i+1} - d_{i+1}) = (s_i - s_{i+1}) - (d_i - d_{i+1}) > 0 \quad \text{for } i = 1, 2, ..., p-1.$$

But $(s_p - d_p) > 0$. Hence we have $(s_1 - d_1) > (s_2 - d_2) > \cdots > (s_p - d_p) > 0$ and $r_1 > r_2 > \cdots > r_p \geq 0$.

Hence every determinant in sum (i) is positive by first part of theorem (2.1) of section (2). [Note that the sum involved in (i) is nonempty]. Hence $F_D(r,s,p) > 0$ for every $D \in [0, R]$. Hence the theorem.

More generally, if $\min\{s_1 - s_2, s_2 - s_3, ..., s_{p-1} - s_p, s_p\} > L$ then as above, it follows that

$$F_D(r,s,p) > 0 \quad \text{for every } D \in [0, L].$$

It may be noted that Abhyankar has posed the following problem.

Find a function $\emptyset(r,s) = \emptyset(r_1, r_2, ..., r_p, s_1, s_2, ..., s_p)$ such that $F_D(r,s,p) = 0$ if and only if $D > \emptyset(r,s)$.

We now prove a few more results in this direction.

Remembering definitions and notation developed in section (3), we first express recursive formula stated in section (3), in an alternative notation.

For this fix $k = [1, 2]$ and put

$$m(k) - a(k,i) = r_i,$$

$$m(k') - a(k',i) = s_i \quad \text{for } i = 1, 2, ..., p$$

and
$$m^*(k) - a^*(k,i) = r_i^*,$$

$$m^*(k') - a^*(k',i) = s_i^* \text{ for } i = 1,2,...,p^*.$$

Incidently by a triplet (r,s,p) we mean strictly decreasing sequences r : $r_1 > r_2 > \cdots > r_p \geq 0$, $s : s_1 > s_2 > \cdots s_p \geq 0$ of integers. Let $u \in J(p) =$ all subsets of $[1,p]$. For any $a \in \text{vec}\,[2,p]$, we have $a \in M[p,b,k,u]$ if and only if
$$s_{i+1}^* + 1 \leq s_i \leq s_i^* - 1 \text{ for } i \in [1,p-1] \cap u$$

and
$$0 \leq s_p \leq s_{p^*}^* - 1 \text{ if } p \in u, \quad s_j = s_j^* \text{ for } j \in [1,p] \setminus u,$$

and
$$r_i = r_i^* - 1 \text{ for } i \in [1,p] \text{ if } r_{p^*}^* \neq 0 \text{ and } p = p^*.$$

On the other hand if $r_{p^*}^* = 0$ and $p = p^* - 1$, then $a \in M[p,b,k,u]$ if and only if $s_{i+1}^* + 1 \leq s_i \leq s_i^* - 1$ for $i \in [1,p] \cap u$, $s_j = s_j^*$ for $j \in [1,p] \setminus u$ and $r_i = r_i^* - 1$ for $i \in [1,p]$.

For convenience, let us say $(r,s,p) = (r_1,r_2,...,r_p;s_1,s_2,...,s_p;) \in \tilde{M}[p,b,k,u]$ if and only if $a \in M[p,b,k,u]$ and put
$$\tilde{M}(p,b,k,U) = \bigcup_{u \in J(p,U)} \tilde{M}[p,b,k,u].$$

With this the recursive formula of section (3) can be rewritten as
$$F_D(r^*,s^*,p^*) = \sum_{U \in [0,p]} \sum_{(r,s,p) \in \tilde{M}(p,b,k,u)} S_U(r,s,p;D).$$

Remark (I) From theorem (3.2), it easily follows that $\tilde{M}(p,b,k,0) = \{r^* - 1, s^*, p)\}$ where $p = p^*$ if $r_{p^*}^* \neq 0$ and $p = p^* - 1$ if $r_{p^*}^* = 0$.

Also $\tilde{M}(p,b,k,1) = \emptyset$ if and only if $s_1^*, s_2^*, ..., s_{p^*}^* = 0$ is a strictly decreasing sequence of consecutive integers if $r_{p^*}^* = 0$ (with $p^* = p+1$). Moreover in this case we have $M(p,b,k,U) = \emptyset$ for every $U \geq 2$.

Remark (II) If k is replaced by k' roles of r and s are interchanged. It may further be noted that $F_D(r,s,p)$ is symmetrical in r and s (Refer to section (8) of [1]).

We used the recursive formula of section (3), to prove the following.

Theorem (4.2) *Given an (r,s,p) and (r^*,s^*,p^*) such that (1) holds and either (1′) holds or (1*) holds, we have $F_1(r^*,s^*,p^*) = 0$ if and only if $r_{p^*}^*, r_2^*, ..., r_{p^*}^*$, $s_1^* s_2^*, ..., s_{p^*}^*$ are strictly decreasing sequences of consecutive integers with $r_{p^*}^* = 0$ or $s_{p^*}^* = 0$.*

Proof: By recursive formula, we have

$$F_1(r^*, s^*, p^*) = F_1(r^* - 1, s^*, p) + \sum_{(r,s,p)\in \tilde{M}(p,b,k,1)} (F_1 + F_0)(r, s, p)$$

where $p^* = p$ if $r_1^* > 0$ and $p^* = p+1$ if $r_{p^*}^* = 0$. Now $F_1(r^*, s^*, p^*) = 0$ implies that $\tilde{M}(p, b, k, 1) = \emptyset$ since $F_0(r, s, p) > 0$. Hence necessary parts follow by Remarks (I) and (II) made above. The converse is clear.

Theorem (4.3) *Assumptions being the same as in theorem (4.2), we have for any $D \in N, F_D(r^*, s^*, p^*) > 0$ if $D \leq \min\{r_1^* - r_2^* - 1, s_1^* - s_2^* - 1\}$.*

Proof: By recursive formula, we note that $F_D(r^*, s^*, p^*) = 0 \Rightarrow F_D(r^* - 1, s^*, p) = 0$ where $p = p^*$ if $r_{p^*}^* > 0$ and $p = p^* - 1$ if $r_{p^*}^* = 0$. From this it easily follows that $F_D(r_1^* - r_2^* - 1, s_1^*, 1) = 0$, so that $r_1^* - r_2^* - 1 < D$ or $s_1^* < D$.

The result follows by symmetry in r^* and s^*.

We conclude this paper with the following theorem whose proof easily follows from the recursive formula.

Theorem (4.4) *Let there be given any (r, s, p) and (r^*, s^*, p^*) such that (1) holds and either (1′) holds or (1*) holds. Then we have the following.*
(a) If $r_{p^*}^* > 0$, $p = p^*$ and s_i^*, s_{i+1}^* are non consecutive for exactly D values of $i \in [1, p-1]$, then

(1) $F_D(r^*, s^*, p^*) > 0$ if $s_{p^*}^* = 0$, and

(2) $F_{D+1}(r^*, s^*, p^*) > 0$ if $s_{p^*}^* > 0$.

(b) If $r_{p^*}^* = 0$, $p^* = p+1$ and s_i^* , s_{i+1}^* are nonconsecutive for exactly D values of $i \in [1, p]$, then $F_D(r^*, s^*, p^*) > 0$.

Acknowledgments: I wish to thank Professor Shreeram S. Abhyankar for his valuable guidance.

15.5 References

[1] S. Abhyankar, *Enumerative Combinatorics of Young Tableaux*, Marcel Dekker, Inc. (1988).

[2] C. DeConcini, D. Eisenbud, and C. Procesi, Young diagrams and determinantal varieties; *Inventiones Mathematicae* **56**, 129-165 (1980).

[3] G. Shrinivas and G. Udpikar, On Hilbert Polynomial of certain determinantal ideal, to appear in *International Journal of Mathematics and Mathematical Sciences.*

16
Correspondences Between Tableaux and Monomials

Sanjeevani V. Vaidya (Joshi)

16.1 Introduction

Let X be an $m(1)$ by $m(2)$ matrix whose entries X_{ij} are independent indeterminates over a field K and let $K[X]$ be the ring of polynomials in these $m(1)m(2)$ indeterminates. A p by p minor of X can be represented by the row indices $1 \leq a(1,1) < a(1,2) < \ldots < a(1,p) \leq m(1)$ and the column indices $1 \leq a(2,1) < a(2,2) < \ldots < a(2,p) \leq m(2)$. Such a pair of strictly increasing sequences of positive integers of same (finite) length may be called a bivector whose length is p and which is bounded by $m = (m(1), m(2))$; such a bivector may be denoted as $a(k,i)_{1 \leq k \leq 2, 1 \leq i \leq p}$; if a' is another bivector whose length is p' and which is bounded by m then we write $a \leq a'$ to mean that $p \geq p'$ and $a(k,i) \leq a'(k,i)$ for $k = 1,2$ and $i = 1,2,\ldots,p'$. A Young bitableau T bounded by m, i.e., a finite sequence of bivectors bounded by m, can be made to correspond to the product of the corresponding minors of X, i.e., to a certain monomial in the minors of X. This monomial is said to be standard if the bitableau T is standard, i.e., if T is a finite nondecreasing sequence of bivectors. Then we have the *Standard Basis Theorem*, which says that the set of standard monomials in the minors of X, corresponding to standard bitableaux bounded by m, gives a K-vector-space-basis of $K[X]$. The said *Standard Basis Theorem* was first proved by Doubilet-Rota-Stein in [13] where they call it the *Straightening Law*. This law has proved to be of much significance in some aspects of Algebraic Geometry and Invariant Theory.

Some other proofs of this law may be found in Abhyankar [3], Abhyankar-Kulkarni [4], Abhyankar-Ghorpade [10], Deconcini-Eisenbud-Procesi [11] and Desarmenian-Kung-Rota [12]. The proof of the Straightening Law in Abhyankar [3] is based on the fact that $\text{stab}(2,m,p,a,V)$ and $\text{mon}(2,m,p,a,V)$ have the same cardinality where, for any positive integer V, by $\text{stab}(2,m,p,a,V)$ we denote a certain finite set of standard bitableaux determined by m, a, V, and by $\text{mon}(2,m,p,a,V)$ we denote the exponent system of a certain finite set of 'indexed' monomials in the $m(1)m(2)$ indeterminates X_{ij} determined by m, a, V. The fact that $\text{stab}(2,m,p,a,V)$ and $\text{mon}(2,m,p,a,V)$ have the same cardinality was proved in Theorem (9.9) of Abhyankar [3] by enumeration, and in Remark (9.10) of Abhyankar [3] it was suggested that a bijective proof of this be found; such a bijective proof has now been given in Abhyankar-Kulkarni [4]. It may be noted that [1] is a precursor of [3] whereas [2] provides a summary of [3]. The said bijective proof given in Abhyankar-Kulkarni [4] suggested that certain two analogous sets costab$(2, m, p, A, V)$ and comon$(2, m, p, A, V)$ also might have the same cardinality where A is a cobivector whose length is p and which is bounded by m, i.e., A is a bisequence of positive integers $A(k,i)_{1\leq k\leq 2, 1\leq i\leq p}$ such that $A(k,i) \leq A(k,i+1)$ for $k = 1,2$ and $i = 1,2,\ldots,p-1$ and $A(k,p) \leq m(k)$ for $k = 1,2$. In Abhyankar-Kulkarni [5] it was shown that indeed these two sets costab$(2, m, p, A, V)$ and comon$(2, m, p, A, V)$ do have the same cardinality, and yet another proof of the Straightening Law was deduced from this.

To generalize the above matter to higher dimensions, let us consider a multimatrix X of any dimension q and size $m = (m(1), m(2), \ldots, m(q))$, i.e., a higher dimensional matrix

$$X = (X_{r(1),\ldots,r(q)})_{1\leq r(k)\leq m(k) \text{ for } k=1,2,\ldots,q}$$

whose entries constitute a system of $m(1)m(2)\ldots m(q)$ independent indeterminates over a field K and let $K[X]$ be the ring of polynomials in these indeterminates. Now a multiminor of size p of X can be represented by a multivector a of width q whose length is p and which is bounded by m, i.e., by a multisequence of positive integers $a(k,i)_{1\leq k\leq q, 1\leq i\leq p}$ such that $a(k,i) < a(k,i+1)$ for $k = 1,2,\ldots,q$ and $i = 1,2,\ldots,p-1$ and $a(k,p) \leq m(k)$ for $k = 1,2,\ldots,q$; if a' is another multivector of width q whose length is p' and which is bounded by m then we write $a \leq a'$ to mean that $p \geq p'$ and $a(k,i) \leq a'(k,i)$ for $k = 1,2,\ldots,q$ and $i = 1,2,\ldots,p'$. Correspondingly, a monomial in the multiminors of the multimatrix X can be reepresented by a tableau T of width q bounded by m, by which we mean a finite sequence $T[1], T[2], \ldots, T[d]$ of multivectors of width q which are bounded by m; the tableau T is said to be standard if $T[1] \leq T[2] \leq \ldots \leq T[d]$. For any positive integer V, by stab(q, m, p, a, V) we denote a certain finite set of standard multitableaux determined by q, m, a, V, and by mon(q, m, p, a, V) we denote the exponent system of a certain finite set of 'indexed' monomials in the $m(1)m(2)\ldots m(q)$ indeterminates $X_{r(1),r(2),\ldots,r(q)}$ determined by q, m, a, V. Likewise by costab(q, m, p, A, V) and comon(q, m, p, A, V) we denote certain two analogous sets where A is a comultivector of width q whose length is p and which is bounded by m, i.e., A is a multisequence of positive integers $A(k,i)_{1\leq k\leq q, 1\leq i\leq p}$ such that $A(k,i) \leq A(k,i+1)$ for $k = 1,2,\ldots,q$ and $i = 1,2,\ldots,p-1$ and $A(k,p) \leq m(k)$ for $k = 1,2,\ldots,q$. Detailed definitions of the sets stab(q, m, p, a, V) and mon(q, m, p, a, V) are given in Section 2 of Abhyankar [3], and the detailed definitions of the sets costab(q, m, p, A, V) and comon(q, m, p, A, V) are given in Section 12 of Abhyankar-Kulkarni [5].

As said above, in Theorem (9.9) of Abhyankar [3] it was proved that, in case of $q = 2$, the two sets $\mathrm{stab}(q, m, p, a, V)$ and $\mathrm{mon}(q, m, p, a, V)$ have the same cardinality. Moreover, in the general case, in Problem (8.42) of Abhyankar [3] it was asked as to for what values of q the two sets $\mathrm{stab}(q, m, p, a, V)$ and $\mathrm{mon}(q, m, p, a, V)$ have the same cardinality. One aim of Abhyankar-Joshi [6] was to establish a certain correspondence from the set $\mathrm{stab}(q, m, p, a, V)$ to the set $\mathrm{mon}(q, m, p, a, V)$ for any positive integer q; it turned out that this correspondence is surjective but not injective for $q = 1$, bijective for $q = 2$, and injective but not surjective for $q > 2$. The said correspondence is obtained by generalizing the RSK correspondence, i.e., the correspondence given by Robinson [16], Schensted [17] and Knuth [15]. The RSK correspondence is based on the procedures of inserting a positive integer in a standard unitableau and deleting a positive integer from it. The deletion procedure described in Definition (4.1) of Abhyankar-Joshi [6] involves deletion along rows. For the bitableau case, this procedure coincides with that given in Section 9 of Abhyankar-Kulkarni [4], where a one-to-one correspondence between $\mathrm{stab}(2, m, p, a, V)$ and $\mathrm{mon}(2, m, p, a, V)$ is established. In Section 3 of [6] we have defined the lexical associate of a monomial in the $m(1)m(2)\dots m(q)$ indeterminates $X_{r(1),r(2),\dots,r(q)}$ for any positive integer q and we have noted that this gives bijections between various types of monomials and lexical vectors. In Section 4 of [6] we have defined the generalized rodeletion for tableaux of any width and establish certain correspondences from sets of various types of tableaux to the sets of special lexical multivectors. In Section 5 of [6] we have obtained the above stated correspondence from $\mathrm{stab}(q, m, p, a, V)$ to $\mathrm{mon}(q, m, p, a, V)$ for any positive integer q. In Section 3 we will summarize the said results about rodeletion from Abhyankar-Kulkarni [4], and Abhyankar-Joshi [6].

As noted above, in Abhyankar-Kulkarni [5] it was proved that, in case of $q = 2$, the two sets $\mathrm{costab}(q, m, p, A, V)$ and $\mathrm{comon}(q, m, p, A, V)$ have the same cardinality and a proof of the Straightening Law was deduced from this. One aim of Abhyankar-Joshi [7] was to establish a certain correspondence from the set $\mathrm{costab}(q, m, p, A, V)$ to the set $\mathrm{comon}(q, m, p, A, V)$ for any positive integer q; it turned out that this correspondence is surjective but not injective for $q = 1$, bijective for $q = 2$, and injective but not surjective for $q > 2$. The said correspondence was also obtained by generalizing the RSK correspondence, i.e., the correspondence given by Robinson [16], Schensted [17] and Knuth [15]. The deletion procedure described in Definition (4.1) of Abhyankar-Joshi [7] was called generalized codeletion because it involves deletion along columns rather than rows. For the bitableau case, this procedure coincides with that given in Section 19 of Abhyankar-Kulkarni [5], where a one-to-one correspondence between $\mathrm{costab}(2, m, p, A, V)$ and $\mathrm{comon}(2, m, p, A, V)$ is established. In Section 3 of [7] we have defined the colexical associate of a monomial in the $m(1)m(2)\dots m(q)$ indeterminates $X_{r(1),r(2),\dots,r(q)}$ for any positive integer

q and we have noted that this gives bijections between various types of monomials and colexical vectors. In Section 4 of [7] we have defined the generalized codeletion for tableaux of any width and established certain correspondences from sets of various types of tableaux to the sets of special colexical multivectors. In Section 5 of [7] we have obtained the above stated correspondence from costab $q, m, p, A, V)$ to comon(q, m, p, A, V) for any positive integer q. In Section 4 we will summarize the said results about codeletion from Abhyankar-Kulkarni [5], and Abhyankar-Joshi [7].

The notion of tableaux over integers can be easily generalized to tableaux over totally ordered sets. In greater detail: Let $m^* = m^*(k)_{1\leq k\leq q}$ be a sequence of totally ordered sets. A multisequence $a(k,i)_{1\leq k\leq q, 1\leq i\ p}$ with $a(k,i) \in m^*(k)$ for $k = 1, 2, \ldots, q$ and $i = 1, 2, \ldots, p$ such that $a(k,i) < a(k, i+1)$ for $k = 1, 2, \ldots, q$ and $i = 1, 2, \ldots, p-1$ may be called a multivector of type (q, m^*) whose length is p; if a' is another multivector of type $(q, *)$ whose length is p' then we write $a \leq a'$ to mean that $p \geq p'$ and $a(k,i) \leq a'(k,i)$ for $k = 1, 2, \ldots, q$ and $i = 1, 2, \ldots, p'$. A finite nondecreasing sequence of multivectors of type (q, m^*) will give a standard tableau of type (q, m^*). For any positive integer V, by stab $q, m^*, p, a, V)$ we denote a certain finite set of standard tableaux of type $(q, *)$ determined by m^*, a, V, and by mon(q, m^*, p, a, V) we denote the exponent system of a certain finite set of 'indexed' monomials in the indeterminates X_y, where y ranges over cub(q, m^*), determined by m^*, a, V, where cub(q, m^*) is the set of all sequences $y = y(k)_{1\leq k\leq q}$ with $y(k) \in m^*(k)$ for $k = 1, 2, \ldots, q$. Likewise by costab $q, m^*, p, A, V)$ and comon(q, m^*, p, A, V) we denote certain two analogous sets, where A is a comultivector of type (q, m^*), whose length is p, i.e., A is a multisequence $A(k,i)_{1\leq k\leq q, 1\leq i\leq p}$ with $A(k,i) \in m^*(k)$ for $k = 1, 2, \ldots, q$ and $i = 1, 2, \ldots, p$ such that $A(k,i) \leq A(k, i+1)$ for $k = 1, 2, \ldots, q$ and $i = 1, 2, \ldots, p-1$. Detailed definitions of stab(q, m^*, p, a, V) and mon(q, m^*, p, a, V) are given in Section 3 of Abhyankar-Joshi [8], and the detailed definitions of costab(q, m^*, p, A, V) and comon(q, m^*, p, A, V) are given in Section 2 of Abhyankar-Joshi [9]. It may be noted that these definitions reduce to the previous definitions in case, for $k = 1, 2, \ldots, q$, the ordered set $m^*(k)$ coincides with the set of all integers between 1 to $m(k)$. As an important technique, given any integer k between 1 and q, to any multivector a of type (q, m^*) and a standard tableau T of type (q, m), in Section 3 of Abhyankar-Joshi [8] we have respectively associated a bivector of type $(2, B_k(m^*))$ and a standard bitableau of type $(2, B_k(m^*))$ where $B_k(m^*)(1) = m^*(k)$ and $B_k(m^*)(2)$ is the set of all sequences of $q-1$ tuples with entries in $m^*(1), m^*(2), \ldots, m^*(k-1), m^*(k+1), \ldots, m^*(q)$.

As noted above, in Theorem (9.9) of Abhyankar [3] it was proved that, in case of $q = 2$, the two sets stab(q, m, p, a, V) and mon(q, m, p, a, V) have the same cardinality where m is a pair of positive integers, and in Theorem (9.16) of Abhyankar-Kulkarni [4], a bijective proof of this was given. One aim of Abhyankar-Joshi [8] was to establish a bijection between stab(q, m^*, p, a, V) and mon(q, m^*, p, a, V) where $q = 2$ and m^* is a pair of any totally ordered sets. The said bijection was also obtained by generalizing the RSK correspondence, i.e., the correspondence given by Robinson [16], Schensted [17] and Knuth [15]. In Sections 8 and 9 of Abhyankar-Kulkarni [4], the notions of insertion and rodeletion for tableaux over integers are described. In Section 6 of Abhyankar-Joshi [8] these insertion and deletion procedures along rows are extended to tableaux over any totally ordered sets. Section 7 of [8], a certain injective correspondence from the set $\mathrm{mon}(q, m^*)$ to the set $\mathrm{pab}(q, m^*)$ is obtained, where $q > 2$ and m^* is any sequence of totally ordered sets whose length is q; it turned out that under this correspondence the image of $\mathrm{mon}(q, m^*)$ properly contains $\mathrm{stab}(q, m^*)$ and the image of $\mathrm{mon}[[q, m^*, V]]$ properly contains $\mathrm{stib}(q, m^*, V)$ for every integer $V > 1$; moreover, given any positive integer p, the image of $\mathrm{mon}(q, m^*, p)$ properly contains $\mathrm{stab}(q, *, p)$ and the image of $\mathrm{mon}[q, m^*, p, V]$ properly contains $\mathrm{stab}[q, m^*, p, V]$ for every integer $V > 1$. In Section 5 we will summarize the said results about generalized roinsertion from Abhyankar-Joshi [8]

As said above, in Abhyankar-Kulkarni [5] it was proved that, in case of $q = 2$, the two sets $\mathrm{costab}(q, m, p, A, V)$ and $\mathrm{comon}(q, m, p, A, V)$ have the same cardinality, where m is a pair of positive integers. One aim of Abhyankar-Joshi [9] was to establish a bijection between $\mathrm{costab}(q, m^*, p, A, V)$ and $\mathrm{comon}(q, m^*, p, V)$, where $q = 2$ and m^* is a pair of any totally ordered sets. The said bijection was also obtained by generalizing the RSK correspondence, i.e., the correspondence given by Robinson [16], Schensted [17] and Knuth [15]. In Sections 18 and 19 of Abhyankar-Kulkarni [5], the notions of coinsertion and codeletion for tableaux over integers are described. In Section 4 of Abhyankar-Joshi [9] these insertion and deletion procedures along columns are extended to tableaux over any totally or red sets. In Section 5 of [9], a certain injective correspondence from the set $\mathrm{mon}(q, m^*)$ to the set $\mathrm{pab}(q, m^*)$ is obtained, where $q2$ and m^* is any sequence of totally ordered sets whose length is q; it turned out that under this correspondence the image of $\mathrm{mon}(q, m^*)$ properly contains $\mathrm{stab}(q, m^*)$ and the image of $\mathrm{mon}[[q, m^*, V]]$ properly contains $\mathrm{stib}(q, m^*, V)$ for every intege $V > 1$; moreover, given any positive integer p, the image of $\mathrm{comon}(q, *, p)$ properly contains $\mathrm{costab}(q, m^*, p)$ and the image of $\mathrm{comon}[q, m^*, p, V]$ properly contains $\mathrm{costab}[q, m^*, p, V]$ for every integer $V > 1$. In Section 6 we will summarize the said results about generalized coinsertion from Abhyankar-Joshi [9].

As a consequence of the generalized rodeletive correspondence, in Section 6 of Abhyankar-Joshi [6] it is proved that for dimension greater than two, the set of standard monomials in the multiminors of the multimatrix X, corresponding to standard multitableaux, though independent (which is proved in [10]), does not span the polynomial ring $K[X]$. Other proofs of this can be found in Abhyankar-Joshi [7], [8], and [9]. This paper is essentially the summary of [14] and it may serve as a brief preview of Abhyankar-Joshi [6], [7], [8], and [9].

16.2 Notation and Terminology

We shall use the notation and terminology introduced in Subsections (2.1) to (2.4) of Abhyankar [3].

16.3 Generalized Rodeletive Correspondence

In this section we shall review some of the definitions and results about rodeletion from Abhyankar-Kulkarni [4] and Abhyankar-Joshi [6]. So let there be fixed any $q \in N^*$ and $k \in [1, q]$.

Definition (3.1) *Given any $m \in N^*(q)$, in Definition (3.1) of Abhyankar-Joshi [6] a subset* $\mathrm{lem}(q, k, m)$ *of* $\mathrm{popre}(q, m)$ *is defined and it is noted that for each $t \in \mathrm{mon}(q, m)$ there exists a unique element* $\mathrm{les}[q, k, m](t)$ *in* $\mathrm{lem}(q, k, m)$ *such that*

$$\mathrm{mos}[\mathrm{les}[q, k, m](t), m] = t$$

where $\mathrm{les}[q, k, m](t)$ *is the lexical associate of (q, k, m, t) moreover, it is noted that this gives the bijective map*

$$\mathrm{les}[q, k, m] : \mathrm{mon}(q, m) \to \mathrm{lem}(q, k, m).$$

In Definition (3.1) of Abhyankar-Joshi [6] a subset $\mathrm{lem}(q, k)$ *of* $\mathrm{popre}(q)$ *is defined and in Definition (4.1) of Abhyankar-Joshi [6] a map*

$$\mathrm{GRD}_{q,k} : \mathrm{stab}(q) \to \mathrm{lem}(q, k)$$

is defined such that for every $T \in \mathrm{stab}(q)$ we have $\mathrm{con}(T) = \mathrm{con}[\mathrm{GRD}_{q,k}(T)]$; consequently given an $m \in N^*(q)$, in Definition (4.6) of [6] a map

$$\mathrm{GRD}_{q,k,m} : \mathrm{stab}(q, m) \to \mathrm{lem}(q, k, m)$$

is obtained by putting $\mathrm{GRD}_{q,k,m}(T) = \mathrm{GRD}_{q,k}(T)$ for all $T \in \mathrm{stab}(q, m)$, where $\mathrm{GRD}_{q,k}(T)$ is the eneralized rodeletion of (q, k, T).

In view of the said definitions, given any $m \in N^*(q)$, in (5.1) of Abhyankar-Joshi [6] a map

$$\mathrm{GMRD}_{q,k,m} : \mathrm{stab}(q, m) \to \mathrm{mon}(q, m)$$

is obtained by putting $\mathrm{GMRD}_{q,k,m}(T) = \mathrm{mos}[\mathrm{D}_{q,k,m}(T), m]$ for all $T \in \mathrm{stab}(q, m)$, where $\mathrm{GMRD}_{q,k,m}(T)$ is the generalized monomial rodeletion of (q, k, m, T)

Remark (3.2) *Let $q = 2$, and $k' = 1$ or 2 according as $k = 2$ or 1. Let there be given any $m \in N^*(2)$. Then the map*

$$\mathrm{MRD}_{k',m} : \mathrm{stab}(2, m) \to \mathrm{mon}(2, m)$$

which is defined in Definition (9.12) of Abhyankar-Kulkarni [4], coincides with the map

$$\mathrm{GMRD}_{q,k,m} : \mathrm{stab}(2, m) \to \mathrm{mon}(2, m).$$

In view of the Remark (3.2), the following theorem follows from (9.13) of Abhyankar-Kulkarni [4].

Theorem (3.3) *Let $q = 2$. Given any $m \in N^*(2)$, the map*

$$\mathrm{GMRD}_{q,k,m} : \mathrm{stab}(2, m) \to \mathrm{mon}(2, m)$$

is bijective.

In the following theorems, we will see that the generalized rodeletive map is surjective but not injective for $q = 1$, and injective but not surjective for $q > 2$. These theorems are proved in Section 5 of Abhyankar-Joshi [6].

Theorem (3.4) *Assume that $q = 1$ and let there be given any $m \in N^*(1)$ such that $m(1) > 1$. Then we have the following.*

(3.4.1) The map

$$\mathrm{GMRD}_{q,k,m} : \mathrm{stab}(1, m) \to \mathrm{mon}(1, m)$$

is surjective but not injective.

(3.4.2) For every integer $V > 1$, the map $\mathrm{GMRD}_{q,k,m}$ induces a map

$$\mathrm{stib}(1, m, V) \to \mathrm{mon}[[1, m, V]]$$

which is surjective but not injective.

Theorem (3.5) *Assume that $q > 2$ and let there be given any $m \in N^*(q)$ such that $m(k_1) \neq 1 \neq m(k_2)$ for some $k_1 \neq k_2$ in $[1, q]$. Then we have the following.*

(3.5.1) The map

$$\mathrm{GMRD}_{q,k,m} : \mathrm{stab}(q, m) \to \mathrm{mon}(q, m)$$

is injective but not surjective.

(3.5.2) For every integer $V > 1$, the map $\mathrm{GMRD}_{q,k,m}$ induces a map

$$\mathrm{stib}(q, m, V) \to \mathrm{mon}[[q, m, V]]$$

which is injective but not surjective.

Theorem (3.6) *Assume that $q > 2$ and let there be given any $m \in N^*(q)$ such that $m(k_1) \neq 1 \neq m(k_2)$ for some $k_1 \neq k_2$ in $[1, q]$ and let there be given any positive integer p. Then we have the following.*

(3.6.1) The map $\mathrm{GMRD}_{q,k,m}$ induces a map

$$\mathrm{stab}(q, m, p) \to \mathrm{mon}(q, m, p)$$

which is injective but not surjective.

(3.6.2) For every integer $V > 1$, the map $\mathrm{GMRD}_{q,k,m}$ *induces a map*

$$\mathrm{stab}[q, m, p, V] \to \mathrm{mon}[q, m, p, V]$$

which is injective but not surjective.

Theorem (3.7) *Assume that $q > 2$ and let there be given any $m \in N^*(q)$. Also let there be given any positive integer p and let $a \in \mathrm{vec}(q, m, p)$ be such that $a(k_1, 1) < m(k_1)$ and $a(k_2, 1) < m(k_2)$ for some $k_1 \neq k_2$ in $[1, q]$. Then we have the following.*

(3.7.1) The map $\mathrm{GMRD}_{q,k,m}$ *induces a map*

$$\mathrm{stab}(q, m, p, a) \to \mathrm{mon}(q, m, p, a)$$

which is injective but not surjective.

(3.7.2) For every integer $V > 1$, the map $\mathrm{GMRD}_{q,k,m}$ *induces a map*

$$\mathrm{stab}(q, m, p, a, V) \to \mathrm{mon}(q, m, p, a, V)$$

which is injective but not surjective.

16.4 Generalized Codeletion

In this section we shall review some of the definitions and results about codeletion from Abhyankar-Kulkarni [5] and Abhyankar-Joshi [7]. So let there be fixed any $q \in N^*$ and $k \in [1, q]$.

Definition (4.1) *Given any $m \in N^*(q)$, in Definition (3.1) of A yankar-Joshi [7] a subset* $\mathrm{colem}(q, k, m)$ *of* $\mathrm{popre}(q, m)$ *is defined and it is noted that for each $t \in \mathrm{mon}(q, m)$ there exists a unique element* $\mathrm{coles}[q, k, m](t)$ *in* $\mathrm{colem}(q, k, m)$ *such that*

$$\mathrm{mos}[\mathrm{coles}[q, k, m](t), m] = t$$

where $\mathrm{coles}[q, k, m](t)$ *is the colexical associate of (q, k, m, t); moreover, it is noted that this gives the bijective map*

$$\mathrm{coles}[q, k, m] : \mathrm{mon}(q, m) \to \mathrm{colem}(q, k, m).$$

In Definition (3.1) of Abhyankar-Joshi [7] a subset $\mathrm{colem}(q, k)$ *of* $\mathrm{popre}(q)$ *is defined and in Definition (4.1) of Abhyankar-Joshi [7] a map*

$$\mathrm{GCD}_{q,k} : \mathrm{stab}(q) \to \mathrm{colem}(q, k)$$

is defined such that for every $T \in \mathrm{stab}(q)$ we have $\mathrm{con}(T) = \mathrm{con}[\mathrm{GCD}_{q,k}(T)]$; consequently given any $m \in N^*(q)$, in Definition (4.6) of [7] a map

$$\mathrm{GCD}_{q,k,m} : \mathrm{stab}(q, m) \to \mathrm{colem}(q, k, m)$$

is obtained by putting $\mathrm{GCD}_{q,k,m}(T) = \mathrm{GCD}_{q,k}(T)$ *of all* $T \in \mathrm{stab}(q,m)$, *where* $\mathrm{GCD}_{q,k}(T)$ *is the* generalized codeletion *of* (q,k,T).

In view of the said definitions, given any $m \in N^*(q)$, *in (5.1) of Abhyankar-Joshi [7] a map*

$$\mathrm{GMCD}_{q,k,m} : \mathrm{stab}(q,m) \to \mathrm{mon}(q,m)$$

is obtained by putting $\mathrm{GMCD}_{q,k,m}(T) = \mathrm{mos}[\mathrm{GCD}_{q,k,m}(T), m]$ *for all* $T \in \mathrm{stab}(q,m)$, *where* $\mathrm{GMCD}_{q,k,m}(T)$ *is the generalized monomial codeletion of* $(q,k,m,$.

Remark (4.2) *Let* $q = 2$, *and* $k' = 1$ *or* 2 *according as* $k = 2$ *or* 1. *Let there be given any* $m \in N^*(2)$. *Then the map*

$$\mathrm{MCD}_{k',m} : \mathrm{stab}(2,m) \to \mathrm{mon}(2,m)$$

which is defined in Definition (19.12) of Abhyankar-Kulkarni [5], coincides with the map

$$\mathrm{GMCD}_{q,k,m} : \mathrm{stab}(2,m) \to \mathrm{mon}(2,m).$$

In view of the Remark (4.2), the following theorem follows from (19.13) of Abhyankar-Kulkarni [5].

Theorem (4.3) *Let* $q = 2$. *Given any* $m \in N^*(2)$, *the map*

$$\mathrm{GMCD}_{q,k,m} : \mathrm{stab}(2,m) \to \mathrm{mon}(2,m)$$

is bijective.

In the following theorems, we will see that the generalized codeletive map is surjective but not injective for $q = 1$, and injective but not surjective for $q > 2$. These theorems are proved in Section 5 of Abhyankar-Joshi [7].

Theorem (4.4) *Assume that* $q = 1$ *and let there be given any* $m \in N^*(1)$ *such that* $m(1) > 1$. *Then we have the following.*

(4.4.1) The map

$$\mathrm{GMCD}_{q,k,m} : \mathrm{stab}(1,m) \to \mathrm{mon}(1,m)$$

is surjective but not injective.

(4.4.2) For every integer $V > 1$, *the map* $\mathrm{GMCD}_{q,k,m}$ *induces a map*

$$\mathrm{stib}(1,m,V) \to \mathrm{mon}[[1,m,V]]$$

which is surjective but not injective.

Theorem (4.5) *Assume that* $q > 2$ *and let there be given any* $m \in N^*(q)$ *such that* $m(k_1) \neq 1 \neq m(k_2)$ *for some* $k_1 \neq k_2$ *in* $[1,q]$. *Then we have the following.*

(4.5.1) The map

$$\mathrm{GMCD}_{q,k,m} : \mathrm{stab}(q,m) \to \mathrm{mon}(q,m)$$

is injective but not surjective.

(4.5.2) For every integer $V > 1$, the map $\mathrm{GMCD}_{q,k,m}$ *induces a map*

$$\mathrm{stib}(q,m,V) \to \mathrm{mon}[[q,m,V]]$$

which is injective but not surjective.

Theorem (4.6) *Assume that $q > 2$ and let there be given any $m \in N^*(q)$ such that $m(k_1) \neq 1 \neq m(k_2)$ for some $k_1 \neq k_2$ in $[1,q]$ and let there be given any positive integer p. Then we have the following.*

(4.6.1) The map $\mathrm{GMCD}_{q,k,m}$ *induces a map*

$$\mathrm{costab}(q,m,p) \to \mathrm{comon}(q,m,p)$$

which is injective but not surjective.

(4.6.2) For every integer $V > 1$, the map $\mathrm{GMCD}_{q,k,m}$ *induces a map*

$$\mathrm{costab}[q,m,p,V] \to \mathrm{comon}[q,m,p,V]$$

which is injective and if $\mathrm{costab}[q,m,p,V] \neq \emptyset$ *then the induced map is not surjective.*

Theorem (4.7) *Assume that $q > 2$ and let there be given any $m \in N^*(q)$. Also let there be given any positive integer p and let $a \in \mathrm{covec}(q,m,p)$ be such that $a(k_1,1) < m(k_1)$ and $a(k_2,1) < m(k_2)$ for some $k_1 \neq k_2$ in $[1,q]$. Then we have the following.*

(4.7.1) The map $\mathrm{GMCD}_{q,k,m}$ *induces a map*

$$\mathrm{costab}(q,m,p,a) \to \mathrm{comon}(q,m,p,a)$$

which is injective but not surjective.

(4.7.2) For every integer $V > 1$, the map $\mathrm{GMCD}_{q,k,m}$ *induces a map*

$$\mathrm{costab}(q,m,p,a,V) \to \mathrm{comon}(q,m,p,a,V)$$

which is injective and if $\mathrm{costab}(q,m,p,a,V) \neq \emptyset$ *then the induced map is not surjective.*

16.5 Generalized Roinsertion

In this section we shall review some of the definitions and results about roinsertion and rodeletion from Abhyankar-Kulkarni [4] and Abhyankar-Joshi [8].

By a total type m of length q, where q is a positive integer, we mean a sequence $m(1), m(2), \ldots, m(q)$ of totally ordered sets. In the subsection (5.1) we will consider the roinsertion for any total type of length 2, and in the Subsection (5.2) we will discuss the roinsertion for any total type whose length is greater than 2.

16.5.1

In this subsection let m be a total type of length 2 and let there be given any $k \in [1, 2]$.

Definition (5.1.1) *Given any $m' \in N^*(2)$, in Definition (8.10) of Abhyankar-Kulkarni [4] a subset* $\mathrm{leb}(k, m')$ *of* $\mathrm{pop}(2, m')$ *is defined and in Definition (8.12) of Abhyankar-Kulkarni [4] a map*

$$R_{k,m'} : \mathrm{leb}(k, m') \to \mathrm{stab}(2, m')$$

is obtained by putting $R_{k,m'}(w) = R_k(w)$ for all $w \in \mathrm{leb}(k, m')$, where $R_k(w)$ is the roinsertion of (k, w).

In Definition (6.1) of Abhyankar-Joshi [8] a subset $\mathrm{leb}(k, m)$ *f* $\mathrm{pre}(2, m)$ *is defined and it is noted that for all $w \in \mathrm{leb}(k, m) \setminus \{\mathrm{vec}_0(2, m)\}$ we have $n(w) \in \mathrm{leb}(k[w])$, where $l[w]$ is a member of $N^*(2)$ and it is the* true length *w, and $n[w]$ is a member of* $\mathrm{popre}(2, l[w], \mathrm{len}(w))$ *a it is the normalization at w. Consequently using (5.2) of [8], in Definition (6.1) of [8] a map*

$$R_{k,m} : \mathrm{leb}(k, m) \to \mathrm{stab}(2, m)$$

is obtained by putting

$$R_{k,m}(w) = \begin{cases} n_w(R_{k,l[w]}(n(w))) & \text{if } w \in \mathrm{leb}(m) \setminus \{\mathrm{vec}_0(2, m)\} \\ \mathrm{stab}_o(2, m) & \text{if } w = \mathrm{vec}_0(2, m) \end{cases}$$

where $R_{k,m}(w)$ is the roinsertion *of (k, m, w).*

In Definition (6.6) of Abhyankar-Joshi [8], it is noted that for each $\mathrm{mon}(2, m)$ *there exists a unique element* $\mathrm{les}[k, m](t)$ *in* $\mathrm{leb}(k, m)$ *such that*

$$\mathrm{mos}[\mathrm{les}[k, m](t), m] = t$$

where $\mathrm{les}[k, m](t)$ *is the lexical associate of (k, m, t); moreover, it is noted that this gives the bijective map*

$$\mathrm{les}[k, m] : \mathrm{mon}(2, m) \to \mathrm{leb}(k, m).$$

In view of the said definitions, in Definition (6.8) of Abhyankar-Joshi [8], a map

$$\mathrm{MR}_{k,m} : \mathrm{mon}(2, m) \to \mathrm{stab}(2, m)$$

is obtained by putting

$$\mathrm{MR}_{k,m}(t) = R_{k,m}(\mathrm{les}[k,m](t))$$

for all $t \in \mathrm{mon}(2,m)$*, where* $\mathrm{MR}_{k,m}(t)$ *is the monomial roinsertion of* (k,m,t).

Definition (5.1.2) *Given any* $m' \in N^*(2)$*, in Definition (9.9) of Abhyankar-Kulkarni [4] a map*

$$\mathrm{RD}_{k,m'} : \mathrm{stab}(2,m') \to \mathrm{leb}(k,m')$$

is defined by putting $\mathrm{RD}_{k,m'}(T) = \mathrm{RD}_k(T)$ *for all* $T \in \mathrm{stab}(2,m')$*, where* $\mathrm{RD}_k(T)$ *is the rodeletion of* (k,T).

In Definition (6.1) of Abhyankar-Joshi [8] it is noted that for all $T \in \mathrm{stab}(2,m) \setminus \{\mathrm{stab}_0(2,m)\}$ *we have* $n(T) \in \mathrm{stab}(2,l[T])$*, where* $l[T]$ *is a member of* $N^*(2)$ *and it is the* true length *of* T*, and* $n[T]$ *is a member of* $\mathrm{popid}(2,l[T],\mathrm{dep}(T))$ *and it is the* normalization *at* T*. Consequently using (6.1.2) of [8], in Definition (6.1) of [8] a map*

$$\mathrm{RD}_{k,m} : \mathrm{stab}(2,m) \to \mathrm{leb}(k,m)$$

is obtained by putting

$$\mathrm{RD}_{k,m}(T) = \begin{cases} n_T(RD_{k,l[T]}(n(T))) & \text{if } T \in \mathrm{tstab}(2,m) \setminus \{\mathrm{stab}_0(2,m)\} \\ \mathrm{vec}_o(2,m) & \text{if } T = \mathrm{stab}_0(2,m) \end{cases}$$

where $\mathrm{RD}_{k,m}(T)$ *is the rodeletion of* (k,m,T).

In view of the said definitions, in Definition (6.13) of Abhyankar-Joshi [8] a map

$$\mathrm{MRD}_{k,m} : \mathrm{stab}(2,m) \to \mathrm{mon}(2,m)$$

is obtained by putting

$$\mathrm{MRD}_{k,m}(T) = \mathrm{mos}[\mathrm{RD}_{k,m}(T),m]$$

for all $T \in \mathrm{stab}(2,m)$*, where* $\mathrm{MRD}_{k,m}(T)$ *is the* monomial rodeletion *of* (k,m,T).

In the following theorems, we will see that the roinsertion and rodeletion maps are inverses of each other and they induce bijections between sets of various types of bitableaux and monomials. These theorem are proved in Section 6 of Abhyankar-Joshi [8].

Theorem (5.1.3) *The maps* $\mathrm{MR}_{k,m} : \mathrm{mon}(2,m) \to \mathrm{stab}(2,m)$ *and* $\mathrm{MRD}_{k,m} : \mathrm{stab}(2,m) \to \mathrm{mon}(2,m)$ *are inverses of each other, i.e., for every* $t \in \mathrm{mon}(2,m)$ *we have* $\mathrm{MRD}_{k,m}(\mathrm{MR}_{k,m}(t)) = t$*, and for every* $T \in \mathrm{stab}(2,m)$ *we have* $\mathrm{MR}_{k,m}(\mathrm{MRD}_{k,m}(T)) = T$*. Hence in particular, both e maps* $\mathrm{MR}_{k,m}$ *and* $\mathrm{MRD}_{k,m}$ *are bijective.*

Theorem (5.1.4) *Given any* $V \in N$*, the maps* $\mathrm{MR}_{k,m}$ *and* $\mathrm{MRD}_{k,m}$ *induce bijective maps* $\mathrm{mon}[[2,m,V]] \to \mathrm{stib}(2,m,V)$ *and* $\mathrm{stib}(2,m,V) \to \mathrm{mon}[[2,m,V]]$ *which are inverses of each other.*

Theorem (5.1.5) *Given any* $p \in N$*, the maps* $\mathrm{MR}_{k,m}$ *and* $\mathrm{MRD}_{k,m}$ *induce bijective maps* $\mathrm{mon}(2,m,p) \to \mathrm{stab}(2,m,p)$ *and* $\mathrm{stab}(2,m,p) \to \mathrm{mon}(2,m,p)$ *which are inverses of each other, and for every* $V \in N$*, the maps* $\mathrm{MR}_{k,m}$ *and* $\mathrm{MRD}_{k,m}$ *induce bijective maps* $\mathrm{mon}[2,m,p,V] \to \mathrm{stab}[2,m,p,V]$ *and* $\mathrm{stab}[2,m,p,V] \to \mathrm{mon}[2,m,p,V]$ *which are inverses of each other.*

Theorem (5.1.6) *Given any* $p \in N$ *and* $a \in \mathrm{vec}(2,m,p)$*, the maps* $\mathrm{MR}_{k,m}$ *and* $\mathrm{MRD}_{k,m}$ *induce bijective maps* $\mathrm{mon}(2,m,p,a) \to \mathrm{stab}(2,m,p,a)$ *and* $\mathrm{stab}(2,m,p,a) \to \mathrm{mon}(2,m,p,a)$ *which are inverses of each other, and for every* $V \in N$*, the maps* $\mathrm{MR}_{k,m}$ *and* $\mathrm{MRD}_{k,m}$ *induce bijective maps* $\mathrm{mon}(2,m,p,a,V) \to \mathrm{stab}(2,m,p,a,V)$ *and* $\mathrm{stab}(2,m,p,a,V) \to \mathrm{mon}(2,m,p,a,V)$ *which are inverses of each other.*

16.5.2

In this subsection let m be a total type of length q where q is some integer greater than 2 and let there be given any $k \in [1,q]$, and let $m^* = B_k(m)$.

Definition (5.2.1) *In Section 3 of Abhyankar-Joshi [8], to any* $t \in \mathrm{mon}(q,m)$ *we have associated a unique element in* $\mathrm{mon}(2,m^*)$ *and we have obtained the bijective map*

$$\mathrm{BM}_{k,m} : \mathrm{mon}(q,m) \to \mathrm{mon}(2,m^*)$$

by putting

$$\mathrm{BM}_{k,m}(t) = B_k(t)$$

for all $t \in \mathrm{mon}(q,m)$*, where* $B_k(t)$ *is the* k^{th} *biassociate of* t*. In the same section of [8], to any* $T \in \mathrm{pab}(q,m)$ *we have also associated a unique element in* $\mathrm{pab}(2,m^*)$ *and we have obtained the bijective map*

$$\mathrm{BT}_{k,m} : \mathrm{pab}(q,m) \to \mathrm{pab}(2,m^*)$$

by putting

$$\mathrm{BT}_{k,m}(T) = B_k(T)$$

for all $T \in \mathrm{pab}(q,m)$*, where* $B_k(T)$ *is the* k^{th} *biassociate of* T*. By*

$$\mathrm{TB}_{k,m} : \mathrm{pab}(2,m^*) \to \mathrm{pab}(q,m)$$

we have denoted the inverse of the bijective map $\mathrm{BT}_{k,m}$*. Using e said maps, in Definition (7.1) of Abhyankar-Joshi [8], a map*

$$\mathrm{GMR}_{k,m} : \mathrm{mon}(q,m) \to \mathrm{pab}(q,m)$$

is obtained by the composition of the maps

$$\text{mon}(q,m) \xrightarrow{\text{BM}_{k,m}} \text{mon}(2,m^*) \xrightarrow{\text{MR}_{2,m^*}} \text{stab}(2,m^*)$$

$$\rightarrow \text{pab}(2,m^*) \xrightarrow{\text{TB}_{k,m}} \text{pab}(q,m)$$

where the unnamed map is the natural injection, and where for any $t \in \text{mon}(q,m)$*, the pretableau* $\text{GMR}_{k,m}(t)$ *is the generalized monomial roinsertion of* (q,k,m,t).

In the following theorems, we will see that the generalized roinsertion map is injective and under this map the images of sets of various types of monomials properly contain sets of standard multitableaux. These theorems are proved in Section 7 of Abhyankar-Joshi [8].

Theorem (5.2.2) *The map* $\text{GMR}_{k,m} : \text{mon}(q,m) \rightarrow \text{pab}(q,m)$ *is injective. Moreover for any integer* V *it induces an injective map* $\text{mon}[[q,m,V]] \rightarrow \text{pib}(q,m,V)$.

Theorem (5.2.3) *Assume that* $\text{card}(m(l)) > 1$ *for every* $l \in [1,q]$. *Then we have the following.*

1. $\text{GMR}_{k,m}(\text{mon}(q,m)) \supseteq \text{stab}(q,m)$ *and* $\text{GMR}_{k,m}(\text{mon}(q,m)) \neq \text{stab}(q,m)$.

2. *Given any integer* $V > 1$, *we have that* $\text{GMR}_{k,m}\text{mon}[[q,m,V]]) \supseteq \text{stib}(q,m,V)$ *and* $\text{GMR}_{k,m}\textit{text}(\text{mon}[[q,m,V]]) \neq \text{stib}(q,m,V)$.

3. *Given any positive integer* p, *we have that* $\text{GMR}_{k,m}(\text{mon}(q,m,p)) \supseteq \text{stab}(q,m,p)$ *and* $\text{GMR}_{k,m}(\text{mon}(q,m,p)) \neq \text{stab}(q,m,p)$.

4. *Given any integer* $V > 1$, *and any positive integer* p, *we have that* $\text{GMR}_{k,m}(\text{mon}[q,m,p,V]) \supseteq \text{stab}[q,m,p,V]$ *and* $\text{GMR}_{k,m}(\text{mon}[q,m,p,V]) \neq \textit{stab}[q,m,p,V]$.

Corollary (5.2.4) *Assume that* $1 < \text{card}(m(l)) < \infty$ *for every* $l \in [1,q]$. *Then we have the following.*

1. *Given any integer* $V > 1$, *we have* $\text{card}(\text{mon}[[q,m,V]]) > \text{card}(\text{stib}(q,m,V))$.

2. *Given any integer* $V > 1$ *and any positive integer* p, *we have* $\text{card}(\text{mon}[q,m,p,V]) > \text{card}(\text{stab}[q,m,p,V]$.

Corollary (5.2.5) *Let there be given any* $m' \in N^*(q)$ *such that* $m'(l) > 1$ *for every* $l \in [1,q]$. *Then we have the following.*

1. *Given any integer* $V > 1$, *we have* card(mon[[q, m', V]] > *card*(stib(q, m', V)).

2. *Given any integer* $V > 1$ *and any positive integer* p, *we have* card(mon[q, m', p, V]) > card(stab[q, m', p, V]).

16.6 Generalized Coinsertion

In this section we shall review some of the definitions and results about coinsertion and codeletion from Abhyankar-Kulkarni [5] and Abhyankar-Joshi [9].

In the subsection (6.1) we will consider the coinsertion for any total type of length 2, and in the subsection (6.2) we will discuss the coinsertion for any total type whose length is greater than 2.

16.6.1

In this subsection let m *be a total type of length 2 and let there be given any* $k \in [1, 2]$.

Definition (6.1.1) *Given any* $m' \in N^*(2)$, *in Definition (18.10) of Abhyankar-Kulkarni [5] a subset* coleb(k, m') *of* pre($2, m'$) *is defined and in Definition (18.12) of Abhyankar-Kulkarni [5] a map*

$$C_{k,m'} : \mathrm{coleb}(k, m') \to \mathrm{stab}(2, m')$$

is obtained by putting $C_{k,m'}(w) = C_k(w)$ *for all* $w \in \mathrm{coleb}(k, m')$, *where* $C_k(w)$ *is the* coinsertion *of* (k, w).

In Definition (4.1) of Abhyankar-Joshi [9] a subset coleb(k, m) *of* pre($2, m$) *is defined and it is noted that for all* $w \in mcoleb(k, m) \setminus \{\mathrm{vec}_0(2, m)\}$ *we have* $n(w) \in \mathrm{coleb}(k, l[w])$ *where,* $l[w]$ *is a member of* $N^*(2)$ *and it is the* true length *of* w, *and* $n[w]$ *is a member of* popre($2, l[w], \mathrm{len}(w)$) *and it is the* normalization *at* w. *Consequently using (5.2) of [8], in Definition (4.1) of [9] a map*

$$C_{k,m} : \mathrm{coleb}(k, m) \to \mathrm{stab}(2, m)$$

is obtained by putting

$$C_{k,m}(w) = \begin{cases} n_w(C_{k,l[w]}(n(w))) & \text{if } w \in \mathrm{colebk}, m) \setminus \{\mathrm{vec}_0(2, m)\} \\ \mathrm{stab}_o(2, m) & \text{if } w = \mathrm{vec}_0(2, m) \end{cases}$$

where $C_{k,m}(w)$ *is the coinsertion of* (k, m, w).

In Definition (4.6) of Abhyankar-Joshi [9], it is noted that for each $t \in$ mon($2, m$) *there exists a unique element* coles[k, m](t) coleb(k, m) *such that*

$$\mathrm{mos}[\mathrm{coles}[k, m](t), m] = t$$

where $\operatorname{coles}[k,m](t)$ *is the colexical associate of* (k,m,t)*; moreover, it is noted that this gives the bijective map*

$$\operatorname{coles}[k,m] : \operatorname{mon}(2,m) \to \operatorname{coleb}(k,m).$$

In view of the said definitions, in Definition (4.8) of Abhyankar-Joshi [9], a map

$$\mathrm{CR}_{k,m} : \operatorname{mon}(2,m) \to \operatorname{stab}(2,m)$$

is obtained by putting

$$\mathrm{CR}_{k,m}(t) = C_{k,m}(\operatorname{coles}[k,m](t))$$

for all $t \in \operatorname{mon}(2,m)$*, where* $\mathrm{CR}_{k,m}(t)$ *is the monomial coinsertion of* (k,m,t).

Definition (6.1.2) *Given any* $m' \in N^*(2)$*, in Definition (19.9) of Abhyankar-Kulkarni [5] a map*

$$\mathrm{CD}_{k,m'} : \operatorname{stab}(2,m') \to \operatorname{coleb}(k,m')$$

is defined by putting $\mathrm{CD}_{k,m'}(T) = \mathrm{CD}_k(T)$ *for all* $T \in \operatorname{stab}(2,m')$*, where* $\mathrm{CD}_k(T)$ *is the codeletion of* (k,T).

In Definition (4.1) of Abhyankar-Joshi [9] it is noted that for all $T \in \operatorname{stab}(2,m) \setminus \{\operatorname{stab}_0(2,m)\}$ *we have* $n(T) \in \operatorname{stab}(2,l[T])$*, where* $l[T]$ *is a member of* $N^*(2)$ *and it is the true length of* T*, and* $n[T]$ *is a member of* $\operatorname{popid}(2,l[T],\operatorname{dep}(T))$ *and it is the* normalization *at* T*. Consequently using (4.1.2 of [9], in Definition (4.1) of [9] a map*

$$\mathrm{CD}_{k,m} : \operatorname{stab}(2,m) \to \operatorname{coleb}(k,m)$$

is obtained by putting

$$\mathrm{CD}_{k,m}(T) = \begin{cases} n_T(CD_{k,l[T]}(n(T))) & \text{if } T \in \operatorname{stab}(2,m) \setminus \{\operatorname{stab}_0(2,m)\} \\ \operatorname{vec}_o(2,m) & \text{if } T = \operatorname{stab}_0(2,m) \end{cases}$$

where $\mathrm{CD}_{k,m}(T)$ *is the codeletion of* (k,m,T).

In view of the said definitions, in Definition (4.13) of Abhyankar-Joshi [9] a map

$$\mathrm{MCD}_{k,m} : \operatorname{stab}(2,m) \to \operatorname{mon}(2,m)$$

is obtained by putting

$$\mathrm{MCD}_{k,m}(T) = \operatorname{mos}[\mathrm{CD}_{k,m}(T), m]$$

for all $T \in \operatorname{stab}(2,m)$*, where* $\mathrm{MCD}_{k,m}(T)$ *is the monomial codeletion of* (k,m,T).

In the following theorems, we will see that the coinsertion and codeletion maps are inverses of each other and they induce bijections between various types of bitableaux and monomials. These theorems are proved in Section 4 of Abhyankar-Joshi [9].

Theorem (6.1.3) *The maps* $\mathrm{CR}_{k,m} : \mathrm{mon}(2,m) \to \mathrm{stab}(2,m)$ *and* $\mathrm{MCD}_{k,m} : \mathrm{stab}(2,m) \to \mathrm{mon}(2,m)$ *are inverses of each other, i.e., for every* $t \in \mathrm{mon}(2,m)$ *we have* $\mathrm{MCD}_{k,m}(\mathrm{CR}_{k,m}(t)) = t$, *and for every* $T \in$ *stab*$(2,m)$ *we have* $\mathrm{CR}_{k,m}(\mathrm{MCD}_{k,m}(T)) = T$. *Hence in particular, both the maps* $\mathrm{CR}_{k,m}$ *and* $\mathrm{MCD}_{k,m}$ *are bijective.*

Theorem (6.1.4) *Given any* $V \in N$, *the maps* $\mathrm{CR}_{k,m}$ *and* $\mathrm{MCD}_{k,m}$ *induce bijective maps* $\mathrm{mon}[[2,m,V]] \to \mathrm{stib}(2,m,V)$ *and* $\mathrm{stib}(2,m,V) \to \mathrm{mon}[[2,m,V]]$ *which are inverses of each other.*

Theorem (6.1.5) *Given any* $p \in N$, *the maps* $\mathrm{CR}_{k,m}$ *and* $\mathrm{MCD}_{k,m}$ *induce bijective maps* $\mathrm{comon}(2,m,p) \to \mathrm{costab}(2,m,p)$ *and* $\mathrm{costab}(2,m,p) \to \mathrm{comon}(2,m,p)$ *which are inverses of each other, and for every* $V \in N$, *the maps* $\mathrm{CR}_{k,m}$ *and* $\mathrm{MCD}_{k,m}$ *induce bijective maps* $\mathrm{comon}[2,m,p,V] \to \mathrm{costab}[2,m,p,V]$ *and* $\mathrm{costab}[2,m,p,V] \to \mathrm{comon}[2,m,p,V]$ *which are inverses of each other.*

Theorem (6.1.6) *Given any* $p \in N$ *and* $a \in \mathrm{covec}(2,m,p)$, *the maps* $\mathrm{CR}_{k,m}$ *and* $\mathrm{MCD}_{k,m}$ *induce bijective maps* $\mathrm{comon}(2,m,p,a) \to \mathrm{costab}(2,m,p,a)$ *and* $\mathrm{costab}(2,m,p,a) \to \mathrm{comon}(2,m,p,a)$ *which are inverses of each other, and for every* $V \in N$, *the maps* $\mathrm{CR}_{k,m}$ *and* $\mathrm{MCD}_{k,m}$ *induce bijective maps* $\mathrm{comon}(2,m,p,a,V) \to \mathrm{costab}(2,m,p,a,V)$ *and* $\mathrm{costab}(2,m,p,a,V) \to \mathrm{comon}(2,m,p,a,V)$ *which are inverses of each other.*

16.6.2

In this subsection let m be a total type of length q where q is some integer greater than 2 and let there be given any $k \in [1,q]$, and let $m^* = B_k(m)$.

Definition (6.2.1) *In Definition (5.1) of Abhyankar-Joshi [9] a map*

$$\mathrm{GCR}_{k,m} : \mathrm{mon}(q,m) \to \mathrm{pab}(q,m)$$

is obtained by the composition of the maps

$$\mathrm{mon}(q,m) \xrightarrow{\mathrm{BM}_{k,m}} \mathrm{mon}(2,m^*) \xrightarrow{\mathrm{CR}_{2,m^*}} \mathrm{stab}(2,m^*)$$

$$\longrightarrow \mathrm{pab}(2,m^*) \xrightarrow{\mathrm{TB}_{k,m}} \mathrm{pab}(q,m)$$

where the unnamed map is the natural injection, and where for any $t \in \mathrm{mon}(q,m)$, *the pretableau* $\mathrm{GCR}_{k,m}(t)$ *is the eneralized monomial coinsertion of* (q,k,m,t).

In the following theorems, we will see that the generalized coinsertion map is injective and under this map the images of sets of various types of monomials properly contain sets of standard multitableaux. These theorems are proved in Section 5 of Abhyankar-Joshi [9].

Theorem (6.2.2) *The map* $\mathrm{GCR}_{k,m} : \mathrm{mon}(q,m) \to \mathrm{pab}(q,m)$ *is injective. Moreover for any integer* V *it induces an injective map* $\mathrm{mon}[[q,m,V]] \to \mathrm{pib}(q,m,V)$.

Theorem (6.2.3) *Assume that* $\mathrm{card}(m(l)) > 1$ *for every* $l \in [1,q]$. *Then we have the following.*

1. $\mathrm{GCR}_{k,m}(\mathrm{mon}(q,m)) \supseteq \mathrm{stab}(q,m)$ *and* $\mathrm{GCR}_{k,m}(\mathrm{mon}(q,m)) \neq \mathrm{stab}(q,m)$.
2. *Given any integer* $V > 1$, *we have that* $\mathrm{GCR}_{k,m}(\mathrm{mon}[[q,m,V]]) \supseteq \mathrm{stib}(q,m,V)$ *and* $\mathrm{GCR}_{k,m}(\mathrm{mon}[[q,m,V]]) \neq \mathrm{stib}(q,m,V)$.
3. *Given any positive integer* p, *we have that* $\mathrm{GCR}_{k,m}(\mathrm{comon}(q,m,\ p)) \supseteq \mathrm{costab}(q,m,p)$ *and* $\mathrm{GC}_{k,m}(\mathrm{comon}(q,m,p)) \neq \mathrm{costab}(q,m,p)$.
4. *Given any integer* $V > 1$, *and any positive integer* p, *if* $\mathrm{costab}[2,m^*,p,V] \neq \emptyset$, *then we have that* $\mathrm{GCR}_{k,m}(\mathrm{comon}[q,m,p,V]) \supseteq \mathrm{costab}[q,m,p,V]$ *and* $\mathrm{GCR}_{k,m}(\mathrm{comon}[q,m,p,V]) \neq \mathrm{costab}[q,m,p,V]$.

Corollary (6.2.4) *Assume that* $1 < \mathrm{card}(m(l)) < \infty$ *for every* $l \in [1,q]$. *Then we have the following.*

1. *Given any integer* $V > 1$, *we have* $\mathrm{card}(\mathrm{mon}[[q,m,V]]) > \mathrm{card}(\mathrm{stib}(q,m,V))$.
2. *Given any integer* $V > 1$ *and any positive integer* p, *if* $\mathrm{costab}[2,m^*,p,V] \neq \emptyset$, *then we have* $\mathrm{card}(\mathrm{comon}[q,p,V]) > \mathrm{card}(\mathrm{costab}[q,m,p,V])$.

Corollary (6.2.5) Let there be given any $m' \in N^*(q)$ such that $m'(l) > 1$ for every $l \in [1,q]$. Then we have the following.

1. Given any integer $V > 1$, we have $\mathrm{card}(\mathrm{mon}[[q,m',V]] > \mathrm{card}(\mathrm{stib}(q,m',V))$.
2. Given any integer $V > 1$ and any positive integer p, if $\mathrm{costab}[2,m',p,V] \neq \emptyset$, then we have $\mathrm{card}(\mathrm{comon}[q,m',p,V]) > \mathrm{card}(\mathrm{costab}[q,m',p,V])$.

16.7 Applications

Throughout this section let there be fixed any integer $q \geq 2$, any $m \in N^*)$, any field R, and any subfield K of R. We shall use the notation and terminology about multimatrices, their multiminors, monomials in multiminors, and determinantish maps from Section 3 of Abhyankar-Ghorpade [10].

The following theorem, which says that the set of standard monomials in the multiminors of the multimatrix X, corresponding to standard ultitableaux bounded by m, is linearly independent over K, is proved in (3.6.1) of Abhyankar-Ghorpade [10].

Theorem (7.1) *If M is a determinantish map and R contains an indeterminate over $K(X)$, then we have the following.*

(7.1.1) For every $V \in N$ we have that the map

$$M[X, \text{stib}, V] : \text{stib}(q, m, V) \to R$$

is K-independent and its image is contained in $K[X]_V$.

(7.1.2) We have that the map

$$M[X, \text{stab}] : \text{stab}(q, m) \to R$$

is K-independent and its image is contained in $K[X]$.

The following theorem, which says that for $q = 2$, the set of standard monomials in the minors of the matrix X, corresponding to standard bitableaux bounded by m, gives a K-vector-space-basis of $K[X]$, is proved in Theorem (20.3) of Abhyankar [3] and in (4.3) of Abhyankar-Ghorpade [10].

Standard Basis Theorem (7.2) *If $q = 2$ and $M(x) = \det x$ for all $x \in \text{smul}(R, q)$, then we have the following.*

(7.2.1) For every $V \in N$ we have that $M[X, \text{stib}, V]$ is a K-basis of $K[X]_V$.

(7.2.2) We have that $M[X, \text{stab}]$ is a K-basis of $K[X]$.

The following theorem, which says that for dimension greater than two, the set of standard monomials in the multiminors of the multimatrix X, corresponding to standard multitableaux bounded by m, does not span $K[X]$, is proved in (6.3) of Abhyankar-Joshi [6]. Other proofs of this can be found in Abhyankar-Joshi [7], [8], and [9].

Theorem (7.3) *Assume that $q > 2$ and $m(k) > 1$ for every $k \in [1, q]$. Also assume that M is a determinantish map and R contains an indeterminate over $K(X)$. Then we have the following.*

(7.3.1) For any integer $V > 1$ we have that the map

$$M[X, \text{stib}, V] : \text{stib}(q, m, V) \to R$$

is not a K-generator of $K[X]_V$.

(7.3.2) We have that the map

$$M[X, \mathrm{stab}] : \mathrm{stab}(q, m) \to R$$

is not a K-generator of $K[X]$.

16.8 References

[1] S. S. Abhyankar, Combinatoire des tableaux de Young, varietes determinantielles et calcul de fonctions de Hilbert, *Rendiconti del Seminario Matematico des Universita et Polytecnico di Torino*, **42**, 1984 pp. 65–88.

[2] S. S. Abhyankar, Determinantal loci and enumerative combinatorics of Young tableaux, *Algebraic Geometry and Commutative Algebra in Honor of Masayoshi Nagata*, **I**, Kinokuniya Company Ltd., Tokyo, 1987, pp. 1–26.

[3] S. S. Abhyankar, *Enumerative combinatorics of Young tableaux* Marcel Dekker, New York, 1988.

[4] by S. S. Abhyankar and D. M. Kulkarni, Bijection between indexed monomials and standard bitableaux, *Discrete Mathematics*, **79**, 1989/90, pp. 48.

[5] S. S. Abhyankar and D. M. Kulkarni, Coinsertion and standard bitableaux, to appear in discrete Mathematics

[6] S. S. Abhyankar and S. B. Joshi Generalized rodeletive correspondence between multitableaux and multimonomials, to appear in *Discrete Mathemati.*

[7] S. S. Abhyankar and S. B. Joshi, Generalized codeletion and standard multitableaux, *Montreal Conference Proceedings, Group Actions and Invariant Theory*, Canadian Mathematical Society **10**, 1989 pp. 1-24.

[8] S. S. Abhyankar and S. B. Joshi, Generalized roinsertion and standard monomials in multiminors of a multimatrix, to appear in *Discrete Mathematics.*

[9] S. S. Abhyankar and S. B. Joshi, Generalized coinsertion and standard monomials in multiminors of a multimatrix, fourthcoming.

[10] S. S. Abhyankar and S. R. Ghorpade, Young tableaux and linear independence of standard monomials in multiminors of a multimatrix, forthcoming.

[11] C. Deconcini and D. Eisenbud and C. Procesi, Young diagrams and determinantal varieties, *Inv. Math.*, **56**, 1980, pp. 129-165.

[12] J. Desarmenian and J. P. S. Kung and G. C. Rota, Invariant theory, Young bitableaux, and combinatorics, *Advances in Mathematics*, **27**, 1978, pp. 63-92.

[13] P. Doubilet and G. C. Rota and J. Stein, Foundations of combinatorics IX: Combinatorial methods in invariant theory, *Studies in Applied Mathematics*, **53**, 1974, pp. 185–216.

[14] S. B. Joshi, Thesis, Purdue University, 1989.

[15] D. E. Knuth, Permutations, matrices and generalized Young tableaux, *Pacific Journal of Mathematics*, **34**, 1970, pp. 709–727.

[16] G. deB Robinson, On the representations of the symmetric group, *American Journal of Mathematics*, **60**, 1938, pp. 745–760.

[17] C. Schensted, Longest increasing and decreasing subsequences, *Canadian Journal of Mathematics*, **13**, 1961, pp. 179–191.

Part VII

Commutative Algebra

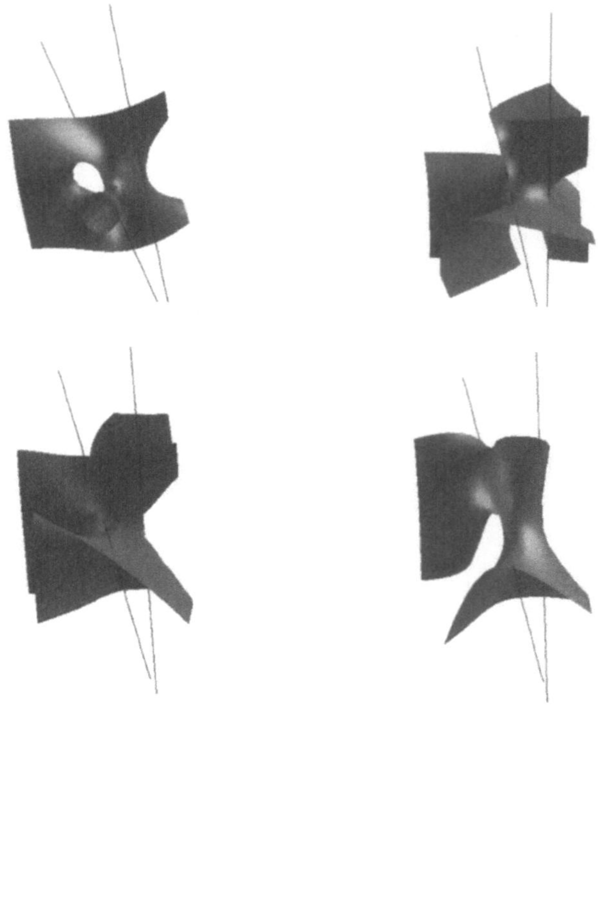

17

Report on the Torsion of the Differential Module of an Algebraic Curve

Robert W. Berger

ABSTRACT There is a conjecture, that the torsionfreeness of the module of differentials in a point of an algebraic or algebroid curve should imply that the curve is non singular at that point. A report on the main results is given.

17.1 Introduction

Let k be a perfect field and R the local ring of a closed point of an algebraic or algebroid curve over k. There is a conjecture that R is regular if and only if the (universally finite) differential module $\Omega_{R/k}$ is torsionfree. The nontrivial part is of course to show that for a singular point the torsion submodule $\tau(\Omega_{R/k})$ of $\Omega_{R/k}$ is not zero. Although a solution for the general case is not in sight there are many special cases which have been treated successfully. In all of these the conjecture has been found to be true. It is the purpose of this paper to give a survey on some of these results with hints concerning the proofs. For simplicity let us assume for the following that R is a reduced complete analytic k-algebra of dimension one with maximal ideal $\mathbf{m}$ and embedding dimension n, which then can be represented in the form $R = k[\![X_1, \ldots, X_n]\!]/I = k[\![x_1, \ldots, x_n]\!]$, where I is a reduced ideal in the formal power series ring $k[\![X_1, \ldots, X_n]\!]$. We will also restrict ourselves to the case *char* $k = 0$, although many of the results are also valid for perfect ground fields. One can distinguish several cases:

17.2 Conditions on the Number of Generators of I

Let $d(R) = \mu_R(I) - (n-1)$ denote the *deviation* of R, where $\mu_R(I)$ denotes the minimal number of generators of I. R is called a *complete intersection* if $d = 0$ and an *almost complete intersection* if $d \leq 1$. In [5] the cases $d \leq 1$ were solved in R is a domain. This was generalized to $d \leq 3$ in the reduced case by Ulrich [33], [34].

Denote by S the integral closure of R in its full ring of quotients K, and let $\mathbf{D} : S \to SDS = \Omega_{S/k}$ and $d : R \to RdR = \Omega_{R/k}$ be the universally finite derivations of S and R over k respectively. Since RdR and SDS are both of rank 1 and SDS is torsionfree (even free), the kernel of the canonical homomorphism $\varphi : RdR \to SDS$ is $T := \tau(\Omega_{R/k})$, so that we have an exact sequence

$$O \to T \to RdR \to SDS \to SDS \,/\, RDR \to 0.$$

Let $x \in R$ be a normalizing parameter and $s = k[\![x]\!]$. Then R and S are finite s-modules and the universally finite derivations $\delta : R \to R\delta R = \Omega_{R/s}$ and $\Delta : S \to S\Delta S = \Omega_{S/s}$ coincide with the universal derivations of R and S over s respectively. $R\delta R$ and $S\Delta S$ are finitely generated torsion modules and therefore have finite lengths as s-modules. Further we have $R\delta R = RdR \,/\, Rdx$ and $Rdx \cap T = 0$ and therefore $SDx \,/\, RDx \cong S \,/\, R$. Let α be the natural map $R\delta R \to S\Delta S$ induced by the inclusion $R \hookrightarrow S$. One gets a commutative diagram with exact rows and columns:

$$\begin{array}{ccccccccc}
 & & & & 0 & & 0 & & \\
 & & & & \downarrow & & \downarrow & & \\
 & & 0 & & T & & ker\ \alpha & & \\
 & & \downarrow & & \downarrow & & \downarrow & & \\
0 & \longrightarrow & Rdx & \longrightarrow & RdR & \longrightarrow & R\delta R & \longrightarrow & 0 \\
 & & \downarrow & & {\scriptstyle\varphi}\downarrow & & {\scriptstyle\alpha}\downarrow & & \\
0 & \longrightarrow & SDx & \longrightarrow & SDS & \longrightarrow & S\Delta S & \longrightarrow & 0 \\
 & & \downarrow & & \downarrow & & \downarrow & & \\
 & & S/R & & SDS/RDR & & Coker\ \alpha & & \\
 & & \downarrow & & \downarrow & & \downarrow & & \\
 & & 0 & & 0 & & 0 & &
\end{array}$$

By the snake lemma one obtains an exact sequence

$$0 \to T \to Ker\ \alpha \to S \,/\, R \to SDS \,/\, RDR \to Coker\ \alpha \to 0$$

and from this for the lengths as s-modules:

$$l(T) = l(Ker\ \alpha) - l(Coker\ \alpha) + l(SDS \,/\, RDR) - l(S \,/\, R).$$

But $l(Ker\ \alpha) - l(Coker\ \alpha) = l(R\delta R) - l(S\Delta S)$, so that we have

Lemma 1.

$$l(T) = l(R\delta R) - l(S\Delta S) + l(SDS\,/\,RDR) - l(S\,/\,R).$$

17.2.1 The Case $\mathbf{d(R) \leq 1}$.

One can represent $R\delta R = F\,/\,U$ where F is a free R-module and U is generated by at most $rank(F) + 1$ elements. Therefore by ([5], Satz 2) one has $l(R\delta R) = l(D(R\delta R)^{-1}\,/\,R)$, where $D(R\delta R)$ denotes the 0th Fitting ideal of $R\delta R$, which is by definition also the 0th Kähler Different $\mathcal{D}_K(R\,/\,s)$ of R over s. By the general theory of the differents one has the following inclusions for the Dedekind, Kähler, and Noether differents $\mathcal{D}_D$, $\mathcal{D}_K$, $\mathcal{D}_N$ and the complementary modules $R^* = \mathcal{C}(R\,/\,s)$ and $S^* = \mathcal{C}(S\,/\,s)$:

$$R \subseteq S \subseteq S^* \subseteq R^* \subseteq \mathcal{D}_D(R\,/\,s)^{-1} = \mathcal{D}_N(R\,/\,s)^{-1} \subseteq \mathcal{D}_K(R\,/\,s)^{-1}.$$

Therefore we have

$$\begin{aligned} l(R\delta R) = l(\mathcal{D}_K(R\,/\,s)^{-1}\,/\,R) &\geq l(R^*\,/\,R) \\ &= l(S^*\,/\,S) + l(R^*\,/\,S^*) + l(S\,/\,R) \\ &= l(S^*\,/\,S) + 2 \cdot l(S\,/\,R) \text{ by duality.} \end{aligned}$$

Now S is a principal ideal ring and $S\Delta S$ is of projective dimensions 1, so $l(S\Delta S) = l(S\,/\,\mathcal{D}_K(S\,/\,s)) = l(S\,/\,\mathcal{D}_D(S\,/\,s)) = l(S^*\,/\,S)$, and therefore $l(R\delta R) - l(S\Delta S) \geq 2 \cdot l(S\,/\,R)$.

If R is a complete intersection one even has $R^* = \mathcal{D}_D(R\,/\,s)^{-1}$ and $\mathcal{D}_K(R\,/\,s) = \mathcal{D}_D(R\,/\,s)$ by ([22], §10 and Appendix G, G1 and G2). Therefore $l(R\delta R) - l(S\Delta S) = 2 \cdot l(S\,/\,R)$ and we obtain

Theorem 1. ([5], Stax 7): *If* $d \leq 1$ *then*

$$l(T) \geq l(SDS\,/\,RDR) + l(S\,/\,R)$$

with equality if R *is a complete intersection. Consequently: If* $T = 0$ *then* R *is regular.*

There is anothe rexpression for $l(T)$ in this case due to Kunz [21]: Look at the natural homomorphism of RdR into the R-module ω_R of regular differentials:

$$c_R : RdR \to R^* \cdot Dx = \omega_R$$

induced by the inclusion $R \hookrightarrow Qout(R)$. Obviously $ker\ c_R = T$, but also $coker\ c_R$ is interesting in this context: One has the includions $R^* \cdot Dx \supseteq S^* \cdot Dx \supseteq RDR \supseteq RDx$ and so

$$\begin{aligned} l(coker\ c_R) &= l(R^* \cdot Dx \,/\, RDR) = l(R^* \,/\, R) - l(RDR \,/\, RDx) \\ &= l(R^* \,/\, S^*) l(S^* \,/\, S) + l(S \,/\, R) + l(SDS \,/\, RDR) \\ &\quad\ -l(SDS \,/\, SDx) - l(SDx \,/\, RDx) \\ &= l(S \,/\, R) + l(S\Delta S) + l(S \,/\, R) + l(SDS \,/\, RDR) \\ &\quad\ -l(S\Delta S) - l(S \,/\, R) \\ &= l(SDS \,/\, RDR) + l(S \,/\, R). \end{aligned}$$

If R is a complete intersection then it follows with Theorem 1 that $l(T) = l(coker\ c_R)$. If R is only an almost complete intersection, Junz shows in [21] that $l(T) - l(coker\ c_R) = l(\tau(I \,/\, I^2))$, where $\tau(\ )$ denotes the torsion submodule. for a complete intersection $\tau(I \,/\, I^2) = 0$ by ([21], Theorem 1), so that we have

Theorem 1′. *If* $d \leq 1$ *then*

$$l(T) = l(coker|c_R) + l(\tau(I \,/\, I^2)).$$

17.2.2 The Case $\mathbf{d(R) \leq 3}$.

In general it is not possible to give an inequality for the length of an R-module using only the 0th Fitting ideal. So the proof from 1.1. cannot be applied. (See the examples in [5].) But Ulrich developed a more complex formula, using a chain of certain determinantal ideals, by which a lwoer bound for the length of an arbitrary finitely generated R-module can be computed ([33], Satz 2, [34], Satz 1).

We start again with Lemma 1, but this time we choose the parameters x so that $SDR = SDx$. This is possible by ([33], Hilfssatz 3). By Lemma 1 we have

$$\begin{aligned} l(T) &= l(R\delta R) - l(S\Delta S) + l(SDS \,/\, SDR) \\ &\quad\ +l(SDR \,/\, RDR) - l(S \,/\, R) \\ &= l(R\delta R) - l(S\Delta S) + \underbrace{l(SDS \,/\, SDx)}_{=S\Delta S} \\ &\quad\ +l(S \cdot RDR \,/\, RDR) - l(S \,/\, R) \text{ by choice of } x. \end{aligned}$$

Hence

Proposition 1. ([33], Satz 3)

$$\begin{aligned} l(T) &= l(R\delta R) - l(S\,/\,R) + l(S \cdot RDR\,/\,RDR) \\ &\geq l(R\delta R) - l(S\,/\,R). \end{aligned}$$

Definition 1. ([33], Def. 3) R has minimal torsion if and only if $l(T) = l(R\delta R) - l(S\,/\,R)$.

Proposition 2. ([33], Bemerkung 3) *If R has minimal torsion then there exists a non zero divisor $y \in \mathbf{m}$ with $y \cdot \mathbf{m} = \mathbf{m}^2$. If R is singular then $T \neq 0$.*

Proof. If R has minimal torsion then by proposition 1 $l(S \cdot RDR\,/\,RDR) = 0$, hence $RDR = S \cdot RDR = S \cdot Dx \cong S$. $\Rightarrow l_R(\mathbf{m}\,/\,\mathbf{m}^2) = n \geq \mu_R(RDR) = \mu_R(S) = l_R(S\,/y\mathbf{m} \cdot S)$. Now there is a non zero divisor $y \in \mathbf{m}$ with $S \cdot \mathbf{m} = S \cdot y$ since S is a principal ideal ring ([33], Hilfssatz 2). Then $l_R(\mathbf{m}\,/\,y \cdot \mathbf{m}) = l_R(S\,/\,S \cdot y) = l_R(S\,/\,S \cdot \mathbf{m}) l_R(\mathbf{m}\,/\,\mathbf{m}^2)$ hence $\mathbf{m}^2 \supseteq y \cdot \mathbf{m} \supseteq \mathbf{m}^2$. By Theorem 8 we have then $T \neq 0$, if R is singular.

Now by the above mentioned length formula of Ulrich, using also the relations between the various differents and the fact that R^* as in 1.1 is a cononical R-ideal, Urich shows in ([33], Staz 6):

Proposition 3. *If $d \leq 3$ then $l(R\delta R) \geq l(S\,/\,R) + l(S^*\,/\,S) \geq l(S\,/\,R)$.*

Now we can prove the torsion conjecture in this case:

Theorem 2. ([33], Satz 6, [34], Satz 3): *If $d(R) \geq 3$ and $T = 0$ then R is regular.*

Proof. By propositions 1 and 3 we have $0 = T \geq l(R\delta R) - l(S\,/\,R) \geq 0$, and hence R has minimal torsion. Then by proposition 2 $T \neq 0$ if R is not regular.

Remark 1. *From position 1 and 3 we also get $0 = l(T) \geq l(S^*\,/\,S) \geq 0$, and therefore $S^* = S$. So S over s is unramified. If k is algebraically closed and R a domain, it follows immediately that $S = s$ and therefore also $R = S$ without using Theorem 7.*

17.3 Exact Differentials, Maximal Torsion and Quasi Homogeneous Singularities

17.3.1 Exact Differentials

A secon class of curve singularities for which the conjecture is true consists of all R such that every differential of RDR is exact:

Theorem 3. ([28], Theorem 3): *Assume* $RDR = DR$. *If* $T = 0$ *then* R *is regular.*

For the proof we need the following lemma ([10], Satz 1), which can be proved by a direct computation:

Lemma 2. *Let* $A := k[\![Z_1, \ldots, Z_q]\!]$ *be a formal power series ring over* k, $t \in \mathbf{N}$, $A_t := A/(Z_1, \ldots, Z_q)^t$. *Then* $dim_k \Omega) A_t/k = (t-1) \cdot \left({q+t-l \atop t} \right)$.

Proof of Theorem 3. Let $x \in \mathbf{m} \setminus \mathbf{m}^2$ be a non zero divisor of R, $\bar{R} := R/x$ and $\bar{d} : \bar{R} \to \bar{R}\bar{d}\bar{R}$ the universally finite derivation of $\bar{R}$ over k. Then $edim\ \bar{R} = edim\ R - 1 = n - 1$, so that we can represent $\bar{R} = P/\lfloor$, $P := k[\![Y_1, \ldots, Y_{n-1}]\!]$ a formal power series ring with maximal ideal $\mathcal{M} = (Y_1, \ldots, Y_{n-1})$ and $\lfloor$ an ideal with $\lfloor \subseteq \mathcal{M}^2$. Denote by $\partial : P/\mathcal{M}^2 \to \Omega_{\frac{P/\mathcal{M}^2}{k}}$ the universally finite derivation of $P/\mathcal{M}^2$ over k. We obtain the following commutative diagram with exact rows of canonical maps:

$$\begin{array}{ccccc} R & \longrightarrow & P/\mathcal{M}^2 & \longrightarrow & 0 \\ \bar{d}\Big\downarrow & & \partial\Big\downarrow & & \\ \bar{R}\bar{d}\bar{R} & \longrightarrow & \Omega_{\frac{P/\mathcal{M}^2}{k}} & \longrightarrow & 0 \end{array}$$

This yields a surjection $\bar{R}\bar{d}\bar{R}/Im\ \bar{d} \to \Omega_{\frac{P/\mathcal{M}^2}{k}}/Im\ \partial \to 0$. Now by assumption we have $0 = T = \ker(RdR \to RDR)$ and $RDR = DR$. Hence $RdR = dR$ and consequently $\bar{R}\bar{d}\bar{R} = RdR/Rdx = \bar{d}\bar{R} = Im\ \bar{d}$. $\Rightarrow Im\partial = \Omega_{\frac{P/\mathcal{M}^2}{k}}$. $\Rightarrow n = dim_k P/\mathcal{M}^2 = dim_k \Omega_{\frac{P/\mathcal{M}^2}{k}} + dim_k Ker\partial \geq \begin{pmatrix} n \\ 2 \end{pmatrix} + 1$ by Lemma 2. $\Rightarrow n \leq 2$. $n = 2$ is not possible, because then R is a plane curve singularity and therefore $T \neq 0$ by 1.1. Then $n = 1$ and consequently R is regular.

When is the condition $RDR = DR$ satisfied?

17.3.2 Maximal Torsion.

Consider the universally finite derivation $S \to SDS$ of S over k. Since S is a direct product of formal power series rings $S_i \cong k_i[\![t_i]\!]$, where k_i is an algebraic extension of k, D is the direct product of the formal derivations $D_i : f(t_i) \mapsto f'(t_i) \cdot D_i t_i$. It follows by formal integration that the D_i are surjective and hence also D is surjective. This induces a surjective k-linear map $\tilde{D} : S / R \to SDS / RDR$, and so $l(S / R) = dim_k(S / R) \geq dim_k(SDS / RDR) = l(SDS / RDR)$. Together with lemma 1 we get

$$l(T) = l(R\delta R) - l(S\Delta S) - \underbrace{[l(S / R) - l(SDS / RDR)]}_{\geq 0} \leq l(R\delta R) - l(S\Delta S).$$

Definition 2. ([28]): R has *maximal torsion* if and only if

$$l(T) = l(R\delta R) - l(S\Delta S)$$

or, equivalently, if and only if

$$l(S / R) = l(SDS / RDR).$$

If R has maximal torsion the conjecture is true:

Theorem 4. ([28], Theorem 1): *If R has maximal torsion then $RDR = DR$. Consequently: If $T = 0$ then R is regular. If k is algebraically closed and R is a domain then also the converse is true:*
If $RDR = DR$ then R has maximal torsion.

Proof. As shown above the k-linear map $D : S \to SDS$ is surjective and therefore

$$\begin{aligned} l(S / R) &= dim_k(S / R) \geq dim_k(S / (R + \ker D)) \\ &= dim_k(SDS / DR) \\ &\geq dim_k(SDS / RDR) = l(SDS / RDR). \end{aligned}$$

If R has maximal torsion then $l(S / R) = l(SDS / RDR)$ and therefore $DR = RDR$.

Assume now that k is algebraically closed and R is a domain. Then $\ker D = k \subseteq R$ and therefore $l(S / R) = l(SDS / DR) = l(SDS / RDR)$ by hpothesis. So R has maximal torsion.

Remark 2. Let k be algebraically cloased and R a domain.

a) Let $x \in \mathbf{m} \setminus \mathbf{m}^2$ be a superficial element of R. Then one can find a uniformizing parameter t for S such that $x = t^{m(R)}$, where $m(R)$ denotes the multiplicity of R. With $s := k[\![x]\!]$ and $S = k[\![t]\!]$ we get $l(S\Delta S) = l(SDt \,/\, SDt^{m(R)}) = m(R) - 1$. Therefore R has maximal torsion if and only if $l(T) = l(R\delta R) - m(R) + 1$. This shows that definition 2 is equivalent to Urich's Definition 2 in [33].

b) Zariski [36] considers the case of an irreducible plane algebroid curve over k and shows that $l(T) \leq 2 \cdot l(S \,/\, R)$. Equality holds if and only if $RDR = DR$, which by Theorem 4 means that R has maximal torsion. He proves that this is the case if and only if the curve can be represented by a quasi homogeneous equatio of the form $Y^p - X^q = 0$ with $(p, q) = 1$.

For upper bounds of the torsion of a plane curve in terms of the characteristic pairs see Azevedo ([1], Propositions 3 and 4).

17.3.3 Quasi Homogeneous Singularities

Definition 3. ([29], 9.8): R is called *quasi homogeneous* if there exists a surjective R-module homomorphism $\Omega_{R/k} \to \mathbf{m}$.

Let $\gamma = (\gamma_1, \ldots, \gamma_n) \in \mathbb{N}_0^n$. A polynomial $F = \sum \alpha_{i_1 \ldots i_n} \cdot X_1^{i_1} \ldots X_n^{i_n}$ with coefficients in k is called quasi homogeneous of type γ, if there is a $d \in \mathbb{N}$ such that for all $\alpha_{i_1 \ldots i_n} \neq 0$ one has $\sum \gamma_j \cdot i_j = d$. In this case d is called the degree of F.

From ([23], Satz 2.1. and Satz 3.1.) one obtains:

Remark 3. If I is generated by polynomials, then R is quasi homogeneous if and only if I is generated by quasi homogeneous polynomials of a fixed type γ. For an irreducible R this is equivalent to R being isomorphic to the analytic semigroup ring $k[\![H]\!]$ for the value semigroup H of R.

Theorem 5. ([28], Theorem 2): *If R is quasi homogeneous and I generated by polynomials then R has maxmal torsion. Consequently: If $T = 0$ then R is regular.*

Proof: (For a different proof of the second assetion without the assumption of I being generated by polynomials see Scheja ([29], Satz 9.8).) By hypothesis there is a R-linear map : $Rd\ R \to \mathbf{m}$, which by ([23], Proof of Satz 2.1), can be chosen such that $(dx_i) = \gamma_i x_i$ with $\gamma_i \in \mathcal{N}$ for $i = 1, \ldots, n$. Since RdR and $\mathbf{m}$ are both of rank 1 and $\mathbf{m}$ is torsionfree we have ker $= T$. We may assume that $x = x_1$ is a superficial element of degree 1 of R. Since $Rdx \cap T = 0$ we have an exact sequence $0 \to T \to RdR \,/\, Rdx \to \mathbf{m} \,/\, Rx \to 0$. Therefore $l(T) = l(RdR \,/\, Rdx) - l(\mathbf{m} \,/\, Rx) = l(R\delta R) - m(R) + 1$. Now by Remark 2 a) R has maximal torsion.

17.3.4 The Value Semigroup

Let, as before, R be a domain and k algebraically closed. Then $S = k[\![t]\!]$ is a discrete valuation ring. Let ν denote the normed valuation with $\nu(t) = 1$. $H = \{\nu(y) | y \in R\backslash 0\}$ is called the value semigroup of R. Since $SDS = SDt$ every $\omega \in RDR$ is of the form $\omega = z \cdot Dt$ with $z \in S$, and so we can define $\nu(\omega) := \nu(z) + 1$. This definition is independent of the choice of t. For all $y \in \mathbf{m}$ with we then have $\nu(Dy) = \nu(y)$ and therefore $\nu(RDR) \supseteq \nu(\mathbf{m})$, but there may be elements $\omega \in RDR$ with $\nu(\omega) \notin \nu(\mathbf{m})$. Yoshino calls them exceptional differentials ([35], Def. 2.3). If $RDR = DR$ then obviously $\nu(RDR) = \nu(\mathbf{m})$. The converse follows from ([36], Proof of Corollary 3). So we obtain

Theorem 6. ([33], Satz 4): *Let k be algebraically closed, and R a domain. Then $\nu(RDR) = \nu(\mathbf{m})$ if and only if $RDR = DR$. Consequently: If $\nu(RDR) = \nu(\mathbf{m})$ and $T = 0$ then R is regular.*

This was also stated in Theorem 4.1 of [35], but there the proof of Proposition 3.3, which is used in the proof, is wrong.

17.4 Conditions on the Embedding Dimension, The Index of Stability, and the Multiplicity

17.4.1 Low Embedding Dimension

In the case of $n = edim\ R = 2$ the ring R represents a plane curve singularity and therefore R is a complete intersection. Then the conjecture is true by Theorem 1. But also in the cases $n = 3$ and $n = 4$ one has the following results by Herzog ([12], Satz 3.2 and Satz 3.3), that are obtained using properties of the Koszul complex for which we refer the reader to [12].

Theorem 7. *a) If $n \leq 3$ and $T = 0$ then R is regular. b) If $n = 4$, R is Goenstein and $T = 0$ then R is regular.*

17.4.2 Low Index of Stability

The main tool in this section is a reduction to the case of $dim\ R = 0$, which was first used by Scheja in [29]. The proofs of the following results are very technical, and so we will mostly contend ourselves with references to the iterature. First we need a technical lemma which generalizes the well known formula for $\mu(RdR)$:

Lemma 3. ([10], Lemma 1a): *Let $B = k[\![X_1, \ldots, X_n]\!] / \dashv$ such that $\dashv \subseteq (X_1, \ldots, X_n)^t$ for a $t \in \mathbb{N}$, $\mathbf{n}$ the maximal ideal of B, and ∂ the universally finite derivation of B over k. Then for $r = t$ and $r = t + 1$ we have*

$$l_B(B\partial\mathbf{n}^r \,/\, \mathbf{n}^r \cdot B\partial B) = \mu_B(\mathbf{n}^r).$$

This lemma together with Lemma 2 is the main tool for proving

Proposition 4. ([10], Satz 2): Let $\bar{R}$ with maximal ideal $\bar{\mathbf{m}}$ be an analytic k-algebra with $dim\ \bar{R} = 0$, and $\bar{d}$ the universally finite derivation of $\bar{R}$ over k. Assume that $\bar{R} = k[\![X_1,\ldots,X_n]\!] \,/\, \dashv$ with $\dashv \subseteq (X_1,\ldots,X_n)^r$ for some $r \in I\!N$. Then

$$l_{\bar{R}}(\bar{R}\bar{d}\bar{R}) \geq (r-1)\cdot \binom{n+r-1}{r} + \mu_{\bar{R}}(\bar{\mathbf{m}}^r) + \mu_{\bar{R}}(\bar{\mathbf{m}}^{r+1}).$$

In particular one has always (with $n = edmi\ \bar{R}$)

$$l_{\bar{R}}(\bar{R}\bar{d}\bar{R}) \geq \frac{1}{2}\cdot n\cdot(n+1) + \mu_{\bar{R}}(\bar{\mathbf{m}}^2) + \mu_{\bar{R}}(\bar{\mathbf{m}}^3).$$

The following easy lemma enables us to apply the preceding results to the torsion problem for one-dimensional analytic k-algebras:

Lemma 4. ([33], Hilfssatz 8): *Let R be a one-dimensional Cohen-Macaulay ring, $y \in \mathbf{m}$ a non zero divisor for R, M a finitely generated R-module with rank r, and $T_y := \{z | z \in M, y\cdot z = 0\}$. Then $l(M \,/\, y\cdot M) = r\cdot l(R\,/y.\,R) + l(T_y)$.*

One can now show the following theorem, which is a generalization of ([34], Satz 7):

Theorem 8. ([10], Satz 4): *Let $n \geq 3$, and let $T_y := \{z | z \in T, y\cdot z = 0\}$. If there exists a non zero divisor $y \in \mathbf{m}$ such that $\mathbf{m}^4 \subseteq R\cdot y$, then:*

$$\text{If } edim(R\,/\,R\cdot y) = n-1, \quad \text{then} \quad l(T_y) = \frac{1}{2}\cdot(n-2)\cdot(n-1) \qquad (17.1)$$

$$\text{If } edim(R\,/\,R\cdot y) = n, \quad \text{then} \quad l(T_y) = \frac{1}{2}\cdot(n-2)\cdot(n+1) \qquad (17.2)$$

Consequently: If there exists a non zero divisor $y \in \mathbf{m}$ with $\mathbf{m}^4 \subseteq R\cdot y$ then: If $T = 0$ then R is regular.

Remark 4. ([10])

1. If one weakens the hypothesis of Theorem 8 to $\mathbf{m}^5 \subseteq R\cdot y$, one can still show $l(T_y) \geq \frac{1}{2}\cdot(n-2)\cdot(n-1) - r(R)$, where $r(R)$ denotes the type of R. So, if R is Gorenstein and $n \geq 4$ one has again $T \neq 0$.

2. The condition $\mathbf{m}^{t+1} \subseteq R\cdot y$ for some t is satisfied for instance if $\mathbf{m}^t$ is stable in the sense of ([14], Definition 1.2). (See ([14], Remark 1.5)).

If $m(R) \leq edim\ R + 1$ then even $\mathbf{m}^3 \subseteq R \cdot y$ ([33], Bemerkung 9b): Take for y a superficial element of degree one. Then $l_R(R / R \cdot y) = m(R)$ and so $l((\mathbf{m}^2 + R \cdot y) / R \cdot y) = l(R / R \cdot y) - l(R / (\mathbf{m}^2 + r \cdot y)) = m(R) - edim\ R \leq 1$. It follows $\mathbf{m} \cdot (\mathbf{m}^2 + R \cdot y) \subseteq R \cdot y$ and therefore $\mathbf{m}^3 \subseteq R \cdot y$.

This is a first example of a condition between the multiplicity and the embedding dimension of R, which will be generalized in Theorem 9.

Obviously the condition $\mathbf{m}^{t+1} \subseteq R \cdot y$ plays an important role. If y is a superficial element of degree 1 then for all large $t \in I\!N$ we have $y \cdot \mathbf{m}^t = \mathbf{m}^{t+1}$.

Definition 4. The minimal $t \in I\!N$ such that there is a superficial element (of degree 1) y with $y \cdot \mathbf{m}^t = \mathbf{m}^{t+1}$ is called the index of stability of R and is denoted by $t(R)$.

17.4.3 Relatively Low Multiplicity

Lemma 5. ([10], Lemma 3): *Let $t(R) \geq 2$. Then for every superficial element x of degree one $l((Rdx + x \cdot RdR) / x \cdot RdR) \geq 2$.*

Using Lemma 4 and 5 together with Proposition 4 one can now derive:

Theorem 9. ([10], Satz 5′ and 5): *Let x be a superficial element of degree one, $\bar{R} := R / R \cdot x$, and $\bar{\mathbf{m}}$ the maximal ideal of $\bar{R}$.*

If $m(R) \leq \frac{1}{2} \cdot n \cdot (n-1) + \mu_{\bar{R}}(\bar{\mathbf{m}}^2) + \mu_{\bar{R}}(\bar{\mathbf{m}}^3) + 1$ and $T = 0$ then R is regular.

More generally: Let $X \in k[\![X_1, \ldots, X_n]\!]$ be a representative for x, and assume that $I \subseteq (X_1, \ldots, X_n)^r + (X)$ for an $r \in I\!N, r \geq 2$. Then If $m(R) \leq (r-1) \cdot \binom{n+r-2}{r} + \mu_{\bar{R}}(\bar{\mathbf{m}}^r) + \mu_{\bar{R}}(\bar{\mathbf{m}}^{r+1}) + 1$ and $T = 0$ then R is regular.

Proof. Assume R not regular ($n \geq 2$). By Lemma 4 we have $l(T_x) = l(RdR / x \cdot RdR) - l(\bar{R}) = l(\bar{R}d\bar{R}) + l(Rdx + x \cdot RDR . x \cdot RdR) - l(\bar{R})$. Since x is superficial we have $l(R) = m(R)$. By hypothesis we can write $\bar{R} = R / R \cdot x = k[\![Z_1, \ldots, Z_{n-1}]\!] / \lfloor$ with $\lfloor \subseteq (Z_1, \ldots, Z_{n-1})^r$ and so by Proposition 4 $l(\bar{R}d\bar{R}) \geq (r-1) \cdot \binom{n+r-2}{r} + l(\bar{\mathbf{m}}^r) + l(\bar{\mathbf{m}}^{r+1})$. Together with Lemma 5 (we may assume $t(R) \geq 2$ because of Theorem 7) we get

$$l(T) \geq l(T_x) \geq (r-1) \cdot \binom{n+r-2}{r} + l(\bar{\mathbf{m}}^r) + l(\bar{\mathbf{m}}^{r+1}) + 2 - m(R),$$

and if the hypothesis of the theorem is satisfied, the righthand side is positive.

Remark 5.

1. In ([18], Theorem 1) Isogawa gives a simple proof in the special case of the above theorem that $m(R) \leq \frac{1}{2} \cdot n \cdot (n-1) - 1$.

2. In ([26], Satz 3.13) Pohl shows that Theorem 9 can be applied for instance in the following situation: Let (with the notation in the Proof of Theorem 9) $a := min\{i|(Z_1, \ldots, Z_{n-1})^i \subseteq \lfloor\}$ and $c := max\{j|(Z_1, \ldots, Z_{n-1})^j \subseteq \lfloor\}$. If $a - c \leq 2$ or R a Gorenstein singularity and $a - c \leq 3$ then the hypothesis of Theorem 9 is satisfied with $r = c$. In particular the first condition is satisfied if R has *maximal Hilbert function* (i.e., $\mu_R(\mathbf{m}^i) = min\{\binom{n+i+1}{i}, m(R)\}$ for all $i \in I\!N$).

Combining Theorem 9 with ([12], Satz 3.2 and Satz 3.3) one gets:

Theorem 10. ([10], Satz 6): *Let R be a domain.*

a) If $m(R) \leq 9$ and $T = 0$ then R is regular.

b) If $m(R) \leq 13$, R Gorenstein, and $T = 0$ then R is regular.

Proof. Let x be a superficial element of degree one. Assume $\mathbf{m}^3 \not\subseteq R \cdot x$ in view of Theorem 8. Using the same notation as in Proposition 4 we then have $\mu(\bar{\mathbf{m}}^2) \geq 1$ and $\mu(\bar{\mathbf{m}}^3) \geq 1$. Now either $m(R) \leq \frac{1}{2} \cdot n \cdot (n-1)\frac{1}{2} \cdot n \cdot (n-1) + \mu(\bar{m}^2) + 3 \leq +\mu(\bar{\mathbf{m}}^3) + 1$, and then R is regular by Theorem 8, or $m(r) > \frac{1}{2} \cdot n \cdot (n-1) + 3$.

On the other hand we have by hypothesis

a) $m(R) \leq 9$ and therefore $n \leq 3$. ([12], Satz 3.3) gives our result.

b) $m(R) \leq 13$ and therefore $n \leq 4$. Now the result follows with ([12], Satz 3.2).

17.5 Conditions on the Linkage Class

In this section let k be algebraically closed. Recall the definition of linkage: Two perfect ideals I and J of the same grade in a Gorenstein local ring are said to be *linked* (or *1-linked*), if there exists an ideal G generated by a regular sequence, $G \subset I$, $G \subset J$, such that $I = G : J$ and $J = G : I$.

The analytic k-algebra R is said to be *linked* (or *1-linked*) to the analytic k-algebra R', if $R = k[\![X_1, \ldots, X_n]\!] / I$ and $R' = k[\![X_1, \ldots, X_n]\!] / J$ with I linked to J.

R is said to be *in the same linkage class* as R' (or *t-lined to* R'), if there exists a sequaence $R = R_0, \ldots, R_t = R'$ such that R_i is linked to R_{i+1} for $i = 0, \ldots, t-1$. One defines an invariant

$$\sigma(R) := l(||^{-1} / R) - l(R / ||)$$

for any canonical ideal $|| \subseteq R$ of R. One has $\sigma(R) \leq l(S / R) =: \delta(R)$ and $\sigma(R) < \delta(R)$ if R is singular ([19], Satz 1, [15], pp. 337–338). If R is Gorenstein (e.g., complete intersection) then $\sigma(R) = 0$.

Theorem 11. ([15], Theorem): *Let R be 1-linked to R'. Then with the otation of 1.1:*

$$l(coker\ c_R) + l(\tau(I / I^2)) - l(T) = \sigma(R')$$

For the proof see [15]. Here we will look at some of the consequences:

It follows immediately that both sides of the equation are even-linkage invariants (that is invariants for t-linkage with t even).

By Theorem 11 one has

$$\begin{array}{rclcl} l(T) = l(coker\ c_R) & + & l(\tau(I / I^2) & - & \sigma(R') \\ l(SDS / RDR) & + & l(\tau(I / I^2) & + & l(S / R) - \sigma(R'). \end{array}$$

Now, if $\sigma(R' = \sigma(R)$ and R is not regular then $l(S / R) - \sigma(R') > 0$ and hence $T \neq 0$. If R and therefore also $R' := R_1$ is in the linkage class of a complete inetersection, it is easily seen that R is also *evenly* linked to a complete intersection ([15], p. 336). Since σ is an even-linkage invariant one has $\sigma(R') = 0$ and therefore by the formula of Theorem 11:

Theorem 12. ([15], Corollary3, [16], Theorem 4.5): *If R is in the linkage class of a complete intersection then*

$$\begin{array}{rcl} l(T) & = & l(coker\ c_R) + l(\tau(I / I^2)) \\ & = & l(S / R) + l(SDS / RDR) + l(\tau(I / I^2)). \end{array}$$

Consequently: If $T = 0$ then R is regular.

There is a second class of rings in which this method gives a result:

Definition 5. ([15]): Let R_0 be a reduced complete analytic k-algebra. R is called a *small extension* of R_0 if there exists a non zero divisor x of the integral closure of R_0 such that $R = R_0[x]$ and $x^2 \in R_0$.

Now let R be a small extension of a complete intersection R_0. It is easily seen that R is linked to itself, and so any R' in the linkage class of R is evenly linked to R. So $\sigma(R') = \sigma(R)$ and the formula of Theorem 11 yields:

Theorem 13. ([15], Corollary 4): *If R is in the linkage class of a small extension of a complete intersection then*

$$\begin{aligned} l(T) &= l(coker\ c_R) + l(\tau(I\,/\,I^2)) - \sigma(R) \\ &> l(SDS\,/\,RDR) + l(\tau(I\,/\,I^2)) \text{ if } R \text{ is not regular.} \end{aligned}$$

Consequently: If $T = 0$ then R is regular.

Quite generally one can show that of two linked singular analytic k-algebras at least one of them must have non trivial torsion of the differential module:

Theorem 14. ([15], Corollary 5 and Theorem on p. 335): *Let $\tau(\Omega_{R/k}) = 0$, R' singular and 1-lined to R. Then $\tau(\Omega_{R'/k}) \neq 0$.*

In order to show this we first need:

Lemma 6. ([15], Corollary 2a): *Let R be 1-lined to a singular R' and let $\delta(R) \geq \sigma(R')a$ then $\tau(\Omega_{R/k}) \neq 0$.*

Proof. By Theorem 11 one has $l(T) = l(SDS\,/\,RDR) + l(\tau(I\,/\,I^2)) + \delta(R) - \sigma(R') \geq l(SDS\,/\,RDR) + l(\tau(I\,/\,I^2))$. It follows that $T \neq 0$ if $SDS \neq RDR$. Assume now that $SDS = RDR$, then also $S \cdot RDR = RDR$. We can choose s as in 1.2. Then by Proposition 1 $l(T) = l(R\delta R) - l(S\,/\,R)$ and therefore R has minimal torsion. By Proposition 2 we then have $T \neq 0$.

Proof of Theorem 14. Let R' be singular and 1-lined to R. Since $\tau(\Omega_{R/k}) = T = 0$ we must have $\sigma(R') > \sigma(R)$ because of Lemma 6. So $\delta(R') \geq \sigma(R') > \delta(R) \geq \sigma(R)$. Applying again Lemma 6 with R and R' interchanged yields $\tau(\Omega_{R'/k}) \neq 0$.

17.6 Smoothability Conditions

Definition 6. R is called *smoothable*, if $R \cong P\,/\,J$, where P is a normal analytic k-algebra and J is generated by a regular sequence of P.

In the terminology of deformation theory that means that R can be defomred into a complete intersection.

Bassein [2] shows in the complex analytic case, generalizing a result of Pinkham [25] for Gorenstein singularities:

Theorem 15. ([2], Theorem 2.4): *If R is smoothable then $l(T) = \delta(R) + dim(SDS\,/\,RDR)$. Consequently: If $T = 0$ then R is regular.*

Buchweitz and Greuel ([3], 6.1) weakened the condition "smoothable" to the codition, that the degree of singularity in the general fiber is at most $\frac{1}{2} \cdot \delta(R)$. In his Dissertation Koch proves a version for an arbitrary ground field k. He generalizes the above results so that in the complex analytic case he only requires that R can be deformed into an almost complete intersection:

Theorem 16. ([20], Korolar 3 to Satz 10): *Let P be excellent and an almost complete intersection in codimension 1, and $R = P \,/\, J$ with J generated by a regular sequence of S. Then $l(T) \geq \delta(R) + l(SDS \,/\, RDR)$. Consequently: If $T = 0$ then R is regular.*

For more results concerning smoothability see [16], where formulas involving the higher (analytic) derived functors $T_i(R \,/\, s, R)$ and $T^i(R \,/\, s, R)$ are obtained. In particular Theorem 12 is proved there by showing that R is smoothable.

17.7 Quadratic Transforms

For simplicity let us assume in this section that R is a domain and k algebraically closed. Instead of looking at T as the Kernel of the natural map $RdR \to SDS$, one can also take an intermediate ring A with $R \subseteq A \subseteq S$ instead of S as long as one can be sure that there is a nontrivial kernel of the induced map of the differential modules. A good canddate for this is the first quadratic transform $R_1 := R\left[\frac{x_2}{x_1}, \ldots, \frac{x_n}{x_1}\right]$ where we assume without loss of generality that x_1 is a superficial element of degree one in R, because a short direct computation shows that whenever $T \neq 0$ then there is also ker $\varphi \neq 0$. The converse is trivial. Therefore:

Proposition 3. ([6], Theorem 1): Let R_1 be the first quadratic transform of R and $\varphi : \Omega_{R/k} \to \Omega_{R_1/k}$ the natural homomorphism induced by the inclusion $R \hookrightarrow R_1$. Then $T \neq 0$ if and only if ker $\varphi \neq 0$.

The hope is that the behavior of the differential modules when going from R to a suitable A might be easier accessible than that when going from R to S. One would think that the length of the torsion should strictly decrase when going from a singular R to its quadratic transform R_1. Indeed Bertin and Carbonne show that this is so in the case of a complete intersection ([3], Theorem 2). If one wants only to show $T \neq 0$ it is enough to show ker$\varphi \neq 0$ for some A. By the same reasoning as in Section 1 we obtain the analog of the formula in Lemma 1:

$$l(\ker \varphi) = l(AD_AA \,/\, RD_AR) + l(R\delta R) - l(A\partial A) - l(A \,/\, R),$$

(with ∂ the universal derivation of A over s and D_A the universally finite derivation of A over k).

Unfortunately, so far there is no method known to achieve this goal for R. But one can at least constuct a ring $\tilde{R}$ with $R \subseteq \tilde{R} \subseteq R_1$ which has the same multiplicity and the same quadratic transforms as R, for which by the above formula one obtains $l(\ker \tilde{\varphi}) \geq \frac{1}{2} \cdot m \cdot (m-1)$, with $\tilde{\varphi}$ denoting the natural homomorphism $\Omega_{\tilde{R}/k} \to \Omega_{R_1/k}$. More precisely:

Theorem 17. ([6], Corollary to Theorem 2): *Let $\tilde{R} = s + \mathbf{m} \cdot R_1$. Then $l(\tau(\Omega_{\tilde{R}/k})) \geq \frac{1}{2} \cdot m \cdot (m-1)$. Moreover R has the same multiplicity m and the same quadratic transforms as R. Consequently: If $\tau(\Omega_{\tilde{R}/k}) = 0$ then $\tilde{R}$ (and therefore also R) is regular.*

Proof. By definition $s = k[\![x]\!]$ and x is an element of minimal value in R. We have $\mathbf{m} \cdot R_1 = x \cdot R_1$, therefore x has minimal value in $\tilde{R}$ as well and hence $m(\tilde{R}) = m(x) = m(R)$. For the maximal ideal $\tilde{\mathbf{m}}$ of R one has $\tilde{\mathbf{m}}^2 = x \cdot \mathbf{m}$, therefore $\tilde{R}$ has maximal embedding dimension $edim(\tilde{R}) = m(\tilde{R}) = m$. Theorem 8 then yields $l(\tau(\Omega_{\tilde{R}/k})) \geq \frac{1}{2} m(m-1)$. A more detailed analysis in [6] shows that this inequality already holds for $ker\tilde{\varphi}$.

Remark 6. The ring R is the glueing of $\mathbf{m} \cdot R_1$ over $\mathbf{m}$ in the sense of Tamone [32].

17.8 Equisingularity

We conclude this paper with some remarks on the torsion and the quisingularity class of R. Let us always assume that R is irreducible and k algebraically closed.

For plane curves there are many equivalent definitions of equisingularity. For a report on the fundamental papes of Zariski [37] and the definitions see [8], Chapters III–V.

The three main equisingularity classes are:

(ES1): R and R' are equisingular if they have the same saturation.

(ES2): R and R' are equisingular if they have the same multiplicity sequence with respect to quadratic transforms.

(ES3): R and R' are equisingular if they have the same value semigroup.

While for plane curves all three concepts are equivalent (and also equivalent to other definitions via characteristic exponents or characteristic pairs), already for curves in 3-space over the complex numbers there is no relation between the three defintions (see [8], Examples 5.3.4 and Remark 5.4.4). But, nevertheless:

Theorem 18. ([26], Satz 2.6): *For every singular R thre is always an R' equisingular to R with $\tau(\Omega_{R'/k}) \neq 0$, no matter which of the three definitions one takes.*

Proof

(ES1): There is a generic plane projection R' of R equisingular to R, and $\tau(\Omega_{R'/k}) \neq 0$ because of $edimR' = 2$.

(ES2): Take for R' the ring $\tilde{R}$ defined in Section 6. By Theorem 17 it is (ES2)-equisingular to R and $\tau(\Omega_{R'/k}) \neq 0$. ($R$ has not only the same multiplicity sequence as R but even the same quadratic transforms.)

(ES3): Take for $R' = k[\![H]\!]$ the analyic semigroup ring for the value semigroup H of R. By defintion R' is (ES3)-equisingular to R and by Remark 3 and Theorem 5 we have $\tau(\Omega_{R'/k}) \neq 0$.

Unfortunately the length of the torsion is not an invariant of the equisingularity class of the curve. For instance Zariski shows in [36] that for a plane curves $l(T) = 2\delta(R)$ is and only if it can be represented by an equation $Y^p - X^q = 0$ with $(p,q) = 1$, but of course there are many other plane curves R' with just one characteristic pair (p,q) (and therefore in the same equisingularity class as R) for which $l(T) < 2\delta(R') = 2\delta(R)$.

17.9 References

[1] A. Azevedo, *The Jacobian Ideal of a Plane Algebroid Curve*, Ph.D. Thesis, Purdue University, 1967.

[2] R. Bassein, On Smoothable Curve Singularities: Local Methods, *Math. Ann.*, **230**, 1977, pp. 273–277.

[3] J. Bertin et. Ph. Carbonne, *Sur le sous-module de torsion du module des différentielles*, C.R. Acad. Sc. Paris, A **277**, 1973, pp. 797–800.

[4] R. Berger, *Über verschiedene Differentenbegriffe*, Sitzungsber. d. Heidelberger Akad. d. Wiss., Math-Naturw. Kl., 1. Abh., Springer-Verlag, 1960.

[5] R. Berger, Differentialmoduln eindimensionaler lokaler Ringe, *Math. Z.*, **81**, 1963, pp. 326–354.

[6] R. Berger, On the torsion of the differntial module of a curve singularity, *Arch. Math.*, **50**, 1988, pp. 526-533.

[7] R. Buchweitz and G. Greul, The Milnor numer and deformations of complex curve singularities, *Inventiones Math.*, **52**, 1980, pp. 241-281.

[8] A. Campillo, *Algebroid Curves in Positive Characteristics*, Springer Lect. Notes in Math., **813**, Springer, Berlin-Heidelberg-New York, 1980.

[9] K. Güttes, *Einige Untersuchungen zum Torsionsproblem bei Kurvensingularitäten*, Dissertation, Saarbrücken, 1988.

[10] K. Güttes, Zum Torsionsproblem bei Kurvensingularitäten, *Arch. Math.*, **54**, 1990, pp. 499-510.

[11] J. Herzog, Eindimensionale fast-vollständige Durchschnitee sind nicht starr, *Manus. Math*, **30**, 1979, pp. 1-20.

[12] J. Herzog, Ein Cohen-Macaulay Kriterium mit Anwendungen auf den Konormalenmodul und den Differentialmodul, *Math. Z.*, **163**, 1978, pp. 149–162.

[13] J. Herzog and E. Kunz, eds., *Der Kanonische Modul eines Cohen-Macaulay-Rings*, Springer Lect. Notes in Math., **238**, Springer, Berlin-Heidelberg-New York, 1971.

[14] J. Herzog and R. Waldi, A note on the Hilbertfunction of a one-dimensional Cohen-Macaulay ring, *Manus. Math.*, **16**, 1975, pp. 251–260.

[15] J. Herzog and R. Waldi, Differentials of linked curve singularities, *Arch. Math.*, **42**, 1984, pp. 335–343.

[16] J. Herzog and R. Waldi, Cotangent functors of curve singularities, *Manuscripta Math.*, **55**, 1986, pp. 307–341.

[17] R. Hübl, A note on the torsion of differential forms, *Arch. Math.*, **54**, 1990 pp. 142-145.

[18] S. Isogawa, On Berger's conjecture about one dimensional local rings, *Arch. Math.*, (to appear).

[19] J. Jäger, Längenberechnung und kanonische Ideale in eindimensionalen Ringen, *Arch. Math.*, **24**, 1977, pp. 504-512.

[20] J. Koch, *Über die Torsion des Differentialmoduls von Kurvensingularitäten*, Dissertation Regensburg, Regensburger Mathematische Schriften, **5**, Fakultät f. Math Univ. Regensburg, 1983.

[21] E. Kunz, The conormal module of an almost complete intersection, *Proc. Amer. Math. Soc.*, **73**, 1979, pp. 15–21.

[22] E. Kunz, *Kähler Differentials*, Advanced Lectures in Mathematics, Vieweg, Braunschweig/Wiesbaden, 1986.

[23] E. Kunz and W. Ruppert, Quasihomogene Singularitäten algebraischer Kurven, *Manus. Math.*, **22**, 1977, pp. 47-61.

[24] E. Kunz and R. Waldi, Regular Differential Forms, *Contemporary Mathematics*, **79**, ASM, 1988.

[25] H. Pinkham, Deformations of algebraic varieties with G_m action, *Astérisque*, **20**, 1974.

[26] Th. Pohl, *Über die Torsion des Differentialmoduls von Kurvensingularitäten*, Dissertation Saarbrücken, 1989.

[27] Th. Pohl, Torsion des Differentialmoduls von Kurvensingularitäten mit maximaler Hilberfunktion, *Arch. Math.*, **52**, 1989, pp. 53–60.

[28] Th. Pohl, Differential Modules with maximal Torsion, *Arch. Math.*, (to appear).

[29] G. Scheja, *Differentialmoduln lokaler analytischer Algebren, Schriftenreihe Math. Inst. Univ., Fribourg*, Univ. Fribourg, Switzerland, 1970.

[30] U. Storch, Zur Längenberechnung von Moduln, *Arch. Math.*, **24**, 1973, pp. 39–43.

[31] G. Tamone, *Sugli incollamenti di ideali primari e la genesi di certi singolarità*, Analizi Funzionale e Applicazione B.U.M.I. (Supplemento) Algebra e Geometria Suppl. **2**, 1980, pp. 243–258.

[32] G. Tamone, Blowing-up and Glueings in one-dimensional Rings, Commutative Algebra *Proc. of the Trento Conference* (s. Greco and G. Valla, eds.), Lect. Notes in Pure and Appl. Math, **84**, Marcel Decker, Inc., New York and Basel, 1983, pp. 321–337.

[33] B. Ulrich, *Torsion des Differentialmoduls und Kotangentenmodul von Kurvensingularitäten*, Dissertation Saarbrücken, 1980.

[34] B. Ulrich, Torsion des differentialmoduls und Kotangentenmodul von Kurvensingularitäten, *Arch. Math.*, **36**, 1981, pp. 510–523.

[35] Y. Yoshino, *Torsion of the differential modules and the value semigrou of one dimensional local rings*, Math. Rep. Toyama Univ. **9**, 1986, pp. 83-96.

[36] O. Zariski, Characterization of plane algebroid curves whose module of differentials has maximum torsion, *Nat. Acad. Sci.*, **56**, 1966, pp. 781–786.

[37] O. Zariski, Studies in Equisingularity I, *Amer. J. Math.*, **87**, 1965, pp. 507–535; II **87**, 1965, pp. 972–1006; III, **90**, 1965, pp. 961–1023.

18

A Quick Proof of the Hartshorne–Lichtenbaum Vanishing Theorem

Markus Brodmann
Craig Huneke

18.1 Introduction

In this note we give a short and elementary proof of the Hartshorne-Lichtenbaum vanishing theorem [8], 3.1 (which we refer to below as "HLVT"). Other authors have given nice proofs of this result, cf. [9], III, 3.1 and [4]. The reason we present another proof is threefold. First, it is an important and basic result, for example HLVT plays a crucial role in the vanishing theorems for local cohomology in [7], and in the simple proof of Falting's connectedness theorem in [2]. Secondly we feel our proof does represent a true simplification, and thirdly the method of proof is probably useful in other contexts and is certainly suggestive of other vanishing theorems. Often a different proof is necessary for a variety of reasons, e.g. G. Lyubeznik [11] recently needed to give a new proof of HLVT (using a local Bertini theorem) to extend it to étale cohomology.

The proof presented here was discovered independently by both authors. We feel the proof is very much in the spirit of Abhyankar; it uses little machinery and relies on "down-to-earth" methods to prove an important result. See [4] for more detail.

18.2 The Proof

If R is a ring, I an ideal of R, and M and R-module, by $H_I^j(M)$ we denote the jth local cohomology of M with respect to I. (See [5] for general information concerning local cohomology.) The completion of any local ring R in this paper is with respect to the powers of its maximal ideal, and will be denoted $\hat{R}$. HLVT is the following theorem:

Theorem 1.1 [H, 3.1]. *Let (R, m) be a local Noetherian ring and let I be an ideal of R. Set $d = \dim(R)$. Then the following statements are equivalent:*

i) $H^d_I(R) = 0$.

ii) For all minimal primes P of $\hat{R}$ *with* $\dim(\hat{R}/P) = \dim(R)$, $\dim(\hat{R}/(I\hat{R}+P)) > 0$.

Consider the following two statements, the first of which is a special case of Theorem 1.1, the second of which is a special case of the first.

(1.2) Let (R,m) be a complete local domain and let $P \in Spec(R)$. Let $d = \dim(R)$ and assume that $\dim(R/P) = 1$. Then $H^d_P(R) = 0$.

(1.3) Let (R,m) be a complete local Gorenstein domain and let $P \in Spec(R)$. Set $d = \dim(R)$ and assume that $\dim(R/P) = 1$. Then $H^d_P(R) = 0$.

It is a standard reduction to see that the hard direction of Theorem 1.1 (namely that ii) implies i)) reduces to proving (1.2). Hartshorne gave such a reduction in his original proof. We sketch a reduction below. Likewise (1.3) is easy to prove by using an argument also found in [8]. For completeness we also sketch this argument.

The main work in the proof of HLVT comes in reducing (1.2) to (1.3). In [8], [9], and [4], (1.2) is reduced to (1.3) by mapping a Gorenstein local ring S *onto* R and then each in different ways showing that (1.3) for S implies (1.2) for R.

The main contribution of this note is to give a quick proof, using very little machinery, that (1.2) can be reduced to (1.3) by finding a Gorenstein ring *inside* R which has all the needed information. With the elementary proof of (1.3) and the proof that (1.2) implies Theorem 1.1, this provides a quick proof of HLVT.

Proof of (1.3) (Cf. [H]). Write $P^n = P^{(n)} \cap J_n$, where J_n is m-primary. As R is a domain, $\cap_{n\geq 0} P^{(n)} = (0)$ and Chevalley's Theorem shows that each J_n contains some $P^{(k)}$, which shows that $\{P^n\}$ and $\{P^{(n)}\}$ are cofinal.

Hence $H^d_P(R) = \lim_{\rightarrow} Ext^d_R(R/P^{(n)}, R) = 0$ since depth $R/P^{(n)} > 0$ and R is Gorenstein of dimension d. (This is immediate from local duality, but a quick proof is as follows: choose an element $x \notin P$, and apply $Hom_R(\ ,R)$ to the exact sequence of multiplication by x on $R/P^{(n)}$. Since $id_R R = d$, one sees that $Ext^d_R(R/P^{(n)}, R)$ is divisible by x, hence is 0 by Nakayama's lemma.)□

Next we sketch the proof that (1.2) implies Theorem 1.1. For this, we may first complete R. If $H^d_I(R) \neq 0$, then it is easy to see this is true modulo some minimal prime of R of maximal dimension. Hence we may assume R is a complete local domain. Choose I maximal such that $H^d_I(R) \neq 0$ but $\dim(R/I) > 0$; if I is not a prime of dimension 1 we may choose $x \notin I$ such that $\dim(R/(I,x)) > 0$; in this case the exact sequence (cf. [1], 3.9)

$$\ldots \rightarrow H^i_{(I,x)}(R) \rightarrow H^i_I(R) \rightarrow H^i_{I_x}(R_x) \rightarrow \ldots$$

at $i = d$ shows that $H^d_{(I,x)}(R) \neq 0$ (as $\dim R_x < d$), contradicting the maximality of I.□

Finally we prove (1.2), and thus Theorem 1.1, by showing that (1.2) reduces to (1.3).

Proof of (1.2) Let k be a coefficient ring of R. If k is not a field, let q denote the characteristic of its residue field. Of course in this case k is a complete DVR with uniformizing parameter q.

Choose $x_1, \ldots, x_d \in P$ such that $P = \sqrt{(x_1, \ldots, x_d)}$. Such $x_1, \ldots, x_d$ can be chosen: first choose $x_1, \ldots, x_{d-1}$ in P such that $ht(x_1, \ldots, x_{d-1}) = d-1$; also note that x_1 can be chosen to be any nonzero element in P. Choose x_d in P which is not in the union of the other minimal primes containing $x_1, \ldots, x_{d-1}$. Clearly $\sqrt{(x_1, \ldots, x_d)} = P$ in this case. In case k is not a field, we may choose $q = x_1$ if $q \in P$. Without loss of generality, we may assume that $x_1, \ldots, x_{d-1}$ is part of a system of parameters, and further if k is not a field and $q \notin P$, that $x_1, \ldots, x_{d-1}, q$ is a full system of parameters. If either k is a field or $q \in P$, choose $y \notin P$ such that $x_1, \ldots, x_{d-1}, y$ is a full system of parameters. If $q \notin P$, set $y = 0$. Set $A = k[[x_1, \ldots x_{d-1}, y]]$. A is a complete regular local ring, and R is a finite A-module. Finally let $B = A[x_d]$. As $A \subseteq B \subseteq R$, B is necessarily finite over A, hence both complete and local. Furthermore B is Gorenstein since it is a hypersurface ring. Set $Q = P \cap B$. Note that $H^d_P(R) = H^d_{(x_1,\ldots,x_d)R}(R)$ as $P = \sqrt{(x_1, \ldots, x_d)}$. As $x_1, \ldots, x_d \in B$, we necessarily get that $Q = \sqrt{(x_1, \ldots x_d)B}$ (by lying over) so that $H^d_Q(B) = H^d_{(x_1,\ldots,x_d)B}(B)$. By (1.3), $0 = H^d_Q(B)$, and we obtain $0 = H^d_{(x_1,\ldots,x_d)B}(B)$ and finally that $0 = H^d_{(x_1,\ldots,x_d)B}(B) \otimes_B R \cong H^d_{(x_1,\ldots,x_d)R}(R) = H^d_P(R)$. (The middle isomorphism follows as tensor is right exact and $H^j_J(M) = 0$ for $j > d$ for any ideal J of B and any B-module M.)□

Acknowledgments: We would like to thank Gennady Lyubeznik and Rodney Sharp for many valuable conversations concerning local cohomology.

18.3 References

[1] M. Brodmann, Einige Ergebnisse aus der Lokalen Kohomologie theorie und Ihre Anwendung *Osnabrücker Schriffen zur Mathematik*, Heft 5, 1983.

[2] M. Brodmann and J. Rung, Local cohomology and the connectedness dimension in algebraic varieties, *Comment. Math. Helvetici*, **61**, 1986, pp. 481–490.

[3] M. Brodmann and R. Sharp, Work in preparation, 1988–.

[4] F. W. Call and R. Sharp, A short proof of the local Lichtenbaum-Hartshorne theorem on the vanishing of local cohomology, *Bull. Lon-*

don Math. Soc., **18**, 1986, pp. 261–264.

[5] A. Grothendieck (notes by R. Hartshorne), *Local Cohomology*, Springer Lecture Notes in Math., 41, Springer-Verlag, 1966.

[6] R. Hartshorne, Cohomological Dimension of Algebraic Varieties, *Annals of Math.* **88**, 1968, pp. 403–450.

[7] C. Huneke and G. Lyubeznik, *On the vanishing of local cohomology modules*, preprint, 1988.

[8] G. Lyubeznik, *The ëtale cohomological dimension of algebraic varieties*, preprint, 1989.

[9] C. Peskine and L. Szpiro, Dimension projective finie et cohomologie locale, *I.H.E.S.*, **42**, 1973, pp. 323–395.

19

Projective Lines Over One-Dimensional Semilocal Domains and Spectra of Birational Extensions

William Heinzer
David Lantz
Sylvia Wiegand

19.1 Introduction

In [7], Nashier asked if the condition on a one-dimensional local domain R that each maximal ideal of the Laurent polynomial ring $R[y, y^{-1}]$ contracts to a maximal ideal in $R[y]$ or in $R[y^{-1}]$ implies that R is Henselian. Motivated by this question, we consider the structure of the projective line $\text{Proj}(R[s,t])$ over a one-dimensional semilocal domain R (the projective line regarded as a topological space, or equivalently as a partially ordered set). In particular, we give an affirmative answer to Nashier's question. (Nashier has also independently answered his question [9].) Nashier has also studied implications on the prime spectrum of the Henselian property in [8] as well as in the papers cited above.

We also investigate the structure of prime spectra of finitely generated birational extensions of $R[y]$ and of blowups of parameter ideals of a two-dimensional Cohen-Macaulay local domain. In each case we note some analogies with $\text{Spec}(R[y])$, which was analyzed in [2].

Since the Henselian property is so crucial to this work, it seems appropriate to thank Professor Abhyankar here for his inspiration and contributions to an earlier paper [1]. In [1] an example was constructed of a non-Henselian local two-dimensional domain D such that D/P is Henselian for each height-one prime ideal P of D.

The present paper is in part an extension and generalization of work in [2]. One of the results of that paper is the following:

Theorem Let R be a countable one-dimensional semilocal domain.

1. If R is not Henselian and has exactly n maximal ideals, then $\text{Spec}(R[y])$ is isomorphic (as topological spaces or partially ordered sets) to

Spec($L[y]$), where L is a localization of the integers $\mathbf{Z}$ outside n distinct nonzero prime ideals.

2. If R is Henselian (which implies R is local), then Spec($R[y]$) is isomorphic to Spec($H[y]$), where H is a Henselization within the complex numbers of $\mathbf{Z}$ localized outside $2\mathbf{Z}$.

In analogy with the affine case given in the Theorem above, we prove in Theorem 2.3 that if R is a countable one-dimensional Noetherian domain with n maximal ideals, then up to homeomorphism or isomorphism, there are exactly two possibilities for Proj($R[s,t]$) if $n = 1$, and only one if $n > 1$. As before, the two cases distinguish between Henselian and non-Henselian rings.

In Section 20.3 we consider certain birational extensions of the polynomial ring $R[y]$, where R is a one-dimensional semilocal domain. For example, if $(R, \mathbf{m})$ is a countable one-dimensional local domain and $f \in R[y] - \mathbf{m}[y]$, then Spec($R[y]$) $\cong$ Spec($R[y, 1/f]$). But if the ideal $fR[y]$ has prime radical and B is a finitely generated R-algebra that is properly between $R[y]$ and $R[y, 1/f]$, we show in Proposition 3.1 that Spec(B) is not homeomorphic to Spec($R[y]$).

Section 20.4 concerns the blowup of a parameter ideal of a two-dimensional Cohen-Macaulay local domain. We show in Proposition 4.1 that affine pieces of this blowup satisfy many of the axioms satisfied by the spectrum of a polynomial ring in one variable over a one-dimensional local domain. Proposition 4.2 gives similar results for the entire blowup.

All rings we consider are commutative and contain a multiplicative identity. The terms "local" and "semilocal" include "Noetherian." The symbol $<$ between sets means proper inclusion.

It will be convenient to set some notation for partially ordered sets from earlier papers:

1.1 Notation For U a partially ordered set, $u \in U$, and T a subset of U,

$$\mathrm{G}(u) = \{w \in U \mid w > u\}\ , \qquad \mathrm{L}(u) = \{w \in U \mid w < u\}\ ,$$

$$L_e(T) = \{w \in U \mid \mathrm{G}(w) = T\}.$$

Note that the set called L(T) in [2] is denoted $L_e(T)$, the "exactly-less-than" set, here.

We will be concerned with partially ordered sets of dimension two with a unique minimal element, specifically the spectra of two-dimensional integral domains. In this context, if P is a height-one prime, then G(P) is the set of maximal ideals containing P, while if T is a set of height-two maximal ideals, then $L_e(T)$ is the set of height-one primes contained in the intersection of the elements of T and not contained in any other maximal ideal of the ring.

Roger Wiegand has conjectured in [10] that the spectrum of any two-dimensional domain that is a finitely generated algebra over $\mathbf{Z}$ is homeomorphic to the spectrum of the polynomial ring $\mathbf{Z}[y]$. It is shown in [10] that if k is a field and A is a two-dimensional domain that is finitely generated as a k-algebra, then $\mathrm{Spec}(A) \cong \mathrm{Spec}(\mathbf{Z}[y])$ if and only if k is contained in the algebraic closure of a finite field. His method was to provide an axiom system characterizing $\mathrm{Spec}(\mathbf{Z}[y])$ up to homeomorphism or isomorphism. Motivated by his result, the following axiom systems were formulated in [2]:

1.2 Definition A partially ordered set U is "$C\mathbf{Z}(n)P$" if it satisfies:

(P0) U is countable.

(P1) U has a unique minimal element u_0.

(P2) U has dimension two.

(P3) There exist infinitely many height-one maximal ideals.

(P4) There exist n height-one nonmaximal "special" elements $u_1, u_2, \ldots u_n$ satisfying: (i) $\mathrm{G}(u_1) \cup \ldots \cup \mathrm{G}(u_n) = \mathrm{height-twoelementsof}U \}$, (ii) $\mathrm{G}(u_i) \cap \mathrm{G}(u_j) = \emptyset$ for $i \neq j$, and (iii) $\mathrm{G}(u_i)$ is infinite for each i, $1 \leq i \leq n$.

(P5) For each height-one nonspecial element u, $G(u)$ is finite.

(P6) For each nonempty finite subset T of $\{$ height-two elements of U $\}$, $L_e(T)$ is infinite.

(Pictorially, a $C\mathbf{Z}(n)P$ partially ordered set looks like this:

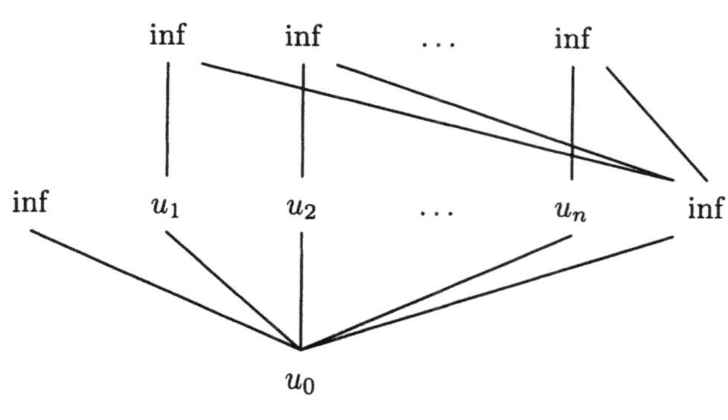

The relationships of the lower right boxed section, determined by (P5) and (P6), are too complicated to display.)

1.3 Definition A partially ordered set U is "CHP" if it satisfies:

(P0)–(P5) Same as $C\mathbf{Z}(1)P$ above.

(P6) For each finite subset T of { height-two elements of U } of cardinality greater than one, $L_e(T) = \emptyset$. For each singleton $t \in$ { height-two elements of U }, $L_e(\{t\})$ is infinite.

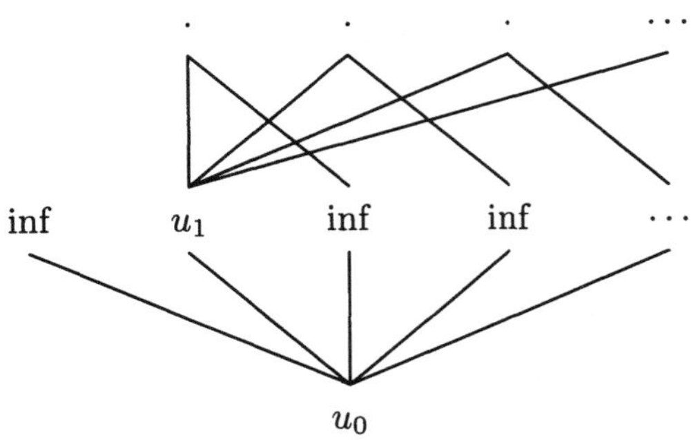

It was shown in [2] that (1) these axiom systems are categorical; (2) if $(R, \mathbf{m}_1, \ldots, \mathbf{m}_n)$ is a countable semilocal one-dimensional domain that is not Henselian, then Spec($R[y]$ is $C\mathbf{Z}(n)P$; and (3) if R is a countable Henselian one-dimensional (local) domain, then Spec($R[y]$) is CHP. We use these facts in the present paper.

R. Wiegand proves in [10] that if D is an order in an algebraic number field, then Spec($D[y]$) $\cong$ Spec($\mathbf{Z}[y]$). A crucial point in this proof is his axiomatic characterization of Spec($\mathbf{Z}[y]$), and the crucial axiom here is (rW5), called (P5) in [10], which states that if $P_1, \ldots, P_r$ are height-one primes and $M_1, \ldots, M_s$ are maximal ideals, then there exists a height-one prime Q such that $Q \subset M_i$, for each $i = 1, \ldots, s$, and if M is a maximal ideal containing Q and some P_i, then M is one of the M_j. If A is a two-dimensional domain that is finitely generated as a $\mathbf{Z}$-algebra and if P is a height-one prime of A, then it is known that every maximal ideal of A/P is the radical of a principal ideal. It follows that Spec(A) satisfies a restricted version of axiom (rW5) where $r = 1$ and $s = 1$. This motivates the following

Question Suppose A is a two-dimensional Noetherian domain having the property that Spec(A) is countable, every height-one prime of A is contained in infinitely many maximal ideals, and for each height-one prime P and each maximal ideal M containing P, there exists a height-one prime Q such that $P + Q$ is M-primary, does it follow that Spec(A) satisfies axiom (rW5) mentioned above?

Our work in this paper is part of an on-going study of the general question: What partially ordered sets arise as the prime spectrum of a Noetherian ring? This question is entirely open, even for two-dimensional rings. It is even unknown how to characterize polynomial rings over one-dimensional countable rings (even polynomial rings in two variables over a countable field).

19.2 The Projective Line Over a One-Dimensional Semilocal Domain

Let $(R, \mathbf{m}_1, \mathbf{m}_2, \ldots, \mathbf{m}_n)$ be a one-dimensional semilocal domain and s, t be indeterminates. In this section, we study the projective line X over R. It will be convenient to use two interpretations of the projective line: (1) $X = \text{Proj}(R[s,t])$, the set of relevant homogeneous primes in the polynomial ring in two indeterminates over R, and (2) X is the union of its affine pieces $\text{Spec}(R[y])$ and $\text{Spec}(R[1/y])$, where $y = s/t$. (The only elements in the second affine piece that are not in the first are the height-one prime $(1/y)R[1/y]$ and the height-two maximals $(\mathbf{m}_i, 1/y)R[1/y]$, and the two pieces intersect in $\text{Spec}(R[y, 1/y])$.) We will refer to homogeneous relevant prime ideals of $R[s,t]$ as points of X. Each height-two point of X has the form $(\mathbf{m}, f(s,t))R[s,t]$ where $\mathbf{m}$ is a maximal ideal of R and f is a homogeneous polynomial of which the image mod $\mathbf{m}$ is irreducible in $(R/\mathbf{m})[s,t]$. In such an f the highest power of at least one of s, t has coefficient not in $\mathbf{m}$; and if only one (say s) has coefficient not in $\mathbf{m}$, then $f(s,t)$ can be taken to be s times an element of $R - \mathbf{m}$. (Warning: If R is not integrally closed, the ideal $f(s,t)R[s,t]$ need not be prime despite the fact that its image in $(R/\mathbf{m})[s,t]$ is a prime ideal.)

In analogy with the axiom systems in [10] and [2], we introduce the following:

2.1 Definition We say that the partially ordered set U is "$\mathbf{PCZ}(n)P$" if it satisfies:

(P0) U is countable.

(P1) U has a unique minimal element u_0.

(P2) U has dimension two.

(P3) Every maximal element has height two.

(P4) There exist n height-one nonmaximal "special" elements $u_1, u_2, \ldots u_n$ satisfying: (i) $\text{G}(u_1) \cup \ldots \cup \text{G}(u_n) = \{\text{height} - \text{two elements of } U\}$, (ii) $\text{G}(u_i) \cap \text{G}(u_j) = \emptyset$ for $i \neq j$, and (iii) $\text{G}(u_i)$ is infinite for each i, $1 \leq i \leq n$.

(P5) For each height-one nonspecial element u, $G(u)$ is finite and $\text{G}(u) \cap \text{G}(u_i) \neq \emptyset$ for each i, $1 \leq i \leq n$.

(P6) For each nonempty finite subset T of height-two elements of U } such that $\{u_1, \ldots, u_n\} \subseteq \bigcup\{L(t) \mid t \in T\}$, $L_e(T)$ is infinite. (Here $L_e(T)$ is the exactly-less-than set.)

2.2 Definition We say that the partially ordered set U is "**P**CH**P**" if it satisfies:

(P0)–(P5) Same as $\mathbf{P}C\mathbf{Z}(1)P$ above.

(P6) For each finite subset T of { height-two elements of U } of cardinality greater than one, $L_e(T) = \emptyset$. For each singleton $t \in$ { height-two elements of U }, $L_e(\{t\})$ is infinite. ($L_e(T)$ as above.)

2.3 Theorem Let R be a countable one-dimensional semilocal Noetherian domain with n maximal ideals. If $n = 1$, then the projective line over R is $\mathbf{P}CHP$ if R is Henselian and $\mathbf{P}C\mathbf{Z}(1)P$ otherwise. If $n > 1$, then the projective line over R is $\mathbf{P}C\mathbf{Z}(n)P$.

The proof of this result will occupy most of this section. We show first that these axiom systems are categorical:

2.4 Lemma Every two partially ordered sets which satisfy the axioms $\mathbf{P}CHP$ are order-isomorphic. The same is true for $\mathbf{P}C\mathbf{Z}(n)P$ for any fixed positive integer n.

Proof We show this for $\mathbf{P}C\mathbf{Z}(n)P$; the argument for $\mathbf{P}CHP$ is similar, and both are only slight adaptations of those of [10] or [2]: Given two posets U, V satisfying $\mathbf{P}C\mathbf{Z}(n)P$, define the order-isomorphism $f : U \to V$ by sending the minimal element u_0 to the minimal element v_0, the n height-one special elements $u_1, \ldots, u_n$ bijectively to the n height-one special elements $v_1, \ldots, v_n$, and for each i, $1 \le i \le n$, the elements of $\mathrm{G}(u_i)$ to the elements of $\mathrm{G}(f(u_i))$, each in any bijective way. Now enumerate the nonspecial height-one elements of U: $u_{n+1}, u_{n+2}, \ldots$, and for $k > n$, enumerate $L_e(f(\mathrm{G}(u_k)))$ in such a way that if $k' < k$ but $\mathrm{G}(u_{k'}) = \mathrm{G}(u_k)$, then $L_e(f(\mathrm{G}(u_k)))$ is enumerated in the same order as $L_e(f(\mathrm{G}(u_{k'})))$. Then inductively define $f(u_k)$ to be the first element of $L_e(f(\mathrm{G}(u_k)))$ that is not of the form $f(u_{k'})$ for some $k' < k$.□

We now begin to show that for a countable one-dimensional semilocal domain R, $X = \mathrm{Proj}(R[s,t])$ is either $\mathbf{P}C\mathbf{Z}(n)P$ or $\mathbf{P}CHP$. Since we are assuming that R is countable, so is $R[s,t]$. The relevant homogeneous primes in $R[s,t]$ are generated by finite subsets, so X is also countable, and (P0) holds. This is the only use we make of the hypothesis of countability on R.

Of course (0) is the unique minimal element of X, so (P1) holds. Since $R[s,t]$ has Krull dimension 3 and the irrelevant maximal ideals $(\mathbf{m}_i, s, t)$ are not elements of X, we see that $\dim(X) = 2$, i.e., (P2) holds.

Axiom (P3) follows from the second assertion in (P5). For (P4), as in the affine case, the "special" elements are the extensions $\mathbf{m}_i[s,t]$ to $R[s,t]$ of the maximal ideals $\mathbf{m}_1, \mathbf{m}_2, \ldots, \mathbf{m}_n$ of R. Since any two of these extensions generate the unit ideal $R[s,t]$, it is clear that no point of X contains two of them; so (P4)*(ii)* holds. Since $\text{Proj}((R/\mathbf{m}_i)[s,t])$ is the (infinite) projective line over the field $R/\mathbf{m}_i$, we also have (P4)*(iii)*.

To see that X satisfies (P4)*(i)* and (P5), we picture X as the union of its affine pieces $\text{Spec}(R[y])$ and $\text{Spec}(R[1/y])$. Since these affine spectra are either $C\mathbf{Z}(n)P$ or CHP ([2], p. 583), we see that each height-two point in X contains one of the special elements, i.e., that (P4)*(i)* holds; and that each nonspecial height-one element is contained in only finitely many height-two points, i.e., that the first part of (P5) holds.

To see that the second part of (P5) holds, assume by way of contradiction that the height-one nonspecial prime P in $R[y]$ is comaximal with the special prime $\mathbf{m}[y]$ in X. We may safely localize all the rings in question at the complement of $\mathbf{m}$ in R, so we assume that R is local and P is a height-one maximal in $R[y]$. Since $yR[y]$ is not maximal, $y \notin P$, so P survives in the localization $R[y, 1/y]$ of $R[y]$ at the powers of y, i.e., $PR[y,1/y] \in \text{Spec}(R[y,1/y])$. There are polynomials $f(y)$ in P and $g(y)$ in $\mathbf{m}[y]$ for which $1 = f(y) + g(y)$, so the coefficients of $f(y)$, except the constant term, are in $\mathbf{m}$, but the constant term is a unit. Hence the extension of P to $R[y,1/y]$ contains $f(0)^{-1}f(y)/y^{\deg(f)}$, a monic polynomial in $1/y$. Let $Q = R[1/y] \cap PR[y,1/y]$ (i.e., the element in $\text{Spec}(R[1/y])$ corresponding to P in $\text{Spec}(R[y])$). Then since Q contains a monic polynomial and meets R in (0), $R[1/y]/Q$ is integral over R, so it has a maximal ideal lying over $\mathbf{m}$, and hence Q is contained in a maximal ideal of $R[1/y]$ that also contains $\mathbf{m}[1/y]$. It follows that the element of X represented by P or Q is not comaximal with the special element represented by the extension of $\mathbf{m}$, the desired contradiction.

We now begin the proof that $X = \text{Proj}(R[s,t])$ satisfies P(6) of $\mathbf{P}C\mathbf{Z}(n)P$ or $\mathbf{P}CHP$. We deal first with the non-Henselian case. Note first that by adjoining to the field of fractions K of R the roots and a deg(f)-th root of the leading coefficient of a dehomogenized version of $f(s,t)$ (i.e., $f(s/t,1)$ or $f(1,t/s)$) to obtain a field L, and letting S be the integral closure of R in L, we have that each of the points of $\text{Proj}(S[s,t])$ lying over a height-two point of X is of the form $(\mathbf{n}, as+bt)S[s,t]$ where $\mathbf{n}$ is a maximal ideal of S and $a, b \in S$, not both in $\mathbf{n}$.

We use the following lemma to deduce the existence of a generic point (in the sense of ([10], Def. 4.7, p. 25)) for a certain subset of $\text{Proj}(R[s,t])$ from the fact that an appropriate set in $\text{Proj}(S[s,t])$ has a generic point.

2.5 Lemma Let $B = \bigoplus_{n=0}^{\infty} B_n$ be a graded ring and $A = \bigoplus_{n=0}^{\infty} A_n$ be a graded subring (in the sense that $A \cap B_n = A_n$ for each n) such that $A \subseteq B$ satisfies the going-up property. (In particular, this holds if B is integral over A.) Let $\mathcal{Q}$ be a set of homogeneous prime ideals in B.

If there exists a homogeneous prime ideal $\mathbf{q}$ of B such that $\mathcal{Q} = \{Q : Q \text{is a homogeneous prime ideal in } B \text{containing} \mathbf{q}\}$, then $\mathbf{p} = \mathbf{q} \cap A$ is a homogeneous (prime) ideal in A, and

$$\{Q \cap A : Q \in \mathcal{Q}\} = \{P : P \text{is a homogeneous prime ideal in } A \text{ containing } \mathbf{p}\}.$$

Proof The homogeneous components of an element of $\mathbf{q} \cap A$ are in both $\mathbf{q}$ and A (the latter because of the uniqueness of the expression of an element of $B = \bigoplus_{n=0}^{\infty} B_n$ as a sum of its homogeneous components); so $\mathbf{p}$ is homogeneous. Any $Q \cap A$, for Q in $\mathcal{Q}$, clearly contains $\mathbf{p}$, so let P be a homogeneous prime of A containing $\mathbf{p}$. By going-up, there is a prime ideal Q_1 of B containing $\mathbf{q}$ and such that $Q_1 \cap A = P$. The homogeneous ideal $\mathbf{q} + PB$ of B is contained in Q_1, so Q_1 contains a minimal prime Q of $\mathbf{q} + PB$. By [10], Proposition 5.11, p. 34, Q is homogeneous, and $\mathbf{q} \subseteq Q$, so $Q \in \mathcal{Q}$. Also, since $PB \subseteq Q \subseteq Q_1$, $Q \cap A = P$. □

We can now verify that, if R is not Henselian, then for a set T satisfying the hypothesis of (P6) of $\mathbf{PCZ}(n)P$, $L_e(T)$ is at least nonempty. Note that by the second assertion of (P5), if the set T does not satisfy the hypothesis of (P6), then $L_e(T)$ is empty.

2.6 Theorem Suppose R is not Henselian. Then for each finite set $M_1, \ldots, M_r$ of height-two points of $X = \mathrm{Proj}(R[s,t])$ such that each maximal ideal of R is contained in at least one M_i, there is a height-one element P of X that is contained in $M_1, \ldots, M_r$ but not in any other height-two point of X.

Proof In view of Lemma 2.5, we may replace R by its integral closure in a finite algebraic extension of its field of fractions K, and the collection $\{M_1, \ldots, M_r\}$ by the (possibly larger) set of points in the projective line over that integral closure that lie over these M_i. Therefore, we may assume that each M_i has the form $(\mathbf{m}, as + bt)R[s,t]$ for some $\mathbf{m}$ maximal in R and some $a, b \in R$, not both in $\mathbf{m}$.

Now we use the fact that R is not Henselian. Since each of the maximal ideals of R is ∞-split ([2], Theorem 1.1), there exists a finite algebraic extension L of K for which the integral closure S of R in L has the property that, for each $\mathbf{m}$ maximal in R, the number of maximal ideals $\mathbf{n}$ of S lying over $\mathbf{m}$ is greater than or equal to the number of M_i containing $\mathbf{m}$. For each $\mathbf{n}$, we pick an $M_i = (\mathbf{n} \cap R, as + bt)R[s,t]$ in such a way that every M_i is picked at least once, and set $N_{\mathbf{n}} = (\mathbf{n}, as + bt)S[s,t]$. Then $N_{\mathbf{n}} \cap R[s,t] = M_i$.

Since the a, b now vary with $\mathbf{n}$, we write them as $a_{\mathbf{n}}, b_{\mathbf{n}}$. By the Chinese Remainder Theorem, there are elements a, b of S for which $a \equiv a_{\mathbf{n}} \mathrm{mod} \mathbf{n}$ and $b \equiv b_{\mathbf{n}} \bmod \mathbf{n}$ for every maximal ideal $\mathbf{n}$ of S. Let $Q = (as + bt)S[s,t]$. Since not both $a_{\mathbf{n}}, b_{\mathbf{n}}$ are in $\mathbf{n}$ for each $\mathbf{n}$, a, b generate the unit ideal in S;

so Q is a prime ideal, that is, $Q \in Y = \mathrm{Proj}(S[s,t])$. Observe that for each maximal ideal $\mathbf{n}$ of S, the polynomial $as + bt$ is in exactly one height-two point of Y containing $\mathbf{n}$ (because the image of $as + bt$ in the polynomial ring $(S/\mathbf{n})[s,t]$ over the field $S/\mathbf{n}$ is a nonzero linear form). Therefore, the set $\{N_{\mathbf{n}} : \mathbf{n} \in \mathrm{Mspec}(S)\}$ is precisely the set of height-two points of Y that contain Q. Since $\{N_{\mathbf{n}} \cap R[s,t] : \mathbf{n} \in \mathrm{Mspec}(S)\} = \{M_1, \ldots, M_r\}$, it follows from Lemma 2.5 that $P = Q \cap R[s,t]$ is contained in $M_1, \ldots, M_r$ but not in any other height-two point of X.□

Next, we argue that, if R is Henselian, then (P6) in $\mathbf{P}CHP$ holds. Suppose R is Henselian (and hence local, with maximal ideal $\mathbf{m}$). Then no two distinct height-two points of X contain the same nonspecial height-one element of X. For, if $y = s/t$ and P is a height-one prime of the polynomial ring $R[y]$ such that $P \cap R = (0)$, then P is contained in a unique maximal ideal of $R[y]$ ([2], Proposition 1.4); if P is not itself maximal, it suffices to observe that P contains a monic polynomial in y and therefore is not contained in the height-two point at infinity for $\mathrm{Spec}(R[y])$ in X (i.e., the prime in $R[1/y]$ corresponding to P in X is not contained in the maximal ideal $(\mathbf{m}, 1/y)R[1/y]$). To see that P contains a monic polynomial in y, consider the domain $R[y]/P = D$, an algebraic extension of R. The integral closure S of R in the field of fractions L of D is a local domain since R is Henselian and a finite intersection of DVR's since R is a one-dimensional local domain. Therefore S is the unique DVR of L containing R. Since D is not a field, it follows that $D \subseteq S$, and hence P contains a monic polynomial in y. Thus we have shown that, for t a height-two element of X, $L_e(\{t\})$ is at least nonempty, since any nonspecial height-one element u contained in t is such that $\mathrm{G}(u) = \{t\}$. (In fact, since a height-two prime in the Noetherian ring $R[y]$ contains infinitely many height-one primes, we get the full strength of the second sentence in (P6) of $\mathbf{P}CHP$ immediately. But the next paragraph treats both Henselian and non-Henselian cases at once.)

Finally, we complete the proof of (P6) in both the Henselian and non-Henselian cases, by showing that if $L_e(T)$ is nonempty, then it is infinite: For a height-one nonspecial element P of $\mathrm{Proj}(R[s,t])$, recall $\mathrm{G}(P) = \{M \in \mathrm{Proj}(R[s,t]) : \mathrm{ht}(M) = 2 \mathrm{and} P \subset M\}$. We contend that, given a finite set $\mathcal{M}$ of height-two points of $\mathrm{Proj}(R[s,t])$ such that $\mathcal{M} = \mathrm{G}(P)$ for some height-one nonspecial element P of $\mathrm{Proj}(R[s,t])$, there are infinitely many height-one nonspecial elements P of $\mathrm{Proj}(R[s,t])$ for which $\mathrm{G}(P) = \mathcal{M}$. To see this, let S be a domain that is a finitely generated integral extension of R such that, in $\mathrm{Proj}(S[s,t])$, there is a finite set of maximal ideals $\mathcal{N}$ such that (1) each maximal ideal of S is contained in exactly one element of $\mathcal{N}$ (i.e., the map $\mathcal{N} \to \mathrm{Mspec}(S) : N \mapsto S \cap N$ is a bijection), (2) there is at least one N in $\mathcal{N}$ lying over each M in $\mathcal{M}$, and (3) each N in $\mathcal{N}$ has the form $N = (S \cap N, a_N s + b_N t)S[s,t]$ with a_N, b_N in S, not both in $S \cap N$. (In the non-Henselian case, we saw in the proof of Theorem 2.6 that such an S exists. In the Henselian case, there is only one M; it contains the unique maximal ideal $\mathbf{m}$ of R, and S can be any extension such that the generator of the image of M in $(R/\mathbf{m})[s,t]$ has a linear factor over the residue field of S.) Then choose a, b in S such that $a \equiv a_N \text{mod}(S \cap N)$ and $b \equiv b_N \text{mod}(S \cap N)$ for each N in $\mathcal{N}$ and note that, if $P = (as + bt)L[s,t] \cap R[s,t]$, where L is the field of fractions of S, then $P \subset M$ iff $M \in \mathcal{M}$. Note that $P = f(s,t)K[s,t] \cap R[s,t]$, where K is the field of fractions of R and f is an irreducible element in $K[s,t]$, unique up to constant multiple, of which $as + bt$ is a factor in $L[s,t]$. Now, the choice of a, b above was determined only up to the (infinite) Jacobson radical J of S; we could add any element of J to either of a, b without changing the resulting $\mathrm{G}(P)$. But since a nonzero element f of $K[s,t]$ has only finitely many nonassociate linear factors over an algebraic closure of K, if we fix a nonzero a and add to b nonzero elements of the Jacobson radical of S, then the prime ideals in $L[s,t]$ generated by the elements $as + bt$ are distinct, and only finitely many of these different primes can give the same P. Thus, there are infinitely many P that give the same $\mathrm{G}(P)$.

The proof of Theorem 2.3 is now complete. We close this section by providing our affirmative answer to Nashier's question.

2.7 Proposition Let $(R, \mathbf{m})$ be a one-dimensional local domain and y an indeterminate. If for every maximal ideal P in $R[y, 1/y]$, either $P \cap R[y]$ is maximal in $R[y]$ or $P \cap R[1/y]$ is maximal in $R[1/y]$, then R is Henselian.

Proof Assume R is not Henselian and let $X = \mathrm{Spec}(R[y]) \cup \mathrm{Spec}(R[1/y])$ be the projective line over R. By the proof of Theorem 2.3, X satisfies (P1)–(P6) of $\mathbf{PC}\mathbf{Z}(1)P$. If P is any height-one element of X that is in $L_e((\mathbf{m}, y)R[y], (\mathbf{m}, 1/y)R[1/y])$, then $PR[y, 1/y]$ is maximal, while both $PR[y]$ and $PR[1/y]$ are nonmaximal.

An alternative proof, not using Theorem 2.3, is the following: Assuming R is not Henselian, by ([6], (43.12)), R has a finite integral extension A that is not local, and the integral closure A' of A is also not local, though it is a semilocal PID. Let $N_1, \ldots, N_n$ be all the maximal ideals of A', and pick an element c of the field of fractions K of A such that $c \in N_1 A'_{N_1}$ and $c \notin A'_{N_i}$ for $2 \le i \le n$. Then since none of the maximal ideals of A' survive in $A'[c, 1/c]$, $A'[c, 1/c]$ is a field. Since it is an integral extension of $R[c, 1/c]$, $R[c, 1/c]$ is also a field. Hence the kernel of the R-homomorphism $R[y, 1/y] \to$K: $y \mapsto c$ is a maximal ideal P. But since $R[c] \subseteq A'_{N_1}$ and $R[1/c] \subseteq A'_{N_2}$, $R[c]$ and $R[1/c]$ are not fields, so neither $P \cap R[y]$ nor $P \cap R[1/y]$ is maximal. □

19.3 Spectra of Birational Extensions of the Affine Line

In this section we establish the following result:

3.1 Proposition Let $(R, \mathbf{m}_1, \ldots, \mathbf{m}_n)$ be a one-dimensional semilocal domain, K its field of fractions, y an indeterminate, $A = R[y]$, $f \in A - \bigcup_{i=1}^{n} \mathbf{m}_i[y]$, and B a finitely generated A-algebra strictly between A and $A[1/f]$. Then Spec(B) satisfies the following axioms from $C\mathbf{Z}(n)P$ or CHP (Definitions 1.2 and 1.3):

(a) (P0) holds if R is countable.

(b) (P1)–(P3) hold without additional hypotheses.

(c) The number m of "special" elements (height-one elements $u_1, \ldots, u_m$ for which (P4)(iii) holds, i.e., G(u_i) is infinite), may be greater than the number n of maximal ideals of R, but it is still finite, and (P4)(i) and (P5) hold (the latter trivially). Any "special" element meets R in a maximal ideal.

(d) If fA has prime radical, then $m > n$ and (P4)(ii) may fail, i.e., the "special" elements need not be comaximal.

3.2 Remark (1) Spec($A[1/f]$) $\cong$ Spec(A), since Spec($A[1/f]$) and Spec(A) both satisfy the axioms for either $C\mathbf{Z}(n)P$ or CHP. The reason for this is that, in localizing A at f, only finitely many height-one primes of A are lost, none of them special, and consequently only finitely many maximal ideals (those containing those height-one nonspecials) are lost.

(2) If B were a non-Noetherian ring strictly between A and $A[1/f]$, then (P6) of both $C\mathbf{Z}(n)P$ and CHP could fail, and the partially ordered set Spec(B) could fail to represent Spec(C) for any Noetherian ring C. For example, if $R = k[x]_{(x)}$, $f = y$, and $B = R[y, x/y, x/y^2, x/y^3, \ldots]$, then B has a height-two maximal ideal $M = yB$, that contains only one height-one prime $P = \bigcap_{n=1}^{\infty} y^n B$; cf. ([3], page 7, Exercise 5). But this phenomenon is impossible in a Noetherian ring: By Krull's Principal Ideal Theorem, every height-two prime ideal M in a Noetherian ring must contain infinitely many height-one primes. (For, if M contained only r height-one primes $P_1, \ldots, P_r$, then for any a in $M - \bigcup_{i=1}^{r} P_i$, the height-two prime ideal M would be minimal over a, a contradiction.)

(3) The stronger hypothesis that B is finitely generated as an algebra over A is used below to insure that the dimension formula holds.

We now begin the proof of Proposition 3.1. If R is countable, then so is B, and since B is also Noetherian, Spec(B) is countable.

Of course, Spec(B) always has unique minimal element (0).

We claim that B has dimension two. Indeed, a bit more generally, if $f \in A - \text{Jac}(R)A$ and $B \subseteq A[1/f]$, then $\dim(A[1/f]) = 2$ and since $A[1/f] = B[1/f]$, $\dim(B) \geq 2$. Since B is also a birational extension of the two-dimensional Noetherian domain A, we have $\dim(B) \leq 2$ so $\dim(B) = 2$.

At most finitely many of the height-one maximals in A (those containing f) extend to the unit ideal in B. Let Q be a prime of B lying over a height-one maximal P in A not containing f. Then $B_Q = A_P$ and $Q = PA_P \cap B$ (since $P \cap R = \emptyset$, so A_P is a localization of $K[y]$ and hence a DVR), and $A/P \subseteq B/Q \subseteq A_P/PA_P = A/P$ (the last equality because P is maximal), and hence Q is a height-one maximal in B. Therefore Spec(B) has infinitely many height-one maximals.

We want to see that the number of height-one primes Q in Spec(B) such that G(Q) is an infinite set is finite: Let Q be one of them. If it meets A in a nonspecial height-one prime P, then, because none of the height-two maximals of B containing Q meet A in P (for, if N is a prime in B such that $N \cap A = P$, then B_N is between the one-dimensional Noetherian domain A_P and its field of fractions and hence has dimension at most one), we get an infinite-to-finite map on the maximal spectra Mspec$(B/Q) \to$ Mspec(A/P), so that at least one of the extensions of maximals in A/P to the Noetherian ring B/Q would have infinitely many minimal primes, a contradiction. Thus Q meets A in either a special height-one prime or a height-two maximal, and in either case it meets R in a maximal ideal $\mathbf{m}$, and hence Q is a minimal prime of $\mathbf{m}B$. But since R is semilocal, so is $\bigcup\{\text{minimalprimesof}\mathbf{m}B\ \} : \mathbf{m} \in \text{Mspec}(R)\}$.

Since $B < A[1/f]$, $fB \neq B$, so fB has at least one minimal prime Q, and if B is Noetherian, then $\text{ht}(Q) = 1$. Since R is Cohen-Macaulay, $A = R[X]$ is Cohen-Macaulay, so every associated prime of fA is of height one. If $P_1, \ldots P_m$ are the associated primes of fA, then

$$A = A[1/f] \cap A_{P_1} \cap \ldots \cap A_{P_m} = B \cap A_{P_1} \cap \ldots \cap A_{P_m} .$$

Suppose that $m = 1$, i.e., that fA has prime radical P (e.g., $f = y$). In this case, since $f \notin \bigcup_{i=1}^{n} \mathbf{m}_i A$, $P \cap R = 0$, so P is contracted from $K[y]$, and hence A_P is a DVR. Assume that the center on A of a prime Q of B is exactly P; then $A_P \subseteq B_Q < K(y)$, and hence (since A_P is a DVR) $A_P = B_Q$. So:

$$B \subseteq A[1/f] \cap B_Q = A[1/f] \cap A_P = A ,$$

a contradiction. Therefore, for each minimal prime Q of fB, $Q \cap A$ properly contains P and hence is a height-two maximal in A. By the dimension formula, e.g., [13, ?] (since A is Cohen-Macaulay, it is universally catenary),

$$1 = \text{ht}(Q) = \text{ht}(Q \cap A) + \text{tr.deg.}(B/A) - \text{tr.deg.}(B/Q)/(A/(Q \cap A)) .$$

Since $\text{ht}(Q \cap A) = 2$, and $\text{tr.deg.}(B/A) = 0$, we see that $\text{tr.deg.}(B/Q)/(A/(Q \cap A)) = 1$, and since B/Q is finitely generated over the field $A/(Q \cap A)$, B/Q has infinitely many maximal ideals. So Q is contained in infinitely many maximal ideals of B, i.e., the closure G(Q) is infinite. But for each maximal ideal $\mathbf{m}$ of R, $\mathbf{m}A[1/f] \cap B$ is also a height-one prime with infinite closure.

Finally, to see that the height-one primes Q of Spec(B) with infinite closure need not be comaximal, we provide an example: Let R be a discrete rank-one valuation domain with maximal ideal $\mathbf{m} = aR$, let $f = y$, and let $B = A[a/y]$. Now yB amd $(a/y)B$ are height-one primes with infinite closures (since $B/yB \cong B/(a/y)B \cong (R/\mathbf{m})[t]$); but they are not comaximal, because $(y, a/y)B$ is a proper ideal of B. This completes the proof of Proposition 3.1.

The example in the last paragraph is somewhat special. We remark that even under the following hypotheses, it is possible that B has exactly two height-one primes with infinite closure, and these two primes are comaximal: Let R be a DVR with $\mathbf{m} = aR$, $A = R[y]$, fA a height-one prime ideal of A such that $fA \cap R = (0)$, $g \in A - fA$ (so that $A < A[g/f]$) such that $(f, g)A < A$ (so that $A[g/f] < A[1/f]$), and $B = A[g/f]$. One such example is obtained by setting $f = y^2 + a^3$ and $g = y$. The two height-one primes of B with infinite closure are $\mathbf{m}B[1/f] \cap B$ and $(a, y)B$; the former contains $a^3/(y^2 + a^3)$, and the latter y, so they are comaximal.

We close this section with two questions suggested by the axiom systems $C\mathbf{Z}(n)P$ and CHP.

Questions 1. If R is not Henselian and $\mathcal{M}$ is a finite set of height-two maximals of B, is there a height-one prime P of B for which $\mathcal{M} = \mathrm{G}(P)$ (i.e., $\mathcal{M}$ is precisely the set of maximal ideals of B that contain P)? We remark that if R is Henselian and P is a height-one prime of B distinct from the finitely many minimal primes of $\mathbf{m}B$, then P is contained in a unique maximal ideal of B. Therefore, if R is Henselian, then there exist such sets $\mathcal{M}$ for which there is no corresponding P.

2. Given a set $\mathcal{M}$ such that $\mathcal{M} = \mathrm{G}(P)$ for one height-one prime P in B, are there infinitely many P for which $\mathcal{M} = \mathrm{G}(P)$?

19.4 Spectra of Parameter Blowups of Two-Dimensional Local Domains

Let $(R, \mathbf{m})$ be a two-dimensional Cohen-Macaulay local domain and let x, y be a system of parameters for R, i.e., the ideal $(x, y)R$ is primary for the maximal ideal $\mathbf{m}$ of R. In this section we examine the "blowup" of the ideal $(x, y)R$, to see how many of the axioms above it satisfies.

We consider first an affine piece $A = R[y/x]$ of the blowup, and we refer to the axiom systems $C\mathbf{Z}(1)P$ and CHP of [2]. Since x, y form a regular sequence, the kernel of the R-algebra homomorphism of the polynomial ring $R[t] \to A$ defined by $t \mapsto y/x$ is the principal ideal $(xt - y)R[t]$, which is contained in $\mathbf{m}R[t]$, a height-two prime ideal of $R[t]$; so $\mathbf{m}A$ is a height-one prime ideal of A. Moreover, $A/\mathbf{m}A \cong (R/\mathbf{m})[t]$, a polynomial ring in one indeterminate over the residue field of R. Thus, the maximal ideals of A containing $\mathbf{m}A$ are in one-to-one correspondence with the maximal ideals of this polynomial ring; in particular, there are infinitely many height-two maximal ideals of A containing $\mathbf{m}A$. On the other hand, for any height-one prime Q of A distinct from $\mathbf{m}A$, $Q \cap R = P$ is a height-one prime in R; since the ideal $(xt - y)R[t]$ is not contained in $PR[t]$, the image of y/x in A/Q is algebraic over R/P, and since this image generates A/Q over R/P, A/Q is a semilocal Noetherian domain of dimension at most one. Therefore, Spec(A) satisfies axiom (P5) of either CHP or $C\mathbf{Z}(1)P$ in [2]. Also, axioms (P1) and (P2) clearly hold for Spec(A), as does (P0) if R is assumed to be countable. Let us observe that there are infinitely many height-one maximal ideals in A: No two of the elements $x - y^{n+1}$, as n varies over the natural numbers, are in the same height-one prime of R; if P is a minimal prime of such an element, then since $x \notin P$, $A \subseteq R_P$ and $PR_P \cap A = Q$ is maximal in A (since in A/Q the image of y/x is the inverse of the image of y^n, an element in the maximal ideal of R/P). Thus, (P3) also holds. To see (P4), all that remains to show is that every height-two maximal N of A meets R in $\mathbf{m}$; so assume that for some N, $N \cap R = P$ has height one. Then the ring of fractions of A with respect to the complement of P in R lies between the one-dimensional Noetherian domain R_P and its field of fractions, so its dimension is at most one; but N survives in this ring of fractions, a contradiction.

Let Q be a height-one prime of A other than $\mathbf{m}A$, and set $P = Q \cap R$. If R/P is Henselian, then A/Q is algebraic over a one-dimensional Henselian local domain and hence is local (cf. ([2], pp. 577–8)). Thus, Q is contained in a unique maximal ideal of A. Suppose that R/P is Henselian for each height-one prime P of R; then each height-one prime of A other than $\mathbf{m}A$ is contained in a unique maximal ideal. If N is a height-two maximal of A, then N is the union of the height-one primes contained in it. Since each of these height-one primes other than $\mathbf{m}A$ is contained in no maximal ideal except N, we see that Spec(A) satisfies axiom (P6) of CHP.

Thus we have shown:

4.1 Proposition Let R be a two-dimensional Cohen–Macaulay local domain, x, y be a system of parameters of R, and $A = R[y/x]$. Then Spec(A) satisfies axioms (P1)–(P5) of [2]. If R is countable and, for each height-one prime P of R, R/P is Henselian, then Spec(A) is CHP.

It is shown in [1] that the hypotheses in Proposition 4.1, including the assumption that R/P is Henselian for each height-one prime P, do not imply that R is Henselian.

So we turn our attention to the case where some R/P is not Henselian, and try to prove (P6) of $C\mathbf{Z}(1)P$. An example relevant to our situation here is the following. Let k be a field and let x, y be indeterminates over k. Let R be the ring $k[y(y-1), y^2(y-1)][[x]]$ localized at the maximal ideal generated by $y(y-1), y^2(y-1)$, and x. Let $f = (x - y^2(y-1))/(y(y-1))$, let $A = R[f]$, and let P be the height-one prime of R generated by x. Then $A \subseteq R_P$. Let $Q = PR_P \cap A$. Since the image of f in A/Q is the same as that of y and since adjoining this element to $R/P \cong k[y(y-1), y^2(y-1)]_{(y(y-1),y^2(y-1))}$ gives a ring with two maximal ideals, we see that Q is contained in precisely two maximal ideals of A. Note that if P' is a height-one prime of R that is distinct from P, then R/P' is complete and therefore Henselian. Therefore, if Q' is a height-one prime of A distinct from both Q and $\mathbf{m}A$, then as we observed above Q' is contained in a unique maximal ideal of A. Therefore in this example Spec(A) is neither CHP nor $C\mathbf{Z}(1)P$. So it is natural to ask:

Question If for each height-one prime P of R the ring R/P is not Henselian, does it follow that Spec(A) satisfies $C\mathbf{Z}(1)P$?

We can provide a first step toward a proof of (P6) of $C\mathbf{Z}(1)P$: For each maximal ideal N of height two of A we show that there exists a height-one prime Q contained in N and not contained in any other maximal ideal of A: If N is a height-two maximal ideal in A, then as we saw above, $\mathbf{m} = N \cap R$. Further above we noted that $A/\mathbf{m}A$ may be identified with the polynomial ring $(R/\mathbf{m})[t]$, where t is the image of y/x. Hence $N = (\mathbf{m}, f)A$, where the image $\overline{f}$ of f in $(R/\mathbf{m})[t]$ is a monic irreducible polynomial. If $\overline{f} = \overline{r_0} + \overline{r_1}t + \ldots + t^n$ for $r_i \in R$, and we set $f = r_0 + r_1(y/x) + \ldots + (y/x)^n$, then N is the unique height-two prime of A that contains f. It follows that there exists a height-one prime Q of A contained in N having the property that N is the unique maximal ideal of A containing Q: Take Q to be a minimal prime of the principal ideal fA.

It seems plausible that, given a height-two maximal ideal N in A, we can find infinitely many height-one primes Q contained in N but not in any other maximal ideal of A. But we wonder whether for every finite set of height-two maximal ideals of A there exists a height-one prime Q of A that is contained in precisely this set of maximal ideals. In certain examples this is the case. For instance, let x, y be indeterminates over a field k, and set $R = k[x, y]_{(x,y)}$ and $A = R[y/x]$. Then using the fact that A is a ring of fractions of $k[x]_{(x)}[y/x]$, we see by Section 2 that Spec(A) satisfies $C\mathbf{Z}(1)P$.

Now let us consider the entire blowup of the ideal $I = (x, y)R$, i.e., $X = \text{Proj}(T)$, where $T = \bigoplus_{n=0}^{\infty} I^n$ is the Rees algebra of I; and refer to the axiom systems $\mathbf{P}C\mathbf{Z}(1)P$ and $\mathbf{P}CHP$. Since X is also the union of its affine pieces Spec($R[y/x]$) and Spec($R[x/y]$), Proposition 4.1 provides some of the answers immediately: If R is countable, then so is X. The poset X has a unique minimal element and dimension two. Every height-two point of X contains the extension of the maximal ideal $\mathbf{m}$ of R, and there are infinitely many height-two points. For a height-one element P of X distinct from the extension of the maximal ideal of R, G(P) is finite.

To show that X satisfies (P3), it suffices to show that if P is a height-one prime of R, then at least one of the rings $R[y/x], R[x/y]$ is contained in R_P, and the center of R_P on at least one of these rings is not a maximal ideal. If $x \in P$, then $y \notin P$, so $R[x/y] \subseteq R_P$, and the center of R_P on $R[x/y]$ is properly contained in $(\mathbf{m}, x/y)R[x/y]$. So we may assume that

$x, y \notin P$, and hence both $R[y/x]$ and $R[x/y]$ are contained in R_P. Assume by way of contradiction that the center of R_P on each ring is maximal, and let z denote the image of y/x in R_P/PR_P. Then the images $(R/P)[z]$ and $(R/P)[1/z]$ of $R[y/x]$ and $R[x/y]$ are both the field R_P/PR_P, so their intersection is again R_P/PR_P. But either z or $1/z$ is in every valuation ring between R/P and its field of fractions R_P/PR_P, so $(R/P)[z] \cap (R/P)[1/z]$ is integral over the one-dimensional domain R/P, the desired contradiction.

Suppose that for each height-one prime P of R, R/P is Henselian. Then as we saw above, a height-one element of Spec($R[y/x]$) distinct from the extension of $\mathbf{m}$ is contained in a unique maximal ideal. So the first sentence of (P6) of $\mathbf{P}CHP$ can fail for X only if there is a height-one prime P of R such that both $R[y/x]$ and $R[x/y]$ are contained in R_P and the center of R_P on $R[y/x]$ is properly contained in a maximal ideal that is lost in $R[x/y]$ and vice versa. Let P be a height-one of R such that both $R[y/x]$ and $R[x/y]$ are contained in R_P and the center of R_P on each is nonmaximal. Again let z denote the image of y/x in R_P/PR_P. Then $(R/P)[z]$ and $(R/P)[1/z]$ are both properly contained in the field of fractions R_P/PR_P of R/P. Since R/P is one-dimensional and Henselian, both z and $1/z$ are integral over R/P, so $(R/P)[z] = (R/P)[1/z]$ (cf. for example [6], (10.5)). Therefore the height-two point in Spec($R[y/x]$) containing P is the same point of X as the one in Spec($R[x/y]$). The second sentence of (P6) of $\mathbf{P}CHP$ also follows, because a height-two maximal of $R[y/x]$ is the union of the height-one primes in it.

Thus we have shown:

4.2 Proposition Let R be a two-dimensional Cohen–Macaulay local domain, x, y be a system of parameters of R, $I = (x, y)R$, and $T = \bigoplus_{n=0}^{\infty} I^n$. Then the blowup Proj($T$) of I satisfies axioms (P1)–(P5) of $\mathbf{P}C\mathbf{Z}(1)P$ or $\mathbf{P}CHP$. If R is countable and, for each height-one prime P of R, R/P is Henselian, then Proj(T) is $\mathbf{P}CHP$.

19.5 References

[1] Shreeram S. Abhyankar, William Heinzer, and Sylvia Wiegand, On the compositum of two power series rings, *Proc. Amer. Math. Soc.*, to appear.

[2] William Heinzer and Sylvia Wiegand, Prime ideals in two-dimensional polynomials rings, *Proc. Amer. Math. Soc.*, **107**, 1989, pp. 577–586.

[3] Irving Kaplansky, *Commutative Rings*, Univ. of Chicago Press, Chicago, 1974.

[4] Ernst Kunz, *Introduction to Commutative Algebra and Algebraic Geometry*, Birkhäuser, Boston, 1985.

[5] Hideyuki Matsumura, *Commutative Algebra, Second Edition*, Benjamin / Cummings, Reading, MA, 1980.

[6] Masayoshi Nagata, *Local Rings*, Interscience, New York/London/Sydney, 1962.

[7] Budh Nashier, Henselian rings and Weierstrass polynomials, *Proc. Amer. Math. Soc.*, to appear.

[8] Budh Nashier, On one-dimensional primes in Laurent polynomial rings over a Henselian ring, *Comm. Algebra*, to appear.

[9] Budh Nashier, Maximal ideals in Laurent polynomial rings, preprint.

[10] Roger Wiegand, The prime spectrum of a two-dimensional affine domain, *J. Pure Appl. Algebra*, **40**, 1986, pp. 209–214.

20
Some Questions on $Z[\sqrt{14}]$

Masayoshi Nagata

20.1 Introduction

There are several definitions of Euclid rings (cf. [1]), and we recall here two of them. Throughout this article, we mean by a ring a commutative ring with identity.

Defintion A A ring R is a Euclid ring if (1) there is a map φ of $R \setminus \{0\}$ into the set $\mathbf{N}$ of natural numbers and (2) for any $a, b \in R$ such that $a \neq 0$, there are $q, r \in R$ such that

$$b = aq + r \quad \text{with either } r = 0 \text{ or } \varphi(r) < \varphi(a).$$

Under the circumstances, we may say that R is euclidean under φ.

Defintion B A ring R is euclidean if (1) R is the ring of integers in a number field of finite degree and (2) R is euclidean under the norm map: $a \longrightarrow N(a) = \sharp(R/aR)$ for $a \neq 0$.

In the case of a imaginary quadratic field, the ring of integers is euclidean in the sense of Defintion B if and only if it is so under Defintion A. The main reason for this is that in $\mathbf{Q}(\sqrt{-m})$, m being a natural number without square factor and $m \geq 13$, the units are ± 1 only and no integer have norm 2 or 3.

In the case of a real quadratic field, there is a fundamental unit, and we feel a big difference. Namely, we feel that the following is plausible:

> The ring of integers of a real quadratic field of class number 1 is a Euclid ring under Defintion A.

$\mathbf{Z}[\sqrt{14}]$ is the ring of integers of $\mathbf{Q}(\sqrt{14})$. Though the class number is 1, $\mathbf{Z}[\sqrt{14}]$ is not euclidean in the sence of Defintion B. Our conclusion at the moment is the following conjecture.

Conjecture Let φ be the function of $E \setminus \{0\}$ into $\mathbf{N}$ such that $\varphi = \varphi_2\varphi_1$ with (1) φ_1 defined as follows: If $a \in E$ $a \neq 0$, then we factor a in the form $a_1(4 + \sqrt{14})^e$ (a_1 not divisible by $4 + \sqrt{14}$), then $\varphi_1(a) = N(a_1)\sqrt{6}^e$ and (2) φ_2 an order preserving map of the image of φ_1 into $\mathbf{N}$. Then $\mathbf{Z}[\sqrt{14}]$ is euclidean under φ.

In this article, we discuss some bases for this conjecture.

20.2 Notation and Terminology

From now on, we denote by E this ring $\mathbf{Z}[\sqrt{14}]$. v_2 denotes $(4+\sqrt{14})$-adic additive valuation on E.

$\begin{bmatrix} m \\ n \end{bmatrix}$ denotes the set $\{(x,y) \in \mathbf{R}^2 \mid |(x-m)^2 - 14(y-n)^2| < 1\}$.

To each point (x,y) on $\mathbf{R}^2$, we associate $x + y\sqrt{14}$. Then the multiplication by a fundamental unit $15 + 4\sqrt{14}$ correspontds to the linear transformation defined by $\begin{pmatrix} 15 & 56 \\ 4 & 15 \end{pmatrix}$. Thus, we often observe linear transformations on the plane defined by $\{\begin{pmatrix} 15 & 56 \\ 4 & 15 \end{pmatrix}^n \mid n \in \mathbf{Z}\}$. We also observe $\mathbf{R}/(2\mathbf{Z}, \mathbf{Z})$. By the orbit of a point (x,y), we mean the set of representatives of $\{\begin{pmatrix} 15 & 56 \\ 4 & 15 \end{pmatrix}^n \begin{pmatrix} x \\ y \end{pmatrix} \mid n \in \mathbf{Z}\}$ modulo $(2\mathbf{Z}, \mathbf{Z})$ in the set $\{(a,b) \mid -1 < a \leq 1, -0.5 < b \leq 0.5\}$.

S denote the set $\{(x,y) \mid x, y \in \mathbf{Q}, v_2(x) = 0, v_2(y) \geq 0, -1 < x \leq 1, -0.5 < y \leq 0.5\}$

20.3 Some questions

Because of our defintion of φ, it is enough that the requirement in Defintion A be satisfied by φ_1. By the nature of φ_1, it is enough to observe the case where a and b are coprime.

As was shown by [2], the complement of $\bigcup_{n,m\in\mathbf{Z}} \begin{bmatrix} m \\ n \end{bmatrix}$ in $\mathbf{R}^2$ consits of points $\{(m+0.5, n+0.5) \mid m, n \in \mathbf{Z}\}$. This means that if a, b are coprime, then $b = aq + r$ with $N(r) < N(a)$ fails only if a is associate with 2 and $b = (2m+1) + (2n+1)\sqrt{14}$ with $m, n \in \mathbf{Z}$. In this case, r with smallest norm is an associate of $3 \pm \sqrt{14}$. Thus, in our observation, factor $4 + \sqrt{14}$ is important. Namely:

Question 1 Assume $a, b \in E$ are coprime. Find all cases where we cannot have $q, r \in E$ such that $b = aq + r$, $N(r) < N(a)$, $v_2(r) \leq v_2(a)$.

If we could answer this question, we should be able to settle our problem. Aiming to have an answer of this question, we tried to see

Question 2 Consider $\bigcup_{m,n\in\mathbf{Z}} \begin{bmatrix} 2m \\ n \end{bmatrix} / (2\mathbf{Z}, \mathbf{Z})$. Find out all the points of S which are not in the this set.

20.4 Observation

(1) Point of S whose orbit consists of a single point.

Linear condition for such a point (x, y) is that $15x+56y \equiv x \pmod{2\mathbf{Z}}$, $4x + 15y \equiv y \pmod{\mathbf{Z}}$. It follows easily that $x \in \mathbf{Z}$, $7y \in \mathbf{Z}$. Such points of S are $\{(1, c/7) \mid c = 0, \pm 1, \pm 2, \pm 3\}$. So the point among them and not in the set $\bigcup_{m,n\in\mathbf{Z}} \begin{bmatrix} 2m \\ n \end{bmatrix} /(2\mathbf{Z}, \mathbf{Z})$ are $(1, 0)$ and $(1, \pm 3/7)$.

(2) Point of S whose orbit consists of two points.

Linear condition on such a point is $15x + 56y \equiv 15x - 56y \pmod{2\mathbf{Z}}$, $4x+15y \equiv -4x+15y \pmod{\mathbf{Z}}$. Because of the nature of S, we have $x \in \mathbf{Z}$, $7y \in \mathbf{Z}$. This implies that these point are those in (1) above. Namely, no orbit of a point of S consist of two points.

(3) Point of S whose orbit consists of three points.

Linear condition on such a point is $449x+1680y \equiv 15x-56y \pmod{2\mathbf{Z}}$, $120x+449y \equiv -4x+15y \pmod{\mathbf{Z}}$. By the nature of S, we see easily that the condition is reduced to $31x \in \mathbf{Z}$, $217y \in \mathbf{Z}$. Naturally, by this condition, points in (1) are included. The writer examined if some point of orbits of these points are in $\begin{bmatrix} 2m \\ n \end{bmatrix}$ with rather small m and suitable n. The result of the examination is that:

As for these pints, they are in the union of $\begin{bmatrix} 2m \\ n \end{bmatrix}$ except for the following 24 points.

$(1, 94/217)$	$(-23/31, 108/217)$	$(23/31, 108/217)$
$(1, -94/217)$	$(23/31, -108/217)$	$(-23/31, -108/217)$
$(1, 108/217)$	$(27/31, 101/217)$	$(-27/31, 101/217)$
$(1, -108/217)$	$(-27/31, -101/217)$	$(27/31, -101/217)$
$(19/31, 101/217)$	$(-23/31, 94/217)$	$(-27/31, -102/217)$
$(-19/31, 101/217)$	$(27/31, -102/217)$	$(23/31, 94/217)$
$(19/31, -101/217)$	$(-27/31, 102/217)$	$(-23/31, -94/217)$
$(-19/31, -101/217)$	$(23/31, -94/217)$	$(27/31, 102/217)$

Here, each row is an orbit. Thus, in view of symmetry, these points are classified into three groups.

(4) Point of S whose orbit consists of 4 points.

Linear condition on these points is $449x + 1680y \equiv 449x - 1680y$, $120x + 449y \equiv -120x + 449y$ (mod$2\mathbf{Z}, \mathbf{Z}$, respectively). Thus, we see that $15x$, $105y \in \mathbf{Z}$. Similar examination as in (3) was done, and the result is that:

As for these points, they are in the union of $\begin{bmatrix} 2m \\ n \end{bmatrix}$ except for the following 4 points, which form an orbit.

$$(13/15, 7/15) \quad (-13/15, 7/15) \quad (-13/15, -7/15) \quad (13/15, -7/15)$$

(5) Other points of S.

Linear conditions for 5 points orbit and 6 points orbit imply

$$899x \in \mathbf{Z}\ (899 = 28 \times 31),\ 6293y \in \mathbf{Z}\ (6293 = 899 \times 7), \text{ and}$$
$$6735x \in \mathbf{Z}\ (6735 = 15 \times 449),\ 47145y \in \mathbf{Z}\ (47145 = 6735 \times 7),$$

respectively. It is too dificult to examine these points by hand computation. But, points having orbits consisting of many points have better chance to be in $\bigcup_{m,n\in\mathbf{Z}} \begin{bmatrix} 2m \\ n \end{bmatrix}$. Therefore, the writer believes that there are few points of S which are not in the union.

20.5 Bases of the conjecture

According to our observation made above, let us suppose that the points of S which are not listed above are in $\bigcup_{m,n\in\mathbf{Z}} \begin{bmatrix} 2m \\ n \end{bmatrix}$. Under this supposition, we can show that the conjecture is true.

Namely, we consider $a, b \in E$ such that they are coprime and a is not a unit.

(1) The case where $v_2(a) \neq 0$.

In this case, $b = aq + r$ implies that $v_2(r) = 0$. Therefore, we can choose r so that $N(r) = \varphi_1(r) < N(a) \leq \varphi_1(a)$ or $N(r) = 5 < 6 = \varphi_1(2)$.

(2) The case where $v_2(a) = v_2(b) = 0$.

Consider b/a and express it in the form $x + y\sqrt{14}$ with $x, y \in \mathbf{Q}$. If (x, y) modulo $(2\mathbf{Z}, \mathbf{Z})$ is not among the points listed above, then by our supposition, there is $\begin{bmatrix} 2m \\ n \end{bmatrix}$ $(m, n \in \mathbf{Z})$ which contains (x, y). Then $q = 2m + n\sqrt{14}$, $r = a((x - 2m) + (y - n)\sqrt{14})$ are good quotient and residue, because $v_2(r) = 0$.

Assume now that the orbit of (x, y) is one listed above. We may replace (x, y) with a convenient point in the orbit or symmetries of the orbit.

i) Point $(1, 0)$. This is the case a divides b.

ii) Point $(1, 3/7)$. The norm of $2 + ((3/7) - 1)\sqrt{14}$ is $|4 - (32/7)| = 4/7$. Thus we have a residue which is associate with 2. Thus $\varphi_1(a) = N(a) = 7 > 6 = \varphi_1(2)$.

iii) Point $(23/31, 108/217)$. The norm of $1 + (23/31) + (108/217)\sqrt{14}$ is $7^{-1}31^{-2}|7 \times 54^2 - 2 \times 108^2| = 2916/6727$. In this case we have a residue $2r$ with $N(r) = 729$. Then $\varphi_1(a) = N(a) = 6727$, $\varphi_1(2r) = 6 \times 729 = 4374$. Thus $2r$ is a good residue.

iv) Point $(27/31, 101/217)$. The norm of $1 + (27/31) + (101/217)\sqrt{14}$ is $7^{-1}31^{-2}|7 \times 58^2 - 2 \times 101^2| = 3146/6727$. In this case, we have a residue $(4 + \sqrt{14})r$ such that $N(r) = 1573$. Thus $\varphi_1(a) = N(a) = 6727$, $\varphi_1((4 + \sqrt{14})r) = \sqrt{6} \cdot N(r) = 3853.04736\cdots$. Thus, this is a good residue.

v) Point $(27/31, 102/217)$. The norm of $1 + (27/31) + (102/217)\sqrt{14}$ is $7^{-1}31^{-2}|7 \times 58^2 - 2 \times 102^2| = 2740/6727$. Thus, we have a residue $2r$ with $N(r) = 685$. Then $\varphi_1(a) = N(a) = 6727$, $\varphi_1(2r) = 6 \cdot N(r) = 4110$. Thus this is a good residue.

vi) Point $(13/15, 7/15)$. The norm of $1+(13/15)+(7/15)\sqrt{14}$ is $15^{-2}|28^2 - 686| = 98/225$. Thus we have a residue $(4 + \sqrt{14})r$ with $N(r) = 49$. Thus $\varphi_1(a) = 225$, $\varphi_1((4+\sqrt{14})r) = \sqrt{6} \cdot N(r) = 120.02499\cdots$. So this is a good residue.

(3) The remaining case is such that $v_2(a) = 0$, $v_2(b) \neq 0$.

This case is reduced to (2) above considering $a + b$ instead of b. This completes the proof of our conjecture under the supposition made above.

Though there may be some more points in S which are not in the union of $\begin{bmatrix} 2m \\ n \end{bmatrix}$ $(m, n \in \mathbf{Z})$, the writer feels that such cases can be handled similarly.

20.6 Remark

At the beginning, the writer tried to see the union of $\begin{bmatrix} 2m \\ n \end{bmatrix}$ $(m, n \in \mathbf{Z})$. Several data on $\begin{bmatrix} 2m \\ n \end{bmatrix}$ were computed by his colleague Dr. T. Nishida of Kyoto University. On one hand the writer is thankful to Dr. Nishida, and on the other hand, the writer saw that even if we take the union of $\begin{bmatrix} 2m \\ n \end{bmatrix}$ with $m, n \in \mathbf{Z}$ and $|m| \leq 20000$, there are too many area not covered by them. Thus the writer is believing that:

> There are some irrational points which are not in the union of all of $\begin{bmatrix} 2m \\ n \end{bmatrix}$ $(m, n \in \mathbf{Z})$.

20.7 References

[1] M. Nagata, On the defintion of a Euclid ring, Adv. Stud. Pure Math., 11 (1987), Kinokuniya, Tokyo, pp. 167-171.

[2] M. Nagata, A pairwise algorithm and its application to $\mathbf{Z}[\sqrt{14}]$, Proc. Alg. Geom. Sem. Singapore 1987, World Sci., Singapore, pp. 69-74.

21

Function Fields of Conics, a Theorem of Amitsur–MacRae, and a Problem of Zariski

Jack Ohm

21.1 Introduction

One of the first and most fundamental results in the theory of non-algebraic field extensions is

> **Lüroth's theorem (1876)**: If $k \subset L \subsetneq k(t) = K$ are field extensions, with t transcendental over k, then there exists $u \varepsilon L$ such that $L = k(u)$.

In other words, a subfield of a simple transcendental extension is again simple transcendental.

The proof of this theorem, although short and elementary, is tricky (cf. [**vdW**, p. 198]). To get a feeling for the difficulties, one might pick at random a couple of elements u_1, u_2 from $k(t)$ and try to exhibit a single generator u for $L = k(u_1, u_2)$. (Incidentally, there is an algorithm for computing u, but we shall not go into that here.) Ever since its discovery, conjectured generalizations of Lüroth's theorem have attracted the efforts of some of the best mathematicians of the day.

For example, one possible generalization allows $k(t)/k$ to be replaced by a finitely generated pure transcendental (or 'rational') extension, giving rise to the

> **Rationality problem**: If $k \subset L \subsetneq k(t_1, \ldots, t_n)$ are field extensions, with $t_1, \ldots, t_n$ algebraically independent over k, does it follow that L/k is also pure transcendental?

The answer is *yes* if the degree of transcendence (abbreviated **dt**) of L/k is 1 (Gordan, Netto, Igusa). In the classical geometric situation of k algebraically closed of characteristic 0, the answer is *yes* if dt $L/k = 2$ (Castelnuovo, Zariski) and *no* if dt $L/k = 3$ (Artin–Mumford, Clemens–Griffiths, Iskovskih–Manin; in geometric terminology, there exists a unirational algebraic threefold which is not rational). If k is not algebraically closed or if $k(t_1, \ldots, t_n)/L$ is not separable, then there are counterexamples with dt $L/k = 2$.

In a special case of the rationality problem known as E. Noether's problem, one additionally assumes that L is the set of elements left fixed by a finite group of permutations of $t_1, ..., t_n$. There are famous counterexamples, by Swan and Voskresenskii, to this problem, and there are some definitive positive results by H.W. Lenstra for an abelian group of permutations that acts transitively.

A different direction of generalization of Lüroth's theorem, one that we shall be interested in here, involves retaining the hypothesis that dt $K/k = 1$ while loosening the requirement that K/k be simple transcendental. What we have in mind is

> **Amitsur's theorem (1955)**: If $k \subset L \subseteq K$ are field extensions and K/k is a regular function field of a conic, then either L/k is simple transcendental or L/k is isomorphic to K/k.

An extension K/k is called regular if K/k is separable and k is algebraically closed in K, and K/k is said to be a function field of a conic if dt $K/k = 1$ and $K = k(x, y)$ where x and y satisfy an irreducible polynomial relation of degree 2 over k.

Amitsur's proof of this theorem comes at the end of a long and difficult paper on central simple algebras [**A1**], and our present objective is to give a direct, field-theoretic proof of it. In fact, we shall prove more, namely, that if K/k is not simple transcendental, then $[K : L]$ is even if and only if L/k is simple transcendental. It is MacRae's proof of this result in [**M**] that provided the key idea for our proof of the combined Amitsur–MacRae theorem.

Aside from its esthetic appeal, Amitsur's theorem is also of interest because it can be used to prove the 1-dim case of the

> **Zariski problem (1949)**: Let K and L be finitely generated extensions of a field k. If a simple transcendental extension of K is k-isomorphic to a simple transcendental extension of L, does it follow that K is k-isomorphic to L?

The answer is *yes* if dt $K/k \leq 1$ and *no* in general, the latter by an astonishing 1985 counterexample of Beauville, Colliot-Thélène, Sansuc, and Swinnerton-Dyer [**BCSS**]. This counterexample is nothing other than a function field of a conic over a ground field k which is a simple transcendental extension of another field k_0, or, in other words, the function field of a surface rationally fibered by conics. (Thus, in studying function fields of conics it is important to allow ground fields that are not necessarily algebraically closed.)

Our goal here has been to give a self-contained, elementary, field-theoretic introduction to these topics, in the hope of making them accessible to a wide audience. Needless to say, the subject can also be approached from other more sophisticated and conceptual (but probably less algorithmic) points of view: for instance, the Brauer group, galois cohomology, K-theory, Severi–Brauer varieties, Hilbert symbols, etc. (See the bibliographic remarks at the end of the paper.) Perhaps the surprising thing is the wealth of complexity that lies hidden in this simplest of algebraic geometric settings.

21.1.1 Summary of the contents

Section 22.1 begins with the definitions of function field, conic, and associated form of a conic. A simple transcendental extension is the simplest example of a function field of a conic, and such extensions play a central role in the theory. The rationality criterion of §22.1 asserts that a function field K/k of a conic $P(X,Y)$ is simple transcendental iff k is algebraically closed in K and $P(X,Y)$ has a k-zero.

In §22.2 we prove that given a regular function field K/k of a conic, there exist nonzero c and f in k such that X^2-cY^2-f (resp. $X^2-XY-cY^2-f$) is a defining conic for K/k if char $k \neq 2$ (resp., char $k = 2$). The notation

$$H_c(X,Y) = \begin{cases} X^2 - cY^2 & \text{if char } k \neq 2 \\ X^2 - XY - cY^2 & \text{if char } k = 2 \end{cases}$$

is used to treat the two cases simultaneously. (On a first reading the reader might mentally replace $H_c(X,Y)$ by $X^2 - cY^2$.)

The proof of the Amitsur–MacRae theorem is given in sections 22.3 and 22.4 and has two main parts. The first of these appears in §22.3: If c is a nonzero element of k and α a root of $H_c(X,1)$ such that $K(\alpha)/k(\alpha)$ is simple transcendental, then there exists a nonzero $f \in k$ such that $H_c(X,Y)-f$ is a defining conic for K/k. Such an α is said to split K/k. It is a consequence of Lüroth's theorem that the same α splits both extensions L/k and K/k of the Amitsur–MacRae theorem, and therefore L/k and K/k have respective defining conics $H_c(X,Y)-f'$ and $H_c(X,Y)-f$. The second part of the proof, in §22.4, involves establishing a relationship between f and f', which can then be used to show that L/k and K/k have a common defining conic and thus are isomorphic. Section 22.4 concludes with a proof that two conics are equivalent if (and only if) their function fields are isomorphic.

Section 22.5 contains a discussion of the Zariski problem, including a proof of the 1-dim case and a description of the counterexample to the general case. The former uses the Amitsur–MacRae theorem. As for the counterexample, it is a 2-dim field extension K/k_0 which is 'stably rational' but not rational and is related to the previous considerations by the property that there is a $t \in K$ such that $K/k_0(t)$ is a function field of a conic.

21.1.2 Notation and conventions

Whenever possible, we shall work inside a universal field extension Ω of a fixed base field k. That is, Ω will be a fixed algebraically closed extension of

k having infinite degree of transcendence over k, and our objects of study will be subfields K,L of Ω which are finitely generated extensions of k. As usual, KL will denote the subfield of Ω generated by K and L. Explicit reference to k and Ω will often be omitted, and when we speak of elements without specifying where they come from, it will be tacitly assumed that they lie in Ω.

By a k-polynomial we mean a polynomial with coefficients in k, and by a k-zero of a polynomial $P(X,Y) \varepsilon k[X,Y]$ we mean an ordered pair (α,β) with $\alpha,\beta \varepsilon k$ such that $P(\alpha,\beta) = 0$.

$\approx$ denotes equivalence of forms (§22.2).

dt K/k = deg of transcendence of K/k.

We use tr. to abbreviate transcendental; char to abbreviate characteristic; $\subset$ for proper containment; and X, Y, Z for indeterminates.

21.2 Function fields

Suppose $P(X)$ is an irreducible polynomial (of deg > 0) in $k[X]$, the polynomial ring in n variables $(X) = (X_1,\ldots,X_n)$. Since $k[X]$ is a unique factorization domain, $P(X)$ generates a prime ideal (P) in $k[X]$. Hence the quotient ring $k[X]/(P)$ is a domain having a canonically constructed quotient field K, which may be regarded as an extension of k by identifying k with its canonical image in K.

If (x) is the image of (X) in K, then $P(x) = 0$ and $K = k(x)$. Moreover, the degree of transcendence of K/k is $n-1$. For, if $P(X)$ has a term with nonzero coefficient involving, say, X_n, then the ideal (P) can contain no polynomial in the other variables $X_1,\ldots,X_{n-1}$ alone, and therefore $x_1,\ldots,x_{n-1}$ are algebraically independent over k and x_n is algebraic over $k(x_1,\ldots,x_{n-1})$.

Since we shall only be interested in field extensions up to isomorphism, a field extension K/k will be said to be *the function field of* $P(X)$ *over* k if K is k-isomorphic to the quotient field of $k[X]/(P)$. The irreducible polynomial $P(X)$ will be called a *defining polynomial* for K/k, and K/k will also be said to be the function field defined by the k-polynomial $P(X)$.

In the sequel we shall focus almost exclusively on the case $n = 2$, i.e., the case of a plane curve; and instead of writing $k[X_1,X_2]$, we shall write $k[X,Y]$. In particular, an irreducible polynomial of deg 2 in $k[X,Y]$ will be called a *k-conic*. Such a polynomial has the form

$$P(X,Y) = aX^2 + bXY + cY^2 + dX + eY + f,$$

where a,b,c,d,e,f are in k and not all of a,b,c are zero.

1.1 Proposition: *Let $P(X,Y)$ be an irreducible polynomial in $k[X,Y]$. A field extension K/k is the function field of $P(X,Y)$ over k if (and only if) dt*

$K/k = 1$ and there exist x, y in K such that $K = k(x, y)$ and $P(x, y) = 0$.

PROOF: Consider the k-homomorphism $\tau : k[X,Y] \longrightarrow k[x,y]$ taking X to x and Y to y. Since $P(x,y) = 0$, τ factors through the canonical homomorphism $k[X,Y] \longrightarrow k[X,Y]/(P) = k[x',y']$; and therefore we have a k–homomorphism $k[x',y'] \longrightarrow k[x,y]$. Since $1 = \text{dt } k(x',y')/k = \text{dt } k(x,y)/k$, it follows that this latter homomorphism is an isomorphism (cf. [**ZS**, p. 101, Theorem 29]) and thus extends to an isomorphism of $k(x',y')$ to $k(x,y)$. □

The above proof also shows that the kernel of the k-homomorphism $\tau : k[X,Y] \longrightarrow k[x,y]$ is (P). It follows that x and y satisfy no nontrivial k-polynomial relationship of deg < deg P; in particular, if $P(X,Y)$ is of deg > 1, then x and y are nonzero.

1.2 The associated form: To any polynomial $P(X,Y)$ of deg > 0 in $k[X,Y]$ there is associated a homogeneous polynomial, or form, $P^h(X,Y,Z)$ in $k[X,Y,Z]$ having the same deg as $P(X,Y)$ and which is uniquely determined by the property that $P(X,Y) = P^h(X,Y,1)$. Conversely, if we start with a homogeneous polynomial $Q(X,Y,Z)$ not divisible by Z, then deg $Q(X,Y,1) =$ deg $Q(X,Y,Z)$, so $Q(X,Y,Z)$ is the associated form of $Q(X,Y,1)$, i.e., $Q^h(X,Y,1) = Q(X,Y,Z)$. Note that any irreducible form other than one of the linear monomials aX, bY, or cZ (and, in particular, any irreducible form of deg > 1) is not divisible by X, Y, or Z.

For example, if $P(X,Y) = aX^2 + bXY + cY^2 + dX + eY + f$, then

$$P^h(X,Y,Z) = aX^2 + bXY + cY^2 + dXZ + eYZ + fZ^2.$$

Such a deg 2 k-form which is also irreducible will be called a *conic k-form.*

Since the factors of a homogeneous polynomial are again homogeneous, $P(X,Y)$ is irreducible in $k[X,Y]$ iff $P^h(X,Y,Z)$ is irreducible in $k[X,Y,Z]$. We shall define *the function field over k of an irreducible k-form* $Q(X,Y,Z)$ of deg > 1 to be the function field over k of $Q(X,Y,1)$ and shall say that $Q(X,Y,Z)$ is a *defining k-form* for this function field. For such a $Q(X,Y,Z)$ the polynomials $Q(X,Y,1)$, $Q(1,Y,Z)$, and $Q(X,1,Z)$ all define the same function field K/k. This is a consequence of 1.1 and the observation that we can write $K = k(x,y)$ with $0 = Q(x,y,1) = Q(1, y/x, 1/x) = Q(x/y, 1, 1/y)$.

1.3 Simple transcendental extensions: Recall that a field extension K/k is called simple tr. if there exists $t \in K$ tr. over k such that $K = k(t)$. Clearly, any two simple tr. extensions of k are k-isomorphic. Moreover, a simple tr. extension is a function field of a conic; for example, $XY - 1$ is a defining conic for $k(t)/k$, as one sees from 1.1 by writing $k(t) = k(t, 1/t)$.

If $P(X,Y)$ is a conic with function field K/k, we can write $K = k(x,y)$ with $P(x,y) = 0$ and either x or y tr. over k. If, say, x is tr. over k and K/k

is simple tr., then $K = k(t)$ and we can furthermore write $x = a(t)/b(t)$, where $a(t)$ and $b(t)$ are polynomials in $k[t]$ of deg ≤ 2. This is a consequence of the fact that $[K : k(x)] \leq 2$ and the following elementary

> **Lüroth Lemma** (cf. [**vdW**, p. 197]): Let k be a field and t be tr. over k, and let $x \in k(t)\backslash k$. If $x = a(t)/b(t)$ with $a(t), b(t)$ relatively prime elements of $k[t]$, then $[k(t) : k(x)] = \max\{\deg a(t), \deg b(t)\}$.

There are many defining conics for a simple tr. extension. For example, if char $k \neq 2$, then $X^2 + Y^2 - 1$ is irreducible and therefore a conic. That its function field K/k is simple tr. may be seen as follows. Consider the Taylor expansion of $X^2 + Y^2 - 1$ at $(0, 1)$:

$$X^2 + Y^2 - 1 = 2(Y - 1) + X^2 + (Y - 1)^2.$$

Let x, y be elements of K such that $x^2 + y^2 - 1 = 0$ and $K = k(x, y)$. Then

$$\begin{aligned} 0 &= x^2 + y^2 - 1; \text{ hence} \\ 0 &= 2[(y-1)/x](1/x) + 1 + [(y-1)/x]^2. \end{aligned}$$

Therefore $1/x \in k((y-1)/x)$ and $K = k((y-1)/x)$. A generalization of this argument yields the implication $(ii) \Rightarrow (iii)$ of the next theorem.

1.4 Theorem (Rationality criterion): *Let K/k be the function field of a k-conic $P(X, Y)$. Then the following are equivalent:*

i) k is algebraically closed in K and there exist α, β, γ, not all zero, in k such that $P^h(\alpha, \beta, \gamma) = 0$.

ii) k is algebraically closed in K and there exist α, β in k such that $P(\alpha, \beta) = 0$.

iii) K/k is simple tr.

PROOF: Let x, y be elements of K such that $K = k(x, y)$ and $P(x, y) = 0$.

$(i) \Rightarrow (ii)$: If $\gamma \neq 0$, then $0 = P^h(\alpha/\gamma, \beta/\gamma, 1) = P(\alpha/\gamma, \beta/\gamma)$; so we may assume $\gamma = 0$. Thus, if $P(X, Y) = aX^2 + bXY + cY^2 + dX + eY + f$, then $0 = P^h(\alpha, \beta, 0) = a\alpha^2 + b\alpha\beta + c\beta^2$. If $a = c = 0$, then $P(\alpha', \beta') = 0$ for $-e/b \neq \alpha' \in k$ and $\beta' = -(d\alpha' + f)/(b\alpha' + e)$; so by symmetry, we may assume $a \neq 0$. Then it follows that $aX^2 + bXY + cY^2 = a(X - rY)(X - r'Y)$, where r, r' are in k. By making the change of variables $X' = X - rY$, $Y' = Y$, we have

$$P(X, Y) = Q(X', Y') = aX'^2 + [a(r - r')X' + dr + e]Y' + dX' + f \quad (21.1)$$

and it suffices to show that $Q(X', Y')$ has a k-zero.

If $a(r - r') \neq 0$, choose $\alpha' = -(dr + e)/a(r - r')$, and solve (22.1) for Y' to find β' such that $Q(\alpha', \beta') = 0$. If $a(r - r') = 0$ and $dr + e \neq 0$, choose α' arbitrarily in k and again use (1) to find β'. If, finally, $a(r - r') = dr + e = 0$, then $Q(X',Y') = aX'^2 + dX' + f$. Since this is a defining polynomial for K/k, it has a root α' in K; and since k is algebraically closed in K, α' is in k. But then $Q(\alpha',1) = 0$.

(ii) $\Rightarrow$ *(iii)*: By rewriting $P(X,Y)$ as a polynomial in $k[X-\alpha, Y-\beta](= k[X,Y])$, we eliminate the constant term and therefore may assume that $P(X,Y)$ has constant term 0. Then

$$\begin{aligned} 0 &= P(x,y) = ax^2 + bxy + cy^2 + dx + ey; \text{ hence} \\ 0 &= a + b(y/x) + c(y/x)^2 + [d + e(y/x)](1/x). \end{aligned}$$

If $d + e(y/x) \neq 0$, we can solve for $1/x$ to conclude $1/x$ is in $k(y/x)$, in which case $K = k(y/x)$. On the other hand, suppose $d + e(y/x) = 0$. If $e \neq 0$, then $y/x = -d/e \,\varepsilon\, k$, and $K = k(x)$. Therefore we may suppose $d = e = 0$. But then $P(X,Y) = aX^2 + bXY + cY^2$, so $P(x,y) = 0$ implies y/x is algebraic over k, hence is in k, since by hypothesis k is algebraically closed in K. Thus, $y \,\varepsilon\, k(x)$ and $K = k(x)$.

(iii) $\Rightarrow$ *(i)*: Suppose $K = k(t)$. Let v be the subdeg valuation of $k(t)/k$, i.e., v is the valuation of K whose valuation ring V is

$$\{f(t)/g(t) \mid f(t), g(t) \,\varepsilon\, k[t] \, and \, g(0) \neq 0\}.$$

Let s be an element of least value from among $x, y, 1$. Then $x/s, y/s, 1/s$ are in V and $P^h(x/s, y/s, 1/s) = P^h(x,y,1) = P(x,y) = 0$. Now let α, β, γ be the k-residues of $x/s, y/x, 1/x$ under the homomorphism $V \longrightarrow k$ defined by sending t to 0. □

It follows from 1.4 that for a function field K/k of a conic with k algebraically closed in K, if one defining conic for K/k has a k-zero, then every defining conic for K/k does. Note also that *(ii)* $\Rightarrow$ *(iii)* becomes false if the hypothesis that k is algebraically closed in K is deleted. For example, take $k =$ the rationals and $P(X,Y) = X^2 + Y^2$. Then $P(0,0) = 0$, but $x/y = \sqrt{-1}$ and $K = k(\sqrt{-1}, y)$.

21.3 The canonical defining conics

2.1 Equivalence of forms: If $P(X,Y,Z)$ is a form in the variables X,Y,Z and $Q(X',Y',Z')$ a form in the variables X',Y',Z', then we shall say that $P(X,Y,Z)$ is *k-equivalent* to $Q(X',Y',Z')$ if there exists an invertible 3x3 k-matrix A and a nonzero element $q \,\varepsilon\, k$ such that

$$\begin{aligned} (X\,Y\,Z) &= (X'\,Y'\,Z')A, \text{ and} \\ P(X,Y,Z) &= qQ(X',Y',Z'). \end{aligned}$$

This is an equivalence relation on the set of k-forms of a given deg in three variables and will be denoted $P(X,Y,Z) \approx Q(X',Y',Z')$. The equality

$(X\,Y\,Z) = (X'\,Y'\,Z')A$ will be referred to as *a change of variables over k*. If a k-form is irreducible over k, then so also is any form k-equivalent to it. Two k-polynomials $P(X,Y)$ and $Q(X',Y')$ will be called k-equivalent if their associated forms are k-equivalent via a change of variables with $Z = Z'$.

2.2 Proposition: *k-equivalent irreducible k-forms of deg $n > 1$ (in three variables) define the same function field over k.*

PROOF: Let $P(X,Y,Z)$ and $Q(X',Y',Z')$ be the two forms and $(X\,Y\,Z) = (X'\,Y'\,Z')A$ be the change of variables such that $P(X,Y,Z) = qQ(X',Y',Z')$. The function field of $P(X,Y,Z)$ is an extension K/k with $K = k(x,y)$ and $P(x,y,1) = 0$. If x',y',z' are defined by $(x\,y\,1) = (x'\,y'\,z')A$ and $A = (a_{ij})$, then $1 = a_{13}x' + a_{23}y' + a_{33}z'$. Hence at least one of x',y',z', say x', must be nonzero; and then $0 = P(x,y,1) = qQ(x',y',z') = qx'^nQ(1,y'/x',z'/x')$. Therefore $k(x,y) = k(x',y',z') = k(y'/x',z'/x')$, the first equality because A is invertible and the second equality because $1/x' \,\varepsilon\, k(y'/x',z'/x')$. Since $Q(1,y'/x',z'/x') = 0$, by 1.1 $k(y'/x',z'/x')/k$ is the function field of $Q(1,Y',Z')$ and hence also of $qQ(X',Y',Z')$. □

We shall prove in 4.4 that the converse to 2.2 holds for $n = 2$.

2.3 Theorem: *Every conic k-form $P^h(X,Y,Z)$ is k-equivalent to one of the following 'canonical' forms:*
i) $X^2 - cY^2 - fZ^2$, *if char* $k \neq 2$;
ii) either $X^2 - cY^2 - fZ^2$ *or* $X^2 - XY - cY^2 - fZ^2$, *if char* $k = 2$.

PROOF: Let $P^h(X,Y,Z) = aX^2 + bXY + cY^2 + dXZ + eYZ + fZ^2$.

Suppose char $k \neq 2$. Let M be the symmetric matrix associated to the quadratic form $aX^2 + bXY + cY^2 + dXZ + eYZ + fZ^2$:

$$\begin{aligned} P^h &= (X\;Y\;Z)\;M\;(X\;Y\;Z)^t \\ &= (X\;Y\;Z)\begin{pmatrix} a & b/2 & d/2 \\ b/2 & c & e/2 \\ d/2 & e/2 & f \end{pmatrix}\begin{pmatrix} X \\ Y \\ Z \end{pmatrix} \end{aligned}$$

where t denotes transpose. It is easily seen that M can be reduced to a diagonal matrix with nonzero entry in the $(\;)_{11}$ position by a sequence of elementary row operations followed by the same sequence of elementary column operations. (In matrix terminology, M is 'congruent' to a diagonal matrix; cf. [**K**, p. 3].) If A is the matrix that effects this sequence of elementary row operations, then a scalar multiple of $P^h = (X'\,Y'\,Z')AMA^t(X'\,Y'\,Z')^t$ is of the desired form.

Suppose char $k = 2$. First, if $b = d = e = 0$, then $P^h = aX^2 + cY^2 + fZ^2$ and (after possibly a permutation of variables) we are done; so we may assume $b \neq 0$. We have

$$P^h = aX^2 + X(bY + dZ) + cY^2 + eYZ + fZ^2.$$

If $X' = X$, $Y' = bY + dZ$, $Z' = Z$, then this is a change of variables since $b \neq 0$, and

$$\begin{aligned} P^h &= aX'^2 + X'Y' + c'Y'^2 + e'Y'Z' + f'Z'^2 \\ &= aX'^2 + Y'(X' + e'Z') + c'Y'^2 + f'Z'^2. \end{aligned}$$

Therefore another change of variables: $X'' = X' + e'Z'$, $Y'' = Y'$, $Z'' = Z'$, yields (using char 2)

$$P^h = a''X''^2 + X''Y'' + c''Y''^2 + f''Z''^2.$$

If $a'' \neq 0$, we are done after multiplying through by a'' and replacing $a''X''$ by X'''. If $a'' = 0$ and $c'' \neq 0$, we can interchange X'' and Y'' (a change of variables) and are in the previous situation. If $a'' = c'' = 0$, then a change of variables of the form $X'' = X'''$, $Y'' = X''' + Y'''$, $Z'' = Z'''$ produces the required form. □

2.4 Separability: Recall (cf. [**We**, p. 14]) that a finitely generated field extension K/k is separable iff every generating set contains a separating transcendence basis. In the case that dt $K/k = 1$ and $K = k(x, y)$, this means that K/k is separable iff either y is separably algebraic over $k(x)$ or x is separably algebraic over $k(y)$.

If x and y satisfy an irreducible polynomial relation over k of deg 2, $ax^2 + bxy + cy^2 + dx + ey + f = 0$, then K/k is inseparable iff char $k = 2$ and $b = d = e = 0$. Thus, if K/k is an inseparable function field of a conic, then char $k = 2$ and *every* defining conic for K/k has the form $aX^2 + cY^2 + f$. Moreover, this shows that the two possibilities of 2.3(ii) are mutually exclusive: If char $k = 2$ and K/k has a defining conic of the form $X^2 - XY - cY^2 - f$ (resp., of the form $X^2 - cY^2 - f$), then K/k is separable (resp., inseparable).

Before continuing, we introduce some notation that will allow us to treat the char $\neq 2$ and char $= 2$ separable cases simultaneously.

2.5 Definition of $H_c(X,Y)$: For $c \in k$, we define $H_c(X,Y)$ by

$$H_c(X,Y) = \begin{cases} X^2 - cY^2 & \text{if char } k \neq 2 \\ X^2 - XY - cY^2 & \text{if char } k = 2. \end{cases}$$

2.6 Corollary: *The function field of a conic k-form $P(X,Y,Z)$ is separable iff there exist c and $f \in k$ such that $P(X,Y,Z)$ is k-equivalent to $H_c(X,Y) - fZ^2$.*

2.7 Regularity: Now recall (cf. [**We**, p. 18]) that a finitely generated field extension K/k is *regular* iff K/k is separable and k is algebraically closed in K. If K/k is a function field of an irreducible polynomial $P(X,Y) \in k[X,Y]$ and $\bar{k}$ is the algebraic closure of k (in Ω), then K/k is regular iff $P(X,Y)$

remains irreducible in $\bar{k}[X,Y]$ (i.e., iff P(X,Y) is *absolutely irreducible*; cf. [**We**, p. 18, Theorem 5]). It follows that if K/k is a regular function field of $P(X,Y)$ and $k'(\subset \Omega)$ is any algebraic extension of k, then $k'K/k'$ is also regular and, by 1.1, a function field of $P(X,Y)$; so the property of being a regular function field of a conic is preserved under algebraic base extension.

2.8 Proposition: *Let c, f be elements of a field k. Then*

i) $X^2 - cY^2 - f$ is irreducible in $k[X,Y]$ iff $2cf \neq 0$ or $\sqrt{c} \notin k$ or $\sqrt{f} \notin k$; and

ii) if char $k = 2$, then $X^2 - XY - cY^2 - f$ is irreducible in $k[X,Y]$ iff $f \neq 0$ or $X^2 - X - c$ is irreducible in $k[X]$.

The proof is a routine calculation and is left to the reader.

2.9 Corollary: *i) $X^2 - cY^2 - f \in k[X,Y]$ is absolutely irreducible (and hence defines a regular function field of a conic) iff $2cf \neq 0$, i.e., iff char $k \neq 2$ and c and f are nonzero.*

ii) If char $k = 2$, $X^2 - XY - cY^2 - f \in k[X,Y]$ is absolutely irreducible (and hence defines a regular function field of a conic) iff $f \neq 0$.

2.10 Corollary: *The function field of a conic k-form P(X,Y,Z) is regular iff there exist nonzero c and $f \in k$ such that $P(X,Y,Z)$ is k-equivalent to $H_c(X,Y) - fZ^2$.*

Proof: The 'if' direction is immediate from 2.9 and 2.2. Only if: By 2.6 there exist c and f in k such that $P(X,Y,Z) \approx H_c(X,Y) - fZ^2$, and by 2.9 $cf \neq 0$ if char $k \neq 2$ and $f \neq 0$ if char $k = 2$. Thus, it only remains to treat the possibility that char $k = 2$, $f \neq 0$, and $c = 0$. In this case $X^2 - XY - fZ^2 \approx X'^2 - X'Y' - fY'^2 - fZ'^2$ via the change of variables $X = X'$, $Y = Y'$, $Z = Z' + Y'$. □

In view of 2.10, we shall make the following

2.11 Definition: A conic of the type $H_c(X,Y) - f$ with $cf \neq 0$ will be called *a regular conic.*

By 2.10 the regular function fields of conics coincide with the function fields of regular conics.

We continue with an observation on the inseparable case. By 2.8(i), in char 2 a k-polynomial of the type $X^2 - cY^2 - f$ is irreducible iff $\sqrt{c} \notin k$ or $\sqrt{f} \notin k$. As for the condition that k be algebraically closed in K, the following may be thought of as the inseparable analogue of 2.10.

2.12 Corollary: *Let K/k be an inseparable function field of a conic $P(X,Y)$*

$= X^2 - cY^2 - f$. *Then k is algebraically closed in K iff* $[k(\sqrt{c}, \sqrt{f}) : k] = 4$.

PROOF: Since $P(X,Y)$ is by definition irreducible in $k[X,Y]$, by 2.8(i) $\sqrt{c} \notin k$ or $\sqrt{f} \notin k$. Therefore $[k(\sqrt{c}, \sqrt{f}) : k] = 2$ or 4. Suppose $[k(\sqrt{c}, \sqrt{f}) : k] = 2$, and hence that $\sqrt{f} \in k(\sqrt{c})$ or $\sqrt{c} \in k(\sqrt{f})$; let us assume the former, the latter case being treated symmetrically. Then there exist $\alpha, \beta \in k$ such that $\sqrt{f} = \alpha + \beta\sqrt{c}$. If we write $K = k(x,y)$ with $x^2 + cy^2 - f = 0$, then (using char 2) we have $x^2 + c(y-\beta)^2 + \alpha^2 = 0$. But y is tr. over k since $x^2 + cy^2 - f = 0$ implies x is algebraic over $k(y)$; so $y - \beta \neq 0$. Therefore $c = [(\alpha + x)/(y-\beta)]^2$, and $\sqrt{c} \in k(x,y)$. Thus k is not algebraically closed in K.

Conversely, suppose there exists $d \in K$ which is algebraic of deg > 1 over k. Then d is also of deg > 1 over $k(y)$, and therefore $k(y,d) = k(y,x) = K$ and d is of deg 2 over k. Since $x \in k(y,d)$ and $P(x,y) = 0$, $P(X,y)$ is reducible in $k(y,d)[X]$, and hence by Gauss's lemma, also in $k(d)[y,X]$. But then $P(X,Y)$ is reducible over $k(d)$, so by 2.8 $\sqrt{c}$ and $\sqrt{f}$ are in $k(d)$. Therefore $[k(\sqrt{c}, \sqrt{f}) : k] = 2$. □

For an example of a function field of a conic which is regular but not simple tr., let k = rationals and $P(X,Y) = X^2 + Y^2 + 1$. By 2.9, $X^2 + Y^2 + 1$ defines a regular function field, and by 1.4 this function field is not simple tr. (The following arithmetic version of this example requires the machinery of Chapters I-IV of [**Se1**] for its proof: If k = the rationals and c is an integer > 1 without repeated prime factors, then the function field of the regular k-conic $X^2 - cY^2 + 1$ is simple tr. iff every prime factor of c is congruent to 1 or 2 mod 4. See also [**Sa3**, p. 62, 4.6].) More examples will be given in §22.5.

Finally, the following technical lemma will be used later.

2.13 Change of variables lemma: *Let c, e_1, e_2 be elements of a field k. Then*

$$H_c(X,Y)H_c(e_1,e_2) = H_c(X',Y'),$$

where

$$\begin{cases} X' = e_1X + ce_2Y \\ Y' = e_2X + dY, \end{cases}$$

with $d = e_1$ if char $k \neq 2$ and $d = e_1 - e_2$ if char $k = 2$. (Note also that $e_1d - ce_2^2 = H_c(e_1,e_2)$.)

PROOF: Direct verification.

Actually, we shall use 2.13 in the following form: *If c, e_1, e_2 are elements of k such that $H_c(e_1,e_2) \neq 0$, then there exists a change of variables such that $(1/H_c(e_1,e_2))H_c(X,Y) = H_c(X',Y')$.*

21.4 Splitting

3.1 Definition: We shall say that an element α algebraic over k *splits* an extension K/k of dt 1 if $K(\alpha)/k(\alpha)$ is simple tr. We shall also say that the field $k(\alpha)$ splits K/k. In general, an extension k' of k will be said to split a finitely generated field extension K/k if dt $k'K/k' =$ dt K/k and $k'K/k'$ is pure tr.

For $0 \neq c \varepsilon k$, $H_c(X,1)$ is a separable polynomial, i.e., its roots are distinct: if α is one root of $H_c(X,1)$, then the roots are $\pm\alpha$ $(= \pm\sqrt{c})$ if char $k \neq 2$ and $\alpha, \alpha - 1$ if char $k = 2$. On the other hand, given any separable algebraic extension k_1/k of deg 2, there exists a nonzero $c \varepsilon k$ such that $k_1 = k(\alpha)$, where α is a root of $H_c(X,1)$.

If K/k is a regular function field of a conic, then by 2.10 there exist nonzero $c, f \varepsilon k$ such that $H_c(X,Y) - f$ is a defining conic for K/k. Since $(\sqrt{f}, 0)$ is a zero of $H_c(X,Y) - f$, it follows from 1.4 that $\sqrt{f}$ splits K/k. It similarly follows from 1.4 that the roots α, α' of $H_c(X,1)$ split K/k, which can also be seen directly as follows: Let $k = k(x,y)$ with $H_c(x,y) - f = 0$. Then $(x - \alpha y)(x - \alpha' y) = f$, so $x - \alpha' y = f/(x - \alpha y) \varepsilon k(x - \alpha y)$. Therefore $(x - \alpha' y) - (x - \alpha y) = (\alpha - \alpha')y \varepsilon k(x - \alpha y)$ and $y \varepsilon k(\alpha, x - \alpha y)$. But then also $x \varepsilon k(\alpha, x - \alpha y)$ and $k(\alpha, x, y) = k(\alpha, x - \alpha y)$. □

Conversely,

3.2 Splitting theorem: *Let K/k be a finitely generated extension of dt 1 with k algebraically closed in K. If c is a nonzero element of k such that the roots of $H_c(X,1)$ split K/k, then*

i) K/k is a separable (hence regular) function field of a conic, and

ii) there exists a nonzero $f \varepsilon k$ such that $H_c(X,Y) - f$ is a defining conic for K/k.

PROOF: By hypothesis, if α is a root of $H_c(X,1)$, then there exists $t \varepsilon K(\alpha)$ such that $k(\alpha, t) = K(\alpha)$. Since $k(\alpha,t)/k$ is separable and $k \subset K \subseteq k(\alpha, t)$, K/k is also separable (cf. [**We**, p. 17, Proposition 19]). Thus, it remains to prove that K/k is the function field of a conic of the form $H_c(X,Y) - f$, with $f \neq 0$.

First we want to quickly dispose of the case that K/k is simple tr.: $H_c(X,Y) + c$ is a regular k-conic, and since $H_c(0,1) + c = 0$, by 1.4 its function field is simple tr. Thus, we may assume to begin with that K/k is not simple tr.

Then α is not in k, and therefore also α is not in K since k is algebraically closed in K. Hence α is (separably) algebraic of deg 2 over K. Note that if we write $t = x + y\alpha$, with $x, y \varepsilon K$, as we can do, then $k(x,y) \subseteq K$ and $k(x,y,\alpha) = K(\alpha)$; and then $K = k(x,y)$. If we now had the additional property that $H_c(x,y) \varepsilon k$, then we would be done by 1.1. Our approach will therefore be to replace t by a suitably chosen element t_1 for which this property holds.

Let * denote the automorphism of $K(\alpha)/K$ which takes α to its k-conjugate α^*. (In char $\neq 2$, $\alpha^* = -\alpha$, and in char $= 2$, $\alpha^* = \alpha - 1$; cf. 3.1.) We have $\alpha\alpha^* = -c$; and for any $z_1, z_2 \varepsilon K$, if $z = z_1 + z_2\alpha$, then $z^* = z_1 + z_2\alpha^*$ and $zz^* = H_c(z_1, z_2)$.

It will suffice to prove the

Claim: There exists $t_1 \varepsilon k(\alpha, t)$ such that $k(\alpha, t_1) = k(\alpha, t)$ and $t_1 t_1^* \varepsilon k(\alpha)$. For then $t_1 = x + \alpha y$ with $K = k(x, y)$ and $H_c(x, y) = t_1 t_1^* \varepsilon k(\alpha)$; and since k is algebraically closed in K, $H_c(x, y) \varepsilon k$, and we are done.

Proof of the *Claim*. Let $k' = k(\alpha)$. Any generator of the simple tr. extension $k'(t)/k'$ is of the form $(m_{11}t + m_{12})/(m_{21}t + m_{22})$, with $m_{ij} \varepsilon k'$ and $m_{11}m_{22} - m_{12}m_{21} \neq 0$ (cf. the Lüroth Lemma of §1). Therefore, to any ordered pair (t_2, t_1) of generators of $k'(t)/k'$, we can associate a nonsingular 2×2 k'-matrix, determined up to a nonzero scalar multiple: write $t_2 = (m_{11}t_1 + m_{12})/(m_{21}t_1 + m_{22})$ and associate the matrix (m_{ij}) to (t_2, t_1); in notation, $(t_2, t_1) \models (m_{ij})$. A computation shows: if $(t_2, t_1) \models M$ and $(t_3, t_2) \models N$, then $(t_3, t_1) \models NM$, and hence also $(t_1, t_2) \models M^{-1}$. (Think of (t_3, t_1) as being the composite $(t_3, t_2)(t_2, t_1)$.)

The *Claim* asserts that there should exist a generator t_1 and a nonzero $\beta \varepsilon k'$ such that $t_1^* = 1/\beta t_1$, or equivalently, such that $(t_1^*, t_1) \models T$, where

$$T = \begin{pmatrix} 0 & 1 \\ \beta & 0 \end{pmatrix}.$$

Let $(t^*, t) \models P$. By regarding (t_1^*, t_1) as the composite $(t_1^*, t^*)(t^*, t)(t, t_1)$, we see that it suffices to specify a nonsingular matrix Q such that Q^*PQ^{-1} is a scalar multiple of T, for then we can define t_1 by $(t_1, t) \models Q$. (Here Q^* denotes the matrix obtained by applying * to the elements of Q.)

Since $t^{**} = t$, we have $(t, t^*) \models P^*$ and $(t, t) \models P^*P$. Therefore $P^*P = \beta I$, where β is a nonzero element of k'. If

$$P = \begin{pmatrix} p & q \\ r & s \end{pmatrix},$$

then, using the relations $r^*p + s^*r = 0$ and $p^*q + q^*s = 0$, one checks that the following specifications give the desired Q:

$$Q = \begin{pmatrix} 1 & 0 \\ p & q \end{pmatrix} \text{ if } q \neq 0, \text{ and } Q = \begin{pmatrix} 0 & 1 \\ r & s \end{pmatrix} \text{ if } q = 0 \text{ and } r \neq 0.$$

It only remains to note that $q = r = 0$ cannot happen. For otherwise, $t^* = pt/s$. Then, with $\gamma = s/p$, we have $t = \gamma t^* = \gamma\gamma^* t$; so $\gamma\gamma^* = 1$ and by Hilbert's Theorem 90 (or also by an easy computation), cf. [**N2**, p. 96], there exists $\beta \varepsilon k(\alpha)$ such that $\gamma = \beta^*/\beta$. Then $t\beta = t^*\beta^*$; hence $t\beta \varepsilon K$. But $k(t\beta) \subseteq K$ and $k(t\beta, \alpha) = k(t, \alpha) = K(\alpha)$ imply $K = k(t\beta)$, a contradiction to our initial assumption that K/k is not simple tr. □

I am grateful to A. Wadsworth [**W1**] for the above proof, and especially for the elementary treatment of 3.2(i). In a previous version of this manuscript, 3.2(i) had been derived by quoting the following facts about genus, which will still, however, be needed in §22.5:

3.4: *Let K/k be a finitely generated extension of dt 1 with k algebraically closed in K.*

*i) Then K/k is of genus 0 iff K/k is a function field of a conic; cf. [**Ar**, p. 302] (or [**E**, p. 139]).*

*ii) If k' is any separable overfield of k (in Ω) such that dt $k'K/k' = 1$, then genus K/k = genus $k'K/k'$; cf. [**C**, p. 99, Theorem 5] or [**Ar**, p. 291].*

3.5 Remarks:

i) 3.2(i) raises the following question. Suppose L/k is a finitely generated separable field extension of dt 1 with k algebraically closed in L (i.e., L/k is a regular function field of dim 1). If there exists an algebraic extension (not necessarily separable) k' of k such that $k'L/k'$ is simple tr., does it follow that L/k is a function field of a conic? The answer is *no* as one sees by the following example from Lang-Tate [**LT**].

Let k be a field of char 2 and c, f be elements of k such that $[k(\sqrt{c}, \sqrt{f}) : k] = 4$. This implies that $X^2 - cY^2 - f$ is a k-conic having (inseparable) function field K/k with k algebraically closed in K (cf. 2.9 and 2.12). Write $K = k(x, y)$ with $x^2 - cy^2 - f = 0$; let $w = y^3 + x, z = y^2$; and consider the field extensions $k \subseteq L = k(w, z) \subseteq K = k(x, y)$. (Thus, $K = k(\sqrt{f} + \sqrt{c}y, y)$ and $L = k(\sqrt{f} + \sqrt{c}y + y^3, y^2)$.) Then $K(\sqrt{c}, \sqrt{f})/k(\sqrt{c}, \sqrt{f})$ is simple tr., and therefore, by Lüroth's theorem, also $L(\sqrt{c}, \sqrt{f})/k(\sqrt{c}, \sqrt{f})$ is simple tr. Moreover, L is separable over k since w, z satisfy the relation $w^2 = z^3 + cz + f$. However, L/k is the function field of the irreducible k-polynomial $W^2 - Z^3 - cZ - f$ and can be seen to have genus 1 [**LT**, p. 623, Theorem 3]; so by 3.4 L/k is not a function field of a conic.

A similar example, in any char $p > 2$, may be obtained as follows (cf. [**T**], [**Sa2**, p. 3], or [**E**, p. 138]). Let k be a field of char $p > 2$ containing an element f such that $f^{1/p}$ is not in k, and let $k \subset k(x, y) = L$ be the (regular) function field of the absolutely irreducible k-polynomial $Y^2 - X^p - f$. If $t = y/(x - f^{1/p})^{(p-1)/2}$, then the relation $t^2 = x - f^{1/p}$ implies $L(f^{1/p}) = k(f^{1/p}, t)$, so $f^{1/p}$ splits L/k. However, it can be seen that L/k has genus $(p-1)/2$, hence is not a function field of a conic.

ii) If K/k is the function field of a regular conic $H_c(X, Y) - f$, we have already noted that $k(\sqrt{c})$ and $k(\sqrt{f})$ both split K/k. (In general, any finite algebraic extension k_1 of k which splits a function field of a regular conic which is not simple tr. must have even deg over k; this is a special case of Springer's theorem, cf. [**L**, p. 69, Exercise 9 and p. 198, Theorem 2.3].) On the other hand, if char $k = 2$ and K/k is the function field of the (inseparable) conic $X^2 - cY^2 - f$, then any splitting field k_1 for K/k must contain the splitting field $k(\sqrt{c}, \sqrt{f})$, since $X^2 - cY^2 - f$ becomes reducible over

k_1 because k_1K/k_1 is simple tr. and therefore separable.

While 3.2 is all that will be needed in §22.4 for the proof of the Amitsur–MacRae theorem, the following more precise version of 3.2(ii) will be needed for the proof of 4.4.

3.6 Theorem (Variation on 3.2(ii)): *Let c, c', f' be nonzero elements of a field k, and suppose the roots of $H_c(X,1)$ split the function field K/k of $H_{c'}(X',Y') - f'Z'^2$ and K/k is not simple tr. Then there exists a nonzero $f \varepsilon k$ such that $H_c(X,Y) - fZ^2$ is k-equivalent to $H_{c'}(X',Y') - f'Z'^2$.*

PROOF: First consider the char $k \neq 2$ case (cf. [**L**, p. 200] for a closely related argument). Then $H_c(\alpha,1) = \alpha^2 - c$, so $\alpha^2 = c$. Since $K(\alpha)/k(\alpha)$ is simple tr., by 1.4 there exist nonzero vectors $\mathbf{a_1} = (a_{11}, a_{12}, a_{13})$ and $\mathbf{a_2} = (a_{21}, a_{22}, a_{23})$ in k^3 such that $\mathbf{a_2} + \mathbf{a_1}\alpha$ is a zero of $H_{c'}(X',Y') - f'Z'^2$. Note that the symmetric matrix M belonging to the quadratic form $H_{c'}(X',Y') - f'Z'^2$ (as in the proof of 2.3) is the diagonal matrix with $1, c', -f'$ on the diagonal.

For any two vectors $\mathbf{a_1} = (a_{11}, a_{12}, a_{13})$ and $\mathbf{a_2} = (a_{21}, a_{22}, a_{23})$ in $k(\alpha)^3$, we define a 'dot product' by

$$\mathbf{a_1} \circ \mathbf{a_2} = (a_{11}, a_{12}, a_{13}) M (a_{21}, a_{22}, a_{23})^t,$$

where ^{t} denotes transpose. We also define $\mathbf{a_1}^2$ to be $\mathbf{a_1} \circ \mathbf{a_1}$. Note that $\mathbf{a_1}^2 = H_{c'}(a_{11}, a_{12}) - f'a_{13}^2$, i.e., $\mathbf{a_1}^2$ is $H_{c'}(X,Y) - f'Z^2$ evaluated at $\mathbf{a_1}$. This dot product is a bilinear map from $k(\alpha)^3 \times k(\alpha)^3$ to $k(\alpha)$: $\mathbf{a_1} \circ (\mathbf{a_2} + \mathbf{a_3}) = \mathbf{a_1} \circ \mathbf{a_2} + \mathbf{a_1} \circ \mathbf{a_3}$ and for all $\gamma \varepsilon k(\alpha)$, $\gamma\mathbf{a_1} \circ \mathbf{a_2} = \mathbf{a_1} \circ \gamma\mathbf{a_2} = \gamma(\mathbf{a_1} \circ \mathbf{a_2})$. Moreover, since M is symmetric, $\mathbf{a_1} \circ \mathbf{a_2} = \mathbf{a_2} \circ \mathbf{a_1}$.

To return to the proof of the char $\neq 2$ case, since $\mathbf{a_2} + \mathbf{a_1}\alpha$ is a zero of $H_{c'}(X',Y') - f'Z'^2$, by the bilinear property of the dot product (and using $\alpha^2 = c$, with $\alpha \not\varepsilon k$) we have

$$0 = (\mathbf{a_2} + \mathbf{a_1}\alpha)^2 = (\mathbf{a_2}^2 + c\mathbf{a_1}^2) + 2\alpha(\mathbf{a_1} \circ \mathbf{a_2}),$$

or equivalently,

$$\begin{cases} 0 & = & (\mathbf{a_2}^2 + c\mathbf{a_1}^2) \\ 0 & = & 2(\mathbf{a_1} \circ \mathbf{a_2}). \end{cases} \tag{21.2}$$

Since $2 \neq 0$, $0 = \mathbf{a_1} \circ \mathbf{a_2}$.

Now choose a third nonzero vector $\mathbf{a_3}$ in k^3 such that $\mathbf{a_1} \circ \mathbf{a_3} = \mathbf{a_2} \circ \mathbf{a_3} = 0$; this is possible because two homogeneous linear equations in three unknowns have a nontrivial solution. Since by hypothesis $H_{c'}(X',Y') - f'Z'^2$ has no nontrivial k-zero, $\mathbf{a_i}^2 \neq 0$, (i=1,2,3). Let A be the matrix with rows $\mathbf{a_1}, \mathbf{a_2}, \mathbf{a_3}$. Then, by definition of the dot product, AMA^t is the matrix whose ij^{th} entry is $\mathbf{a_i} \circ \mathbf{a_j}$ and thus is a diagonal matrix with $\mathbf{a_1}^2, \mathbf{a_2}^2, \mathbf{a_3}^2$ on the diagonal. Moreover, AMA^t is invertible implies A is invertible. Therefore $(X'\,Y'\,Z') = (X\,Y\,Z)A$ is a change of variables, and

$$\begin{aligned} X'^2 - c'Y'^2 - f'Z'^2 &= (X'\,Y'\,Z')M(X'\,Y'\,Z')^t \\ &= (X\,Y\,Z)AMA^t(X\,Y\,Z)^t \\ &= \mathbf{a_1}^2X^2 + \mathbf{a_2}^2Y^2 + \mathbf{a_3}^2Z^2. \end{aligned}$$

But by (22.2), $\mathbf{a_2}^2 = -c\mathbf{a_1}^2$, so if we let $d = \mathbf{a_1}^2$ and $f = \mathbf{a_3}^2/\mathbf{a_1}^2$, we have the desired equality $H_{c'}(X', Y') - f'Z'^2 = d(H_c(X, Y) - fZ^2)$.

Now let us turn to the char $k = 2$ case. Then $H_c(\alpha, 1) = \alpha^2 - \alpha - c$. As in the previous case, for any vector $\mathbf{a_1}$ in $k(\alpha)^3$ it still makes sense to define $\mathbf{a_1}^2$ to be $H_{c'}(X', Y') - f'Z'^2$ evaluated at $\mathbf{a_1}$, i.e., $\mathbf{a_1}^2 = H_{c'}(a_{11}, a_{12}) - f'a_{13}^2 = a_{11}^2 - a_{11}a_{12} - c'a_{12}^2 - f'a_{13}^2$ (although one can no longer define a dot product in general).

By hypothesis there exist nonzero $\mathbf{a_1}, \mathbf{a_2} \,\varepsilon\, k^3$ such that $\mathbf{a_2} + \alpha\mathbf{a_1}$ is a zero of $H_{c'}(X', Y') - f'Z'^2$, or equivalently, such that

$$0 = \mathbf{a_2}^2 + \alpha^2\mathbf{a_1}^2 + \alpha(a_{21}a_{12} + a_{11}a_{22}),$$

or, since by definition of α, $\alpha^2 = \alpha + c$,

$$\begin{cases} 0 = \mathbf{a_2}^2 + c\mathbf{a_1}^2, \\ 0 = \mathbf{a_1}^2 + (a_{21}a_{12} + a_{11}a_{22}). \end{cases} \tag{21.3}$$

Since $H_{c'}(X', Y') - f'Z'^2$ has no nontrivial k-zero, $\mathbf{a_1}^2 \neq 0$; and therefore also $(a_{21}a_{12} + a_{11}a_{22}) \neq 0$. Thus, if we let $\mathbf{a_3} = (0\,0\,1)$, then the matrix A whose rows are $\mathbf{a_1}, \mathbf{a_2}, \mathbf{a_3}$ has nonzero determinant and hence is invertible. Moreover, if $(X'\,Y'\,Z') = (X\,Y\,Z)A$, then $H_{c'}(X', Y') - f'Z'^2 = X'^2 - X'Y' - c'Y'^2 - f'Z'^2 = \mathbf{a_1}^2X^2 + \mathbf{a_2}^2Y^2 + \mathbf{a_3}^2Z^2 + (a_{21}a_{12} + a_{11}a_{22})XY$. Taking into account the equalities (22.3), we have the desired result with $d = \mathbf{a_1}^2$, $f = -\mathbf{a_3}^2/d$. □

I thank H.W. Lenstra for telling me how to prove the char $\neq 2$ case of 3.6 and for pointing out the reference to Lam's book.

21.5 The Amitsur–MacRae theorem

4.1 Theorem: *Let $k \subseteq L \subset K$ be field extensions, and assume K/k is a regular function field of a conic and not simple tr. If $[K{:}L]$ is even, then L/k is simple tr.; and if $[K{:}L]$ is odd, then L is k-isomorphic to K.*

PROOF: Since K/k is regular, by 2.10 there exist nonzero c, f in k such that $H_c(X, Y) - f$ is a defining conic for K/k. Write $K = k(x, y)$ with $H_c(x, y) - f = 0$. If α is a root of $H_c(X, 1)$, then by 3.1 $K(\alpha) = k(\alpha, t)$, where $t = x - \alpha y$. Since $k(\alpha) \subset L(\alpha) \subseteq k(\alpha, t)$, by Lüroth's theorem $L(\alpha)/k(\alpha)$ is simple tr. Hence by 3.2, L/k is a regular function field of a conic and there exists a nonzero $f' \,\varepsilon\, k$ such that $H_c(X, Y) - f'$ is a defining conic for L/k. Therefore we can write $L = k(x', y')$ where x', y' are elements of L such that $H_c(x', y') - f' = 0$. Again by 3.1, $L(\alpha) = k(\alpha, u)$, where $u = x' - \alpha y'$.

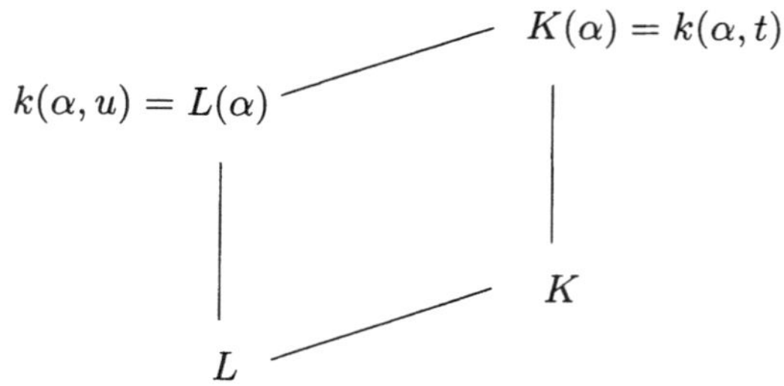

Since K/k is not simple tr., α is not in k and therefore is algebraic of deg 2 over k; and since k is algebraically closed in K, α is not in K and therefore is also algebraic of deg 2 over K. Let * denote the automorphism of $K(\alpha)/K$ defined by taking α to its K-conjugate α^*. (In char $\neq 2$, $\alpha^* = -\alpha$, and in char $= 2$, $\alpha^* = \alpha - 1$; cf. 3.1.) Thus, $\alpha\alpha^* = -c$; and for any z_1, $z_2 \,\varepsilon\, K$, if $z = z_1 + z_2\alpha$, then $z^* = z_1 + z_2\alpha^*$. We shall define $N(z)$ by

$$N(z) := zz^* = H_c(z_1, z_2).$$

Note, in particular, that $N(t) = tt^* = H_c(x, y) = f$ and $N(u) = uu^* = H_c(x', y') = f'$.

At this point we need a 'norm lemma for simple tr. extensions', which contains the crux of the proof of 4.1; its proof is a refinement of a computation found in the paper [**M**] of MacRae.

4.2 Lemma: *Let $k_1(t)/k_1$ be a simple tr. field extension, let * be an automorphism of k_1, let f be a nonzero element of k_1 which is left fixed under *, and extend * to an automorphism of $k_1(t)$ by defining $t^* = f/t$. For any element $z \,\varepsilon\, k_1(t)$, define $N(z) = zz^*$. If u is an element of $k_1(t)\backslash k_1$ such that $N(u) \,\varepsilon\, k_1$, then there exists a nonzero element $\beta \,\varepsilon\, k_1$ such that $N(u) = N(\beta t^m)$, where $m = [k_1(t) : k_1(u)]$.*

Once one has this lemma, the proof of 4.1 concludes as follows. Let $k_1 = k(\alpha)$. By 4.2, there exists $0 \neq \beta \,\varepsilon\, k_1$ such that $N(u) = N(\beta t^m)$, where $m = [k_1(t) : k_1(u)] = [K(\alpha) : L(\alpha)] = [K : L]$. Therefore $f' = N(u) = N(\beta t^m) = N(\beta)N(t)^m = N(\beta)f^m$.

If m is even, then $f' = N(\beta)f^m = N(\beta f^{m/2}) = H_c(e_1, e_2)$, for some $e_1, e_2 \,\varepsilon\, k$, which by 1.4 implies L/k is simple tr.

Now suppose m is odd. Then $f' = N(\beta)f^m = N(\gamma)f$, where $\gamma = \beta f^{(m-1)/2}$.

Claim: The defining conic $H_c(X', Y') - f'$ for L/k is k-equivalent to the defining conic $H_c(X, Y) - f$ for K/k (and then we are done by 2.2). For, there exist $b_1, b_2 \varepsilon k$ such that $\gamma = b_1 + b_2\alpha$, and then $N(\gamma) = H_c(b_1, b_2)$, which is nonzero since $f' \neq 0$. But then

$$\begin{aligned} H_c(X', Y') - f' &= H_c(X', Y') - N(\gamma)f \\ &= H_c(b_1, b_2)[(1/H_c(b_1, b_2))H_c(X', Y') - f] \\ &= H_c(b_1, b_2)[H_c(X'', Y'') - f], \end{aligned}$$

where X'', Y'' are given by the 'change of variables lemma' 2.13. □

PROOF of 4.2: To simplify the notation, let $g = N(u)(= uu^*)$ and $f = N(t)(= tt^*)$. We can write

$$u = dt^n F(t)/G(t),$$

where $d \varepsilon k_1$; t, $F(t)$, $G(t)$ are monic pairwise relatively prime elements of $k_1[t]$; and n is an integer. The equality $uu^* = g$ then yields

$$dd^* f^n F(t) F^*(t^*) = gG(t)G^*(t^*), \tag{21.4}$$

where $F^*(Z)$ denotes the polynomial obtained by applying * to the coefficients of $F(Z)$. We want to rewrite this expression as a polynomial expression in t.

Let $F(t) = f_0 + f_1 t + \cdots + t^r$, where $f_i \varepsilon k_1$. Then

$$\begin{aligned} F^*(t^*) &= f_0^* + f_1^* t^* + \cdots + t^{*r} \\ &= f_0^* + f_1^*(f/t) + \cdots + (f/t)^r \\ &= (f_0^*/t^r)[t^r + (f_1^*/f_0^*)ft^{r-1} + \cdots + f^r/f_0^*]. \end{aligned}$$

Let

$$\hat{F}(t) = t^r + (f_1^*/f_0^*)ft^{r-1} + \cdots + f^r/f_0^*,$$

so $F^*(t^*) = (f_0^*/t^r)\hat{F}(t)$. Similarly, if $G(t) = g_0 + g_1 t + \cdots + t^s$, we can write $G^*(t^*) = (g_0^*/t^s)\hat{G}(t)$, where $\hat{G}(t) = t^s + (g_1^*/g_0^*)ft^{s-1} + \cdots + f^s/g_0^*$. Substituting in (22.4), we have

$$dd^* f^n f_0^* t^s F(t)\hat{F}(t) = gg_0^* t^r G(t)\hat{G}(t). \tag{21.5}$$

The constant term of $\hat{F}(t)$ is f^r/f_0^* and of $\hat{G}(t)$ is f^s/g_0^*. Therefore, equating terms of least deg in (22.5) yields

$$dd^* f^n f_0^* t^s f_0(f^r/f_0^*) = gg_0^* t^r g_0(f^s/g_0^*).$$

It follows that

$$\begin{aligned} r &= s, \text{ and} \\ dd^* f^n f_0 &= gg_0. \end{aligned} \tag{21.6}$$

Similarly, equating the terms of maximal deg in (22.5) yields $dd^* f^n f_0^* = gg_0^*$. Thus, by cancelling in (22.5), we have $F(t)\hat{F}(t) = G(t)\hat{G}(t)$. Since these are monic polynomials and $F(t)$ and $G(t)$ are relatively prime in

$k_1[t]$, it follows that $F(t) = \hat{G}(t)$ and $G(t) = \hat{F}(t)$. Equating constant terms in these last equalities, we then have $f_0 = f^r/g_0^*$ and $g_0 = f^r/f_0^*$. Substituting for f_0 in (22.6) yields

$$dd^* f^{n+r}/g_0 g_0^* = g,$$

or equivalently,

$$N((d/g_0)t^{n+r}) = g = N(u). \tag{21.7}$$

Similarly, substituting for g_0 in (6) yields

$$N(df_0 t^{n-r}) = N(u).$$

By writing $t^{n-r} = t^{2(n-r)}t^{-n+r}$ and using $N(t^2) = f^2 = ff^* = N(f)$, we can rewrite this in the form

$$N((df_0 f^{n-r})t^{-n+r}) = N(u). \tag{21.8}$$

In view of (22.7) and (22.8), it only remains to note that the equality $u = dt^n F(t)/G(t)$ implies $[k_1(t) : k_1(u)] = n + r$ if $n > 0$ and $= -n + r$ if $n < 0$, which follows from the Lüroth Lemma of §22.2. □

4.3 Comments on Theorem 4.1.

i) The assertion of 4.1 that L/k is either simple tr. or isomorphic to K is due to Amitsur [**A1**, p. 42, Corollary 11.3], and that L/k is simple tr. iff $[K{:}L]$ is even to MacRae [**M**]. (The crucial norm equality of 4.2 between u and a power of t undoubtedly reflects a corresponding result in the Brauer group; cf. [**A1**, pp. 34 and 40].)

The history of Theorem 4.1 should also credit an old unpublished manuscript of M. Fried (quoted as "Brauer groups and Jacobians" in [**M**]), which contains the outline of a geometric proof, using the Brauer group, of the Amitsur-MacRae theorem, and some further wide-ranging results and speculations on the lattice of subfields of the function field of a curve. (For an idea of some of the contents of this manuscript, see [**F**, p. 126, Example 1.1.b].) For example, Fried has proved that, under the hypothesis of 4.1, for every integer $m \geq 1$ there exists a subfield L of K such that $[K{:}L] = m$. (This also follows from our proof: Start with $u = t^m = x' + \alpha y'$ and let $L = k(x', y')$ and $f' = N(u)$.)

Finally, it is worthy of note that MacRae and Samuel [**MS**] have made a study of index 2 subfields of function fields of genus 1.

ii) If the assumption that K/k is regular is omitted from 4.1, virtually nothing can be said about the subextensions L/k of K/k. For instance, if K/k is an inseparable function field of a conic, we have seen in 3.5 that L/k need not even be a function field of a conic. Or, if k is not algebraically closed in K, then $K = k(\alpha, t)$ for some α algebraic of deg 2 over k and t tr. over k; and we know by 3.1 that an arbitrary regular function field of a conic is a subextension of such an extension.

However, one might reasonably ask if the following generalization of 4.1 holds: Suppose K/k and L/k are regular function fields of conics and not simple tr., and suppose there exists an element α separably algebraic of deg 2 over k which splits both K/k and L/k. (This is the setting at the start of the proof of 4.1, but there one has the additional property that L is a subfield of K.) Does it follow that K is k-isomorphic to L? The answer is *no*, as the following example shows. Let k_0 be any field not containing $\sqrt{-1}$ and t, t' be elements algebraically independent over k_0; let $k = k_0(t, t')$; and let K/k be the function field of the regular k-conic $X^2 + Y^2 - t$ and L/k the function field of the regular k-conic $X^2 + Y^2 - t'$. Then by 3.1 $\sqrt{-1}$ splits K/k and L/k, and $\sqrt{t}$ splits K/k. On the other hand, $\sqrt{t}$ does not split L/k, as we shall see in 5.4 below; so K/k and L/k are not isomorphic extensions.

In the proof of 4.1, we have actually reached an apparently stronger conclusion than isomorphism, namely, that any defining conic forms for K/k and L/k are equivalent. However, as a consequence of the next theorem, which is the converse to Proposition 2.2, the distinction evaporates. I am indebted to A. Wadsworth [**W1**], who, in commenting on a preliminary version of this paper, sent me an elementary proof, using quaternion algebras, of this converse (for K/k regular and char $k \neq 2$) and suggested that the result be included. After reading his proof, I realised that the regular case (which is the crux of the matter; see (ii) below) also is a consequence of taking $L = K$ in the proof of 4.1. Incidentally, *for k algebraically closed* there is a corresponding statement for any two nonsingular plane curves of the same deg ≥ 4 (cf. [**MH**, p. 71, Proposition 6]); and for char $k \neq 2$ there are some interesting generalizations of 4.4 to quadratic forms in more than 3 variables (cf. [**W2**] and [**Kn**]).

4.4 The equivalence theorem: *If K/k is the function field of the conic k-form $P(X, Y, Z)$ and L/k the function field of the conic k-form $Q(X', Y', Z')$, then K is k-isomorphic to L (if and) only if $P(X, Y, Z)$ is k-equivalent to $Q(X', Y', Z')$.*

PROOF: We clearly may assume $K = L$. The proof will involve consideration of four disjoint cases:

i) K/k is simple tr.

ii) K/k is separable and k is algebraically closed in K (i.e., K/k is regular), but K/k is not simple tr.

iii) K/k is separable and k is not algebraically closed in K.

iv) K/k is inseparable.

In (i), (ii), and (iii), by 2.6 there exist c, c' f, $f' \,\varepsilon\, k$ such that $P(X, Y, Z)$ is k-equivalent to $H_c(X, Y) - fZ^2$ and $Q(X', Y', Z')$ to $H_{c'}(X', Y') - f'Z'^2$. Moreover, in (i) and (ii), c, c', f, f' are nonzero by 2.9.

Proof of (i): By 1.4 there exist e_1, e_2 in k such that $f = H_c(e_1, e_2)$, and then by 2.13 $H_c(X, Y) - fZ^2 \approx H_c(X, Y) - Z^2$. If char $k \neq 2$, then $H_c(X, Y) - Z^2 = X^2 - cY^2 - Z^2 = H_1(X, Z) - cY^2 \approx X^2 - Y^2 - Z^2$, where the equivalence is again by 2.13. Since the same applies to Q, we are done. On the other hand, if char $k = 2$, then the following sequence of equivalences works: $H_c(X,Y) - Z^2 = X^2 - XY - cY^2 - Z^2 \approx cX^2 - XY - Y^2 - cZ^2 = c(X-Z)^2 - XY - Y^2 \approx cZ^2 - XY - Y^2 = cZ^2 - Y(X-Y) \approx cZ^2 - YX \approx Z^2 - YX$, where the equivalences are by obvious changes of variables.

Proof of (ii): By 3.6, there exists $f'' \varepsilon k$ such that $H_{c'}(X',Y') - f'Z'^2$ is k-equivalent to $H_c(X',Y') - f''Z'^2$ (i.e., we may assume $c = c'$). Once one has this, the proof of 4.1 (for $m = 1$) shows that $H_c(X,Y) - fZ^2$ is k-equivalent to $H_c(X',Y') - f''Z'^2$.

Proof of (iii): By 2.9 we may assume $f' = f = 0$, so that $H_c(X,Y)$ and $H_{c'}(X',Y')$ are the two defining conics for K/k. Let k' denote the algebraic closure of k in K. If α is a root of $H_c(X,1)$ and t a tr. over k, then $H_c(\alpha t, t) = 0$; so by 1.1 $K = k(\alpha t, t)$. But then $k' = k(\alpha)$. Similarly, if α' is a root of $H_{c'}(X,1)$, then $k' = k(\alpha')$. Write $\alpha = a + b\alpha'$, where a, $b \varepsilon k$. Since $H_c(a+bX',1)$ vanishes at α' and $H_{c'}(X',1)$ is the irreducible polynomial for α' over k, there exists $d \varepsilon k$ such that $H_c(a+bX',1) = dH_{c'}(X',1)$. But this equality implies $H_c(aY'+bX',Y') = dH_{c'}(X',Y')$; so $X = aY'+bX'$, $Y = Y'$, $Z = Z'$ is the required change of variables.

Proof of (iv): By 2.4 we may assume $X^2 - cY^2 - fZ^2$ and $X'^2 - c'Y'^2 - f'Z'^2$ are the two defining polynomials for K/k. Either $[k(\sqrt{f}, \sqrt{c}) : k] = 2$ or $[k(\sqrt{f}, \sqrt{c}) : k] = 4$. Suppose first the former.

By the symmetry of c and f we may further assume $k(\sqrt{f}, \sqrt{c}) = k(\sqrt{c}) = k(\sqrt{c'})$. Write $\sqrt{f} = a + b\sqrt{c}$, with a, b, εk. Since char $k = 2$, $X^2 - cY^2 - fZ^2 = (X - aZ)^2 - c(Y - bZ)^2 \approx X^2 - cY^2$. By two repetitions of this argument, we obtain $X^2 - cY^2 \approx X'^2 - c'Y'^2 \approx X'^2 - c'Y'^2 - f'Z'^2$.

Now suppose $[k(\sqrt{f}, \sqrt{c}) : k] = 4$. We have noted in 3.5(ii) that $k(\sqrt{f}, \sqrt{c})$ is the unique minimal splitting field for K/k, so it follows that $k(\sqrt{f}, \sqrt{c}) = k(\sqrt{f'}, \sqrt{c'})$ and hence also that either $k(\sqrt{f}, \sqrt{c}) = k(\sqrt{f'}, \sqrt{c})$ or $k(\sqrt{f}, \sqrt{c}) = k(\sqrt{c'}, \sqrt{c})$. By the symmetry between c and f we may assume the former, and then $k(\sqrt{f}, \sqrt{c}) = k(\sqrt{f'}, \sqrt{c}) = k(\sqrt{f'}, \sqrt{c'})$; so again by the symmetry of c and f, it is only necessary to modify one coefficient at a time and hence to see that $X^2 - cY^2 - fZ^2 \approx X^2 - cY^2 - f'Z^2$. We can first write

$$\sqrt{f'} = \gamma - \beta\sqrt{f}, \text{ where } \gamma, \beta \,\varepsilon\, k(\sqrt{c}) \text{ and } \beta \neq 0,$$

and then write

$$\begin{aligned} \beta &= 1/(b_1 - b_2\sqrt{c}) \\ \gamma/\beta &= d_1 - d_2\sqrt{c}, \end{aligned}$$

with b_1, b_2, d_1, $d_2 \,\varepsilon\, k$. Then the change of variables

$$(X\,Y\,Z) = (X'\,Y'\,Z') \begin{pmatrix} b_1 & b_2 & 0 \\ cb_2 & b_1 & 0 \\ d_1 & d_2 & 1 \end{pmatrix}$$

yields

$$X^2 - cY^2 - fZ^2 = (1/\beta^2)(X'^2 + cY'^2 + f'Z'^2).$$

□

21.6 The Zariski problem

This problem was raised by Zariski at a Paris colloquium in 1949. It was first discussed, from a classical geometric standpoint, in a paper [**S**] of B. Segre, and then later (1967) from an algebraic standpoint in the paper [**N**] of Nagata.

Let K and L be fields finitely generated over a common subfield k, and let x, y be elements (in Ω) such that x is tr. over K and y tr. over L.

Zariski problem: *Does $K(x) = L(y)$ imply K is k-isomorphic to L?*

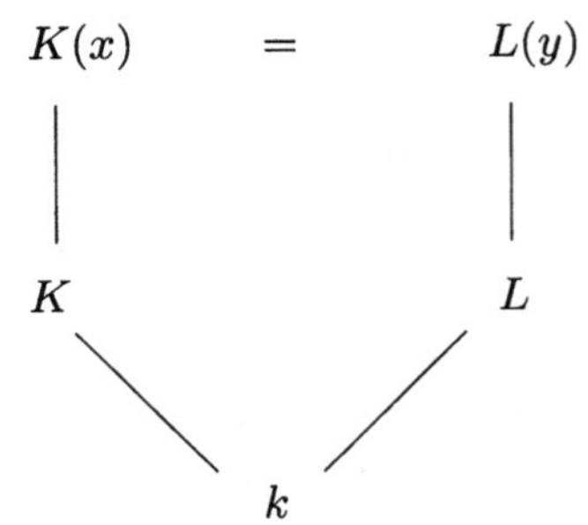

An equivalent formulation is: If X is an indeterminate, does $K(X)$ is k-isomorphic to $L(X)$ imply K is k-isomorphic to L? For a trivial example, one could take $K = k(y)$ and $L = k(x)$. If dt K/k (= dt L/k) $= n$, we shall refer to this as *the n-dim Zariski problem.*

The 0-dim Zariski problem has an affirmative answer, since then $K = L =$ algebraic closure of k in $K(x)$. As an application of 4.1, we want to prove

5.1 Theorem: *The 1-dim Zariski problem has an affirmative answer.*

Before giving the proof, let us make a few additional observations on the n-dim problem. The case that $x = y$ and k is infinite was shown by Samuel [**Sa1**], with a very simple argument, to have an affirmative answer. The 2-dim Zariski problem seems to still be an open question for arbitrary base field k, although an affirmative answer can be given for k algebraically closed of char 0 according to Nagata [**N**, p. 90]. The most important general result is the counterexample to the problem [**BCSS**], which we shall discuss shortly. There is also an *n-dim generalized Zariski problem*, where x is replaced by m algebraically independent elements over K and y by m

algebraically independent elements over L. (See [**O**] for some related considerations.)

We now present the proof of 5.1, following the outline of the paper [**D**] of Deveney. Assume the hypothesis of the 1-dim Zariski problem. First, by the following theorem, K is k-isomorphic to a subfield of L and L is k-isomorphic to a subfield of K.

5.2 Embedding theorem: *Let L be a finitely generated extension of a field k, let $y_1, \dots, y_m$ be algebraically independent elements over L, and let K be a subfield of $L(y_1, \dots, y_m)$ containing k. Then K is k-isomorphic to a subfield of L if (and only if) dt $K/k \leq$ dt L/k.*

(This result is due to Chevalley–Shimura for L/k pure tr. and k infinite, to Roquette for k infinite, and to Ohm for k arbitrary. Its proof is short and elementary; cf. [**O**]).

Theorem 5.1 will now be a consequence of Amitsur's theorem (cf. 4.1) provided we first verify that K/k is a regular function field of a conic. We may assume $K \neq L$, and therefore, by two applications of Lüroth's theorem, also that $K(x) = KL = L(y)$. By replacing k by $K \cap L$, we may further assume that $k = K \cap L$.

With these assumptions, K/k and L/k are regular. For, KL/K and KL/L are simple tr., hence separable; and this implies $KL/(K \cap L)$ is separable (cf. [**Wa**, p. 39] for a simple proof), and therefore also $K/(K \cap L)$ and $L/(K \cap L)$ are separable. Moreover, since K and L are algebraically closed in KL, $K \cap L$ is algebraically closed in KL.

Finally, since $LK/L = L(y)/L$ is simple tr., by 3.4 K/k (and by symmetry L/k) is a function field of a conic. □

We shall describe next the counterexample to the Zariski problem of [**BCSS**] and relate it to some of our previous considerations. The verification of the crucial properties of the example, however, requires extremely sophisticated arguments from algebraic geometry and is far beyond the scope of the present exposition.

A finitely generated field extension K/k_0 is called *rational* if K/k_0 is pure tr. In other words, K/k_0 is rational iff there exist $x_1, \dots, x_n$ in K such that $K = k_0(x_1, \dots, x_n)$ and $x_1, \dots, x_n$ are algebraically independent over k_0, or equivalently, iff if dt $K/k_0 = n$ and K/k_0 can be generated by n elements. An extension K/k_0 will be called *subrational* if there exists a rational extension K'/k_0 such that $K \subseteq K'$. (By 5.2, 'subrational' is equivalent to what is commonly called 'unirational'; cf. [**O**].)

5.3 Hypothesis for the remainder of §22.5: *Let k_0 be a field of characteristic different from 2, let t be an element tr. over k_0, let $f(t)$ be a* *nonzero polynomial in* $k_0[t]$, *and let* c *be a nonzero element of* k_0.

Then $P(X,Y) = X^2 - cY^2 - f(t)$ is a regular $k_0(t)$-conic and defines a regular function field $K/k_0(t)$ (cf. §22.3); and we can write $K = k_0(t)(x,y)$ with $P(x,y) = 0$. The field extension $K/k_0 = k_0(t,x,y)/k_0$ has dt 2 and is said to be 'a function field of a surface fibered by conics' (or 'the function field of a Chatelet surface'). Note that K may be alternatively described as the quotient field of $k_0[t,X,Y]/(X^2 - cY^2 - f(t))$, so K/k_0 is the function field of the k_0-polynomial $X^2 - cY^2 - f(t)$.

5.4 Proposition: *If deg* $f(t)$ *or subdeg* $f(t)$ *is odd* (> 0), *then the function field of* $X^2 - cY^2 - f(t)$ *over* $k_0(t)$ *is simple tr. (if and) only if* $\sqrt{c} \in k_0$.

PROOF: By 1.4 $X^2 - cY^2 - f(t)$ has a $k_0(t)$-zero (p, q), and we may assume p, q are in $k_0[t]$.

Suppose deg $f(t)$ is odd. Consider the equality $p^2 - cq^2 - f = 0$ as a polynomial equality in $k_0[t]$. Since the deg in t of f is odd, it follows that deg $p =$ deg q, and if α is the leading coefficient of p and β the leading coefficient of q, then $\alpha^2 - c\beta^2 = 0$.

If subdeg $f(t)$ is odd, a symmetric argument on the terms of least deg applies. □

For an application of 5.4, note that the following regular conics have function fields that are not simple tr.: $X^2 + Y^2 - t$, or $X^2 + Y^2 - t(t-1)$, or $X^2 + Y^2 - t(t^2 - 2)$, over $k_0(t)$ for any field k_0 which does not contain $\sqrt{-1}$.

If $\sqrt{c} \in k_0$, by 1.4 $K/k_0(t)$ is simple tr. and hence K/k_0 is rational. On the other hand, if deg $f(t) = 3$ and $\sqrt{c} \notin k_0$, then by 5.4 $K/k_0(t)$ is not simple tr., or equivalently, K/k_0 cannot be generated by two elements one of which is t. Nonetheless, K/k_0 might still be rational:

5.5 Theorem ([**BCSS**]): *Suppose* $f(t)$ *has deg 3, distinct roots, and discriminant* c *with* $\sqrt{c} \notin k_0$. *Then the function field* K/k_0 *of the* k_0*-polynomial* $X^2 - cY^2 - f(t)$ *is rational iff* $f(t)$ *has a root in* k_0.

Recall that the discriminant of a polynomial $f(t)$ of deg 3 with roots r_1, r_2, r_3 is the element of k_0 given by

$$[(r_3 - r_2)(r_3 - r_1)(r_2 - r_1)]^2;$$

and if $f(t)$ has the form $t^3 + f_1 t + f_0$, then the discriminant equals $-27f_0^2 - 4f_1^3$ (cf. [**Sa3**, p. 41]).

The counterexample to the Zariski problem is now obtained by choosing $f(t)$ to be irreducible of deg 3 with discriminant c such that $\sqrt{c} \notin k_0$. Then K/k_0 is not rational by 5.5, but it is proved in [**BCSS**] that if one adjoins to K three elements u, v, w algebraically independent over K, then

$K(u, v, w)/k_0$ *is* rational! (In geometric terminology, K/k_0 is 'stably rational' but not rational.) This yields a counterexample to the 2-dim *generalized* Zariski problem, and hence also a counterexample to the n-dim Zariski problem for some (unknown — to me) n such that $2 \leq n \leq 4$.

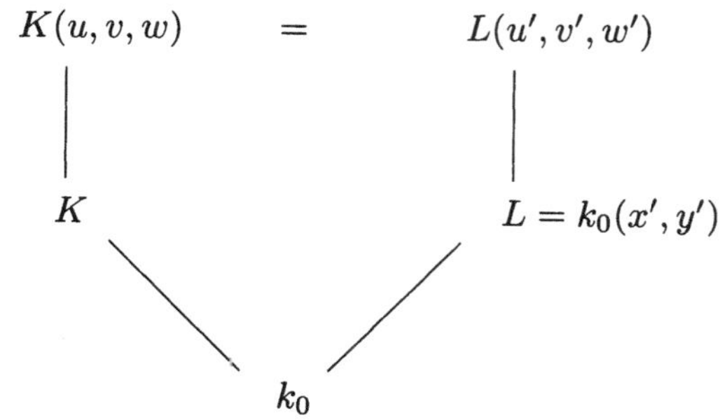

For a specific example, let k_0 = the rationals and $f(t) = t^3 - 2$. Then $c = -108$, and therefore $P(X, Y) = X^2 + 108Y^2 - (t^3 - 2)$. Note that this last expression can be simplified to $X^2 + 3Y'^2 - (t^3 - 2)$ by replacing $6Y$ by Y'.

Incidentally, the field k_0 of the example cannot be chosen algebraically closed, by 5.4. In fact, in view of the Castelnuovo-Zariski theorem mentioned in the introduction, such an example (dt 2 and subrational but not rational) cannot exist over an algebraically closed field k_0 of char 0. However, the authors of [**BCSS**] show that by choosing k_0 to be a simple tr. extension of, say, the complexes C and *then* applying the above construction one obtains a counterexample with base field C.

The "if" assertion of 5.5 is easy to prove:

PROOF (cf. [**Mo**]): Suppose r_1, r_2, r_3 are the roots of $f(t)$ and that r_3 is in k_0.

Since we know dt $k_0(t, x, y)/k_0 = 2$, in order to prove $k_0(t, x, y)/k_0$ is rational, it will suffice to prove there exist x', $y' \,\varepsilon\, k_0(t, x, y)$ such that $k_0(t, x, y) = k_0(x', y')$.

We can write $f(t) = \alpha(t - r_3)(t^2 - \beta t + \gamma)$, for some $\alpha, \beta, \gamma \,\varepsilon\, k_0$. By completing the square, we have

$$t^2 - \beta t + \gamma = (t - \beta/2)^2 - (\beta^2 - 4\gamma)/4.$$

But $\beta^2 - 4\gamma = (r_2 - r_1)^2 = c/[(r_3 - r_1)(r_3 - r_2)]^2 = c/(r_3^2 - \beta r_3 + \gamma)^2$, so then

$$t^2 - \beta t + \gamma = u(t)^2 - cv^2,$$

where $u = (t - \beta/2)$ and $v = 1/2(r_3^2 - \beta r_3 + \gamma)$. Thus,

$$\begin{aligned} x^2 - cy^2 &= \alpha(t - r_3)(u^2 - cv^2), \text{ or} \\ [1/(u^2 - cv^2)](x^2 - cy^2) &= \alpha(t - r_3). \end{aligned}$$

Now, by Lemma 2.13, there exists a $k_0(t)$-change of variables such that this last equation becomes

$$(x'^2 - cy'^2) = \alpha(t - r_3).$$

Therefore $t \varepsilon k_0(x', y') = k_0(t, x, y)$. □

What if deg $f(t) < 3$? That is, is it possible to find a nonrational, stably rational example like the above, but with deg $f(t) < 3$? The following proposition shows that the answer is *no.*

5.6 Proposition: *Under the hypothesis 5.3, if deg $f(t) = 0, 1$, or 2, then K/k_0 is subrational implies K/k_0 is rational.*

PROOF:

i) deg $f(t) = 0$, i.e., $f \varepsilon k_0$. Then $P(X,Y) = X^2 - cY^2 - f$, and $K = k_0(x, y, t)$ with t tr. over $k_0(x, y)$. But $k_0(x, y)/k_0$ is subrational implies $k_0(x, y)/k_0$ is simple tr., by the generalized Lüroth theorem (mentioned in the introduction – apply 5.2 plus Lüroth's theorem for a proof); so a fortiori $k_0(x, y, t)/k_0$ is subrational implies $k_0(x, y, t)/k_0$ is rational.

ii) deg $f(t) = 1$. Then $x^2 - cy^2 - (f_1t + f_0) = 0$, with $f_1 \neq 0$; so $t \varepsilon k_0(x, y)$ and $K/k_0 = k_0(x, y)/k_0$ is rational (without even using the subrational hypothesis).

iii) deg $f(t) = 2$. In this case we may assume $P(X,Y) = X^2 - cY^2 - (f_2t^2 + f_0)$, with $f_2 \neq 0$. If $f_0 = 0$, then $K = k_0(x/t, y/t, t)$, where $(x/t)^2 - c(y/t)^2 - f_2 = 0$, and we are in case (i); so we may assume $f_0 \neq 0$. Then 5.6 follows from

5.7 Lemma: *Let $a_0, a_1, \ldots, a_n$ ($n \geq 2$) be nonzero elements of a field k_0 of char $\neq 2$. Then $a_0 + a_1X_1^2 + \cdots + a_nX_n^2$ is irreducible in $k_0[X_1, \ldots, X_n]$, and therefore defines a function field K/k_0; and the following are equivalent:*

i) $a_0X_0^2 + a_1X_1^2 + \cdots + a_nX_n^2$ has a nontrivial k_0-zero.

ii) $a_0 + a_1X_1^2 + \cdots + a_nX_n^2$ has a k_0-zero.

iii) K/k_0 is rational.

iv) K/k_0 is subrational.

PROOF: The irreducibility assertion is an easy computation (in fact, it is the same computation as one makes for n = 2).

(*i*) ⇒ (*ii*): Let $(\alpha_0, \alpha_1, \ldots, \alpha_n)$ be the given nontrivial k_0-zero. If $\alpha_0 \neq 0$, then $(\alpha_1/\alpha_0, \ldots, \alpha_n/\alpha_0)$ is the required zero. Therefore it remains to consider the case that $\alpha_0 = 0$. By symmetry, we may assume $\alpha_1 = 1$. Since $a_1 + a_2\alpha_2^2 + \cdots + a_n\alpha_n^2 = 0$ and $a_1 \neq 0$, there exists $j \varepsilon \{2, \ldots, n\}$ such that $a_j\alpha_j \neq 0$. Substituting $a_1 = -(a_2\alpha_2^2 + \cdots + a_n\alpha_n^2)$ in $a_0 + a_1X_1^2 + \cdots + a_nX_n^2$, we obtain

$$a_0 + a_2(X_2^2 - \alpha_2^2X_1^2) + \cdots + a_j(X_j^2 - \alpha_j^2X_1^2) + \cdots + a_n(X_n^2 - \alpha_n^2X_1^2).$$

We claim there exist β_j and $\beta_1 \varepsilon k_0$ which satisfy the polynomial relation $a_0 + a_j(X_j^2 - \alpha_j^2 X_1^2) = 0$. This is because the linear equations

$$\begin{cases} X_j - \alpha_j X_1 &= -a_0/a_j \\ X_j + \alpha_j X_1 &= 1 \end{cases}$$

have coefficient determinant $2\alpha_j \neq 0$ and therefore have a solution. Now let $\beta_i = \alpha_i \beta_1$ for $i \varepsilon \{2, \ldots, \hat{j}, \ldots, n\}$. Then $(\beta_1, \ldots, \beta_n)$ is the required k_0-zero. (See, for example, [**Se1**, p. 33, Corollary 1] for a more conceptual proof of $(i) \Rightarrow (ii)$.)

$(ii) \Rightarrow (iii)$: Since dt $K/k_0 = n-1$, we must show that K/k_0 can be generated by $n-1$ elements. Write $K = k_0(x_1, \ldots, x_n)$, where $a_0 + a_1x_1^2 + \cdots + a_nx_n^2 = 0$, and let $(\alpha_1, \ldots, \alpha_n)$ be the k_0-zero given by (ii). Then

$$\begin{aligned} 0 &= a_0 + a_1x_1^2 + \cdots + a_nx_n^2 \\ &= a_1x_1'^2 + \cdots + a_nx_n'^2 - 2(a_1\alpha_1x_1' + \cdots + a_n\alpha_nx_n'), \end{aligned}$$

where $x_i' = x_i + \alpha_i$. If $x_i' = 0$ for some i, then $K = k_0(x_1, \ldots, \hat{x}_i, \ldots, x_n)$ and we are done, so we may assume $x_i' \neq 0$ for all i. Moreover, since $0 = a_0 + a_1\alpha_1^2 + \cdots + a_n\alpha_n^2$ and $a_0 \neq 0$, there exists a j such that $a_j\alpha_j \neq 0$. Therefore

$$\begin{gathered} 0 = [a_1(x_1'/x_j')^2 + \cdots + a_j + \cdots + a_n(x_n'/x_j')^2] - \\ 2(1/x_j')[a_1\alpha_1(x_1'/x_j') + \cdots + a_j\alpha_j + \cdots + a_n\alpha_n(x_n'/x_j')]. \end{gathered}$$

If $[a_1\alpha_1(x_1'/x_j') + \cdots + a_j\alpha_j + \cdots + a_n\alpha_n(x_n'/x_j')] \neq 0$, then we can solve for $1/x_j'$ and are done. If, on the other hand, this expression equals 0, then we can similarly use it to solve for x_j' in terms of the other (x_i')'s and are again done.

$(iii) \Rightarrow (iv)$: Trivial.

$(iv) \Rightarrow (i)$: It is easily seen that any rational extension of k_0 has a valuation (of rank equal to the transcendence deg) whose residue field is k_0, and therefore the same is true, by restriction, for a subrational extension K/k_0. If we divide through the equality $0 = a_0 + a_1x_1^2 + \cdots + a_nx_n^2$ by an x_i of least value for this valuation and apply the residue map to the resulting expression, we obtain the desired k_0-zero. □

It is to be hoped that the counterexample to the Zariski problem will someday achieve the elementary status of the rest of this article, either by a simplification of its proof or by a shift in the meaning of 'elementary'.

21.7 Bibliographic remarks and References

There are some up-to-date surveys available for the problems mentioned in the introduction: for example, [**MT**] and [**Co2**]; both are written in the language of algebraic geometry (as are many of the papers on this subject) and contain extensive bibliographies. Swan's paper [**S**] gives an expert account of Noether's problem and its relationship to galois theory, and [**G**] covers related topics. References for generalizations of Lüroth's theorem can be found in [**O**], which treats the foundations of the theory of ruled and rational field extensions of arbitrary transcendence degree and may be considered a companion piece to the present work.

In 1982 Amitsur [**A2**] returned to his 1955 paper in order "to present the major ideas of the old 1955 paper, in a somewhat more modern language." See also [**A3**] and its bibliography.

For a sampling of the powerful methods of Colliot-Thélène and his collaborators, one could start with [**Co1**] and the short, informative paper [**Cor**], or trace backward from the bibliography of [**Co2**] or of the monumental [**CSS**].

To get an idea of the connections between function fields of conics and some other areas, I recommend the following:

i) algebraic K-theory—Wadsworth's admirable exposition [**W**] of the recent breakthrough theorem of Merkurjev. See also, for example, [**R**] and [**Su**].

ii) quaternion algebras, central simple algebras, and the Brauer group—the classics [**L**] and [**Se1**], and also the detailed recent text [**Ke**] of Kersten.

iii) Galois cohomology and Severi-Brauer varieties—Chapter X of [**Se2**], for a broad overview.

iv) Computing—a series of very readable introductory papers of Abhyankar-Bajaj; see [**AB**] for references.

21.8 References

[**AB**] S.S. Abhyankar and C. Bajaj, Computations with algebraic curves; Symbolic and algebraic computation, ISSAC '88; Lecture notes in computer science, No. 358, Springer, 1989, pp. 274-284.

[**A1**] S.A. Amitsur, Generic splitting fields of central simple algebras, Annals of Math. 62 (1955), 8-43.

[**A2**] S.A. Amitsur, Generic splitting fields; Brauer groups in ring theory and algebraic geometry (F. van Oystaeyen and A. Verschoren editors), Lecture Notes in Math. 917, Springer, Berlin, 1982, pp. 1-24.

[**A3**] S.A. Amitsur, Division algebras – a survey, Contemporary Math. 13 (1982), 3- 26.

[**Ar**] E. Artin, Algebraic numbers and algebraic functions, Gordon and Breach, New York, 1967.

[**BCSS**] A. Beauville, J.-L. Colliot-Thélène, J.-J. Sansuc, and Sir P. Swinnerton-Dyer, Varietes stablement rationnelles non rationnelles, Annals of Math.

121 (1986), 283-315.

[C] C. Chevalley, Introduction to the theory of algebraic functions of one variable, Math. Surveys 6, American Math. Soc., New York, 1951.

[Co1] J.-L. Colliot-Thélène, Les grands themes de Francois Chatelet, L'Enseignement Math. 34 (1988), 387-405.

[Co2] J.-L. Colliot-Thélène, Arithmetic des varietes rationnelles et problemes birationnels, Proc. Int. Congress Math. Berkeley, 1986, 641-653.

[CSS] J.-L. Colliot-Thélène, J.-J. Sansuc, and Sir P. Swinnerton-Dyer, Intersections of two quadrics and Chatelet surfaces, J. fur die reine u. angew. Math.; Part I, vol. 373 (1986), 37-107; Part II, vol. 374 (1986), 72-165.

[Cor] D. Coray, The Hasse principle for pairs of quadratic forms, in: Journees arithmetiques 1980, ed. J.V. Armitage, London Math. Soc. Lecture Notes Series 56 (1982), 237-246.

[D] J. Deveney, Ruled function fields, Proc. Amer. Math. Soc. 86 (1982), 213-215.

[E] M. Eichler, Introduction to the theory of algebraic numbers and functions, Academic Press, New York, 1966.

[F] M. Fried, A note on principal ideals and smooth curves, J. of Algebra 74 (1982), 124-139.

[G] W.-D. Geyer, The automorphism group of the field of all algebraic numbers, Atas da 5a escola de algebra, IMPA vol. 11, Rio de Janeiro, 1978, 167-199.

[K] I. Kaplansky, Linear algebra and geometry, Chelsea, New York, 1974.

[Ke] I. Kersten, Brauergruppen von Korpern, Aspekte der Math. Band D6, Vieweg & Sohn, Braunschweig/Wiesbaden, 1990.

[Kn] M. Knebusch, Generic splitting of quadratic forms I, Proc. London Math. Soc. 33 (1976), 65-93; II, ibid. 34 (1977), 1-31.

[L] T.Y. Lam, The algebraic theory of quadratic forms, Benjamin, Reading, Mass., 1980.

[LT] S. Lang and J. Tate, On Chevalley's proof of Lüroth's theorem, Proc. Amer. Math. Soc. 3 (1952), 621-624.

[M] R.E. MacRae, On rational points on conics, Proc. Amer. Math. Soc. 67 (1977), 38-40.

[MS] R.E. MacRae and P. Samuel, Subfields of index 2 of elliptic function fields, Conference on commutative algebra, Lawrence, Kansas 1972, 171-193; Lecture notes in math. 311, Springer, Berlin, 1973.

[MT] Yu.I. Manin and M.A. Tsfasman, Rational varieties: algebra, geometry and arithmetic, Russian Math. Surveys 41:2 (1986), 51-114.

[MH] T.T. Moh and W. Heinzer, On the Lüroth semigroup and Weierstrass canonical divisors, J. of Algebra 77 (1982), 62-73.

[Mo] L. Moret-Bailly, Varietes stablement rationnelles non rationnelles, Sem. Bourbaki, 1984-85, no. 643, 14 pp.

[N1] M. Nagata, A theorem on valuation rings and its applications, Nagoya Math. J. 29 (1967), 85-91.

[N2] M. Nagata, Field theory, Dekker, New York, 1977.

[O] J. Ohm, On ruled fields, Sém. de théorie des nombres, Univ. Bordeaux 1 (1988-1989), 27-49.

[R] L. Roberts, K_1 of a curve of genus zero, Trans. Amer. Math. Soc. 188 (1974), 319-324.

[Sa1] P. Samuel, Some remarks on Lüroth's theorem, Memoirs of College of Sci. of Univ. of Kyoto 27 (1953), 223-224.

[Sa2] P. Samuel, Lectures on old and new results on algebraic curves, Tata Institute, Bombay, 1964.

[Sa3] P. Samuel, Algebraic theory of numbers, Hermann, Paris, 1970.

[S] B. Segre, Sur un problem de M. Zariski, Coll. international d'algebre et de theorie des nombres (Paris, 1949) 135-138, C.N.R.S., Paris, 1950.

[Se1] J.-P. Serre, A course in arithmetic, Springer, New York, 1973.

[Se2] J.-P. Serre, Local fields, Springer, New York, 1979.

[Su] A.A. Suslin, The quaternion homomorphism for the function field on a conic, Soviet Math. Doklady 26 (1982), 72-77.

[Sw] R. Swan, Noether's problem in galois theory; Emmy Noether in Bryn Mawr (B. Srinivasan and J. Sally, editors), Springer, 1983, pp. 21-40.

[T] J. Tate, Genus change in inseparable extensions of function fields, Proc. Amer. Math. Soc. 3 (1952), 400-404.

[vdW] B.L. van der Waerden, Modern algebra I, Unger, New York, 1949.

[W] A. Wadsworth, Merkurjev's elementary proof of Merkurjev's theorem, Contempory Math. Vol. 55, Part II (1986), 741-774.

[W1] A. Wadsworth, Letter, May 6, 1990.

[W2] A. Wadsworth, Similarity of quadratic forms and isomorphism of their function fields, Trans. Amer. Math. Soc. 208 (1975), 352-358.

[Wa] W. Waterhouse, The structure of inseparable field extensions, Trans.

Amer. Math. Soc. 211 (1975), 39-54.

[We] A. Weil, Foundations of algebraic geometry, Amer. Math. Soc. Colloq. Pub. 29, Providence, 1962.

[ZSI] O. Zariski and P. Samuel, Commutative algebra, vol. I, van Nostrand, Princeton, 1955.

22

Gradings of Polynomial Rings

Peter Russell

22.1 Introduction

These brief notes have their origin in a conversation I had with Abhyankar during a pleasant walk along the Wabash a few years ago. While I tried to explain some ideas on *torus actions on affine spaces*, Abhyankar, in that inimitable way he has to concentrate one's mind, feigned innocence as to such fancy notions, but was quite happy to listen when I proposed talking about *gradings of polynomial rings* instead. The two topics cover exactly the same ground from two different points of view, equally valid and valuable. Torus actions also provide an uncomplicated introduction to the larger subject of reductive group actions. I gave my conference talk hoping that such an introduction in the readily understood language of gradings would be appreciated by some of the non-experts in the rather diverse audience Abhyankar's birthday was bound to bring together. This is a slightly expanded version of my talk. It is entirely expository in nature and not meant for the specialists.

22.2 The Question

We fix an algebraically closed field k (of arbitrary characteristic for the basic results, but usually restricted to characteristic 0 when the going gets tough). Let

$$A = k^{[n]}$$

be a k-algebra isomorphic to the polynomial algebra in n variables over k. This pedantic way of putting things is meant to indicate that one should avoid thinking of A as coming endowed with a specific set of variables. Let

$$X = \mathbf{A}_k^n$$

be affine n-space over k,

$$\begin{aligned} \mathbf{G}_m &= \text{multiplicative group of } k, \\ G &= \mathbf{G}_m^r, \text{ the } r-\text{dimensional torus.} \end{aligned}$$

An *action* of G on X is given by a regular map (morphism)

$$\begin{array}{llll} \varphi : G \times X & \longrightarrow & X \\ (g,x) & \longrightarrow & gx \end{array} \tag{1.1}$$

with the usual properties: $g(g'x) = (gg')x$ $(g, g' \in G, x \in X)$, $ex = x$ ($e =$ neutral element of $G, x \in X$).

In algebraic terms this translates into a homomorphism

$$\begin{array}{llll} \alpha = \varphi^* : A & \longrightarrow & A[t_1, t_1^{-1}, ..., t_r, t_r^{-1}] \\ f & \longmapsto & \sum f_\gamma t^\gamma \end{array} \tag{1.2}$$

Here $\gamma = (\gamma_1, \cdots, \gamma_r)$ ranges over $\mathbf{Z}^r, t = (t_1, \cdots, t_r)$ is a system of coordinates for G and, by convention, $t^\gamma = t_1^{\gamma_1} \cdots t_r^{\gamma_r}$.

The fact that φ is an action is readily seen to be equivalent to

$$\text{for } f \in A \text{ and } \delta \in \mathbf{Z}^r, \alpha(f_\delta) = f_\delta t^\delta, \text{ and } \sum_{\gamma \in \mathbf{Z}^r} f_\gamma = f. \tag{1.3}$$

We therefore have

$$A = \bigoplus_{\gamma \ in \mathbf{Z}^r} A_\gamma, \text{ with } A_\gamma = \{f_\gamma | f \in A\} \quad \text{a } k-\text{subspace of } A \text{ and,}$$
$$\text{for } \gamma, \delta \in \mathbf{Z}^r, \quad A_\gamma A_\delta \subset A_{\gamma+\delta}. \tag{1.4}$$

This, by definition, means that we have a $\mathbf{Z}^r$-grading of A. As usual, elements of A_γ will be called homogeneous of weight γ. Conversely, given 1.4, we use 1.2 to define an action of G on X. So if we fix variables $x_1, ..., x_n$ for A and put $F_i(x,t) = \alpha(x_i) = \sum x_{i\gamma} t^\gamma$, we have $\tau\xi = (F_1(\xi, \tau), ..., F_n(\xi, \tau))$ for $\tau = (\tau_1, ..., \tau_r) \in G$, $\xi = (\xi_1, ..., \xi_n) \in X$.

Familiar gradings of A, which we call *standard*, are those in which a set $x_1, ..., x_n$ of variables is given and each x_i is homogeneous of a certain weight γ_i. We pose the basic

> **QUESTION:** Are all gradings of A standard? That is, given a grading of A, is A isomorphic, as a graded algebra, to $k[x_1, \cdots, x_n]$ with a standard grading, or, is A generated by *some* set of homogeneous variables?

In terms of torus actions on X, we are asking whether all such are linearizable, that is, given by an action on the linear forms in some set of variables. (This is enough, since such an action is then diagonalizable.) Not much is known about our question. We will provide some answers in special cases below.

Let us remark that A is, of course, generated as a k-algebra by finitely many homogeneous elements and hence a quotient, in the graded sense, of a standardly graded polynomial algebra. This fact is only mildly useful.

22.3 Homogeneous Maximal Ideals

Let $A = \bigoplus_{\gamma \in \mathbf{Z}^r} A_\gamma$ be a grading. We put

$$\Gamma = \{A_\gamma \mid A_\gamma \neq 0\}. \tag{2.1}$$

Γ is a semi-group, called the semi-group associated with the grading. We will assume henceforth that Γ generates $\mathbf{Z}^r$ or, equivalently, that the corresponding action is effective (only the neutral element $e \in G$ acts trivially on all of X). We put

$$A^+ = \bigoplus_{\gamma \neq 0} A_\gamma \quad \text{and } I^+ = A^+ A. \tag{2.2}$$

Of particular interest to us are homogeneous (or graded) ideals of A. Clearly these correspond to G-invariant closed subsets of X, and in particular *homogeneous maximal* ideals are in one - one correspondence with the fixpoints for G. One sees easily

Lemma: *Any ideal containing A^+ is homogeneous. Conversely, a homogeneous maximal ideal contains A^+.*

Corollary: $V(I^+)$ *(the set of zeros of I^+) is the set of fixpoints for the action of G on X.*

A basic non-trivial result is

Theorem [3]: *I^+ is a proper ideal.*

If $k = \mathbf{C}$, there is a more precise result. (A formulation is possible also in positive characteristic).

Theorem (see [4]): *$V(I^+)$ has the homology of a point, and hence is non-empty.*

Known proofs of these results proceed by noting that $V(I^+)$ is also the fixpoint set for the action on X of a suitably large subgroup of G of q-power order, q a prime. (If $y_1, ..., y_s$ are homogeneous generators for I^+, with y_i of weight $\gamma_i = (\gamma_{i1}, ..., \gamma_{in})$, it suffices to choose q^m larger than $\max\{\gamma_{ij}\}$ and to consider elements in G whose order divides q^m.) One method, due to Shafarevich (see [5]) proceeds by noting that the relation $1 \in I^+$ already holds over a ring R finitely generated over the prime ring of k, hence, after reduction modulo a suitable maximal ideal of R, over a finite field K. Moreover, we may choose q prime to char K. A straight-forward counting argument now leads to a contradiction. A second approach is via Smith theory (see [4]), which gives precise information (on any manifold) on the mod q-homology of the fixpoint set of an automorphism of prime order q.

Let M be a homogeneous maximal ideal. Both

$$\begin{aligned} A' &= gr_M A = \bigoplus_m M^m/M^{m+1} \text{and} \\ \hat{A} &= M-\text{adic completion of } A \end{aligned}$$

inherit a grading from A, readily seen to be *standard.* (M/M^2 is a direct sum of homogeneous pieces.) Moreover, their semi-groups are the same as that for A (see [7]). We may choose variables $x_1, ..., x_n$ for A in M such that $\xi_i = x_i + M^2$ is homogeneous, of weight γ_i say, and lift the ξ_i to homogeneous variables η_i of the same weight in $\hat{A}$. Then clearly

$$\Gamma \text{ is generated by } \gamma_1, ..., \gamma_n. \tag{2.7}$$

Also, $I^+\hat{A}$ is generated by the η_i with weight $\gamma_i \neq 0$, and it follows that $V(I^+)$ is non-singular. (This is a quite general property for actions of reductive groups on non-singular varieties [5]). Since $V(I^+)$ is connected by 2.6, we now find

Theorem: I^+ *is prime. It would be nice to prove this in a less roundabout way. To see that the result is, most likely, quite delicate one may look at the following two easy, but I think in no way artificial, examples:*

(i) $R = k[x, y, z]$ with $xy = z(z-1)$ and **Z**-graded such that x has weight 1, y weight -1 and z weight 0. Here $V(I^+)$ has two elements.

(ii) $R = k[x, y, u, v]$ with $xy - uv = 1$ and **Z**-graded such that x, u have weight 1 and y, v have weight -1. Now $V(I^+)$ is empty.

22.4 Maximal Homogeneous Ideals

A *maximal homogeneous ideal* P, i.e. an ideal P maximal among homogeneous ideals, is easily seen to be prime. These correspond to minimal closed G-invariant subsets of X (necessarily irreducible), the *closed orbits.* In fact, $V(P)$ is the G-orbit of any point in it. However, in general orbits need not be closed: If M is maximal and P maximal among homogeneous ideals in M, then $V(P)$ is the closure of the orbit of the point $V(M)$. It is possible however, that there is a homogeneous $P' \supsetneq P$, $P' \not\subset M$.

We note that A_0 is a sub k-algebra of A. In terms of the group action, A_0 is the sub-algebra of G-invariant functions in A. One defines the *categorical quotient* X/G by requiring that A_0 be its algebra of regular functions. Since $A_0 \hookrightarrow A$ is "universal" for graded homomorphisms of trivially graded k-algebras into A, this means no more and no less than that $X \to X/G$ is "universal" for equivariant regular maps from X to varieties with trivial G-action. The geometric meaning of A_0, or X/G, is particularly conspicuous in our case. For one sees immediately

Lemma:

(i) If $I \subset A_0$ is an ideal, then $IA \cap A_0 = I$.

(ii) If $P \subset A$ is a maximal homogeneous ideal, then $P \cap A_0$ is maximal.

(iii) If P, P' are maximal homogeneous in A and $P \cap A_0 = P' \cap A_0$, then $P = P'$.

Hence the maximal ideals of A_0 are in one - one correspondence with the maximal homogeneous ideals in A, i.e. X/G parametrizes the closed orbits in X. (This is a general property of reductive group actions on affine varieties.) As a typical example, consider $A = k[x, y, z]$, $\mathbf{Z}$-graded with x, y of weight 1 and z of weight -1. Then $A_0 = k[xz, yz]$ and the maximal homogeneous ideals are $M = (x, y, z)A$ and the ideals $P_{\alpha,\beta} = (xz - \alpha, yz - \beta)A$, $\alpha, \beta \in k$ and not both zero. Note that if $J \subset A$ is an ideal and $J \supset zA$ or $J \supset (x, y)A$, then $J \cap A_0 = M \cap A_0$, i.e. the quotient map $X \to X/G$ contracts the (x, y)-plane and the z-axis to a point.

A_0 is obviously normal. Again, if $k = \mathbf{C}$, one has an important topological result.

Theorem [14]: *X/G is contractible.*

COROLLARY [14]: If dim $A_0 = 1$, then A_0 is a polynomial algebra.

Let us note that X/G may be singular if $n \geq 3$.

22.5 Some Answers

It follows immediately from 2.7 that

Proposition: $r \leq n$. This is a rather special property of tori. It is, for instance, possible to find effective actions of the additive group $\mathbf{G}_a^r$ on, say, $\mathbf{A}_k^2$ for any $r \geq 0$.

It turns out that gradings are the harder to study the larger the codimension $n - r$. (In a sense this also extends to the case $r = 0$. It has proved useful, even in investigating gradings of polynomial algebras, not to assume that A is polynomial, but rather to deduce this from the fact that it is graded, and other properties (see [10], [11], for instance). The philosophy then is that the larger the codimension, the more "other" properties are needed.) The earliest positive answers to our question were given by

Theorem [3]: *If $r = n$ or $r = n - 1$, then the grading is standard.*

For codimension 2 there is only a partial result.

Theorem [11]: *Suppose char $k = 0$ and $r = n - 2$. Assume in addition that I^+ is not maximal. Then the grading is standard.*

The proof is by induction n and based on 4.2 as well as the Epimorphism Theorem of Abhyankar and Moh [1]. (This is the reason why characteristic zero is needed). This theorem appears as the case $n = 2$ of the following result, of some interest even if the grading is known to be standard.

Lemma [11]: *Suppose char* $k = 0$ *and* $r \geq n - 2$. *Let* $F \in A$ *be homogeneous. Suppose* A/F *is a polynomial algebra over* k *and* $I^{+}(A/F)$ *is not maximal. Then* F *is a variable in* A *(i.e.* $A = k[F, F_2, ..., F_n]$ *for some* $F_2, ..., F_n \in A$*).*

Another special result starts with 3.3.

Theorem [13]: *Suppose char* $k = 0$ *and dim* $A_0 = 1$. *Then the grading is standard.*

In a quite different line of investigation one puts restrictions not on A_0 (hence X/G) or I^{+} (hence the fixpoint set), but on the semi-group Γ of the grading.

Definition — Lemma: *The grading* $A = \bigoplus A_\gamma$ *is unmixed if it satisfies the following equivalent condition (see [7]):*

(i) $A_\gamma \neq 0 \Rightarrow A_{-\gamma} = 0$.

(ii) After suitable choice of basis in $\mathbf{Z}^r$, *we have* $\Gamma \subset \mathbf{Z}^r_+$.

(iii) A^+ *is an ideal and hence* $A_0 \simeq A/I^+$.

One shows by a quite straightforward argument, using a *graded Nakayama lemma* [7], that, in the case of an unmixed grading, A is the symmetric algebra of a finitely generated projective module P over A_0. Since A_0 is a retract of the polynomial algebra A, the Quillen-Suslin results [16],[18] apply, i.e. P is free. Hence

Theorem [7]: *In the case of an unmixed grading,* A *is a polynomial algebra over* A_0. *This shows that in the unmixed case, our question is equivalent to the famous*

Cancellation problem: If A is as polynomial algebra over a subalgebra A_0 of A, is A_0 a polynomial algebra over k?

This is true for dim $A_0 \leq 2$ (see [6],[15],[17]), but nothing else appears to be known about this problem. But we have

Corollary: An unmixed grading with dim $A_0 \leq 2$ is standard.

The last condition of 4.5 means that the only closed orbits are fixpoints. Such actions are called *fixpointed*, and 4.7 has been generalized in [2] for such actions in the case of arbitrary reductive groups.

It is clear from the above that all gradings of $k^{[1]}$ and $k^{[2]}$ are standard, as are those, in case char $k = 0$, of $k^{[3]}$ that are of codimension 1 or 0, or are unmixed, or have non-maximal I^{+}. The "next" case, that of a *mixed* **Z**-grading of $k^{[3]}$, with $I^{+} = M$ maximal has, of course, received a lot of attention. Let us introduce variables

$$z, x, y \text{ with } M = (x, y, z)A$$

and z, x, y homogeneous mod M^2. Then all weights (mod M^2) of z, x, y are non-zero, and we may assume that they are

$$-a, b, c \text{ with } a, b, c > 0.$$

Suppose for a moment that z, x, y are homogeneous of the above weights. Consider the group ω of a-th roots of unity acting on x and y with weights b and c mod a and let

$$B = k[x, y]^{\omega}$$

be the algebra of invariants. Then, as one sees easily, $A_0 \simeq B$. A bit artificially, maybe, in terms of the grading, we can think of A as $\mathbf{Z}/a\mathbf{Z}$-graded by collecting together homogeneous pieces A_γ with indicies congruent mod a. Then $z - 1$ is homogeneous and we have $A_0 \simeq (A/z-1)_0$. It turns out that if we let $F \in A_{-a}$ be a generator for I^{-}, the ideal generated by the A_γ with $\gamma < 0$, then in general (see [9])

$$A/F \simeq k^{[2]} \text{ and } A_0 \simeq (A/F-1)_0.$$

An "epimorphism theorem" is, so far, lacking in dimension 3, and one does not have much information on the zero locus of $F - 1$. However, the completion of A_0 at $A_0 \cap M$ is isomorphic to that of B at the origin, i.e. is the complete local ring of a cyclic quotient singularity. All other localizations of A_0 are regular, and the field of quotients of A_0 is purely transcendental over k. The aim then would be to

(i) find good conditions under which an algebra A_0 with such properties is isomorphic to the "standard" global model B (see [8]), and

(ii) to conclude from $A_0 \simeq B$ that $A/F-1$ is isomorphic to $k^{[2]}$.

Remark: There are no difficulties with step (ii) in case a, b, c are pairwise relatively prime (since then the action of ω on the zero locus of $F - 1$ has an isolated fixpoint). It is claimed in [10], proposition 2.3, that reduction to this case is always possible. We would like to take this opportunity, however, to point out that the argument given there has a gap, and that this question remains open.

The following is part of a more general result too technical to explain here in full.

Proposition (see [10],[12]): *Suppose that a, b, c are pairwise relatively prime and char $k = 0$. Then the grading is standard in the following cases:*

(i) $b \equiv -c \bmod a$, in particular for $a \leq 2$,

(ii) $b \equiv c \bmod a$ and $a \geq 9$.

22.6 References

[1] Abhyankar, S.S. and Moh, T.T., Embeddings of the line in the plane, J. Reine Angew. Math. 276 (1975), 148-166.

[2] Bass, H. and Haboush, W., Linearizing certain reductive group actions, Trans. Am. Math. Soc. Vol.292, No.2, (1985), 463-481.

[3] Bialynicki-Birula, A., Remarks on the action of an algebraic torus on k^n, I and II, Bull. Acad. Pol. Sci., 14 (1966), 177-181 and 15 (1967), 123-125.

[4] Floyd, E.E., On periodic maps and the Euler characteristic of associated spaces, Trans. Am. Math. Soc. 72 (1952), 138-147.

[5] Fogarty, J., Fixed point schemes, Amer. J. Math. 95 (1973), 35-51.

[6] Fujita, T., On Zariski problem, Proc. Japan Acad., 55A (1979), 106-110.

[7] Kambayashi, T. and Russell, P., On linearizing algebraic torus actions, J. Pure and Applied Algebra 23 (1982), 243-250.

[8] Koras, M., A characterization of $\mathbf{C}^2/\mathbf{Z}_a$, to appear.

[9] Koras, M. and Russell, P., $\mathbf{G}_m$-actions on $\mathbf{A}^3$, Canadian Math. Soc. Conference Proceedings, Vol. 6 (1986), 269-276.

[10] Koras, M. and Russell, P., On linearizing "good" $\mathbf{C}^*$-actions on $\mathbf{C}^3$, Canadian Math. Soc. Conference Proceedings, Vol.10 (1989), 93-101.

[11] Koras, M. and Russell, P., Codimension 2 torus actions on affine n-space, Canadian Math. Soc. Conference Proceedings, Vol. 10 (1989), 103-110.

[12] Koras, M. and Russell, P., Mixed $\mathbf{C}^*$-actions on $\mathbf{C}^3$ with isolated fixpoint, in preparation.

[13] Kraft, H. and Schwarz, G., Reductive group actions on affine space with one-dimensional quotient, Canadian Math. Soc. Conference Proceedings, Vol.10 (1989), 125-132.

[14] Kraft, H., Petrie, T. and Randall, J., Quotient varieties, Adv. Math. 74, No. 2(1989), 145-162.

[15] Miyanishi, M. and Sugie, T., Affine surface containing cylinderlike open sets, J. Math. Kyoto Univ. 20 (1980), 11-42.

[16] Quillen, D., Projective modules over polynomial rings, Invent. Math. 36 (1976), 167-171.

[17] Russell, P., On affine-ruled rational surfaces, Math. Ann. 255 (1981), 287-302.

[18] Suslin, A.A., On projective modules over polynomial rings, Math. USSR Sbornik 22 (1974) 595-602.

23 Rigid Hilbert Polynomials for m-Primary Ideals

Judith D. Sally

23.1 Introduction

Which Hilbert polynomials for an m-primary ideal I in a d-dimensional, ($d > 0$), local Cohen-Macaulay ring (R, m) determine Heilbert function of I? For example, if we denote the Hilbert function giving the length of R/I^n by $H_I(n)$ and the corresponding polynomial by $p_I(X)$, then any m-primary ideal I having Hilbert polynomial

$$p_I(X) = \lambda \binom{X+d-1}{d}, \quad \text{for some positive integer } \lambda,$$

has Hilbert function $H_I = \lambda \binom{n+d-1}{d}$ for all $n > 0$ and, in addition, I must be generated by d elements.

In this note we give some more examples of Hilbert polynomials which determine their Hilbert function and, in addition, force I to have some nice properties. We begin by extending a definition [1] of Elias and Valla to m-primary ideals. Recall that a polynomial function is a function $F : Z \to Z$ such that there is a polynomial $p_F(X) \in Q[X]$ with $p_F(n) = F(n)$ for $n \gg 0$ in Z.

Definition. A polynomial function F is *rigid* for a class C of M-primary ideals in a class of local Cohen-Macaulay rings if, for all $I \in C$ with $p_I = p_F$, it follows that $H_I = F$. A polynomial $p(X) \in Q[X]$ is rigid for C if there is a polynomial function F rigid for C with $p_F(X) = p(X)$.

Note that the definition of rigid polynomial does not necessarily imply that $p_F = F$.

Thus the polynomial $\lambda \binom{X+d-1}{d}$ is rigid for the class C of m-primary ideals in the class of d-dimensional local Cohen-Macaulay rings. In addition, any I in C with $p_I(X) = \lambda \binom{X+d-1}{d}$ is generated by d elements and has associated graded ring, $gr_I R = R/I \oplus I/I^2 \oplus I^2/I^3 \oplus \ldots$, a polynomial ring in d variables over R/I.

The polynomial $7X - 8$ is not rigid for maximal ideals in 1-dimensional local Cohen-Macaulay rings. For, if k is a field, the maximal ideal m in

$$k[[t^7, t^8, t^9, t^{10}, t^{11}]]$$

has $p_m(X) = 7X - 8$ and $H_m(2) = 6$. The maximal ideal m' in

$$k[[t^7, t^9, t^{15}, t^{17}, t^{19}, t^{20}]]$$

has $p_{m'}(x) = 7X - 8$ and $H_{m'}(2) = 7$.

Before giving other examples of rigid polynomials, we recall some facts about the coefficients of Hilbert polynomials We will denote the length of an R module A by $\lambda(A)$. When we say that an m-primary ideal I has length λ, we mean $\lambda(R/I) = \lambda$. As is well known, $p_I(X)$ can be written in the form

$$\begin{aligned} p_I(X) &= e_0 \binom{X+d-1}{d} - e_1 \binom{X+d-2}{d-1} \\ &+ e_2 \binom{X+d-3}{d-2} - \ldots + (-1)^d e_d \end{aligned}$$

where the e_i are in Z. e_0 is, of course, the multiplicity of I. Northcott [6] proved that

$$e_1 \geq e_0 - \lambda(R/I) \geq 0.$$

Narita [5] proved that $e_2 \geq 0$. It is known that the higher e_i's can be negative if no additional hypotheses are put on i, cf. [13].

Work of Huneke [8] and, independently, Ooishi [7] give another rigid polynomial which, in addition, forces nice behavior of I. They proved

(I) The condition $e_1 = e_0 - \lambda(R/I)$ implies that

(i) $e_i = 0$ for $i \geq 2$

(ii) $H_I(n) = p_I(n)$ for all $n \geq 1$

(iii) $gr_I R$ is Cohen-Macaulay.

Thus the polynomial

$$p_I(X) = e_0 \binom{X+d-1}{d} - (e_0 - \lambda) \binom{X+d-2}{d-1} \qquad (*)$$

is rigid for the class $C_{\lambda,d}$ of m-primary ideals of length λ in d-dimensional local Cohen-Macaulay rings.

There is a companion result [8] to that of Huneke and Ooishi. Namely,

(II) The conditions $e_1 = e_0 - \lambda(R/I) + 1$ and $e_2 \neq 0$, if $d > 1$, imply that

(i) $e_2 = 1$, if $d > 1$, and $e_i = 0$ for $i > 2$

(ii) $H_I(n) = p_I(n)$ for all $n > 1$

(iii) depth $gr_I R \geq d - 1$.

Thus the polynomial

$$p_I(X) = e_0 \binom{X+d-1}{d} - (e_0 - \lambda + 1) \binom{X+d-2}{d-1} + \binom{X+d-3}{d-2} \qquad (**)$$

is rigid for $C_{\lambda,d}$. The nonrigid polynomial $7X - 8$ shows that the condition $e_1 - e_0 - \lambda(R/I) + 2$ does not give rigid polynomials for $C_{1,1}$.

23.2 Rigidity of Polynomials

Actually, the rigidity of the polynomials (*) and (**) is easy to prove and is a much weaker result that (I) and (II). The rigidity of (*) and (**) follows from the extension to m-primary ideals of an argument which Elias and Valla give in [1].

Recall that if F is a numerical function, then ΣF is defined by $(\Sigma F)(n) = \sum_j^n = 1\, F(j)$ and ΔF by $\Delta F(n) = F(n) - F(n-1)$. By [9], L3, L4 pp. 285-286, if x is a superficial element for I which is also a regular element, then

1. $H_{I/xR}(n) - \Delta H_I(n) = \lambda(I^n : xR)$ for all $n \geq 1$
2. $H_{I/xR}(n) = \Delta H_I(n)$ for $n \gg 0$
3. $H_{I/xR}(n) = \Delta H_I(n)$ for all $n \geq 1$ if and only if x maps to a nonzero divisor in $gr_i R$.

A result analogous to (1.1) in [1] easily follows. Here we state exactly what is needed in our context.

Proposition. Let F be a numerical function which is rigid for the class $C_{\lambda,d}$ of m-primary ideals of length λ in d-dimensional local Cohen-Macaulay rings. The following statements hold.

(a) ΣF is rigid for the class $C_{\lambda,d+1}$.

(b) If $I \in C_{\lambda,d+1}$ and $p_I = p_{\Sigma F}$, then depth $gr_I R \geq 1$.

(c) If for all i in $C_{\lambda,d}$ with $H_I = F$, $gr_I R$ is Cohen-Macaulay,

then the same is true for all $I \in C_{\lambda,d+1}$ with $H_I = \Sigma F$.

Proof: Let I be an ideal in $C_{\lambda,d+1}$ with $P_i = p_{\Sigma F}$. We may assume that R/m is an infinite field and let x be a superficial element for I and a regular element. By (2) for $n \gg 0$, $H_{I/xR}(n) = \Delta H_I(n) = F(n)$, and since p_F is rigid for $C_{\lambda,d}$, $H_{I/xR} = F$. Thus, by (1), $\Delta H_I(j) \leq F(j)$ for all $j \geq 1$. But for $n \gg 0$, $\sum_{d=1}^{n} \Delta H_I(j) = H_I(n) = (\Sigma F)(n) = \sum_{j=1}^{n} F(j)$. Thus $\Delta H_I(j) = F(j)$ for all j, and $H_I = \Sigma F$. This proves (a). (b) and (c) follow from (3).

Let us calculate the polynomial associated to the function ΣF for any polynomial function F. Suppose

$$p_F(X) = f_0 \binom{X+d-1}{d} - f_1 \binom{X+d-2}{d-1} + \ldots + (-1)^d f_d.$$

Set

$$\Sigma p_F(X) = f_0 \binom{X+d}{d+1} - f_1 \binom{X+d-1}{d} + \ldots + (-1)^d f_d X.$$

Then we have,

$$p_{\Sigma F} = \Sigma p_F + \sum_{j=1}^{\infty} (F(j) - p_F(j)).$$

Proof of Rigidity of Polynomial (*): By applying the Proposition and the equation above, to show that (*) is rigid for $C_{\lambda,d}$, it suffices to prove rigidity for $C_{\lambda,1}$ and to prove $H_I(n) = p_I(n)$ for all $n \geq 1$ when $d = 1$. This follows easily from (1.9) in [11]. Suppose $I \in C_{\lambda,1}$, and $p_I(X) - (e_0 - \lambda)$, where $\lambda = \text{length}(R/I)$. By (1.9) in [11],

$$\lambda(I^k/I^{k+1}) \leq e_0 \text{ for all } k$$

and equality holds if and only if $I^{k+1}R(U) = xI^k R(U)$, where x is a minimal reduction of $IR(U) = IR[U]_{mR[U]}$, U an indeterminate. Therefore,

$$H_I(n) = \lambda + e_0 - j_1 + e_0 - j_2 + \ldots + e_0 - j_{n-1},$$

where $j_i \geq 0$ and $j_i = 0$ implies $j_{i+1} = 0$. Thus, for $n \gg 0$,

$$H_I(n) = \lambda + (n-1)e_0 - \sum j_i = ne_0 - (e_0 - \lambda + \sum j_i).$$

It follows that $\sum j_i = 0$ and, since $j_i \geq 0$ for all i, $j_i = 0$ for all i. Therefore, $H_I(n) = p_I(n)$ for all $n \geq 1$.

Proof of the Rigidity of Polynomial (**): In this case it suffices to prove that (**) is rigid for $C_{\lambda,1}$ and, for $d = 1$, that $H_I(1) - p_I(1) = 1$ and $H_I(n) = p_I(n)$ for all $n > 1$. Since $p_I(X) = e_0 X - (e_0 - \lambda + 1)$, in the above notation we have $\sum j_i = 1$ so that $j_1 = 1$ and $j_i = 0$ for $i > 1$. It follows that $H_I(n) = p_I(n)$ for All $n > 1$ and $H_I(1) - p_I(1) = \lambda - (e_0 - (e_0 - \lambda + 1)) = 1$. Thus p_I is rigid for $C_{\lambda,1}$, $\sum p_I + 1$ is rigid for $C_{\lambda,2}$ and $\sum\sum \cdots (\sum p_I + 1)$ is rigid for $C_{\lambda,d}$.

We conclude this note with an example. The m-primary ideal (t^4, t^5, t^6) in the 1-dimensional local domain $k[[t^4, t^5, t^6, t^7]]$ is in $C_{2,1}$ and has Hilbert polynomial $p_I(X) = 4X - 3$ which is rigid for $C_{2,1}$. However, depth $gr_I R = 0$, for $t^7 \notin I$ and $t^7 I \subseteq I^2$.

23.3 References

[1] Elias, J. and Valla, G., *Rigid Hilbert Functions*, preprint.

[2] Huneke, C., Hilbert Functions and Symbolic Powers, *Mich. Math. J.* **34** (1987), 293–318.

[3] Lipman, J., Stable Ideals and Arf Rings, *Amer. J. of Math.* **XCIII** (1971), 649–685.

[4] Marley, T., The coefficients of the Hilbert Polynomial and the Reduction Number of an Ideal, *J. London Math. Soc.* **40** 2 (1989), 1–8.

[5] Narita, M., A Note on the Coefficients of Hilbert Characteristic Functions in Semi-REgular Local Rings, *Proc. Camb. Phil. Soc.* **59** (1963), 267–279.

[6] Northcott, D.G., A Note on the Coefficients of the Abstract Hilbert Function, *J. London Math. Soc.* **35** (1960), 209–214.

[7] Ooishi, A., Genera and Arithmetic Genera of Commutative Rings, *Hiroshima Mth. J.* **17** (1987), 47–66.

[8] Sally, J.D., *Hilbert Coefficients and Reduction Number 2*, preprint.

[9] Zariski, O. and Samuel, P., *Commutative Algebra, Vol. 2*, Springer-Verlag, New York-Heidelberg-Berlin, 1960.

24
One-Dimensional Local Rings with Finite Cohen–Macaulay Type

Roger Wiegand

24.1 Introduction

Let $(R, \mathbf{m})$ be a local ring of dimension one. Recall that a non-zero finitely generated R-module M is said to be a maximal Cohen-Macaulay module (MCM module for short) provided $\mathbf{m}$ contains a non-zero-divisor on M. Equivalently, the simple module $R/\mathbf{m}$ does not embed in M. We say R has finite Cohen-Macaulay type (finite CM type) provided R has, up to isomorphism, only finitely many indecomposable MCM modules.

Earlier studies of one-dimensional rings of finite CM type, [9], [2], [7], [6], [10], [12], [13], have imposed various restrictions on the ring R, all of which imply the following:

- **(red)** R is reduced, and
- **(fin)** the integral closure $\tilde{R}$ of R (in its total quotient ring) is a finitely generated R-module.

Assuming, temporarily, that **(red)** and **(fin)** are satisfied, we know [12] that if R has finite CM type , then R satisfies the following conditions introduced by Drozd and Roĭter in [2]:

- **(DR1)** $\tilde{R}$ is generated by 3 elements as an R-module, and
- **(DR2)** $(\mathbf{m}\tilde{R} + R)/R$ is a cyclic R-module.

Conversely, if R satisfies **(DR1)** and **(DR2)** and, in addition, the residue field(s) of $\tilde{R}$ are separable over $R/\mathbf{m}$, then R has finite CM type. These results were proved by Drozd and Roĭter [2] and Green and Reiner [7] for (localizations of) orders in algebraic number fields, and by Greuel and Knörrer [7] (characteristic 0) and Kiyek and Steinke [10] (characteristic p) for curve singularities over an algebraically closed field.

It is well-known (see, for example, [13]) that a ring R satisfying **(red)** and **(fin)** has finite CM type if and only if the completion $\hat{R}$ has finite CM type. In [13] the equicharacteristic reduced complete local rings of finite CM type were classified by means of explicit parametrizations (under the mild separability hypothesis mentioned above). The classification works in all characteristics, and the residue field need not be algebraically closed.

In Theorem 1 of this note we give necessary and sufficient conditions for finite CM type without imposing conditions (**red**) and (**fin**). The analysis follows two separate paths, depending on whether or not R is a Cohen-Macaulay ring (i.e., an MCM R module). If R is Cohen-Macaulay, it turns out that (**red**) and (**fin**) are *necessary* conditions for finite CM type. We will see that in all cases R has finite CM type if and only if the completion $\hat{R}$ does.

Regarding the separability of the residue field extensions induced by the inclusion $R \to \tilde{R}$, we see from (**DR1**) that the only difficulty is caused by purely inseparable extensions of degree 2 or 3. In Theorem 2 we handle the degree 3 extensions, leaving only the case of purely inseparable extensions of degree 2 to complete the classification of one-dimensional local rings of finite CM type.

24.2 Necessary and Sufficient Conditions

Theorem 1 *Let $(R, \mathbf{m})$ be a one-dimensional local ring, and let N be the nilradical of R. Then R has finite CM type if and only if R/N has finite CM type and $\mathbf{m}^i N \cap = (0)$ for some $i \geq 1$. These conditions imply that the integral closure of R/N is finitely generated.*

In view of the last statement and the fact that the rings of finite CM type satisfying (**red**) and (**fin**) are essentially classified, we now have a description of virtually all one-dimensional local rings of finite CM type.

The proof of Theorem 1 will be broken down into a series of easy steps. Throughout, R denotes a local ring with maximal ideal $\mathbf{m}$, and $\tilde{R}$ is the integral closure of R in its total quotient ring.

Lemma 1 *If $\tilde{R}$ is not finitely generated as an R-module, then R has infinitely many non-isomorphic ideals.*

Proof There is a strictly ascending chain

$$R = R_0 \subset R_1 \subset R_2 \subset \dots$$

of subrings of $\tilde{R}$, each finitely generated as an R-module. Each R_i is isomorphic to an ideal of R (clear denominators), and I claim that the R_i are non-isomorphic as R-modules. Note that for any R-submodule X with $R \subseteq X \subseteq \tilde{R}$, the ring $E(X) := \{\rho \in \tilde{R} \mid \rho(X) \subseteq X\}$ is determined by the R-isomorphism class of X. Since $E(R_i) = R_i$, the claim follows.

Lemma 2 *If R is Cohen-Macaulay and $\tilde{R}$ is finitely generated as an R-module, then R is reduced.*

Proof Choose a non-zero-divisor $t \in \mathbf{m}$. If x were a non-zero nilpotent element, then

$$Rx \subset Rt^{-1}x \subset Rt^{-2}x \subset \dots$$

would be a strictly ascending chain of R-submodules of $\tilde{R}$.

Proposition 1 Assume R is Cohen-Macaulay. If either (**red**) or (**fin**) fails, then R has infinitely many non-isomorphic ideals, and hence has infinite CM type.

Proof The lemmas imply that R has infinitely many non-isomorphic ideals, and each non-zero ideal is an MCM module since R is Cohen-Macaulay. There is an integer s (the Goldie dimension of R) such that each ideal of R is a direct sum of at most s indecomposable ideals. If, now, R had finite CM type, there would be only finitely many non-isomorphic indecomposable ideals, and hence only finitely many isomorphism classes of ideals.

Lemma 3 *Let I be an ideal of R such that R/I is one-dimensional. If R has finite CM type, so has R/I.*

Proof Every MCM R/I-module is an MCM R-module.

Now we are ready to prove Theorem 1. Suppose first that R has finite CM type. Then so has R/N by Lemma 3. Since R/N is Cohen-Macaulay, its integral closure is finitely generated by Proposition 1. We need to show that $\mathbf{m}^i N \cap = (0)$ for some i. If R is Cohen-Macaulay, then $N = (0)$ by the proposition, and there is nothing to prove. If R is *not* Cohen-Macaulay, then $\mathbf{m}$ is an associated prime of (0). Let

$$(0) = I \cap J_1 \cap \dots \cap J_r$$

be an irredundant primary decompostion, with $\sqrt{I} = \mathbf{m}$ and $\sqrt{J_i} = P_i$. It will suffice to show that $I \cap N = (0)$. The P_i are the minimal primes of R, and since the J_i are primary, the residue rings R/J_i are Cohen-Macaulay and one-dimensional. By Lemma 3 and the proposition, each R/J_i is reduced, that is, $J_i = P_i$. Therefore $I \cap N = (0)$.

Conversely, suppose R/N has finite CM type and $I \cap N = (0)$ for some $\mathbf{m}$-primary ideal I. I claim that $NX = 0$ for every MCM R-module X. To see this, let $v \in N, x \in X$. Select $t \in I$ but outside every minimal prime ideal of R. Since $\mathbf{m}$ is not an associated prime of X, t is a non-zero-divisor on X. Since $tvx = 0x = 0$, we have $vx = 0$, and the claim is proved. Now we know that every MCM R-module is actually an R/N-module. Since R/N has finite CM type, so does R.

Examples Let $R_1 = k[[x,y]]/(x^2, xy)$. Then (x) is the nilradical, $R_1/(x) = k[[y]]$ has finite CM type, and $(x,y)^2 \cap (x) = (0)$. Therefore R_1 has finite CM type, by Theorem 1. On the other hand, $R_2 := k[[x,y]]/(x^3, x^2y)$ has infinite CM type, since every power of the maximal ideal meets the nilradical (x) non-trivially.

The following restatement of Theorem 1 gives an explicit representation of the non-Cohen-Macaulay rings of finite CM type:

Corollary 1 Let R be an arbitrary commutative ring. Then R is a one-dimensional local ring of finite CM type if and only if there is a Cartesian square (pullback diagram)

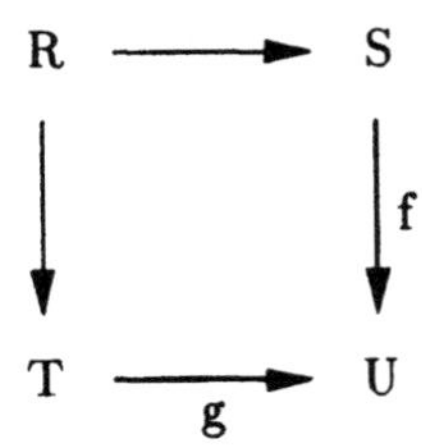

in which

1. S is a one-dimensional, local Cohen-Macaulay ring of finite CM type (necessarily reduced and with finitely generated integral closure),
2. T is an Artinian local ring,
3. f and g are surjective ring homomorphisms, and
4. $U \neq 0$.

Proof If $(R, \mathbf{m})$ has finite CM type, we easily obtain the requisite Cartesian square using Theorem 1: Take $S = R/N, T = R/\mathbf{m}^i, U = R/(N + \mathbf{m}^i)$, and let all maps be the natural ones.

Conversely, given a Cartesian square satisfying (1) - (4), we see that R is a subdirect product of S and T. Therefore R has dimension one, and Eakin's theorem [4] implies that R is Noetherian. Also, R is local because $U \neq 0$. Let $\mathbf{m}$ be the maximal ideal of R.

Let $I = \ker(R \to S)$ and $J = \ker(R \to T)$. Then $J \cap I = (0)$, and J is $\mathbf{m}$-primary because $R/J = T$ is Artinian. Since S is reduced and one-dimensional, I contains the nilradical of R. On the other hand, since T is Artinian it follws easily that I is nil. By Theorem 1, R has finite CM type.

Our earlier example of a non-reduced ring with finite CM type was a homomorphic image of $k[[x,y]]/(xy)$, a *reduced* ring with finite CM type.

Using Corollary 1 one easily obtains rings of finite CM type that cannot be lifted to reduced rings of finite CM type. For example, take $S = k[[t]], T = k[[x_1, \ldots, x_n]]/(x_1, \ldots x_n)^2$, and let f and g be the natural maps to the residue field $U = k$. Suppose, now, that L is a one-dimensional reduced local ring of finite CM type mapping onto R. Then L is Cohen-Macaulay and hence has finitely generated integral closure by Proposition 1. It follows from (**DR1**) that the embedding dimension of L is at most 3. (See the discussion of multiplicity in [5].) Now R is easily seen to have embedding dimension $n+1$ (trivially it is at least n), so by taking n large enough we obtain a contradiction.

Corollary 2 Let $(R, \mathbf{m})$ be a one-dimensional local ring. Then R has finite CM type if and only if the completion $\hat{R}$ has finite CM type.

Proof We observe first, using, e.g., ([13],9.B), that R is Cohen-Macaulay if and only if $\hat{R}$ is Cohen-Macaulay. Let N be the nilradical of R.

If R has finite CM type, then R/N has finite CM type, and its integral closure is finitely generated, by Theorem 1. By ([13],1.2), $\hat{R}/\hat{N} = (R/N)\hat{}$ has finite CM type. Since $(R/N)\hat{}$ is Cohen-Macaulay (by our opening remarks) it is reduced, by Proposition 1. Therefore $\hat{N}$ is exactly the nilradical of $\hat{R}$. Since $\mathbf{m}^i N \cap = (0)$ for some i, we have $\hat{\mathbf{m}}^i \cap \hat{N} = (0)$, and $\hat{R}$ has finite CM type, by Theorem 1. Now assume $\hat{R}$ has finite CM type. Then $(R/N)\hat{} = \hat{R}/\hat{N}$ has finite CM type by Lemma 3. If the integral closure of R/N is finitely generated, it follows from ([13],1.2) that R/N has finite CM type. On the other hand, if the integral closure of R/N is not finitely generated, then R/N has an infinite family of non-isomorphic ideals. By faithfully flat descent, ([4], 2.5.8), the extensions of these ideals to $(R/N)\hat{}$ are still non-isomorphic, contradicting the fact that $(R/N)\hat{}$ has finite CM type. Thus R/N has finite CM type. Finally, since $\hat{N}$ is contained in the nilradical of $\hat{R}$, Theorem 1 shows that $\hat{\mathbf{m}}^i \cap \hat{N} = (0)$ for some i, and it follows that $\mathbf{m}^i \cap N = (0)$. By Theorem 1 R has finite CM type.

It seems that there should be a direct combinatorial argument (using the Krull-Schmidt theorem over $\hat{R}$) that proves Corollary 2 without going through our structure theorem. I have, however, been unable to find such an argument. The difficulty is in deciding which MCM $\hat{R}$-modules come from R-modules.

In order to complete the classification of all one-dimensional local rings of finite CM type, it is enough, in view of Theorem 1 (or Corollary 1) to classify those that satisfy (**red**) and (**fin**). Assuming now that these conditions are satified, we know [12] that finite CM type implies (**DR1**) and (**DR2**), and that the converse holds unless $\tilde{R}$ has a residue field that is purely inseparable of degree 2 or 3 over $R/\mathbf{m}$. It seems reasonable to conjecture that finite CM type is equivalent to (**DR1**) and (**DR2**) in general. The rest of the paper is devoted to a proof of the conjecture in the case of residue field growth of degree three.

24.3 Degree 3 Extensions

Theorem 2 *Assume* $(R, \mathbf{m}, k)$ *is reduced and has finite normalization. If* R *has finite CM type, then* R *satisfies* (**DR1**) *and* (**DR2**). *Conversely, if* R *satisfies* (**DR1**) *and* (**DR2**) *and no residue field of* $\tilde{R}$ *is purely inseparable of degree* 2 *over* k, *then* R *has finite CM type.*

Proof By the remarks above, only the last statement needs to be proved; and we may assume that $\tilde{R}$ has a residue field K of degree 3 over k. (It does not matter whether or not K/k is separable.) By (**DR1**), R is local and its maximal ideal is $\mathbf{m}\tilde{R}$.

Suppose first that R is seminormal, that is, the conductor $\underline{c} := (R : \tilde{R})$ is a radical ideal of R. The conductor square for R is then

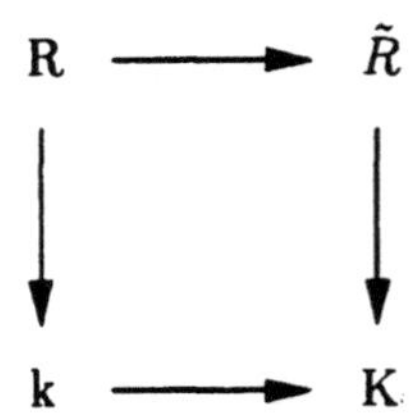

Proceeding exactly as in §3 of [12], we see that the ring B' of [12],3.4) is $K[X]/(f)$ (where f is the minimal polynomial for some element of $K - k$). In particular, B' is a principal ideal ring, which is all that is needed to make the arguments of [12] work. (If K/k is purely inseparable, the only case not covered by ([12], 3.2), $B' \cong K[T]/(T^3)$, and $A' = K$.) In any case, the pair of rings (A', B') has finite representation type [12] by the results of Green and Reiner [7]. Therefore R has finite CM type by ([13], 3.3, 1.9).

If R is *not* seminormal, we will show that R is Gorenstein, and that there is *exactly one* ring S (namely, the seminormalization of R) strictly between R and $\tilde{R}$. If, now, M is an indecomposable MCM R−module, M is reflexive by ([1], 6.2). Then, by ([1], 7.2), either $M \cong R$ or else M is an S-module. Since S satisfies all the assumptions imposed on R and is seminormal, S has finite CM type. It follows that R has finite CM type as well. All we need to complete the proof is the following lemma:

Lemma 4 *Let* $(R, \mathbf{m}, k)$ *satisfy* (**DR1**) *and* (**DR2**), *and assume* $\tilde{R}$ *is local, with residue field* K *of degree* 3 *over* k. *If* R *is not seminormal, then* R

is Gorenstein, and there is exactly one ring S strictly between R and $\tilde{R}$. Moreover, S is a seminormal local ring satisfying (**DR1**) *and* (**DR2**) *and having residue field k.*

Proof All hypotheses and conclusions can be interpreted in terms of the bottom line of the conductor square

$$\begin{array}{ccc} R & \longrightarrow & \tilde{R} \\ \big\downarrow & & \big\downarrow{\scriptstyle \pi} \\ A & \longrightarrow & B \end{array}$$

For simplicity of notation, we let $\mathbf{m}$ and J denote, respectively, the maximals ideals of A and B. We know $J \neq 0$ since R is not seminormmal. Our first (and major) task is to show that $J^2 = 0$.

Condition (**DR1**) implies that $J = \mathbf{m}B$. Select $t \in \mathbf{m} - J^2$, and note that $J = Bt$, as B is a principal ideal ring. It now follows from (**DR2**) that $\dim_k(A+J)/(A+J^2) \leq 1$. Since $(A+J^2)/(\mathbf{m}+J^2)$ is one-dimensional, we have

$$\ell_A(A+J)/(\mathbf{m}+J^2) \leq 2, \tag{1}$$

where ℓ_A denotes the length as an A module. On the other hand, $(A+J)/J \cong A/(J \cap A) = k$, so $\ell_A B/(A+J) = \ell_A B/J - 1 = 2$. Combining this equation with (1) we have

$$\ell_A B/(\mathbf{m}+J^2) \leq 4. \tag{2}$$

Now $\ell_A B/J^2 = 6$, and $\dim_k(At+J^2)/J^2 = 1$, so $\ell_A B/(At+J^2) = 5$. Comparing this with (2), we have

$$\mathbf{m} \not\subseteq At + J^2. \tag{3}$$

Select $f \in \mathbf{m} - (At+J^2)$, and write $f = bt$ where $b \in B - (A+J)$. The image of b in B/J is in $K - k$, so $B = A + Ab + Ab^2 + J$; and since $J = \mathbf{m}B$, Nakayama's lemma implies that B is generated as an A-module by 1, b and b^2. Now A contains $t^2, bt^2 = ft$ and $b^2t^2 = f^2$; therefore $A \supseteq Bt^2$. Since $(B : A) = 0$, $t^2 = 0$, whence $J^2 = 0$.

To show that R is Gorenstein, we appeal to ([8], Korollar 3.7), which states that $\ell_A B \geq 2\ell_A A$, with equality if and only if R is Gorenstein. Since $\ell_A B = 6$, it is enough to show that $\ell_A A \geq 3$, that is, $\mathbf{m}$ is not principal.

But this is clear from (3), since $t \in \mathbf{m} - \mathbf{m}^2$. As a bonus, we learn that $\ell_A A = 3$, whence $\mathbf{m}$ is 2-generated.

Since $\ell_A B/A = 3 = \ell_A B/J$, it follows that $\ell_A(A+J)/A = 1$. Therefore the ring $A + J$ is strictly between A and B. This means the ring $S := \pi^{-1}(A+J)$ is strictly between R and $\tilde{R}$. The residue field of S is $(A+J)/J = k$, and S is seminormal since the conductor of S in its integral closure $\tilde{R}$ is J. Clearly S satisfies (**DR1**) and (**DR2**).

To show that S is unique, it sufffices to show that $A + J$ is the only ring strictly between A and B. Suppose $A \subsetneq C \subsetneq B$, and let U be the radical of C. If $C/U \cong K$, then $C = B$ by Nakayama's lemma. The only other possibility is that $C/U \cong k$. But then $A \subsetneq C = A + U \subseteq A + J$, and since $\ell_A(A+J)/A = 1$ (as shown above), C must be either A or $A + J$.

The remaining piece of the problem of classifying all one-dimensional local rings of finite CM type is to decide whether (**DR1**) and (**DR2**) imply finite CM type in the situation described by Theorem 2. In fact, using methods similar to those in the proof of Theorem 2 and Lemma 4, one can reduce the problem to the following special case: $\tilde{R}$ has exactly two maximal ideals, with residue fields K and k, where K/k is purely inseparable of degree 2. Moreover, letting $\underline{c}$ be the conductor of R in $\tilde{R}$, one can assume that $R/\underline{c}$ is a principal ideal ring and that $\tilde{R}/\underline{c} = B \times k$ (where B is local with residue field K). Unfortunately, since this test ring R is not seminormal, the arguments we used in characteristic 3 do not allow us to conclude that R has finite CM type.

Acknowledgments: I am grateful to Lawrence Levy and Sylvia Wiegand for several very enlightening conversations about the material in this paper.

24.4 References

[1] H. Bass, On the ubiquity of Gorenstein rings, *Math. Z.*, **82**, 1963, pp. 8–28.

[2] Ju. A. Drozd and A. V. Roĭter, Commutative rings with a finite number of indecomposable integral representations, Russian, *Izv. Akad. Nauk. SSSR Ser. Mat.*, **31**, 1967, pp. 783–798.

[3] P. Eakin , The converse to a well known theorem on Noetherian rings, *Math. Ann.*, **177**, 1968, pp. 278–282.

[4] A. Grothendieck and J. Dieudonné, Éléments de Géométrie Algébrique IV, *Publ. Math. IHES*, **24**, 1967,

[5] C. Greither, On the two generator problem for the ideals of a one dimensional ring, *J. Pure Appl. Algebra*, **24**, 1982, pp. 265–276.

[6] G.-M. Greuel and H. Knörrer, Kurvensingularitäten und torsionfreie Moduln, *Math. Ann.*, **270**, 1985, pp. 417–425.

[7] E. Green and I. Reiner, Integral representations and diagrams, *Michigan Math. J.*, **25**, 1978, pp. 53–84.

[8] J. Herzog and E. Kunz, Der kanonische Modul eines Cohen-Macaulay Rings, *Lect. Notes in Math.*, **238**, 1971, Springer-Verlag.

[9] H. Jacobinski, Sur les ordres commutatifs, avec un nombre fini de réseaux indécomposables, *Acta Math.*, **118**, 1967, pp. 1–31.

[10] K. Kiyek and G. Steinke, Einfache Kurvensingularitäten in beliebiger Charakteristik, *Arkiv Math.*, **45**, 1985, pp. 565–573.

[11] H. Matsumura, *Commutative Algebra*, Second Edition, Benjamin, Reading, Massachusetts, 1980.

[12] R. Wiegand , Noetherian rings of bounded representation type, *Commutative Algebra, Proceedings of a Microprogram*, held June 15–July 2, 1987, Springer-Verlag, New York, 1989, pp. 497–516.

[13] R. Wiegand , Curve singularities of finite Cohen-Macaulay type, *Arkiv Math.*, to appear.

Part VIII

Computational Algebraic Geometry

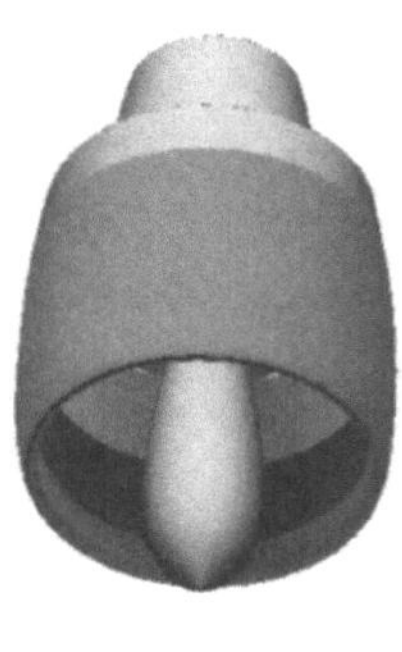

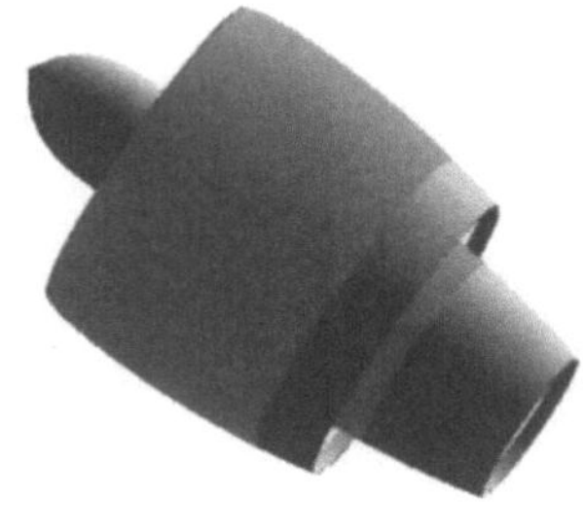

25
Some Applications of Constructive Real Algebraic Geometry

Chandrajit L. Bajaj

ABSTRACT In this short article we summarize a number of recent applications of constructive real algebraic geometry to geometric modelling and robotics, that we have been involved with under the tutelage of Abhyankar.

25.1 Introduction

Macaulay, Tarski and Seidenberg [16, 12, 18] set the tone for current day researchers of constructive methods in algebraic geometry over real closed fields. Constructive methods are clearly at the heart (and soul) of Abhyankar's papers in algebraic geometry and amply evident in his teaching[1]. In this short article we summarize a number of recent applications of constructive real algebraic geometry to geometric modeling and robotics, that we have been involved with under the tutelage of Abhyankar. First is parameterizations, useful for computing intersections, sweeps, offsets etc., required in robotic software simulation systems. Here we consider constructive methods for both local and global real parameterizations of curves and surfaces. Next we look at intersections between curves and surfaces which are fundamental for solid modeling systems based on Boolean set operations. Finally we look at surface fitting with algebraic surface patches, a technique used for both complicated interactive geometric design as well as scattered data fitting.

25.2 Global Parameterization

Certain classes of algebraic curves and surfaces admit both parametric and implicit representations. Algebraic curves and surfaces are the most common representations for curved objects in geometric modeling. Algebraics satisfy *polynomial* equations, usually with rational coefficients. A *rational* algebraic curve or surface is one whose points can be represented as rational functions in some parameters. Each form has certain benefits and drawbacks. The parametric form is better for rapid display and interactive control; the implicit form defines a half-space naturally and is suited

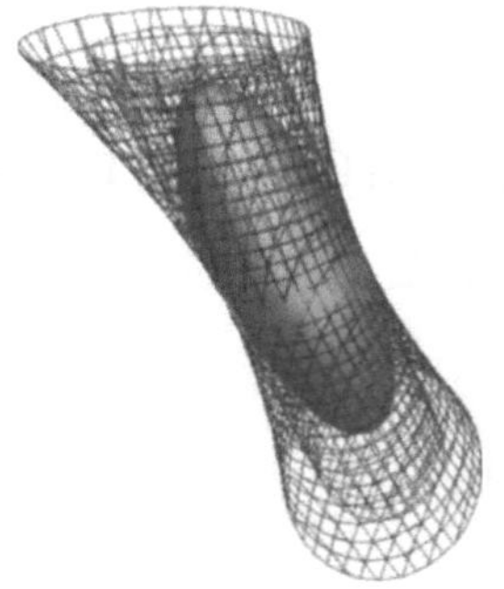

FIGURE 1. Global Parameterization of Quadrics using Finite Precision Arithmetic

for modeling. The class of all algebraics is also much larger than the class of rational algebraics. Having dual forms are highly useful in geometric modeling since they combine the strengths of the two representations[7].

In [13] we consider the problem of computing the rational parameterization of an implicit curve or surface in a finite precision domain. Known algorithms for this problem are based on classical algebraic geometry, and assume exact arithmetic involving algebraic numbers[2, 3, 4, 5, 6]. In this work, we investigate the behaviour of parametrization algorithms in a finite precision domain and derive succinct algebraic and geometric error characterizations. We then indicate numerically robust methods for parameterizing curves and surfaces which yield no error in extended finite precision arithmetic and alternatively, minimize the output error under fixed finite precision calculations

For example, one can obtain succinct bounds on the geometric error incurred in parameterizing quadratic surfaces (quadrics) by mapping (in fixed precision arithmetic) either the constant coefficient or one the squared term coefficients to infinity. The sign of the discriminant, among other quantities, distinguishes amongst the various quadric surfaces. Essentially, perturbing the constant coefficient preserves the center and orientation, although the quadric could degenerate from a hyperboloid of one sheet to a cone to a double-sheeted hyperboloid. Perturbing the highest order coefficients could cause an ellipsoid to change to a cylinder to a one-sheeted hyperboloid, for example, in addition to changing its orientation and center (Figure 1). Since the geometric errors find their extrema along the axes when the center and orientation are fixed, we can bound the errors easily in this case. We simply state the results, for brevity.

Let two quadrics that differ only in their constant coefficient be given and let d_x, d_y, d_z be the distances from the origin to the unperturbed quadric (some may not be finite). Given a number $\epsilon > 0$ that also satisfies $\epsilon < \min(d_x, d_y, d_z)$, and a difference in the constant coefficients of a quantity

d_e, if the geometric perturbations p_x, p_y, p_z are to satisfy

$$\max(|p_x|, |p_y|, |p_z|) \quad < \quad \epsilon$$

then it suffices to choose d_e such that

$$|d_e| < \epsilon \cdot \min(d_x \cdot |\lambda_1|, d_y \cdot |\lambda_2|, d_z \cdot |\lambda_3|)$$

where expressions for λ_i are the roots of a cubic polynomial $\phi(\lambda)$ whose coefficients are expressions in the coefficients of the quadrics. The quadric can be put in standard form in terms of the roots of $\phi(\lambda)$, allowing the the quantities d_x, d_y, d_z to be efficiently calculated.

WIth the parameterization of singular cubic curves, algebraic number computation is unnecessary for exact rational parameterization. Every rational cubic with rational coefficients has a rational singular point. Such a cubic can be parameterized by a pencil of lines through the singularity, which then intersect the cubic at exactly one other point. The coordinates of the latter point parameterized by the slope of the line give parameter functions for the cubic curve. The parameter functions are given as closed form formulas in the parameter t, the coefficients of the curve, and the coordinates (b, c) of the singularity, as shown below:

$$\begin{aligned} X(t) \quad &= a_{30}bt^3 - (3a_{30}c + a_{20})t^2 - \\ &\quad (2a_{21}c + a_{12}b + a_{11})t - (2a_{03}b + a12c + a_{02}) \\ Y(t) \quad &= -((2a_{30}c + a_{21}b + a_{20})t^3 + \\ &\quad (a_{21}c + 2a_{12}b + a_{11})t^2 + (3a_{03}b + a_{02})t - a_{03}c) \\ W(t) \quad &= a_{30}t^3 + a_{21}t^2 + a_{12}t + a_{03} \end{aligned}$$

Therefore, if extended precision rational arithmetic is allowed, one can parameterize an irreducible rational cubic curve without error and without algebraic number computation, by computing the singular point exactly, and substituting the coordinates in the above formula.

25.3 Local Parameterization

In [14, 15] we use a combination of both algebraic and numerical techniques to construct C^1-continuous, piecewise (m, n) rational ϵ-approximation of real algebraic curves of degree d. For example, Figure 3, shows a C^1 continuous $(3,3)$-rational approximation of the curve $(x^2 + y^2)^3 - 4x^2y^2 = 0$ for values of $\epsilon = 0.1, 0.05, 0.025$. At singular points we use the classical Weierstrass Preparation Theorem and Newton power series factorizations, based on the technique of Hensel lifting[1]. These, together with modified rational Padé approximations, are used to efficiently construct locally approximate, rational parametric representations for all real branches of an algebraic curve.

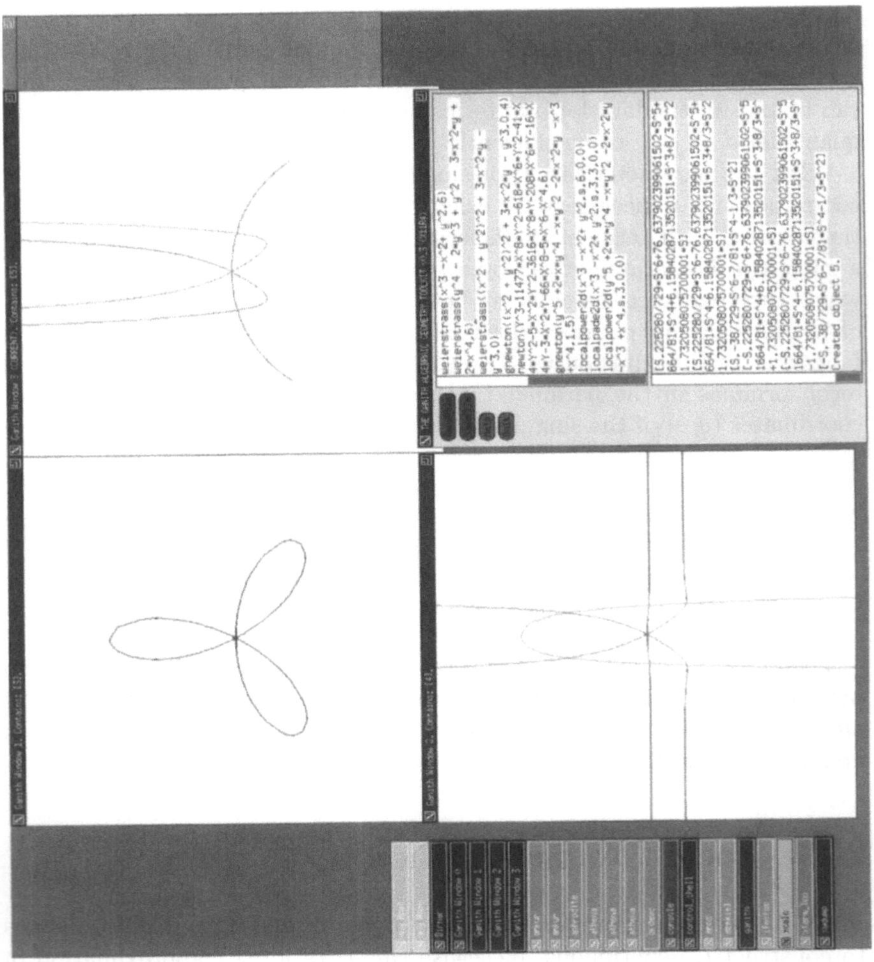

FIGURE 2. Resolution of Singularities: Newton and Weierstrass Factorizations

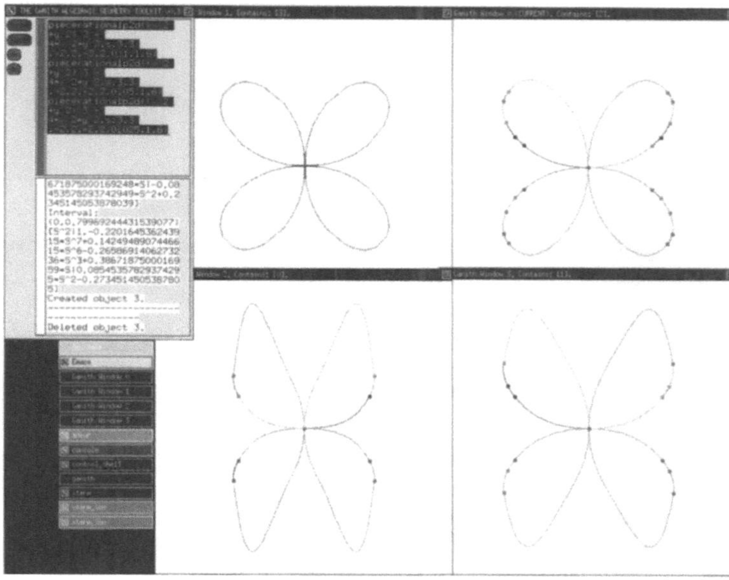

FIGURE 3. Piecewise Rational Approximations of Real Algebraic Plane Curves

A Weierstrass power series factorization is of the form $f(x,y) = g(x,y)\underbrace{(y^e + a_{e-1}(x)y^{e-1} + \cdots + a_0(x))}_{h(x,y)}$ where $g(x,y)$ is a unit power series, i.e., $g(0,0) \neq 0$ while $h(x,y)$ is a "distinguished" polynomial in y with coefficients $a_i(x)$, $i = 0 \ldots e-1$ being non-unit power series, i.e., $a_i(0) = 0$. In Figure 2 the lower right picture shows the two Weierstrass power series factors (truncated to degree 14 in x) of the plane curve $f(x,y) = (x^2 + y^2)^2 + 3x^2y - y^3 = 0$: the unit $g(x,y) = y + 121632404x^14 + 5254746x^12 + 237526x^10 + 11477x^8 + 618x^6 + 41x^4 + 5x^2 - 1 = 0$ which represents the part of the curve away from the origin, and the "distinguished" polynomial in y $h(x,y) = y^3 + (-121632404x^14 - 5254746x^12 - 237526x^10 - 11477x^8 - 618x^6 - 41x^4 - 5x^2)y^2 - (35422240x^14 - 1556448x^12 - 72080x^10 - 3616x^8 - 208x^6 - 16x^4 - 3x^2)y - 498162x^14 - 23038x^12 - 1153x^10 - 66x^8 - 5x^6 - x^4 = 0$ which represents the part of the curve at the origin.

The "distinguished" factor $h(x,y)$ is again shown in the upper left picture of Figure 2 where it is split via Newton factorization into real linear factors of the type $h(x,y) = \Pi_{i=1}^{e}\,(y - \eta_i((t)))$ with $t^m = x$ and m a positive integer and $\eta_i((t))$ a real power series or real meromorphic series.

Besides singular points we obtain an adaptive selection of simple points about which the curve approximations yield a small number of pieces yet achieve C^1 continuity between pieces.

The rational approximation algorithms have been implemented in its entirety as part of GANITH, an X-11 based interactive algebraic geometry

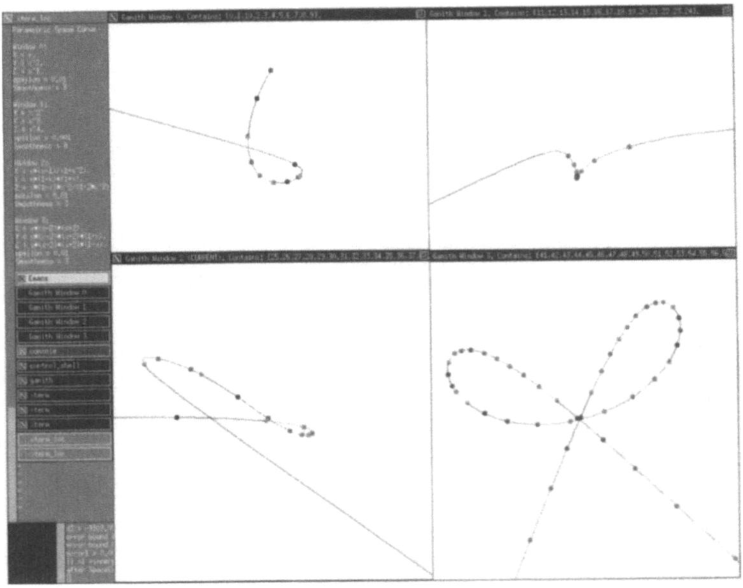

FIGURE 4. Piecewise Rational Approximation of Real Algebraic Space Curves

toolkit, using Common Lisp for the symbolic computation and C for all numeric and graphical computation. The Hensel power series computations as well as its use in Weierstrass and Newton factorizations are based on a robust implementation of the fast euclidean HGCD algorithm. Rational Padé approximants are also computed based on the same HGCD algorithm. Power Series are stored as truncated sparse polynomials, as are the polynomials representing the original algebraic curves, in recursive canonical form. In this form, a polynomial in the variables $x_1, \ldots, x_n$ is represented either as a constant, or as a polynomial in x_n whose coefficients are (recursively) polynomials in the remaining variables $x_1, \ldots, x_{n-1}$. A strength of this form (for purposes of implementation) is that multivariates "look like" univariates, making it easy to modify algorithms for univariate polynomials to handle multivariates.

Floating point coefficients are allowed in the input curve representations, which are then converted to rational numbers for the GCD and power series computations. In Newton factorizations, user options are provided to compute only real branch factorizations. This is achieved by not allowing complex conjugate roots of the appropriate univariate polynomial, to split in the base case of the Henselian computation. Singularity computations and intersection with the bounding box are done in GANITH using multivariate resultants and based on the method of birational maps [8]. Details of this are given in the next section. Examples from the software implementation, are shown in Figures 3 and 4.

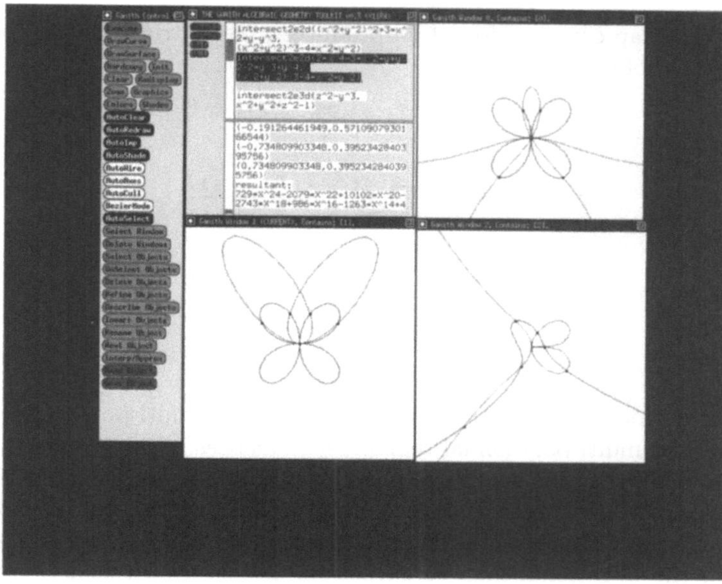

FIGURE 5. Computation of Curve-Curve Intersections

25.4 Intersection

The set of solutions (or *zero* set $Z(S)$) of a collection S of polynomial equations

$$S_1 : f_1(x_1, ..., x_n) = 0$$
$$\cdots$$
$$S_m : f_m(x_1, ..., x_n) = 0 \tag{25.1}$$

is referred to as an *algebraic set.* Algebraic curves and surfaces are algebraic sets of dimension 1 and 2 respectively. Problems dealing with zero sets $Z(S)$, such as the intersection of curves and surfaces, or the decision whether a surface contains a set of curves, are often first versed in an ideal-theoretic form and then solved using Gröbner basis manipulations. In [8] we present an alternative technique based on constructing bi-rational mappings between algebraic varieties and hypersurfaces. questions of intersection and parameterization of algebraic varieties. The bi-rational mapping technique deals directly with the *zero sets* of polynomial equations (rather than just the combinatorial structure of the polynomials), and provides simpler solutions to questions of intersection and parameterization of algebraic varieties.

Given m independent equations in n variables (25.1), let S be the algebraic variety of dimension $n - m$ defined by these equations. Then the bi-rational map construction of [8] produces a new "triangulated" polynomial system of equations

$$\tilde{f}(x_1, ..., x_{n-m+1}) = 0$$

$$x_{n-m+2} = \frac{h_{2m-4}(x_1, \ldots, x_{n-m+1})}{h_{2m-3}(x_1, \ldots, x_{n-m+1})}$$

$$\cdots$$

$$x_{n-1} = \frac{h_2(x_1, \ldots, x_{n-2})}{h_3(x_1, \ldots, x_{n-2})}$$

$$x_n = \frac{h_0(x_1, \ldots, x_{n-1})}{h_1(x_1, \ldots, x_{n-1})} \tag{25.2}$$

This bi-rational map construction is based on the multi-polynomial resultant[16] and multi-polynomial remainder sequences.

Cases of intersection computation of interest in geometric and solid modeling are those of plane curve-curve intersections, surface-surface intersections and three algebraic surface intersections [7]. All these are special cases of the bi-rational map construction. The two prevalent representations of algebraic curves in geometric modeling are the implicit and the rational parametric. Both implicitly and parametrically defined algebraic plane curve-curve intersections reduce to the special case of (25.1) for $n = 2$ and $m = 2$. The common intersection points (x_1, x_2) of the two curves are then obtained from the special case "triangulated" system (25.2) by first computing the zeros of the univariate polynomial $\tilde{f}(x_1) = 0$ and then substituting these into $x_2 = \frac{h_0(x_1)}{h_1(x_1)}$. Examples from the software implementation in GANITH, are shown in Figure 5.

Implicitly defined algebraic surface-surface intersections reduce to the special case of (25.1) for $n = 3$ and $m = 2$. Points (x_1, x_2, x_3) on the common intersection space curve of the two surfaces are then obtained from the special case "triangulated" system (25.2) by first computing points on the plane curve $\tilde{f}(x_1, x_2) = 0$ and then substituting these into $x_3 = \frac{h_0(x_1,x_2)}{h_1(x_1,x_2)}$. Parametrically defined algebraic surface-surface intersections reduce to the special case of (25.1) for $n = 4$ and $m = 3$. Points (x_1, x_2, x_3, x_4) on the common intersection space curve of the two surfaces are then obtained from the special case "triangulated" system (25.2) by first computing points on the plane curve $\tilde{f}(x_1, x_2) = 0$ and then substituting these into $x_3 = \frac{h_0(x_1,x_2)}{h_1(x_1,x_2)}$ and $x_4 = \frac{h_2(x_1,x_2)}{h_3(x_1,x_2)}$. An example from the software implementation in GANITH, involving a sphere and a quartic algebraic surface is shown in Figure 6.

Implicitly defined three algebraic surfaces intersection reduces to the special case of (25.1) for $n = 3$ and $m = 3$. Common intersection points (x_1, x_2, x_3) of the three surfaces are then obtained from the special case "triangulated" system (25.2) by first computing the zeros of the univariate polynomial $\tilde{f}(x_1) = 0$ and then substituting these into $x_2 = \frac{h_0(x_1)}{h_1(x_1)}$, and

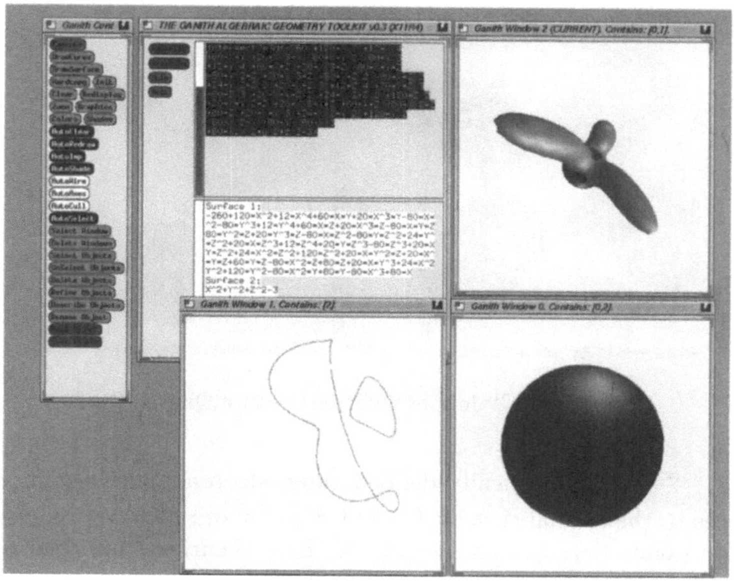

FIGURE 6. Computation of Surface-Surface Intersections

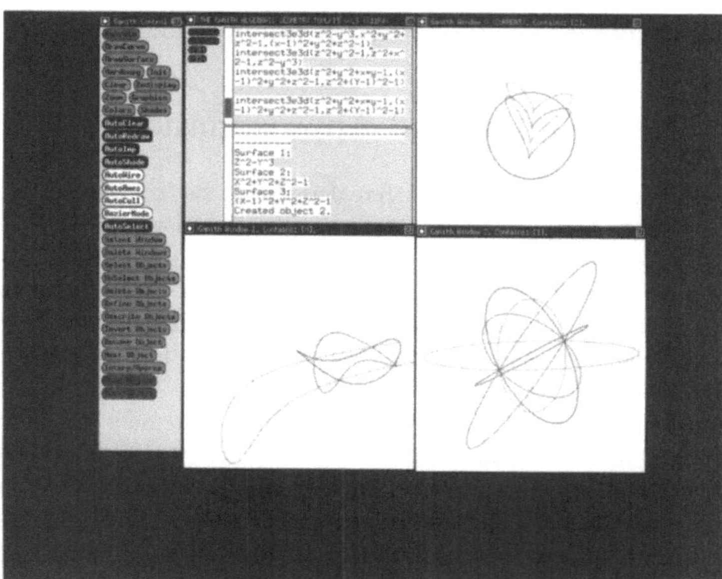

FIGURE 7. Computation of Intersections of Three Surfaces

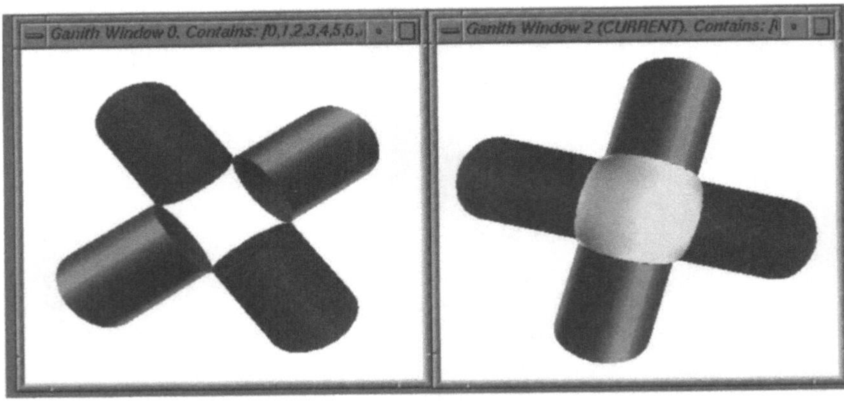

FIGURE 8. C^1 Join of Cylinders with a Quartic Surface

$x_3 = \frac{h_2(x_1)}{h_3(x_1)}$. Parametrically defined three algebraic surfaces intersection reduces to the special case of (25.1) for $n = 6$ and $m = 6$. Common intersection points $(x_1, x_2, x_3, x_4, x_5, x_6)$ of the two surfaces are then obtained from the special case "triangulated" system (25.2) by first computing the zeros of the univariate polynomial $\tilde{f}(x_1, x_2) = 0$ and then substituting these into $x_2 = \frac{h_0(x_1)}{h_1(x_1)}$, $x_3 = \frac{h_2(x_1)}{h_3(x_1)}$, $x_4 = \frac{h_4(x_1)}{h_5(x_1)}$, $x_5 = \frac{h_6(x_1)}{h_7(x_1)}$, and $x_6 = \frac{h_8(x_1)}{h_9(x_1)}$. Examples from the software implementation in GANITH, are shown in Figure 7. The three surface intersection points are shown as the common intersections of the space curves for each pair of surfaces.

25.5 Interpolation and Approximation

The generation of a mesh of smooth real algebraic surface patches or *splines* that interpolate or approximate *triangulated space data* is one of the central topics of geometric design. Prior work on splines have traditionally worked with a given planar triangulation using a piecewise polynomial function basis or over triangulations in three dimensions using parametric surface patches. Little work has been done on spline basis for implicitly defined real algebraic surfaces.

I report briefly on some ongoing work in this extremely interesting and fundamental area of research for geometric design [9, 10]. In [11] we show how low degree blending and joining algebraic surfaces can be computed via C^1 interpolation and least-squares approximation. The algebraic surface fitting scheme reduces to the solution of a finite system of linear equations, based on a proper normalization of the coefficients of the surface. Both the finiteness bound and the linear equations are derived from various invocations of Bezout's theorem. In the example shown in Figure 8 and implemented in GANITH, the joining surface of the four cylindrical sur-

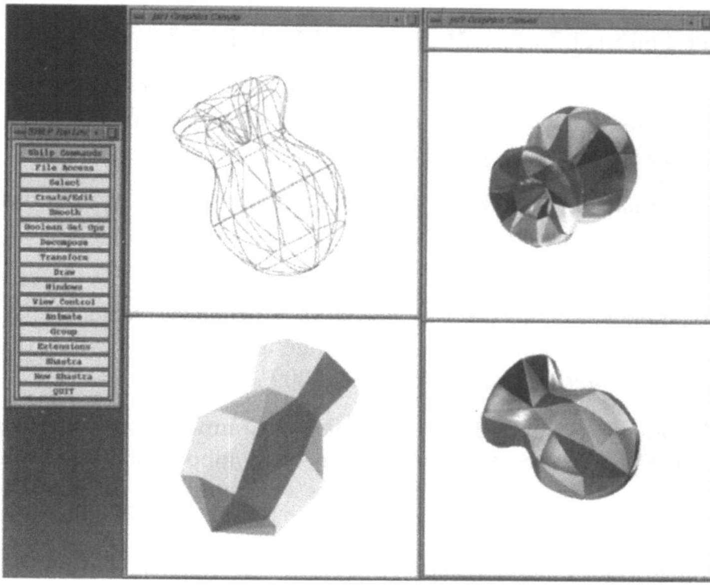

FIGURE 9. Smoothing of a Polyhedron with Triangular Interpolatory Splines

faces is computed by a C^1 interpolation of the four circular cross-sections of the cylinders and the gradient vectors along these cross-sections. Least-squares approximation from a variable radius sphere centered at the mid point of the junction helps select the desired bulge of the joining surface.

In [12] we consider an arbitrary spatial triangulation $\mathcal{T}$ consisting of vertices (x_i, y_i, z_i) in $\mathbb{R}^3$ (or more generally a simplicial polyhedron $\mathcal{P}$ when the triangulation is closed), with possibly "normal" vectors at the vertex points. An algorithm is given to construct a C^1 continuous mesh of low degree real algebraic surface patches S_i , which respects the topology of the triangulation $\mathcal{T}$ or simplicial polyhedron $\mathcal{P}$, and C^1 interpolates all the vertices (x_j, y_j, z_j) in $\mathbb{R}^3$. The technique uses a single implicit surface patch for each triangular face of $\mathcal{T}$ of $\mathcal{P}$, i.e. no local splitting of triangular faces. Each triangular surface patch has local degrees of freedom which are used to provide local shape control. This is achieved by use of weighted least squares approximation from points (x_k, y_k, z_k) generated locally for each triangular patch from the original patch data points and normal directions on them. Examples of this smoothing process are shown in Figure 9 and implemented in GANITH.

25.6 References

[1] Abhyankar, S., (1990) "Algebraic Geometry for Scientists and Engineers", *Mathematical Surveys and Monographs*, Number 35, American

Mathematical Society, Providence, Rhode Island.

[2] Abhyankar, S., and Bajaj, C., (1987) "Automatic Parameterization of Rational Curves and Surfaces I: Conics and Conicoids", *Computer Aided Design*, 19, 1, pp. 11 - 14.

[3] Abhyankar, S., and Bajaj, C., (1987) "Automatic Parameterization of Rational Curves and Surfaces II: Cubics and Cubicoids", *Computer Aided Design*, 19, 9, pp. 499 - 502.

[4] Abhyankar, S., and Bajaj, C., (1988) "Automatic Parameterization of Rational Curves and Surfaces III: Algebraic Plane Curves", *Computer Aided Geometric Design*, 5, (1988), 309 - 321.

[5] Abhyankar, S., and Bajaj, C., (1988) "Computations with Algebraic Curves", (Proc. of Intl. Symposium on Symbolic and Algebraic Computation), (ISSAC8), *Lecture Notes in Computer Science*, No. 358, Springer-Verlag, (1989), 279 - 284.

[6] Abhyankar, S., and Bajaj, C., (1989) "Automatic Parameterization of Rational Curves and Surfaces IV: Algebraic Space Curves", *ACM Transactions on Graphics*, 8, 4, (1989), 324 - 333.

[7] Bajaj, C., (1989) "Geometric Modeling with Algebraic Surfaces" *The Mathematics of Surfaces III*, ed., D. Handscomb, Oxford University Press, pp. 3 - 48.

[8] Bajaj, C., (1990) "Geometric Computations with Algebraic Varieties of Bounded Degree", Proc. of the Sixth ACM Symposium on Computational Geometry, Berkeley, California, 148-156.

[9] Bajaj, C., (1992) "Surface Fitting with Implicit Algebraic Surface Patches" *Topics in Surface Modeling*, edited by H. Hagen, SIAM Publications, (1992), in press.

[10] Bajaj, C., (1992) "The Emergence of Algebraic Curves and Surfaces in Geometric Design", *Directions in Geometric Computing 1991*, edited by R. Martin, Information Geometers Press, United Kingdom (1992), in press.

[11] Bajaj, C. and Ihm, I., (1992), "Algebraic Surface Design with Hermite Interpolation" *ACM Transactions on Graphics*, 19, 1, (1992), 61-91.

[12] Bajaj, C., and Ihm, I., (1992), "C^1 Smoothing of Polyhedra with Implicit Algebraic Splines", *Computer Graphics*, 26, 2, SIGGRAPH'92, 79 - 88.

[13] Bajaj, C., and Royappa, A., (1992) "Parameterization in Finite Precision", Proc. of the Graphics Interface'92 Conference, (1992), Vancouver, Canada, 29-36.

[14] Bajaj, C. and Xu, G., (1991) "Piecewise Rational Approximations of Real Algebraic Curves", Submitted to *Advances in Computational Mathematics.*

[15] Bajaj, C. and Xu, G., (1992) "NURBS Approximation of Surface-Surface Intersection Curves", Submitted to *Advances in Computational Mathematics.*

[16] Macaulay, F. (1916), "The Algebraic Theory of Modular Systems", *Cambridge University Press.*

[17] Seidenberg, A., (1974) "Constructions in Algebra", *Transactions of the AMS*, 197:273–313.

[18] Tarski, A., (1948) "A Decision Method for Elementary Algebra and Geometry", *University of California Press*, Berkeley.

26 An Improved Sign Determination Algorithm

John Canny

ABSTRACT Recently there has been a lot of activity in algorithms that work over real closed fields, and that perform such calculations as quantifier elimination or computing connected components of semi-algebraic sets. A cornerstone of this work is a symbolic sign determination algorithm due to Ben-Or, Kozen and Reif [1]. In this paper we describe a new sign determination method based on the earlier algorithm, but with two advantages: (i) It is faster in the univariate case, and (ii) In the general case, it allows purely symbolic quantifier elimination in pseudo-polynomial time. By purely symbolic, we mean that it is possible to eliminate a quantified variable from a system of polynomials no matter what the coefficient values are. The previous methods required the coefficients to be themselves polynomials in other variables. Our new method allows transcendental functions or derivatives to appear in the coefficients.

26.1 Introduction

Real quantifier elimination algorithms have enjoyed a revival in interest in the last few years, and an important impetus was the parallel sign determination algorithm in [1]. Two of the latest quantifier elimination algorithms, [10] and [7] have very good theoretical complexity. Indeed the bounds in [10] are near the best that can be hoped for the important case of problems with a bounded number of quantifier alternations. Their work builds on a large body of earlier papers, dating from Tarski's original paper [13], and following a line of papers that include [1], [2] [4], [5], [6], [12], [5], and others.

Both [7] and [10], like the author's earlier paper [2], make use of a sign-determination lemma due to Ben-Or, Kozen and Reif [1]. This lemma, henceforth called "BKR", takes a univariate polynomial $p(x)$, and polynomials $q_1(x), \ldots, q_n(x)$, and returns k sign sequences $\sigma \in \{-, 0, +\}^n$, where k is the number of real roots of $p(x)$. Each sign sequence σ corresponds to a particular root α of $p(x)$, in such a way that $\sigma_i = \text{sign}(q_i(\alpha))$. Indeed BKR is an indispensible component of these works, providing two benefits: (i) The ability to work over any real closed field, (ii) The possibility of efficient (NC) parallel implementation. Recently, [9] gave an ingenious parallel algorithm for approximating all the roots of a univariate polynomial. His algorithm can easily be used to parallelize earlier algorithms, like [4], so that (ii) can now be achieved without BKR. But the benefits of being able to work over arbitrary real closed fields should not be underestimated, and BKR remains the most effective tool for this. For example, infinitesimal extensions of the reals, which are themselves real closed fields, are extensively used in recent algorithms for computing connected components of semi-algebraic sets [8], [6], [3].

BKR also seems to perform well in implementations, especially if the polynomial $p(x)$ has few real roots. Although definitive testing has not been done as yet, in some recent implementations, it is often faster to determine the signs of polynomials symbolically using BKR than to compute roots of $p(x)$ numerically and then substitute to find the signs of the $q_i(x)$'s. Since any comparison can be reduced to a sign test of a single polynomial, sign determination is a "universal" calculation, or at least as general as root-finding and substitution.

Given the importance of BKR, it is natural to look for improvements and simplifications. This paper is motivated in part by a desire to improve BKR, and in part by a feeling that a more uniform algorithm (fewer potential branches in the calculation) should be possible. For the most part, these goals have been realised. In BKR, the number of potential branches is exponential. More precisely, BKR makes all its decisions based on sign queries of various polynomials that it computes along the way in the coefficients of $p(x)$ and the $q_i(x)$'s. The number of potential queries in BKR is exponential. By a "branch" we mean one of these queries, rather than the strict sense of a branch in a computation tree.

In this paper, we present a sign determination algorithm which has only a pseudo-polynomial number of potential queries. This makes possible a very simple recursive algorithm for sign determination in the multivariate case, where $p(x)$ and $q_i(x)$ have coefficients which depend on other variables. We simply compute all the potential query polynomials, and return a formula that depends only on their signs. This cannot be done efficiently with BKR, and [7] and [10] use clever subterfuges. Both papers sacrifice generality for efficiency, however. They work only if the coefficients of $p(x)$ and $q(x)$ are themselves polynomials in other variables. Our new method does not have this limitation. It works for any $p(x)$ and $q_i(x)$, without restriction on the coefficients, e.g. they may be exponential functions or derivatives of other variables.

Our algorithm is also simple to implement. Like BKR, it works by solving linear systems $Ax = b$ to find the sign combinations. Whereas BKR recursively computes the matrix A by tensoring, our new algorithm allows A to be computed directly. We also get better bounds on the degree of the query polynomials.

26.2 Sign Determination

If $f(x)$ and $g(x)$ are polynomials, let the Sturm sequence of f and g be denoted $r_0, r_1, \ldots, r_k$, where $r_0(x) = f(x)$, $r_1(x) = g(x)$, and the intermediate remainders are computed via

$$r_{i+1} = q_i r_i - r_{i-1} \tag{26.1}$$

where q_i is the quotient of the polynomial division of r_{i-1} by r_i. In practice, pseudo-remainders are often used so that the r_i have integer coefficients. For a real value v, let $\mathrm{SA}(f, g, v)$ denote the number of changes in sign in the sequence $r_0(v), r_1(v), \ldots, r_k(v)$.

Let $\mathrm{SC}(f, g)$ denote the quantity $\mathrm{SA}(f, g, +\infty) - \mathrm{SA}(f, g, -\infty)$. The classical Sturm theorem states that $\mathrm{SC}(f, f')$ equals the number of real roots of $f(x)$. The most general form of the theorem states that $\mathrm{SC}(f, f'g)$ gives the sum of the number of real roots of $f(x)$ where $g(x) > 0$, minus the number where $g(x) < 0$, common roots making no difference to the count. A useful application of this result is to consider $\mathrm{SC}(f, f'g^2)$ which counts real roots of f where $g \neq 0$. This observation was used by [11] to simplify the BKR sign determination lemma.

Lemma 2.1 [11] *Let r^+ denote the set of roots of $f(x)$ where $g(x)$ is positive, and similarly for r^0 and r^-. Then the numbers of roots in each set satisfy the following identity*

$$\begin{bmatrix} 1 & 1 & 1 \\ 1 & -1 & 0 \\ 1 & 1 & 0 \end{bmatrix} \begin{bmatrix} |r^+| \\ |r^-| \\ |r^0| \end{bmatrix} = \begin{bmatrix} \mathrm{SC}(f, f') \\ \mathrm{SC}(f, f'g) \\ \mathrm{SC}(f, f'g^2) \end{bmatrix} \tag{26.2}$$

Now we can apply this lemma to find the signs of a system of polynomials $q_1(x), \ldots, q_n(x)$ at the roots of a single polynomial $p(x)$. We start by applying the above lemma to $p(x)$ and $q_1(x)$. We can rewrite the above identity in matrix form $H_1 R_1 = S_1$, where H_1 is the 3×3 matrix, R_1 is the vector of root counts for q_1 and S_1 is the vector of Sturm query counts for q_1:

$$\begin{bmatrix} 1 & 1 & 1 \\ 1 & -1 & 0 \\ 1 & 1 & 0 \end{bmatrix} \begin{bmatrix} |r_1^+| \\ |r_1^-| \\ |r_1^0| \end{bmatrix} = \begin{bmatrix} \mathrm{SC}(p, p') \\ \mathrm{SC}(p, p'q_1) \\ \mathrm{SC}(p, p'q_1^2) \end{bmatrix} \tag{26.3}$$

There is a more general identity which we can write as $H_n R_n = S_n$ where we define H_n inductively as

$$H_{n+1} = \begin{bmatrix} H_n & H_n & H_n \\ H_n & -H_n & 0 \\ H_n & H_n & 0 \end{bmatrix} \tag{26.4}$$

and where R_{n+1} is the 3^{n+1} vector obtained by listing the elements of R_n with r_{n+1}^+ appended to each conjunction, then the elements of R_n with r_{n+1}^- appended, and finally a last copy of R_n with r_{n+1}^0. Similarly, S_{n+1} is obtained by listing first a copy of S_n, then another copy of S_n with q_{n+1} multiplying the derivative in each Sturm sequence, and then a third copy of S_n with q_{n+1}^2. We can write these definitions schematically as

$$R_{n+1} = \begin{bmatrix} R_n \cap r_{n+1}^+ \\ R_n \cap r_{n+1}^- \\ R_n \cap r_{n+1}^0 \end{bmatrix} \qquad S_{n+1} = \begin{bmatrix} S_n \\ S_n \cdot q_{n+1} \\ S_n \cdot q_{n+1}^2 \end{bmatrix} \qquad (26.5)$$

and it is easy to verify that $H_{n+1}R_{n+1} = S_{n+1}$. It is also easy to see that H_{n+1} is non-singular. Simply add the middle rows to the last rows to give a block diagonal matrix whose determinant is non-zero if the determinant of H_n is.

Of course, the problem with doing things this way is that we need to invert a $3^n \times 3^n$ matrix to compute the signs of n univariate polynomials. One of the major contributions of [1] is a clever method to avoid all this calculation. They observed that if $p(x)$ has k real roots, then at most k of the elements of R_n can be non-zero, since the elements of R_n form a partition of the roots of $p(x)$.

Suppose then, that one has calculated the at most m sign sequences of $q_1, \ldots, q_n$, which will be counted in $\leq m$ non-zero elements of R_n. When we come to compute R_{n+1} we can use the fact that there are three possible signs for q_{n+1} at each root, hence only $\leq 3m$ possible non-zero elements in R_{n+1}. For the other elements we can immediately fill in zeros, and in fact we can delete the corresponding columns of H_{n+1}, giving a $3^n \times 3m$ submatrix. To solve the system, we need only find $3m$ independent rows of this submatrix, and solve a $3m \times 3m$ system. Finding these rows might potentially take a long time, but [1] show that one can use m independent rows from R_n from the previous step, so the whole calculation takes polynomial time (actually, the description above is for a simplified sequential version of their algorithm, the original [1] uses divide-and-conquer to run in NC).

This is fine for univariate problems, since one never needs to make more than $3mn$ Sturm queries. But it is not known in advance which queries will be made, so the number of *potential* queries is still 3^n. This causes problems when one tries to use the algorithm symbolically. This wasnt fully appreciated in [1], and while their univariate analysis is correct, the multivariate generalization given there is not. The present work was originally motivated by an attempt to correct this error in the simplest way possible, namely by finding good a priori bounds on the number of potential Sturm queries.

It seems that this should be possible for the following reason: The number of potential Sturm queries corresponds to the number of rows of H_n that might ever be used during the algorithm. Since the basic operation is to find $3m$ independent rows given $3m$ columns, a natural step is to find a subset of rows of H_n with the property that any $3m$ columns are linearly independent. Then whatever columns we need to work with, we can still be guaranteed to find independent rows within our new submatrix.

Ideally, this submatrix would have a polynomial number of rows. We will not succeed in this, but we do not miss by much. The number of rows turns out to be pseudo-polynomial in m. This is enough to give us some interesting new bounds for more general quantifier elimination.

Definition Our new matrix is denoted $K_{n,m}$ which indicates that it has 3^n columns and any m of them are linearly independent. Thus it is defined only if $m \leq 3^n$. If $m = 1$, $K_{n,m}$ is just the first row of H_n, i.e. a row of all ones. If $m > 3^{n-1}$, then $K_{n,m}$ is just H_n. Otherwise, it is defined recursively as follows:

$$K_{n,m} = \begin{bmatrix} K_{(n-1),m} & K_{(n-1),m} & K_{(n-1),m} \\ K_{(n-1),\lfloor \frac{m}{2} \rfloor} & -K_{(n-1),\lfloor \frac{m}{2} \rfloor} & 0 \\ K_{(n-1),\lfloor \frac{m}{3} \rfloor} & K_{(n-1),\lfloor \frac{m}{3} \rfloor} & 0 \end{bmatrix} \tag{26.6}$$

and now we must show that this definition does indeed give us m independent columns whichever ones we choose.

Lemma 2.2 Any m columns of $K_{n,m}$ as defined above are linearly independent.

Proof For $m > 3^{n-1}$ the result is immediate, since H_n is non-singular. So we assume it is true for all values of n less than or equal to our given n, and for all values of m less than or equal to the given m.

Now consider the definition (26.6), and pick any set of m columns. We suppose that there exists a linear combination of these columns which is zero, and show that all the coefficients are zero. We do this by considering "initial sections" of the columns, which are the rows in the first group of rows in (26.6). Specifically, let $c_{i_1}, \ldots, c_{i_m}$ be the columns, and suppose that

$$\sum_{j=1}^{m} \lambda_{i_j} c_{i_j} = 0 \tag{26.7}$$

Now consider just the initial sections of those columns, whose rows lie in $K_{(n-1),m}$. Let c'_{i_j} denote the initial section of c_{i_j}. Since up to three columns may have the same initial section, the number q of distinct c''s is between $\frac{m}{3}$ and m, depending on how many columns match in initial sections. The linear combination (26.7) of columns from the previous paragraph gives us

$$\sum \lambda_{i_j} c'_{i_j} = 0 \tag{26.8}$$

But the c''s are just columns of $K_{(n-1),m}$ and any m columns of $K_{(n-1),m}$ are linearly independent. So any column section that appears once in (26.8) must have its $\lambda = 0$.

If exactly two column sections correspond, $c'_{i_1} = c'_{i_2}$, then their coefficient in (26.8) is $(\lambda_{i_1} + \lambda_{i_2})$, and it must be zero as before, since there is no nontrivial linear combination of at most m columns of $K_{n-1,m}$. Similarly, if $c'_{i_1} = c'_{i_2} = c'_{i_3}$, then $(\lambda_{i_1} + \lambda_{i_2} + \lambda_{i_3}) = 0$.

Now we explore further the case where columns correspond in initial sections, since we have not yet been able to show that their λ_{i_j} coefficients are zero. Let c''_{i_j} be the *middle* section of the column c_{i_j}, i.e. consisting of rows of $K_{(n-1),\lfloor \frac{m}{2} \rfloor}$. The identity (26.7) implies an identity on middle sections, but the λ_{i_j}'s may be multiplied by $+1$, -1 or 0, depending on where the column occurs. Since the λ_{i_j}'s are zero for c_{i_j}'s with unique initial sections, we can drop these immediately from the sum of middle sections. What remains are pairs and triples with the same initial section, but these also have the same *middle* section, up to a sign change, or a zero middle section. In each case each pair or triple contributes one distinct column to the sum of middle sections. It follows that the number of distinct middle sections appearing in the sum is at most $\lfloor \frac{m}{2} \rfloor$.

Now we use the fact that any $\lfloor \frac{m}{2} \rfloor$ middle sections are linearly independent, since they are columns of $K_{n-1,\lfloor \frac{m}{2} \rfloor}$. If c_{i_1} and c_{i_2} are two columns whose initial sections agree, the coefficient of their middle section is either $(\lambda_{i_1} - \lambda_{i_2})$ if they lie in the first two groups of columns, or $\pm\lambda_{i_1}$, assuming λ_{i_2} lies in the third group of columns. There are at most $\lfloor \frac{m}{2} \rfloor$ such middle sections, so all their coefficients must be zero. From considering initial sections, we know that $(\lambda_{i_1} + \lambda_{i_2}) = 0$, and now we know that either $(\lambda_{i_1} - \lambda_{i_2}) = 0$, or $\lambda_{i_1} = 0$. In either case, it follows that both λ_{i_1} and λ_{i_2} must be zero.

Finally, we must show that the coefficients are zero for triples of columns whose initial sections agree. Again we delete from the sum all coefficients of columns whose initial sections are unique or match only one other column, since we have determined that these must be zero. This leaves us with at most $\lfloor \frac{m}{3} \rfloor$ triples, and each contributes one final section to the sum of final sections. Since any $\lfloor \frac{m}{3} \rfloor$ columns of $K_{(n-1),\lfloor \frac{m}{3} \rfloor}$ are linearly independent, the corresponding coefficients must all be zero. So if c_{i_1}, c_{i_2} and c_{i_3} are columns with the same initial section, we find that $(\lambda_{i_1} + \lambda_{i_2} + \lambda_{i_3}) = 0$ by considering initial sections, $(\lambda_{i_1} - \lambda_{i_2}) = 0$ by considering middle sections, and $(\lambda_{i_1} + \lambda_{i_2}) = 0$ by considering final sections. Thus $\lambda_{i_1} = \lambda_{i_2} = \lambda_{i_3} = 0$.

Since we have shown that all the λ_{i_j}'s are zero in any zero linear combination of m columns of $K_{n,m}$, no non-trivial combination exists. Therefore any m columns of $K_{n,m}$ are linearly independent. □

Observation $K_{n,m}$ is a submatrix of H_n, consisting of rows of H_n.

This observation is easily proved by induction using the recursive definitions of H_n and $K_{n,m}$ in equations (26.4) and (26.6).

Next we make a simple observation which provides good bounds on the degrees of the polynomials in the Sturm queries. Since $K_{n,m}$ is a submatrix consisting of rows of H_n, the product $K_{n,m}R_n$ is a subvector of S_n. Let $\Gamma_{n,m}$ be this subvector, which consists of rows of S_n which correspond to the rows of $K_{n,m}$. Then

$$\Gamma_{n,m} \quad = \quad K_{n,m}R_n \tag{26.9}$$

and we observe the following:

Lemma 2.3 The maximum number of q_i's occuring in any Sturm query in $\Gamma_{n,m}$ is $\lfloor \log_2(m) \rfloor$.

Proof The proof is by induction on n and m. Now $\Gamma_{n,m}$ can be recursively defined analogously to (26.5) using (26.6) as:

$$\Gamma_{n+1,m} = \begin{bmatrix} \Gamma_{n,m} \\ \Gamma_{n,\lfloor \frac{m}{2} \rfloor} \cdot q_{n+1} \\ \Gamma_{n,\lfloor \frac{m}{3} \rfloor} \cdot q_{n+1}^2 \end{bmatrix} \tag{26.10}$$

and the recursion leads eventually to one of the two base cases (i) if $m = 1$, then $\Gamma_{n,m} = S_0$, or (ii) the case $m > 3^{n-1}$, when $\Gamma_{n,m} = S_n$. We prove the bound for both of these cases, and then by induction for the intermediate $\Gamma_{n,m}$.

For the first base case, $m = 1$, none of the q_i's appear in a Sturm query, and the bound holds. For the second case, since $\Gamma_{n,m} = S_n$, n is exactly the maximum number of q_i's occuring in Sturm queries. Now n is less than $\log_3(m)+1$, and since it is integer, it is bounded by $\lfloor \log_3(m)+1 \rfloor$, which is less than or equal to $\lceil \log_2(m) \rceil$ for $m \geq 2$. Since $m = 1$ was already dealt with, this shows that the bound holds for all base cases.

Now for the inductive step, we have only to inspect (26.10) to notice that if the inductive hypothesis holds for all cases on the right-hand side of (26.10), then it holds for the left-hand side. The polynomial q_{n+1} is only added to Sturm query vectors $\Gamma_{n,\lfloor \frac{m}{2} \rfloor}$ and $\Gamma_{n,\lfloor \frac{m}{3} \rfloor}$ which have at most $\lceil \log_2(m) - 1 \rceil$ polynomials, by hypothesis. Since the recursion (26.10) produces a finite tree of intermediate $K_{n,m}$'s with leaves which are base cases, and since the bound holds for a parent whenever it holds for the descendents, this completes the proof. □

Corollary 2.4 The number of rows in $\Gamma_{n,m}$ or $K_{n,m}$ is

$$\binom{n}{\lceil \log_2(m) + 1 \rceil} 2^{\lceil \log_2(m) \rceil} \quad = \quad (n^{O(\log m)}) \tag{26.11}$$

Proof Each entry in $\Gamma_{n,m}$ is a Sturm query, and can be uniquely indexed by (i) the set of (at most $\log(m)$) q_i's that occur in it and (ii) the vector of exponents of these q_i's which are all either 1 or 2. The bound comes from first counting the number of subsets of at most $\log(m)$ q_i's, which is $\binom{n}{\lceil\log_2(m)+1\rceil}$ and then counting the number of possible exponent vectors for each, which is $2^{\lceil\log_2(m)\rceil}$. □

26.3 A Sign-Determination Algorithm

In this section we show how to turn the results of the previous section into a sign-determination algorithm. Although there are similarities with BKR, there is a major difference. Instead of constructing a submatrix of H by tensoring submatrices from the previous step, our method computes a submatrix of K directly from the set of possible sign sequences. The algorithm works inductively as follows (d is the degree of $p(x)$):

1. We assume at the ith step that the algorithm knows for each sign sequence of $q_1, \ldots, q_{i-1}$, how many roots of $p(x)$ produce this sign sequence. Most of these are zero, and the algorithm only stores the sign sequences with at least one root, and the number of these is $m_{i-1} \leq d$.

2. There are $3m_{i-1}$ possible sign sequences for $q_1, \ldots, q_i$, and each of these defines a column of the matrix $K_{i,3m_{i-1}}$. These columns are linearly independent, so there are $3m_{i-1}$ rows which together with the specified columns, define a square submatrix J_i of $K_{i,3m_{i-1}}$.

3. We solve the $3m_{i-1} \times 3m_{i-1}$ system corresponding to this matrix, to find the actual sign sequences of $q_1, \ldots, q_i$, and repeat the above steps for $i = 1, \ldots, n$.

The first task then, is to give a procedure that accepts a list of m columns of the matrix $K_{n,m}$, and returns a list of m rows, such that the resulting $m \times m$ matrix is non-singular. This procedure needs to run in polynomial time in n and m and not the size of $K_{n,m}$.

The second task, which is very easy, is to determine the entries of this submatrix of $K_{n,m}$. Again this must be in polynomial time in m, so we cannot aford to construct all of $K_{n,m}$. This task reduces to determining the value of a single element of $K_{n,m}$ given its row and column indices.

26.3.1 *Computing rows given columns of K*

To simplify our representation of K, we use the fact that $K_{n,m}$ is a submatrix of H_n, consisting of rows of H_n. To reference a row of $K_{n,m}$, we use the row index of the corresponding row of H_n. So $(K_{n,m})_{i,j} = (H_n)_{i,j}$. In this representation, $K_{n,m}$ has many missing rows, but the row indices are always produced by the algorithm we are about to describe, and we ensure that it always returns a row index that does lie in $K_{n,m}$.

Each column index is a trinary number that corresponds in an obvious way with a particular sign sequence. The sign sequence $\text{sign}(q_1), \ldots, \text{sign}(q_n)$ encodes as the trinary number $t_n t_{n-1} \cdots t1$, where

$$t_i = \begin{cases} 0 & \text{if } \text{sign}(q_i) = + \\ 1 & \text{if } \text{sign}(q_i) = - \\ 2 & \text{if } \text{sign}(q_i) = 0 \end{cases} \tag{26.12}$$

Let the trinary column indices corresponding to the m sign sequences be $c_1, \ldots, c_m$. The first observation we make is that we dont need to pass the algorithm the value of n. Each c_i is an n-digit trinary number, but it is possible that all the leading digits are zero, so they look like $(n-1)$-digit or shorter numbers. If this happens, it is perfectly OK to treat them as shorter numbers, because the corresponding set of rows would come from the matrix $K_{(n-1),m}$, even if we had treated them as n-trit numbers. A close inspection of the results of the previous section will confirm this.

Algorithm Rows(L) The input $L = (c_1, \ldots, c_m)$ is a list of m column indices, and the output is a list of m row indices.

1. If $m = 0$ return the empty list. If $m = 1$, return a list of zero (the index of the first row).

2. Otherwise, we determine n by computing $n = 1+\max(\lfloor \log_3(c_i) \rfloor)$, $i = 1, \ldots, m$, which is the maximum number of digits in any column index.

3. We compute from L, three lists L_1, L_2 and L_3 whose total length is m. These lists contain elements of L reduced modulo 3^{n-1}. The set L_1 consists of all values $c_i \text{mod} 3^{n-1}$, whereas L_2 and L_3 consist of elements which are duplicated or occur three times. Specifically, we have

$$\begin{aligned} L_1 &= \{k \mid k = c_i \text{mod} 3^{n-1} \text{ for some } i\} \\ L_2 &= \{k \mid k = c_i \text{mod} 3^{n-1} \text{ for at least 2 distinct values of } i\} \\ L_3 &= \{k \mid k = c_i \text{mod} 3^{n-1} \text{ for 3 distinct values of } i\} \end{aligned} \tag{26.13}$$

4. We then call our algorithm recursively on each non-empty list, and return the list:

$$\text{Rows}(L_1) \circ (3^{n-1} + \text{Rows}(L_2)) \circ (2 * 3^{n-1} + \text{Rows}(L_3)) \tag{26.14}$$

where $\circ$ denotes concatenation of lists, and where the plus sign between a number and a list means that the number is to be added to every element of the list. The two additions effectively shift the rows returned by Rows(L_2) and Rows(L_3) to the middle and bottom bands of $K_{n,m}$ respectively.

The proof of correctness for the algorithm mimics the lemmas of the previous section very closely, and we do not repeat it here. The algorithm is easily seen to be polynomial time, because the recursion depth is at most n, and the total number of columns in *all* calls at any level is m, so the total number of calls is not more than nm.

26.3.2 Computing elements of K

To compute the elements of the submatrix of $K_{n,m}$ consisting of a particular set of m rows and columns, we need only show that there is an efficient method for computing the element $(K_{n,m})_{r,c}$ given the row and column indices r and c. Since we chose the row indices to correspond to elements of H_n, this element is the same as $(H_n)_{r,c}$. Finally, note that we do not even need to know the value of n explicitly, because $(H_n)_{r,c} = (H_{n-1})_{r,c}$ if r and c are both less than 3^{n-1}.

Algorithm Helt(r,c) The input is a pair of trinary integers r and c, and the output is the correponding element of H_n, which is in $\{-1, 0, 1\}$.

1. Determine $n = 1 + \max(\lfloor \log_3(r) \rfloor, \lfloor \log_3(c) \rfloor)$, the size of the smallest H_n containing this element.
2. Let the trinary expansions of r and c be $r_n r_{n-1} \cdots r_1$ and $c_n c_{n-1} \cdots c_1$ respectively. Compute and return the value of the product

$$\prod_{i=1}^{n} (H_1)_{r_i, c_i} \tag{26.15}$$

   where H_1 is the 3×3 matrix defined earlier.

This expression for the value of $(H_n)_{r,c}$ is obtained by unrolling a recursive algorithm based on the definition in (26.4).

26.3.3 Complexity Analysis

Let m denote the number of real roots of $p(x)$, then m is not more than d, the degree of $p(x)$. The algorithm takes n steps to compute the signs of all q_i's and each step involves inverting a $3m \times 3m$ matrix and the calculation of $3m$ Sturm queries. Each Sturm query involves $p(x)$ and a product of at most $2\log_2(m)$ of the q_i's. The number of arithmetic operations for each

Sturm query is $O(d \log m \log^2(d \log m))$, which simpifies to $O(d \log m \log^2 d)$ because d is at least as large as m. So the total time per step for the Sturm calculation is

$$O(md \log m \log^2 d) \tag{26.16}$$

or $O(d^2 \log^3 d)$ in the worst case when the number of roots is roughly d. The cost of inverting the matrix is $O(m^{2.376})$ and so the overall running time is:

$$O(n(md \log m \log^2 d + m^{2.376})) \tag{26.17}$$

The cost of inverting the matrix dominates if m is close to d. However, the actual number of real roots is often much less than d, and is $O(\log d)$ for a random polynomial. So the Sturm query time will often be the larger in practice, especially if naive algorithms are used.

26.4 Conclusions

In this paper we described a new algorithm for determining the signs of a collection of polynomials $q_1, \ldots, q_n$ at the roots of a polynomial $p(x)$. The idea of the algorithm was to find a certain minimal submatrix of the Hadamard matrix H_n which was sufficient for the sign sequence calculations. We showed that such a matrix exists, denoted $K_{n,m}$, and that it has a pseudo-polynomial number of rows. We also showed that each Sturm query involves at most a logarithmic number of q_i's. Finally we described our method for sign determination, the heart of which is an algorithm for computing a non-singular submatrix of $K_{n,m}$ directly from a list of column indices.

Our algorithm has better sequential complexity than the original BKR method, and it has the important property that the number of potential Sturm queries is pseudo-polynomial, rather than exponential as in BKR. This latter property allows us to do quantifier elimination for formulae constructed from polynomials with arbitrary coefficients. This quantifier elimination algorithm is the subject of a forthcoming paper.

26.5 References

[1] M. Ben-Or, D. Kozen, and J. Reif., The complexity of elementary algebra and geometry, *J. Comp. and Sys. Sci.*, 32:251–264, 1986.

[2] J. Canny, Some algebraic and geometric computations in pspace, in *ACM Symposium on Theory of Computing*, pages 460–467, 1988.

[3] J. Canny, Finding connected components of semi-algebraic sets, Technical report, University of California, Berkeley, 1990, manuscript.

[4] G. Collins, Quantifier elimination for real closed fields by cylindrical algebraic decomposition, *Springer LNCS*, 33:135–183, 1975.

[5] D.Y. Grigor'ev, Complexity of deciding tarski algebra, *Journal of Symbolic Computation*, 5:65–108, 1988.

[6] D.Y. Grigor'ev and N.N. Vorobjov, Solving systems of polynomial equations in subexponential time, *Journal of Symbolic Computation*, 5:37–64, 1988.

[7] J. Heintz, M.F. Roy, and P. Solerno, Complexité du principe de tarski-seidenberg, *Comptes Rendu du l'Acad. de Sciences Paris*, 309:825–830, 1989.

[8] J. Heintz, M.F. Roy, and P. Solerno, Single-exponential path finding in semialgebraic sets, in *Conference on Algebraic Algorithms and Error-Correcting Codes*, 1990.

[9] C.A. Neff, Specified precision polynomial root isolation is in nc, in *IEEE Conference on Foundations of Computer Science*, 1990.

[10] J. Renegar, On the computational complexity and geometry of the first-order theory of the reals, parts i, ii and iii, Technical Report 854,855,856, Cornell University, Operations Research Dept., 1989.

[11] M.F. Roy and A. Szpirglas, Complexity of computation on real algebraic numbers, *Journal of Symbolic Computation*, 10(1):39–52, 1990.

[12] A. Seidenberg, Constructions in algebra, *Trans. AMS*, 197:273–313, 1974.

[13] A. Tarski, *A Decision Method for Elementary Algebra and Geometry*, University of California Press, Berkeley, 1948.

27
Decomposition Algorithms in Geometry

Bernard Chazelle
Leonidas Palios

27.1 Introduction

Decomposing complex shapes into simpler components has always been a focus of attention in computational geometry. The reason is obvious: most geometric algorithms perform more efficiently and are easier to implement and debug if the objects have simple shapes. For example, mesh-generation is a standard staple of the finite-element method; partitioning polygons or polyhedra into convex pieces or simplices is a typical preprocessing step in automated design, robotics, and pattern recognition. In computer graphics, decompositions of two-dimensional scenes are used in contour filling, hit detection, clipping and windowing; polyhedra are decomposed into smaller parts to perform hidden surface removal and ray-tracing.

Methods can be classified into two broad categories: In *image space* techniques the underlying space is discretized, sometimes in a hierarchical manner (e.g., quad-trees, oct-trees) and the discretization guides the decomposition algorithms. In this survey we will focus entirely on *object space* methods, which assume that the underlying universe is continuous. These methods focus entirely on the objects themselves and not so much on the universe in which they are embedded. Image space methods tend to be more hardware-oriented, as they can often be hardwired into special-purpose processors; in practice, they are also more directly tied to the needs of computer graphics and image processing. Object space methods are more universal; they make fewer assumptions about the world and and often enjoy a richer mathematical structure. On the other hand they might be less efficient on particular problems for which the hardware lends itself to a particular discretization scheme.

This paper reviews a large number of state-of-the-art decomposition methods. It does not pretend to be a comprehensive survey. Instead, it focusses its attention on methods that perhaps may not always be the solutions of choice in practice but reveal some of the most important current ideas in the design of decomposition algorithms.

27.2 The Two-Dimensional Case

The notion of decomposition related to a certain two-dimensional planar scene can be understood either as the partition of the plane (or a portion of it) into simple pieces, or as the explicit description of the plane tessellation induced by the elements of the scene. The former interpretation leads to problems involving convex partitions (and in particular triangulations), and related optimization questions; the latter leads to problems pertaining to line arrangements, Voronoi diagrams, or line segment and curve intersections.

27.2.1 Polygons

A *polygon* is a subset of the plane that is bounded by a finite set of nonintersecting closed polygonal lines. The segments that constitute these polygonal lines are the *edges* of the polygon, while their endpoints are its *vertices.* A polygon is *simple* if no two edges share any point other than a common endpoint. A simple polygon P is *convex* if it is a convex set, i.e., the line segment connecting any two of its points lies entirely in P, and is *star-shaped* if there exists a point p in P such that the line segment connecting p to any other point of P lies in P. In our discussion below, we will restrict our attention to simple polygons, and so we will omit the modifier "simple".

Triangulation

By *triangulation* of a polygon we mean its partition into non-overlapping triangles whose vertices belong to the vertex set of the given polygon. The triangulation problem has long been one of the central problems in two-dimensional computational geometry; not only does it have a number of applications in computer graphics, computer aided design and manufacturing, and robotics, but it is a very common preprocessing step in a large number of algorithms operating on polygons or planar subdivisions.

Convex polygons can be easily triangulated in linear time by drawing the diagonals that connect a given vertex of the polygon to all its nonadjacent vertices. The same procedure does not apply, however, in the general case, as such diagonals may cross the boundary of the polygon, or lie completely outside it.

An $O(n \log n)$ time algorithm:
The first subquadratic algorithm for the triangulation of an n vertex polygon was given by Garey, Johnson, Preparata and Tarjan [23] in 1978. Their algorithm works in two phases: during the first one, the polygon is *regularized*, meaning that it is partitioned into pieces that are monotone with respect to a fixed direction; then, in the second phase, each monotone piece is decomposed into triangles, thus completing the triangulation. A

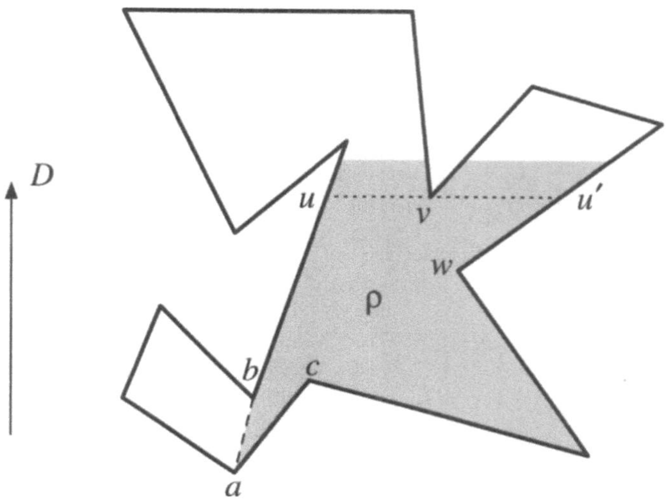

FIGURE 1. Decomposition into Regions Monotone with respect to D.

polygonal chain is said to be *monotone* with respect to a direction D if any line normal to D intersects the chain in no more than one point. A polygon is *monotone* with respect to D if its boundary can be decomposed into two monotone chains (with respect to D). In other words, the source of evil in a non-monotone polygon is the existence of vertices whose incident edges form a chain that is not monotone with respect to D.

The regularization phase fixes a direction D, and employs an algorithm of Lee and Preparata [33] to "resolve" these vertices by adding diagonals incident upon them. It proceeds by sweeping the polygon twice along D, moving in opposite directions. At the general step, the situation is as shown in Figure 1; we assume that the sweepline is moving from top to bottom, in which case only "bad" vertices like b and v, but not c, are resolved. (For instance, vertex b has already been resolved through the diagonal ab.) As a result, the interior of the polygon is split into a number of regions which will eventually yield the monotone partition. Let v, belonging to such a region ρ be the next "bad" vertex to be processed. The line normal to D that passes through v intersects the boundary of ρ in at least two points, two among which, say u and u', define the shortest line segment containing v. If σ is the part of the boundary of ρ that is delimited by u and u' and does not contain v, then v can be resolved by adding a diagonal connecting it to the lowermost vertex of σ (vertex w in the case of Figure 1).

Except for the initial sorting of the vertices of the polygon along D, the

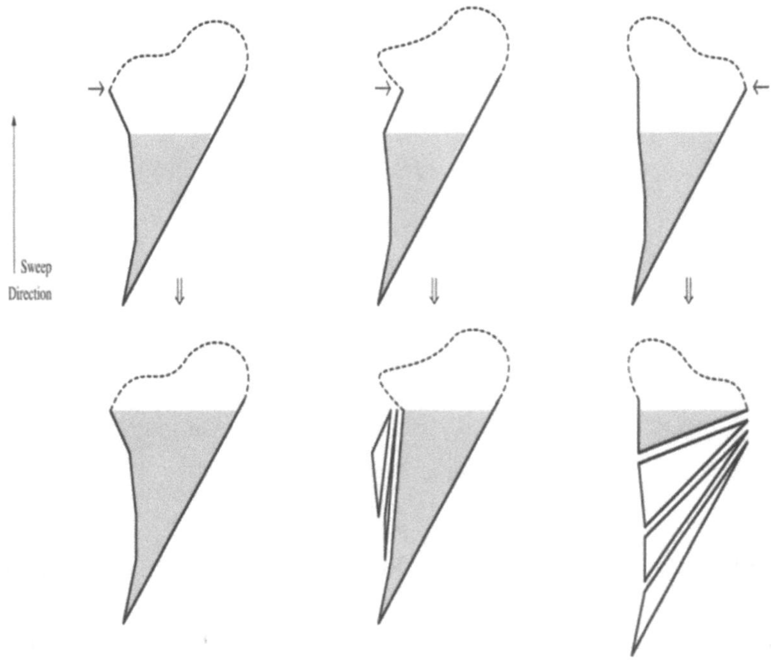

FIGURE 2. Triangulating a Monotone Polygon.

algorithm spends $O(\log n)$ time in each vertex, as it performs at most one Insert, or Delete operation into a balanced tree structure, such as a 2-3 tree, that holds all the information about the monotone regions in which the polygon has been partitioned. This accounts for a total of $O(n \log n)$ time spent in this phase. In fact, Lee and Preparata's algorithm is slightly more general than to our description suggests, and may introduce diagonals that lie outside the polygon; these can be easily located and deleted in linear time, however, or the regularization procedure may be made to ignore them.

The final partition is achieved by triangulating each monotone piece that the regularization phase produces. This can be done in time linear in the number of vertices of the piece for a total of $O(n)$ time. We sweep the piece with a line normal to the direction of monotonicity, maintaining in a stack vertices that form a concave chain. If the next vertex to be processed forms a concave angle with the chain, it is pushed on the stack, thus maintaining the invariant. Otherwise, vertices are popped from the stack (and triangles are reported) until either the vertex forms a concave chain with the rest of the chain, or the latter is reduced to a single vertex. The cases that may arise are shown in Figure 2.

The search for faster algorithms shifted the attention of researchers toward either special classes of polygons, or the discovery and use of the parameters that characterize the complexity of the polygon. In the context of the former trend, Toussaint and Avis [43] presented a linear time triangulation algorithm for edge-visible (a polygon P is edge-visible if there is an edge e of P such that for each point p of P one can always find a point q in e with the line segment pq lying entirely in P) or monotone separable polygons. See also the survey paper of Toussaint [42]. Advances were also made by Hertel and Mehlhorn [27], Fournier and Montuno [21], and Chazelle and Incerpi [11]. In particular, Hertel and Mehlhorn's algorithm has a running time of $O(n + r \log r)$, where r is the number of *cusps* of the polygon, i.e., the vertices of the polygon whose incident edges form an interior angle exceeding π. It is based on the sweep-line paradigm, and combines the two phases of the previous algorithm into one. The idea is to select and sort only a subset of the polygon's vertex set of size proportional to the number of cusps, and then process them in order; all other vertices are reached as we move along the boundary of the polygon and require constant processing time each.

Fournier and Montuno discovered an $O(n \log n)$ time sweep-line method for computing the horizontal (or vertical) *visibility map* of a polygon, which yields its triangulation in linear time postprocessing. The map is obtained by drawing horizontal segments, *chords*, inside the polygon, starting from each vertex and extending all the way across the polygon. The idea is to sort the vertices and process each of them in order, updating the sorted list of active trapezoids stored in a balanced tree.

The algorithm by Chazelle and Incerpi relies on the fact that the notion of visibility map can be extended to polygonal chains, where the horizontal lines that yield the decomposition may extend to infinity. It runs in $O(n \log s)$ time, where s is the *sinuosity* of the polygon, i.e., the number of times its boundary alternates between spirals of opposite orientation. Unlike its predecessors, this algorithm does not sort or sweep the plane. Instead, it computes visibility maps bottom-up by following the mergesort paradigm: the maps of smaller chains of the polygon's boundary are merged together two-by-two in a balanced manner. To achieve linear or even sublinear time, the merging relies on taking shortcuts along the added chords, thus bypassing long portions of the polygon's boundary. As before, a linear time postprocessing step converts the visibility map into a triangulation.

It should be noted that both the sinuosity s and the number of cusps r of a polygon can be as large as $\Omega(n)$, so that none of the algorithms presented so far beats the $\Theta(n \log n)$ time barrier in the worst case. As a matter of fact, this can be shown to be optimal in the case where the polygon contains holes, by using a straightforward reduction from sorting. However, the question of whether the triangulation of a simple polygon without holes was indeed as hard as sorting still remained open for many years. This question was finally settled negatively in 1986 by Tarjan and Van Wyk [41], who presented an $O(n \log\log n)$ time algorithm for the problem.

$O(n \log\log n)$ **time algorithms:**

The visibility map is computed by a clever combination of a linear time *Jordan sorting algorithm* for simple curves (Hoffman, et al. [28]), *divide-and-conquer*, and the fast processing of *finger search trees.* The basic idea in the algorithm is to slice the polygon into smaller pieces by means of a horizontal line L passing through one of its vertices v, and then to apply the technique recursively in the subpolygons produced. The Jordan sorting procedure helps to order along the horizontal line its intersection points with the boundary of the polygon. It is not difficult to see, however, that this approach is doomed if all intersection points are to be reported, as it may be the case that a quadratic number of such points exists. The key is to consider only the *essential* intersection points, i.e., those separating teeth-like portions of the boundary. As a result, a total of $O(n)$ visible pairs are reported. Of course, the Jordan sorting routine has to be appropriately updated to handle this partial information, yielding the *Jordan sorting with error correction.* Apart from that, the balanced divide-and-conquer (that is, making sure that each of the subpolygons that result from the slicing has no more than a constant fraction of the features of the sliced polygon) is used to guarantee that the total number of boundary segments in all regions is $O(n \log n)$. The techniques above in conjunction with the fast access, insertion and concatenation operations in finger search trees result in an $O(n \log\log n)$ time algorithm. Despite the fact that the algorithm has been implemented [32], its intricacy makes it impractical and is only of theoretical interest.

Kirkpatrick, Klawe and Tarjan [31]recently discovered an other $O(n \log\log n)$ algorithm that avoids the use of finger searching. Like Tarjan and Van Wyk's algorithm, it is also recursive in nature and uses a balanced divide-and-conquer approach. To add consistency to the visibility map, it introduces the notion of a *wrap-around partition* of a polygon (or chain): that is the decomposition of the plane induced by horizontal rays through the vertices of the polygon that extend to the left and right until they hit the polygon again. This last condition is always met, since it is assumed that the plane is replaced by an infinite vertical cylinder so that rays extending to the far right emerge again from the far left. Also, the boundary of the polygon is "thickened" so as to make it into a thin snake-like region.

The algorithm works as follows: for $k \geq (n/2)^{2/3}$, the boundary of the given polygon is partitioned into segments of length between k and $k/2$, yielding its *k-uniform partition.* Then the entire polygon is partitioned into *chunks* with the addition of one horizontal chord per segment. The chord associated with a certain segment emanates from a *special point* that splits the segment into two parts containing at least $k/36$ of the vertices of the segment each. Additionally, the endpoints of the chord should split the boundary of the polygon into two chains of size at least $k/36$ each. The special points are found easily, if $k/36$ edges are chopped off from either end of each segment and points are selected from the remaining parts. For a polygon of t vertices, there are $O(t/k)$ special points to be found; locating them and drawing the corresponding horizontal chords takes $O(t)$ time (remember that $k \geq (t/2)^{2/3}$), using standard planar point location techniques. Furthermore, the wrap-around partitions of those chunks that are bounded by at most 2 horizontal chords require another $O(t)$ time. The wrap-around partitions of the remaining chunks are computed by recursive application of the above procedure. Note that each of these chunks is of size at most $t^{2/3}$, while the total number of all their vertices is less than a constant fraction of t. Finally, when all the partial partitions are available, the wrap-around partition of the entire polygon can be constructed applying a linear time merging algorithm similar to [11]. So, each recursive step takes time linear in the size of the current polygon, which amounts to a total of $O(n \log\log n)$ time for the entire algorithm. At least on a conceptual level, this algorithm is much simpler than the Tarjan and Van Wyk method.

$O(n \log^* n)$ expected time probabilistic algorithms:
Clarkson, Tarjan and Van Wyk [16] gave an $O(n \log^* n)$ Las Vegas algorithm, based on random sampling. (A randomized algorithm is *Las Vegas* if it always returns the correct answer but its running time is a random variable.) The algorithm starts by picking a random sample of the polygon edges of the appropriate cardinality, and constructs their visibility map. Then, each of the remaining edges is intersected with the plane partition, and the intersection points of each region are Jordan sorted along its boundary. The information collected during Jordan sorting (the *family trees*) allows the refinement of each region into subregions induced by the non-sampled edges of the polygon. As long as there is some vertex of the polygon whose associated horizontal line has not been drawn, the algorithm recurs on the region enclosing it. The time analysis of the algorithm relies on the observation that, for a random sample of size r out of s noncrossing line segments, with probability greater than some fixed constant, the number of features refining each region is $O(r(\log r)/s)$, while the descriptive size of the entire refined partition of the plane does not exceed $O(s)$. As a result, the Jordan sorting and the region refinement at each level of recursion will take time linear to the size of the input polygon. Finally, selecting a random sample of size $r = s/\log s$ each time the algorithm is applied on a polygon of size s ensures that the vertical visibility decomposition can be computed in $O(r \log r) = O(s)$ time too. It further guarantees that for a polygon of n vertices the recursion depth will be $O(\log^* n)$, achieving the stated time complexity.

Recently, Clarkson and Cole [14] and independently Seidel [40] have announced considerable simplifications of the algorithm which circumvent the use of Jordan sorting.

A Linear Time Algorithm:
An optimal algorithm was recently discovered by Chazelle [7]. The starting observation is that the Chazelle-Incerpi bottom-up method computes too much intermediate information. To speed up the merging process, therefore, only a sample of the chords must be kept around. This sample must be carefully chosen so that the resulting submap represents a balanced approximation of the full-fledged visibility map. Merging is now more difficult because of the missing information, so various oracles are introduced to make it possible. One of these oracles is a primitive that allows us to shoot a ray from within any one of the computed regions and discover which edge (if any) of a given piece of the polygon is hit by the ray. Implementing this oracle requires the implementation of Lipton and Tarjan's planar separator theorem. A major difficulty is that the quality (i.e., the evenness) of the approximation scheme provided by submaps decays as we keep merging submaps together. To repair the damage the algorithm takes advantage of the fact that the duals of submaps have tree structures. It applies a normalization procedure that is best described as a geometric analog of the rotations needed to keep binary trees balanced under dynamic operations. Once a submap of the whole polygon is available, a refinement procedure is called upon to add all the missing chords. This is achieved in a top-down phase that makes use of the auxiliary structures built during the previous merges. Although the algorithm is not a good candidate for practical implementation, it is believed that heuristics based on it will turn out to be useful in practice.

Voronoi Diagrams—Skeletons.

Several geometric problems involve the computation and often the minimization of distances between a given set S of objects. Working on these problems is usually greatly facilitated if for a particular object in S one knows the locus of the points that are closer to it than any other member of S. Such a subset of the plane is called the *Voronoi region* of that object. It is clear from the definition that no pair of Voronoi regions intersect. The tessellation of the plane into Voronoi regions associated with the members of a set S is called the *Voronoi diagram* of S.

Voronoi diagrams are commonly defined with respect to point sets; extensions to segments of straight lines or curves and polygons have also been studied. In the case of polygons, the Voronoi diagram forms their so-called *internal* or *external skeletons.* The internal skeleton of a figure can be more formally defined as the locus of the centers of all the maximal inscribed circles, while the external skeleton of a set of figures coincides with the internal skeleton of their complement [30]. Examples of the Voronoi diagram of a point set, and the internal and external skeletons of two polygons are shown in Figure 3. Finally, it should be noted that the Voronoi diagram of a polygon coincides with the Voronoi diagram of its edges, so that an algo-

FIGURE 3. Voronoi Diagrams of (a) Eight Points, and (b) two quadrilaterals.

rithm operating on line segments can deal with the case of polygons too. Several $O(n \log n)$ algorithms for the computation of the Voronoi diagram of a point set have been proposed, as for example those of Guibas and Stolfi [26], and Fortune [20]. The latter is a sweep-line algorithm, while the former uses divide-and-conquer approach with a linear time merging step of two Voronoi diagrams. Let us mention that another linear time algorithm for merging two Voronoi diagrams follows directly from Chazelle's linear time algorithm for computing the intersection of two convex polytopes in 3-space [6].

The two algorithms that we describe below are applicable not only to point sets, but to sets including line segments or pieces of curves as well. They are both based on the divide-and-conquer paradigm, and they build upon the notion of the *contour* of two sets of the above mentioned objects in the plane. The contour of two such sets P and Q is defined as the locus of all the points in the plane whose minimum distances from both P and Q are equal. If P and Q contain only points, their contour is a set of simple (possibly closed) polygonal lines; if, however, line segments or pieces of curves are allowed in P and Q, the contour may include parabolic, hyperbolic, or elliptic arcs. It should be obvious that in order to merge the Voronoi diagrams of P and Q, we need only to compute their contour, since the Voronoi diagram of $P \cup Q$ coincides with the Voronoi diagram of either P or Q within each region that the contour defines. So, a linear time algorithm to find the contour of two sets guarantees linear time merging of their Voronoi diagrams, resulting in an $O(n \log n)$ time algorithm for the Voronoi diagram computation problem.

Kirkpatrick's algorithm:

In 1979, Kirkpatrick [30] proposed a divide-and-conquer algorithm for the construction of the Voronoi diagram of a point set. The algorithm computes the contour of two point sets using their *Euclidean minimum spanning trees*. The Euclidean minimum spanning trees (EMST) possess two very useful properties: (i) such a tree is a subgraph of the *Delaunay triangulation*, i.e., the dual of the Voronoi diagram of the underlying point set, and can be constructed from it in linear time, and (ii) considering two point sets P and Q on the plane, each edge of the EMST of P or Q intersects either zero or two edges of the contour of P and Q. The latter property can be (and actually is) used to allow the location of points on the contour, and to test whether the entire contour has been computed. Let us briefly outline the merging procedure. We assume that the Voronoi diagrams of two point sets P and Q need to be merged. First, their EMSTs are computed, which are subsequently augmented by a special point far away from the members in P and Q, so that it does not interfere with the edges already in the EMSTs. Clearly, one of the two edges upon which the special point is incident in the trees crosses an edge of the contour, and the point of intersection can in fact be located by moving along the edge in the Voronoi diagrams of P and Q. As soon as this starting point has been located, the corresponding contour component can be traced by following the perpendicular bisector of the current closest pair formed by one point from either point set. If the boundary of a Voronoi region is crossed in any of the partial Voronoi diagrams, the closest pair information is updated. The walk continues until either the starting point is met, meaning that the particular contour component is a closed polygonal line, or the current edge does not cross any other region boundary, and thus extends to infinity. Finally, the contour has been computed in its entirety after each edge of one of the EMSTs has been checked for intersection with the contour and either two or zero such points have been found. In order to estimate the cost of the contour tracing, Kirkpatrick introduces the notion of *spokes*. These are line segments that connect each point of a point set P with the vertices of the Voronoi region enclosing it in the Voronoi diagram of P. The introduction of spokes in a Voronoi diagram restricts the descriptive size of each region to a constant. Then, the cost of tracing a path is equal to the number of spokes and region boundaries crossed, which implies that the total time for the contour construction is linear in the combined size of the two point sets. The Voronoi computation can subsequently be finished off by clipping the partial Voronoi diagrams with the contour components. As the merging is carried out in linear time, the overall time required to compute the Voronoi diagram of n points is $O(n \log n)$.

Kirkpatrick then turned his attention to a generalized Voronoi diagram, that extends the known definition to both points and open line segments (line segments without their endpoints). In this case, the Voronoi regions may be bounded by line segments, half-lines, and parabolic segments. The previous algorithm can be extended to allow the construction of generalized Voronoi diagrams. As we are this time dealing both with points and open line segments, EMSTs are not applicable. Instead, *pseudo-minimum spanning trees*, defined next, can be used. Let us assume that a constant number of visibility directions have been specified, and the visible pairs within a set P of points and open line segments have been computed. Then, if the closest points between the members of any such pair are not endpoints, artificial endpoints are introduced at their locations. The minimum spanning tree of the set of points and of the enlarged set of endpoints of the line segments in P forms its pseudo-minimum spanning tree. The algorithm for the generalized Voronoi diagram computation can be then described as follows: for a given set of points and open line segments, its pseudo-minimum spanning tree is computed, and is separated into two connected components each of which contains at least a constant fraction of the elements in the initial set. After the generalized Voronoi diagrams of the two subsets that correspond to these connected components have been constructed, they are merged in the same way that was used in the standard Voronoi diagram construction, with the exception that EMSTs are replaced by pseudo-minimum spanning trees. The algorithm again runs in $O(n \log n)$ time.

Yap's algorithm:

Eight years later, C.K. Yap [44] proposed a different algorithm for the construction of the Voronoi diagram of n circular and straight line segments that intersect only at their endpoints. As mentioned earlier, in this case the Voronoi diagram consists of straight line, parabolic, hyperbolic and elliptic segments. The algorithm is based on the divide-and-conquer paradigm, too: vertical lines are introduced through the segment endpoints decomposing the plane into *slabs*; then the Voronoi diagram of the entire arrangement is computed by constructing the Voronoi diagram of each slab, and merging pairs of adjacent slabs in a tree-like fashion. If, however, segments contribute pieces in many slabs, this approach may take as much as $\Omega\left(n^2\right)$ time if the Voronoi diagram in each slab is computed in its entirety. The way to avoid that is by computing only the essential part in each slab, whose size is proportional to the number of segment endpoints lying in the slab. In particular, segments that cross both the bounding vertical edges of a slab subdivide it into regions called *quads*. The computation of the Voronoi diagram in the slab is then restricted to only those quads that contain at least one endpoint of some segment. This complicates the merging process, which can still, however, be carried out in linear time, resulting in an overall $O(n \log n)$ running time. It must be noted that, as in the algorithm of Kirkpatrick, the merging process introduces spokes splitting the Voronoi regions into subregions of no more than four sides, and also relies on the contour computation.

Optimization Problems.

Several decomposition problems, such as polygon triangulation, admit more than one valid solution. Quality criteria can then be used to select the best one among them, thus giving rise to optimization questions. For the polygon decomposition problem, such criteria usually involve the number of pieces produced, the total length of the cuts introduced, or the minimum angle between two edges. Interestingly, the existence of "holes" in the polygons makes a big difference in the complexity of the problems. We analyze these two cases separately.

Hole-free Polygons:

The most natural optimization question regarding a given polygon involves the minimization of the number of simpler pieces in which it can be decomposed. Such pieces may be convex, monotone, or star-shaped polygons, triangles, or trapezoids. Nearly all algorithms suggested for the resolution of these questions invariably employ dynamic programming.

Partition into the minimum number of convex polygons:

Chazelle and Dobkin [8] in 1979 presented an algorithm for achieving a partition of a polygon into the minimum number of convex pieces, establishing that this problem can indeed be solved in polynomial time. They introduce the notion of X-patterns that can be used to resolve several notches (i.e., vertices with reflex angles) at the same time. In particular, an X_k-pattern is an interconnection of k notches which removes all reflex angles at the notches involved without creating any new ones (Figure 4). A very important observation is that if the r notches of a polygon are resolved by m X-patterns and some additional line segments, the number of convex pieces produced is at most $r+1-m$; so, the more X-patterns, the fewer convex pieces. Unfortunately, X-patterns have many degrees of freedom, thus making the minimization computation extremely time-consuming. Chazelle and Dobkin introduce a more constrained version of the X-patterns, the Y-patterns. A Y_k-pattern is an X_k-pattern such that no edge joins two Steiner points. For example, the X_3-pattern of Figure 4 is a Y_3-pattern, whereas the X_5-pattern of the same figure is not a Y_5-pattern. The crucial property of Y-patterns is that all X_k-patterns, except for $k = 4$, can be transformed into Y-patterns. The desired set of X_4- and Y-patterns is constructed by applying dynamic programming. The running time of the algorithm is $O\left(n + r^3\right)$, where r is the number of notches of the polygon.

Partition into the minimum number of trapezoids:

Asano and Asano [1] gave an $O(n^3)$ time algorithm, for the case where the pieces of the partition are restricted to trapezoids with their parallel edges parallel to a given direction D. The algorithm makes use of the *minimally effective* diagonals. These are diagonals that either are parallel to D and

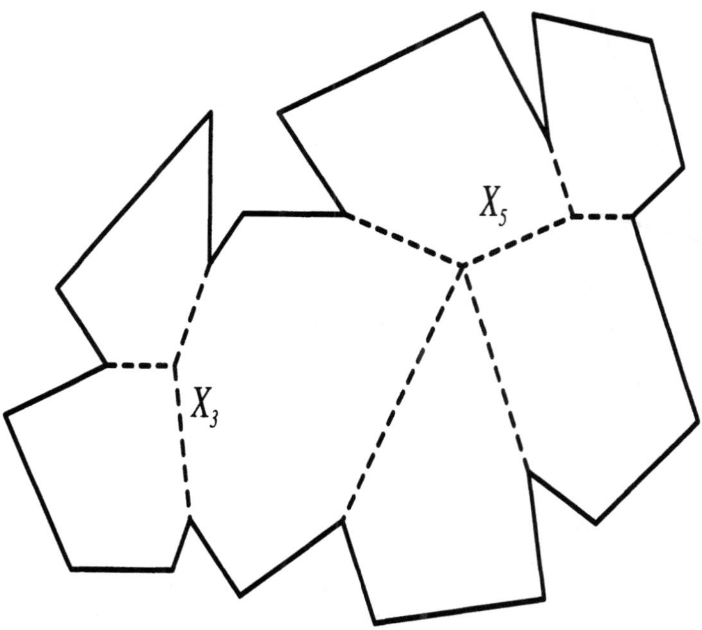

FIGURE 4. A Decomposition using X-patterns.

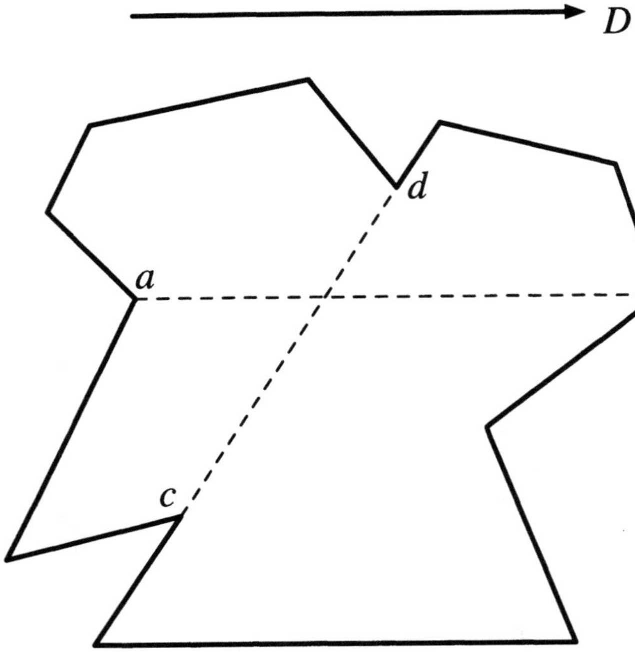

FIGURE 5. The Diagonals *ab* and *cd* are Minimally Effective.

connect two vertices of the polygon, or join two vertices that are incident upon collinear edges (Figure 5).
Since each such diagonal saves exactly one trapezoid in the final decomposition, the largest subset of non-crossing such diagonals ensures a minimum-size decomposition. The algorithm starts by identifying the set of minimally effective diagonals, and reduces the problem of selecting the maximum-cardinality non-intersecting subset among them to finding the maximum independent set of a *circle graph*. The nodes of this graph correspond to the diagonals, while the intersection information is captured in the edge set. The latter problem can be solved by using an algorithm of Gavril [24]. The key observation is that one can reduce the task of finding a maximum independent set of a circle graph of m nodes to m iterations of a routine for finding a maximum independent set on an interval graph. Gavril gave a quadratic time routine for the latter problem. After the appropriate diagonals have been selected, they are added into the polygon. Each subpolygon thus produced can be easily split further into trapezoids by drawing line segments parallel to D emanating from its vertices. The most

expensive component of the algorithm is the maximum independent set routine. Since there may be as many as $O(n)$ minimally effective diagonals it accounts for $O(n^3)$ time in the worst case, which is also the worst-case time performance of the entire algorithm.

The time complexity was improved to $O(n^2)$ in a later paper of Asano, Asano, and Imai [2], where a different routine for the selection of the maximum independent set of an interval graph is used. This routine involves a sorting step and then proceeds in linear time. In this way, all n iterations can be carried out in a total of $O(n^2)$ time.

The algorithms above assume that *Steiner points* are allowed in the decomposition, that is points other than the vertices of the given polygon may be used as vertices of the pieces in the decomposition. For the case in which Steiner points are disallowed, Greene [25] gave an $O(r^2n^2)$ time algorithm for the partition of a polygon into the minimum number of convex pieces. Later, Keil [29], improving on an earlier algorithm, achieved a worst-case time performance of $O(r^2 n \log n)$. The algorithm is based on the observation that each of the diagonals in the partition must be incident upon at least one notch of the polygon. So, the algorithm starts by locating all the notches and finding the vertices of the polygon that are visible from them. Subsequently, a dynamic programming routine computes the minimum decompositions of the subpolygons that the previously computed visible pairs define. The subpolygons are processed in order of increasing size. Unfortunately, the number of valid minimum decompositions may be extremely large; the key is to partition them into equivalence classes and retain a single representative for each such class. This allows the computation of the minimum piece convex partition of a polygon of n vertices and r notches in $O(r^2 n \log n)$ time. The technique can be extended to deal with the partition of a polygon into the minimum number of star-shaped, spiral, or monotone pieces ([29]).

Finally, one may ask for a partition of a polygon that minimizes the total length of the cuts. In the case of a convex partition, and under the additional restriction that Steiner points are disallowed, Keil's dynamic programming approach yields an algorithm requiring $O(r^2 n^2 \log n)$ time in the worst case [29].

Polygons with holes:

In contrast to the previous paragraph, all minimization questions discussed become NP-hard when applied to polygons containing holes. Reductions were drawn from 3SAT or planar-3SAT both known to be NP-complete ([22], [34]).

In the case where Steiner points are allowed in the decomposition, Lingas [35] gave a reduction from planar-3SAT showing that the partition of a polygon in the minimum number of convex pieces is NP-hard. The problem remains NP-hard if the convex pieces are restricted to triangles. Furthermore, O'Rourke and Supowit [38] showed that covering a polygon containing holes with the minimum number of convex pieces (that may be overlapping) is NP-hard by reducing 3SAT to it. Finally, Lingas et al. [36] showed that the problem of minimum edge-length partition of a polygon with holes is NP-hard using a transformation from planar-3SAT.

Disallowing Steiner points does not help in obtaining polynomial time algorithms for these questions. Instead, Lingas [35] showed that the convex partition problem remains NP-complete. NP-completeness results have also been established by O'Rourke and Supowit [38] for the minimum piece covering of a polygon with holes, and by Keil [29] for the minimum edge length convex partition.

27.2.2 Line Segments and Curves.

The problem of line segment intersection can be phrased as follows: Given a set of n line segments, report all their pairwise intersections. Let k denote the number of these intersections. This counts intersections that are combinatorially but not necessarily geometrically distinct. For example, if three segments intersect in one point, k is $\binom{3}{2} = 3$. For line segments, the quantity k may range from a constant up to $O\left(n^2\right)$. As a result, the complexities of the presented algorithms are expressed in terms of both n and k. Although most recent work has been restricted to line segments, extensions to curves have also been described.

An $O\left((n + k)\log n\right)$ time algorithm:
The first non-naive algorithm for line segment intersection is due to Bentley and Ottmann [4]. It runs in $O\left((n + k)\log n\right)$ time and requires $O(n+k)$ space. The algorithm employs the sweep-line paradigm: the endpoints of the segments are sorted according to their x-coordinates; then a vertical line sweeps the plane, stopping at each endpoint or intersection point detected during the sweep. Newly discovered intersection points are inserted into a priority queue using their x-coordinates as ranks. At all times, a balanced binary tree stores the ordered list of segments that are crossed by the vertical sweep-line. Processing a left endpoint involves inserting the corresponding line segment into the tree, and checking it for intersection with its neighbors above and below; processing a right endpoint involves deleting the corresponding line segment from the tree and checking its immediate neighbors for intersection. If the tests for intersection discover intersection points, the latter are inserted in order in the list of points. Processing an intersection point involves switching the order of the intersecting line segments in the tree, and checking no more than two pairs of line segments for intersections, again inserting them (if any) appropriately. The processing of an endpoint or intersection point takes $O(\log n)$ time, so that the total running time is $O\left((n + k)\log n\right)$. The same approach can be extended to curves that are monotone in the x-direction provided that their intersections among themselves and with vertical line segments can be computed

effectively.

A modification suggested by Brown [5] reduces the space requirement of the previous algorithm from $O(n+k)$ to $O(n)$. The idea is to reduce the number of intersection points stored in the ordered list of points pending processing down to $O(n)$. The solution is to maintain at most one intersection point per line segment, which will of course be the one closest to the sweep-line. The other intersections detected can be ignored at that time, for they will be rediscovered and reported later.

An $O(n \log n + k)$ time algorithm:

The first time-optimal algorithm for the line segment intersection problem was proposed by Chazelle and Edelsbrunner [9]. Although it, too, employs the sweep-line paradigm, a number of enhancements and additional techniques are used to guarantee the $O(n \log n + k)$ running time. The algorithm processes the segments in increasing abscissae of their leftmost endpoints, and maintains their (vertical) visibility map. Processing a segment involves locating its left endpoint in the map, in which it is inserted by traversing the regions that it intersects. The problem that arises is that a segment may cross many visibility chords, and the incurred cost may be prohibitive. To overcome that difficulty, Chazelle and Edelsbrunner split the segments into $O(\log n)$ pieces each, using a segment tree. Although visibility chords are attached to these artificial endpoints, as if the pieces resulting from the same initial line segment were independent, each piece is still marked with the index number of the segment it came from. This fragmentation of the line segments implies that many subsegments may have endpoints with the same x-coordinate, so that the visibility chords at these artificial endpoints consist of a large number of small edges, making it costly to scan them all every time such a chord is traversed. So, pointers are associated with these edges enabling fast location of the immediately previous and next non-vertical edges of the neighboring regions.

The location of the region of the visibility map in which the left endpoint of a segment falls, is carried out with the help of the *sweep tree*, a balanced tree that stores the edges that the current sweep-line intersects. It should be noted at this point that unlike most sweep-line algorithms, the sweep-line is not assumed to be straight or even monotone for that matter. In order to describe how the sweep tree is used, the concept of *x-walks* needs to be presented. Let us assume for simplicity that no three line segments intersect at the same point, and for a point p in a segment s let us consider the following path: starting from p, we proceed along s from left to right, up to either the right endpoint of s, in which case the path ends, or the intersection point of s with a segment s', in which case the path continues along s' from left to right. The path that is thus traced is an x-walk starting at p. Figure 6 displays the different x-walks defined on an arrangement of line segments. These walks partition the set of edges of the arrangement into equivalence classes, where two edges belong to the same class, if they belong to the same x-walk. If we want to find the relative position of a point p with respect to the edges that the vertical line through p intersects, we need not have stored these edges explicitly. We only need one representative of each of the equivalence classes in which these edges belong. Putting it another way, the sweep tree invariant is that no edge in the tree lies completely to the right of the vertical line through the current point, and for each edge e that this line intersects there exists an x-walk starting from some edge in the tree and leading to e. This is due to the lazy way in which the tree is updated. Namely, suppose that a segment s is to be inserted in the visibility map. If s is not the leftmost piece of an initial segment t (resulting from the segment tree fragmentation) then its position can be found from the previous piece of t, by means of an $O(n)$ size *appearance table*. Otherwise, the tree is traversed starting at its root. If the edge e stored in some node t of the tree that is visited during the traversal intersects the vertical line through the left endpoint of s, then the test above/below between s and e can be readily carried out, and a decision is made regarding which child of t must be visited next. If not, then starting from e an x-walk is traced until an edge intersecting the current vertical line is found, which reduces to the previous case; alternatively, the x-walk may terminate, in which case the tree node is deleted, and a new tree traversal according to the above guidelines starts at the root of the tree. As a matter of fact, the contents of the tree lag well behind the current vertical line, as only the absolutely necessary part of the tree is updated during insertions. Note that if the right endpoint of a line segment is reached no deletion operation is issued in the tree; deletion may happen later as a side effect of some insertion.

Finally, moving from region to region of the visibility map requires the ability to locate the intersection point of a segment and such a region. This can be carried out by walking along the boundary of the region until the point is found (if it exists). A better way, however, is to use *dovetailing*, where two walks proceeding concurrently in opposite directions along the boundary are initiated. The walks stop as soon as one of them locates the point of intersection, or they cross in which case no such point exists. Dovetailing serves also another purpose; it adds vertical *shortcut* edges in some regions, which speeds up subsequent walks along their boundaries.

Although the algorithm is not very complicated (it was implemented with no great difficulty), its time analysis requires an intricate amortization argument. The time analysis of the algorithm is rather complicated. Since the algorithm produces the complete description of the map of the fragmented line segments, the space required is $O(n \log n + k)$, but Chazelle and Edelsbrunner note that it can be easily reduced to $O(n+k)$. To obtain an $O(n)$ bound remains an open problem.

Probabilistic Algorithms:

In 1989, Clarkson and Shor [15] described three randomized algorithms, each running in $O(n \log n + k)$ expected time. We will describe the most

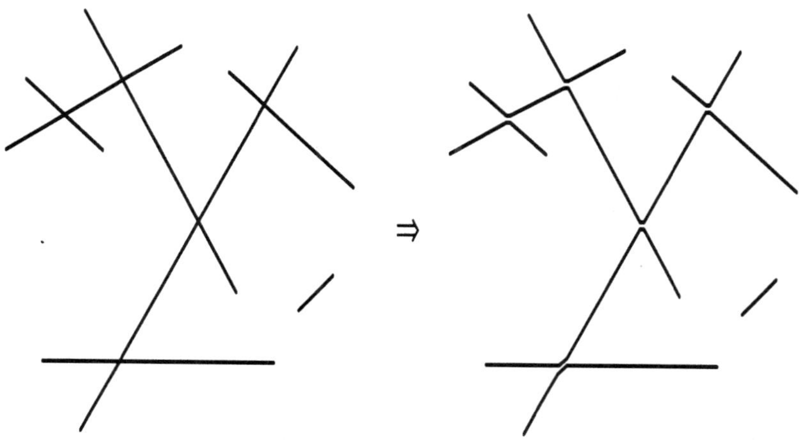

FIGURE 6. A Set of Seven Line Segments and its x-walks.

space efficient one, requiring only $O(n)$ space in the worst case. Let us outline before, however, one of the other two algorithms that is used in the space optimal algorithm. This algorithm processes the segments in random order, and maintains their visibility map in one of the standard planar subdivision representations. In addition, a "conflict" graph is maintained storing pairs of trapezoids in the map and segments that have not been processed yet. In particular, each trapezoid t is associated with a list of segments with an endpoint lying in t, whereas each segment is associated with the trapezoids that contain its endpoints. The addition of a new segment s is carried out as follows: the trapezoid that contains an endpoint of s is found using the conflict graph; then walking along s, we move from trapezoid to trapezoid that s intersects. Trapezoids may be split into no more than four pieces, some of which may merge with adjacent ones into larger cells, as visibility chords may shorten due to the presence of s. In either case, the planar reconfiguration is followed by appropriate updating of the conflict graph. Processing n line segments in this way requires $O(n \log n + k)$ expected time and $O(n + k)$ space.

We are ready now to present the algorithm that requires linear space in n. Clearly, being entitled to only that much space, one cannot store the visibility map of the line segments in its entirety. Instead, divide-and-conquer is applied, until the size of the subproblems becomes $O\left(\sqrt{n}\right)$, the space sufficing for the entire map of each such subproblem. The algorithm begins by picking a "good" random sample of segments of size roughly $\sqrt{n}$;

a sample R of size r is considered *good* if any trapezoid in the visibility map of R intersects $O(n(\log r)/r)$ line segments, while the total number of intersections over all trapezoids is $O(n)$. For each such trapezoid T, the list of line segments S_T that intersect it is constructed, and the algorithm recurs on S_T. A second recursive step on the trapezoids that a good random sample of S_T defines brings the number of line segments conflicting with a particular trapezoid down to $O\left(n^{1/4}(\log n)^{3/2}\right)$. Then, the algorithm that we outlined earlier in this paragraph can be applied. The question that remains is how to pick a good sample, and compute the lists of intersections. To do that, two methods running in parallel are used. The first one works in a brute-force fashion: picks a random sample, constructs its visibility map, and computs the sizes of the intersection lists for each cell in the visibility map. If this computation implies that the sample is not good, a new sample is chosen, and the process is repeated. Otherwise, the chosen sample is adequate, and the actual lists of intersections for the trapezoids are constructed. The second method picks a random sample too, and applies the algorithm that we outlined earlier processing the segments in the sample first. Again, if at any time during the processing of the non-sampled segments, the number of conflicts disqualifies the chosen sample from being good, a new sample is selected, and the method is restarted.

The time analysis of the algorithm relies on two important observations pertaining to a random sample of size r out of a set of n line segments that have a total of k intersections. The first provides a bound of $O\left(kr^2/n^2\right)$ on the expected number of intersections among the members of the sample. The second states that (i) with probability at least 3/4, the total number of conflicts is $O(n+kr/n)$, and (ii) with probability at least $1-1/n^{10}$, each individual trapezoid conflicts with $O(n(\log n)/r)$ segments. The conjunction of these two observations guarantees that a good random sample is found in $O(n\log n+k)$ time; if $k > n\sqrt{n}$ the first method succeeds, otherwise it is the second one that succeeds. Since the two methods run in parallel, the total time for the computation of the conflict lists is $O(n\log n+k)$, and the stated time complexity follows by a two-level induction.

A similar randomized algorithm for the line segment intersection problem has been proposed by Mulmuley [37]. Its expected running time is $O(n\log n+k)$ and its space requirement is $O(n+k)$. Mulmuley assumes the existence of a rectangular box, the *window*, that encloses all the line segments. The algorithm computes the visibility map of the segments clipped within the window. Note that visibility chords, called *attachments*, are drawn through all the endpoints and points of intersection. A key point, is that each face of the partition is associated with only those of the vertices on its boundary that witness a tangent discontinuity. For example, in the Figure 7, the face F_1 is considered to have only u, v, w and z as vertices. In this way, each face has no more than 4 vertices. Initially, the partition is nothing but a collection of $2n+2$ strips induced by the attachments through the endpoints of the line segments. Then, the line segments are

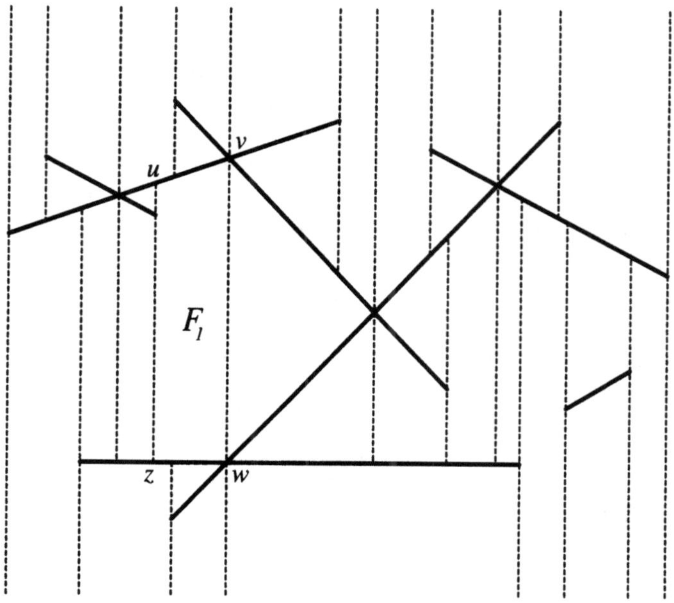

FIGURE 7. The Visibility Map of a Set of Seven Line Segments.

picked, one at a time, and are added to the partition, causing faces to split or merge, and attachments to contract. For a particular segment s, one of its endpoints is used to determine the first face F that s crosses. Then, the boundary ∂F of F is visited until the point of intersection of s and ∂F is found. Now F needs to be split. Furthermore, unless the point of intersection is the other endpoint of s, an attachment needs to be built at that point, and the same process needs to be pursued in the next face that s enters. If the crossing occurs on an attachment, this next face can be located in constant time; otherwise, the segment bounding the two faces is followed until a point of intersection or an endpoint is found. Note that only these points contain the necessary adjacency information to allow the crossing. The time bound of $O(n \log n + k)$ follows from an upper bound of $O(n \log m + k)$ on the expected number of points of attachment visited during the face transitions in the processing of all n line segments, and an upper bound of $5k + 3n + 2n \ln m$ on the expected number of face splits, where $m = O(n)$ is the *average span length*, i.e., the average number of segments that a vertical line through some endpoint intersects.

27.3 The Three-dimensional Case

While two-dimensional decomposition problems are reasonably well-understood, the same cannot be said of their three-dimensional counterparts. For on thing the complexity of the problems tends increases substantially as we move from two to three dimensions. A classical example is the fact that although any polygon can be partitioned into triangles without using Steiner points, not all polytopes can be partitioned into tetrahedra if no Steiner points are allowed. Actually, to decide whether a polytope admits such a partition is NP-complete, as was shown recently by Ruppert and Seidel [39].

Allowing Steiner points, however, reduces the complexity of the problem to polynomial in the size of the polytope. Indeed, Chazelle and Palios [12] have given an $O\left((n + r^2)\log r\right)$ time algorithm for partitioning an n-vertex polytope with r reflex edges into simplices. The algorithm proceeds in two phases: in the first one, the *pull-off* phase, the size of the polytope is reduced to $O(r)$; in the second, the *fence-off* phase, the resulting polytope is partitioned into at most a quadratic number of cylindrical pieces. Central to the pull-off phase is the notion of the *cup* of a vertex that is not incident upon a reflex edge. Let H_v be the convex hull of the vertices adjacent to such a vertex v. The facets incident upon v touch the boundary of H_v and separate it into two patches, one of which, say π, lies between the other patch and v. The cup of v is the polytope bounded by π and the facets of the polytope incident upon v (Figure 8). The idea is that if the cup of v is not hindered (i.e., no vertex of the polytope lies in the interior of the cup(v) or on the cup's boundary that is not adjacent to v), then vertex v and its cup can be removed from the polytope. In fact, there exist at most $2r$ vertices whose cups are hindered. If one subtracts also the vertices that are incident upon reflex edges, which are no more than $2r$ in number, one is left with at least $n - 4r$ vertices whose cups can be removed from the polytope. If among these vertices we always select and remove those that have degree bounded above by some prespecified constant, and iterate on what is left, the tetrahedralization of the removed cups produces a number of tetrahedra linear in n. The entire phase is carried out in $O\left((n + r^2)\log r\right)$ time. In the fence-off phase, vertical *fences* are erected through each edge of the triangulated boundary of the polytope, partitioning it into cylindrical pieces. A triangulation of one of the bases of such a piece is used to yield its decomposition into tetrahedra. This phase runs in $O\left(r^2 \log r\right)$ time, and produces $O\left(r^2\right)$ tetrahedra. Summarizing, the running time of the entire algorithm is $O\left((n + r^2)\log r\right)$, and the total number of tetrahedra produced is $O\left(n + r^2\right)$, which is optimal in the worst case.

The application of decomposition algorithms brought up two important issues: that of the robustness of the algorithm due to finite precision arith-

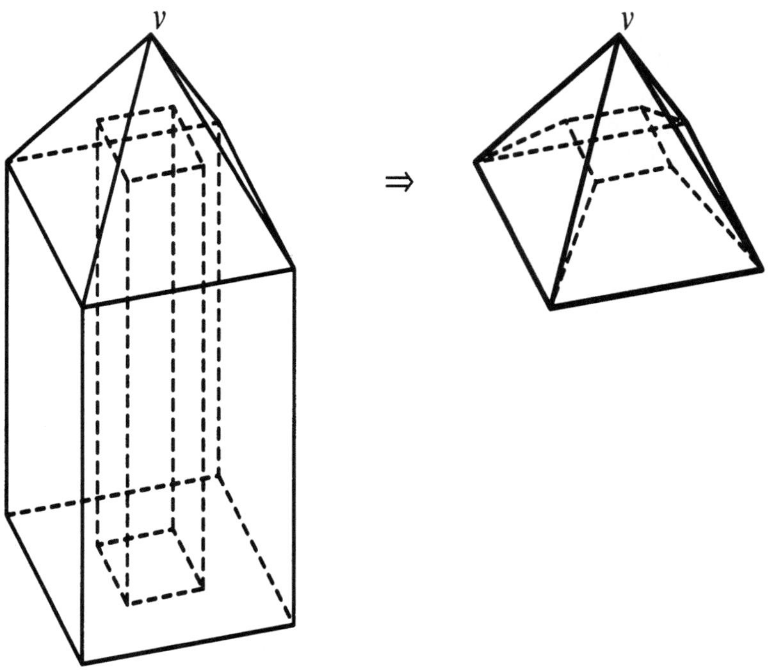

FIGURE 8. A polytope and the cup of vertex v.

metic, and of the "quality" of the produced decomposition. The question of robustness has been receiving considerable attention, and researchers have proposed several sets of primitives to deal with numerical innacuracies both in a general setting and for particular problems. For the problem of partitioning a nonconvex polytope into convex pieces, a numerically robust algorithm was presented by Bajaj and Dey [3]. Finally, in a recent paper, Dey, Bajaj, and Sugihara [19] described an algorithm that triangulates the convex hull of a point set in three dimensions avoiding the creation of flat or long and thin tetrahedra. The algorithm is an extension of an earlier algorithm of Chew [13] for the two-dimensional case, and is based on constrained Delaunay Triangulations.

The idea of a cylindrical decomposition is also exploited by Collins [17] in order to eliminate quantifiers from a prenex formula in first-order logic over the reals. Collins' idea is to partition $\mathbf{R}^d$, where d is the number of variables, into a regular cell complex whose faces are sign-invariant with respect to the polynomials of the formula. The faces of the decomposition are built by lifting cylindrically faces of a base decomposition defined recursively in dimension $d-1$. Each face of the decomposition of $\mathbf{R}^{d-1}$ is made the base of a cylinder, say, in direction x_1, which is then cut off by the real zerosets of the polynomials: in this way, when regarded as polynomials in x_1 (with coefficients in the ring of $(d-1)$-variate polynomials with rational coefficients) the real roots of the polynomials are delinated along any line parallel to x_1 that cuts a given face of the base decomposition. The entire construction produces $O\left(n^{2^d-1}\right)$ cells, and requires $O\left(n^{2^{d+6}}\right)$ time, where n is the number of polynomials in the given formula, and d is the number of variables involved. It is worth noting that the doubly-exponential time bound of Collins' construction is essentially the best one can hope to achieve for quantifier elimination, as shown by Davenport and Heintz [18]. Of course, this is not saying that any sign-invariant cell complex must suffer from this asymptotic blow-up.

Indication to the contrary was given by Chazelle, Edelsbrunner, Guibas, and Sharir [10], who described an algorithm for constructing a sign-invariant stratification of size $O\left(n^{2d-2}\right)$ in $O\left(n^{2d-1}\log n\right)$ time. This is a sign-invariant partition of $\mathbf{R}^d$ into smooth manifolds of constant complexity. It is not a regular cell complex, however, as the cells do not "glue" together; in particular the intersection of the closures of two adjacent cells need not be itself a cell of the stratification. Another drawback is that, although the size is singly exponential in d (which takes us closer to the obvious $\Omega\left(n^d\right)$ lower bound and the Thom-Milnor upper bound of $O\left(n^d\right)$ on the number of connected components in the complement of the varieties defined by the polynomials), the degrees of the intermediate polynomials needed to define the stratification are doubly-exponential in d.

Like Collins', the method has a recursive structure. Let F be a collection $\{f_1, \ldots, f_n\}$ of polynomials with rational coefficients. The sign-invariant stratification of $\mathbf{R}^d$ for F is constructed by lifting cells from a base decomposition. The difference with Collins' method is that we use several base decompositions, so as to provide a pool of cells from which we can choose the ones to be kept in the final stratification. Each cell has a *roof* and a *ceiling*, so we consider all pairs of polynomials to identify these roof/ceiling pairs. Let $\bigvee f_i$ denote the real zeroset of f_i. For each pair $i \leq j$, we project onto $\mathbf{R}^{d-1}$ the intersection of the variety $\bigvee f_i \times f_j$ with each of the $\bigvee f_k$. We also project the silhouettes of all the varieties (i.e., the critical sets of their projection maps). By expressing these projections as subresultants, we can treat them by means of polynomials in $d-1$ variables. This allows us to proceed recursively and thus produce a sign-invariant cell decomposition of $\mathbf{R}^{d-1}$. Next, we lift this decomposition cylindrically into a cell decomposition of $\mathbf{R}^d$. Finally, we use the variety $\bigvee f_i \times f_j$ to chop off the vertical cylinders into cylindrical cells. This provides us with all the cells of the final stratification whose roof/ceiling pair is defined by f_i and f_j (and many other useless ones, of course). We repeat the entire procedure for all pairs f_i, f_j, which gives us a total of $\binom{n+1}{2}$ intermediate cell decompositions of $\mathbf{R}^d$. Next, we examine every cell of every cell decomposition in turn, and keep only those that are free of intersections with any variety $\bigvee f_k$. These candidate cells might still be intersecting, so we add one final selection criterion based on the indices of their defining polynomials. This gives us a collection of mutually disjoint cells which, together with their "upper" boundaries, partition $\mathbf{R}^d$ into the desired sign-invariant stratification.

What is the use of a sign-invariant stratification of the sort we just described? It is shown in [10] how it can be used to perform efficient point location among real-algebraic varieties. Given n polynomials f_1, f_2, etc., we wish to build a data structure such that given any point x in d-space we can determine whether it is the root of any of these polynomials in logarithmic time. If the answer is no, in practice we often want a little more information, such as the name of the cell of the stratification in which x lies.

Such a fast point location algorithm gives us the ability to solve almost any multidimensional searching problem in logarithmic time, using a polynomial amount of storage. This also gives us a means to speed up the solution of a whole class of optimization problems (e.g., finding the longest segment in a polygon, finding whether there is any intersection between two sets of blue and red curves). More generally, suppose that we are given a set Z of n a d-variate polynomial f. In $O(n^d)$ time, it is easy to test whether some subset of Z forms a d-tuple that is a zero of f. The point location technique allows us to perform this test in $O(n^{d-\varepsilon})$ time, for some small $\varepsilon > 0$. Lowering this upper bound substantially is an open problem of major importance.

27.4 Concluding Remarks

The last word on decomposition problems has hardly been said. Which method to use in practice for a given application area remains by and large an open problem. Trade-offs between efficiency and simplicity must be resolved by programmers on an ad hoc basis and a taxonomy to guide us through the maze of decomposition techniques remains to be established. A number of fundamental theoretical questions are still open, especially ones regarding nonlinear higher-dimensional shapes. Also, the issue of robustness, such as how to deal with singularities, degeneracies, and round-off errors, has only begun to be explored, and despite noteworthy efforts, much unknown remains in that area, too.

27.5 References

[1] T. Asano, and T. Asano, Partitioning Polygonal Regions into Trapezoids, *Proc. 24th Annual IEEE Symposium on Foundations of Computer Science* (1983), 233–241.

[2] T. Asano, T. Asano, and H. Imai, Partitioning a Polygonal Region into Trapezoids, *Journal of the ACM* 33 (1986), 290–312.

[3] C.L. Bajaj, and T.K. Dey, Convex Decompositions of Polyhedra and Robustness, *SIAM Journal on Computing* 21 (1992), 339–364.

[4] J.L. Bentley, and T. Ottmann, Algorithms for Reporting and Counting Geometric Intersections, *IEEE Transactions on Computing* C-28 (1979), 643–647.

[5] K.Q. Brown, Comments on "Algorithms for Reporting and Counting Geometric Intersections," *IEEE Transactions on Computing* C-30 (1981), 147–148.

[6] B. Chazelle, An Optimal Algorithm for Intersecting Three-Dimensional Convex Polyhedra, *Proc. 30th Annual IEEE Symposium on Foundations of Computer Science* (1989), 586–591.

[7] B. Chazelle, Triangulating a Simple Polygon in Linear Time, *Proc. 31st Annual Symposium on Foundations of Computer Science* (1990), to appear in *Discrete and Computational Geometry.*

[8] B. Chazelle, and D.P. Dobkin, Optimal Convex Decompositions, *Computational Geometry*, North Holland, (1985), 63-133.

[9] B. Chazelle, and H. Edelsbrunner, An Optimal Algorithm for Intersecting Line Segments in the Plane, *Proc. 29th Annual IEEE Symposium on Foundations of Computer Science* (1988), 590–600.

[10] B. Chazelle, H. Edelsbrunner, L.J. Guibas, and M. Sharir, A Singly Exponential Stratification Scheme for real Semi-Algebraic Varieties and its Applications, *Lecture Notes in Computer Science* 372 (1989), 179–193.

[11] B. Chazelle, and J. Incerpi, Triangulation and Shape-Complexity, *ACM Transactions on Graphics* 3 (1984), 135–152.

[12] B. Chazelle, and L. Palios, Triangulating a Nonconvex Polytope, *Discrete and Computational Geometry* 5 (1990), 505–526.

[13] L.P. Chew, Guaranteed-Quality Triangulation Meshes, Tech. Report, Dept. of Computer Science, Cornell University.

[14] K.L. Clarkson, and R. Cole, in preparation.

[15] K.L. Clarkson, and P.W. Shor, Applications of Random Sampling in Computational Geometry, II, *Discrete and Computational Geometry* 4 (1989), 387–421.

[16] K.L. Clarkson, R.E. Tarjan, and C.J. Van Wyk, A Fast Las Vegas Algorithm for Triangulating a Simple Polygon, *Discrete and Computational Geometry* 4 (1989), 423–432.

[17] G.E. Collins, Quantifier Elimination for Real Closed Fields by Cylindric Algebraic Decomposition, *Lecture Notes in Computer Science* 33 (1975), 134–183.

[18] J. Davenport, and J. Heintz, Real Quantifier Elimination is Doubly Exponential, *Journal of Symbolic Computation* 5 (1988), 29–35.

[19] T.K. Dey, C.L. Bajaj, and K. Sugihara, On Good Triangulations in Three Dimensions, *International Journal of Computational Geometry and Applications* 2 (1992), 75–95.

[20] S.J. Fortune, A Sweepline Algorithm for Voronoi Diagrams, *Algorithmica* 2 (1987), 153–174.

[21] A. Fournier, and D.Y. Montuno, Triangulating Simple Polygons and Equivalent Problems, *ACM Transactions on Graphics* 3 (1984), 153–174.

[22] M.R. Garey, and D.S. Johnson, *Computers and Intractability: A Guide to the Theory of NP-Completeness*, Freeman, 1979.

[23] M.R. Garey, D.S. Johnson, F.P. Preparata, and R.E. Tarjan, Triangulating a Simple Polygon, *Information Processing Letters* 7 (1978), 175-179.

[24] F. Gavril, Algorithms for a Maximum Clique and a Maximum Independent Set of a Circle Graph, *Networks* 3 (1973), 261–273.

[25] D.H. Greene, *The Decomposition of Polygons into Convex Parts*, manuscript, Xerox PARC, 1982.

[26] L.J. Guibas, and J. Stolfi, Primitives for the Manipulation of General Subdivisions and the Computation of Voronoi Diagrams, *ACM Transactions on Graphics* 4 (1985), 74–123.

[27] S. Hertel, and K. Mehlhorn, Fast Triangulation of a Simple Polygon, *Lecture Notes in Computer Science* 158 (1985), 207–218.

[28] K. Hoffman, K. Mehlhorn, P. Rosenstiehl, and R.E. Tarjan, Sorting Jordan Sequences in Linear Time Using Level-Linked Search Trees, *Information and Control* 68 (1986), 170–184.

[29] J.M. Keil, Decomposing a Polygon into Simpler Components, *SIAM Journal on Computing* 14 (1985), 799–811.

[30] D.G. Kirkpatrick, Efficient Computation of Continuous Skeletons, *Proc. 20th Annual IEEE Symposium on Foundations of Computer Science* (1979), 18–27.

[31] D.G. Kirkpatrick, M.M. Klawe, and R.E. Tarjan, Polygon Triangulation in $O(n \log \log n)$ Time with Simple Data Structures, *Proc. 6th Annual ACM Symposium on Computational Geometry* (1990), 34–43.

[32] A. Knight, J. May, J. McAffer, T. Nguyen, and J.-R. Sack, A Computational Geometry Workbench, *Proc. 6th Annual ACM Symposium on Computational Geometry* (1990), 370.

[33] D.T. Lee, and F.P. Preparata, Location of a Point in a Planar Subdivision and its Applications, *SIAM Journal on Computing* 6 (1977), 594–606.

[34] D. Lichtenstein, Planar Formulae and their Uses, *SIAM Journal on Computing* 11 (1982), 329–343.

[35] A. Lingas, The Power of Non-Rectilinear Holes, *Lecture Notes in Computer Science* 140 (1982), 369–383.

[36] A. Lingas, R. Pinter, R. Rivest, and A. Shamir, Minimum Edge Length Partitioning of Rectilinear Polygons, *Proc. 20th Annual Allerton Conference on Communication, Control, and Computing* (1982), 53–63.

[37] K. Mulmuley, A Fast Planar Partition Algorithm, II, *Proc. 5th Annual ACM Symposium on Computational Geometry* (1989), 33–43.

[38] J. O'Rourke, and K.J. Supowit, Some NP-Hard Polygon Decomposition Problems, *IEEE Transactions on Information Theory* IT-29 (1983), 181–190.

[39] J. Ruppert, and R. Seidel, On the Difficulty of Tetrahedralizing 3-Dimensional Non-Convex Polyhedra, *Proc. 5th Annual ACM Symposium on Computational Geometry* (1989), 380–392.

[40] R. Seidel, in preparation.

[41] R.E. Tarjan, and C.J. Van Wyk, An $O(n \log \log n)$-time Algorithm for Triangulating a Simple Polygon, *SIAM Journal on Computing* 17 (1988), 143–178.

[42] G.T. Toussaint, Pattern Recodnition and Geometrical Complexity, *Proc. 5th International Conference on Pattern Recognition* (1980), 1324–1347.

[43] G.T. Toussaint, and D. Avis, On a Convex Hull Algorithm for Polygons and its Application to Triangulation Problems, *Pattern Recognition* 15 (1982), 23–29.

[44] C.K. Yap, An $O(n \log n)$ Algorithm for the Voronoi Diagram of a Set of Simple Curve Segments, *Discrete and Computational Geometry* 2 (1987), 365–393.

28
Single Exponential Path Finding in Semi-algebraic Sets, Part II: The General Case

Joos Heintz
Marie-Francoise Roy
Pablo Solerno

ABSTRACT This paper is devoted to the following result. Let S be a semi-algebraic subset of $\mathbf{R}^n$; one can decide in single exponential time whether two points of S belong to the same semi-algebraically connected component of S, and if they do, one can find a semi-algebraic path connecting them. This paper is the sequel to [HRS 4] in which the result is proved in the particular but fundamental case of a bounded regular hypersurface.

28.1 Introduction

28.1.1 The statement

Let us introduce some notations. We shall denote by $\mathbf{R}$ a real closed field, by $\mathbf{A}$ a subring of $\mathbf{R}$, by S a semi-algebraic (over $\mathbf{A}$) subset of $\mathbf{R}^n$ and by x and x' points of S. We suppose that S, x, x' are described by boolean combinations of polynomial inequalities over polynomials $F_1, ..., F_s$ of $\mathbf{A}[X_1, ... X_n]$ such that $D = \sum degF_i$.

The notion of algorithm we use is the notion of arithmetic network over $\mathbf{A}$ (see [Ga]), and by sequential and parallel complexity of the algorithm (or sequential and parallel time) we refer to the size and depth of the arithmetic network. An admissible algorithm is an algorithm with sequential time $D^{n^{O(1)}}$ and with parallel time $n^{O(1)}\log(D)^{O(1)}$.

The statement of our theorem is precisely the following :

Main Theorem: It is possible to decide whether x and x' belong to the same semi-algebraically connected component of S and if they do to construct a semi-algebraic path connecting them by an admissible algorithm.

Since it is possible to construct in admissible time at least one point in every connected component of S (see [HRS 1,2,3]), the algorithm also permits to compute the exact number of connected components of S in admissible time.

In a forthcoming paper ([HRS 5]) we shall give a new proof of the description of connected components of S in admissible time (for this result, see [CGV]). One consequence of the present paper (cf. §29.4) is that it will be enough to describe the connected components of a smooth bounded algebraic hypersurface.

28.1.2 Comments

This result is in the line of various papers devoted to the complexity of algorithms in real algebraic geometry in single exponential time ([GV 1], [G], [Ca 1], [R], [HRS 1,2,3]). A similar theorem was already known for doubly exponential complexity in n (see for example [Co] or [SS]). The doubly exponential complexity in these papers comes from the use of iterated projections. The first results with single exponential complexity may be found in [Ca 2], where the essential geometric idea, which we are also using , consists in constructing curves on the semi-algebraic set directly, rather than on its iterated projections. There are nevertheless significant differences between Canny's results and ours:

-we have no restrictive hypothesis (such as general position) on the semi-algebraic set,

-we make no use of the theory of stratifications.

28.1.3 The tools of the proof.

We use the following tools:

- some notions of differential topology and topology of semi-algebraic sets,

- some real algebraic geometry over arbitrary real closed fields, dealing in particular with infinitely small (or large) elements and Puiseux series (cf. [GV 1], [G], [HRS 1,2,3]). In particular, we need the notion of semi-algebraic connected components of a semi-algebraic set in an arbitrary real closed field (cf. [BCR]).

-the quantifier elimination method of [HRS 2,3] (or of [R]) in various situations; such as the computation of the closure, the interior, the diameter of a semi-algebraic set, the distance between two semi-algebraic sets and the image of a semi-algebraic function.

28.1.4 Some notations.

Let $x := (x_1, ..., x_n)$ and $y := (y_1, ..., y_n)$ be two points of $\mathbf{R}^n$, we write

$$| x - y | := \sqrt{(x_1 - y_1)^2 + \cdots + (x_n - y_n)^2}$$

for their euclidean distance. If r is a positive element of $\mathbf{R}$ we write

$$B(x, r) := \{y \in \mathbf{R}^n \mid | x - y | < r\}, \bar{B}(x, r) := \{y \in \mathbf{R}^n \mid | x - y | \leq r\}$$

for the open ball and closed ball of radius r centered at x.
For $S \subset \mathbf{R}^n$, we write $\overline{S}$ for its closure in the euclidean topology.

28.2 Some auxiliary results.

28.2.1 About infinitesimals

Let ε be a variable, and let us denote by $\mathbf{R}\widehat{(\varepsilon)}$ the *field of Puiseux series in ε with coefficients in* $\mathbf{R}$. Its elements are 0 and series of the form

$$\sum_{i \geq i_0, i \in \mathbf{Z}} a_i \varepsilon^{\frac{i}{q}}$$

with $i_0 \in \mathbf{Z}$, $a_i \in \mathbf{R}$, $a_{i_0} \neq 0$ and $q \in \mathbf{N}$ (see [W]). One defines $a_k = 0$ for $k < i_0$.

The field $\mathbf{R}\widehat{(\varepsilon)}$ is real closed (see [BCR]). The element ε is "infinitely small positive" (positive and less than all the positive elements of $\mathbf{R}$).The elements of $\mathbf{R}\widehat{(\varepsilon)}$ bounded over $\mathbf{R}$ form a *valuation ring* denoted by $\mathbf{V}(\varepsilon)$; one denotes by *eval* the function from $\mathbf{V}(\varepsilon)$ to $\mathbf{R}$ which maps $\sum_{i \geq i_0, i \in \mathbf{Z}} a_i \varepsilon^{\frac{i}{q}}$ (with $i_0 \in \mathbf{N}$, $a_{i_0} \neq 0$) to the number a_0.

The language of real closed fields with constants in $\mathbf{A}$ is denoted by $\mathcal{L}(\mathbf{A})$.
Let $\Phi(X_1, ..., X_n, Y)$ be a quantifier free formula of $\mathcal{L}(\mathbf{A})$, and let S be a semi-algebraic set defined by

$$S := \{(x, y) \in \mathbf{R}^{n+1} \mid \Phi(x_1, ..., x_n, y)\}.$$

We define:

$$S(\varepsilon) := \{x \in (\mathbf{R}\widehat{(\varepsilon)})^{\,n} \mid \Phi(x_1, ..., x_n, \varepsilon)\}$$

(so that $S(\varepsilon)$ is a semi-algebraic set of $(\mathbf{R}\widehat{(\varepsilon)})^{\,n}$ defined by a quantifier free formula of $\mathcal{L}(\mathbf{A}[\varepsilon])$) and

$$P := \{(x, y) \in \mathbf{R}^{n+1} \mid \Phi(x_1, ..., x_n, y) \wedge y > 0 \}.$$

With these notations, we are going to prove that the *eval* map corresponds, as is intuitively clear, to taking the limit when ε tends to 0.

Proposition 1: Let us suppose that there exists $r \in \mathbf{R}$ such that $S(\varepsilon) \subset B(0, r)$, then $eval(S(\varepsilon)) = \{x \in \mathbf{R}^n \mid (x, 0) \in \overline{P} \}$. In particular, it is possible to compute $eval(S(\varepsilon))$ from $S(\varepsilon)$ in admissible time.

Proof: Let $\mathbf{R}'$ be the field of algebraic Puiseux series (Puiseux series algebraic over polynomials), *eval'* the restriction of *eval* to $\mathbf{V}(\varepsilon) \cap \mathbf{R}'$ and $S(\varepsilon)' = S(\varepsilon) \cap \mathbf{R}'^n$ (then $S(\varepsilon)'$ is a semi-algebraic set of $\mathbf{R}'^n$). We shall first prove the following lemma.

Lemma : Let us suppose that there exists $r \in \mathbf{R}$ such that $S(\varepsilon) \subset B(0,r)$, then $eval(S(\varepsilon)) = eval'(S(\varepsilon)')$.

Proof of the lemma: Let us observe first that if $w \in \mathbf{R}(\varepsilon)^{\widehat{}}$ is a positive element, then there exists $w' \in \mathbf{R}'$ such that $0 < w' < w$ (if $w \notin \mathbf{V}(\varepsilon)$ one may chose $w' = 1$ and if $w = \sum_{i \geq i_0, i \in \mathbf{Z}} a_i \varepsilon^{\frac{i}{q}}$ with $a_{i_0} > 0$ and $i_0 \geq 0$ one may choose $w' = \varepsilon^{\frac{i_0+1}{q}}$). It is enough to prove that for any $z \in S(\varepsilon)$ there exists $z' \in S(\varepsilon)'$ such that $eval(z) = eval'(z')$. This is proved by induction on the number of variables defining $S(\varepsilon)$.
If $n = 1$ the result is clear from the above observation since $S(\varepsilon)$ is a union of intervals with endpoints in $\mathbf{R}'$.
If $n > 1$, the case where the condition is $F_1 > 0, F_2 > 0, ..., F_s > 0$ is immediate since this is an open condition. Otherwise, z verifies a polynomial conjunction, $F_1 = 0, F_2 > 0, ..., F_s > 0$, with polynomials in the variables $X_1, ..., X_n$. Using cylindrical algebraic decomposition (or saucissonnage cf. for example [BCR] Th.2.3.1) in the variable X_n (possibly after a linear change of variable), we can prove that if π is the projection forgetting the last coordinate, then $\pi(z)$ belongs to a semi-algebraic set $W(\varepsilon)$ defined by polynomials $G_1, ..., G_t \in \mathbf{A}[\varepsilon][X_1, ..., X_{n-1}]$, such that there exists a continuous semi-algebraic function $\xi : W(\varepsilon) \rightarrow \mathbf{R}(\varepsilon)^{\widehat{}}$ with $z = (\pi(z), \xi(\pi(z)))$ and $(\pi(w), \xi(\pi(w))) \in S(\varepsilon)$ for all $w \in W(\varepsilon)$. By continuity of ξ, for every rational number $\frac{p}{q}$ there exists another rational number $\frac{p'}{q'}$ such that $| \pi(z) - w | \leq \varepsilon^{\frac{p'}{q'}}$ implies $| (\pi(z), \xi(z)) - (w, \xi(w)) | \leq \varepsilon^{\frac{p}{q}}$. By the induction hypothesis, there exists $w \in \mathbf{R}'^{n-1} \cap \bar{B}(\pi(z), \varepsilon^{\frac{p}{q}}) \cap W(\varepsilon)$, such that $eval(w) = eval(\pi(z))$. Then it is clear that $w' := (w, \xi(w))$ verifies $eval(w') = eval(z)$. □

Let us return to the proof of the proposition.
Using the lemma, it remains to prove that $eval'(S(\varepsilon)') = \{x \in \mathbf{R}^n \mid (x, 0) \in \overline{P}\}$. Let $z = (\xi_1(\varepsilon), ..., \xi_n(\varepsilon)) \in S(\varepsilon)'$, where, for all $1 \leq i \leq n$, $\xi_i(\varepsilon)$ is an algebraic Puiseux series, hence defines a continuous function $\xi_i(t)$ defined over an interval $(0, \delta)$ in $\mathbf{R}(\varepsilon)^{\widehat{}}$ ([BCR] Ex.7.3.6.). Thus, for all $t \in \mathbf{R}(\varepsilon)^{\widehat{}}$, with $t > 0$ small enough, we obtain $(\xi_1(t), ..., \xi_n(t), t) \in P'$, where P' is the subset of $(\mathbf{R}(\varepsilon)^{\widehat{}})^{n+1}$ described by $\Phi(X, Y) \wedge (Y > 0)$. Therefore,

$$\lim_{t \rightarrow 0} (\xi_1(t), ..., \xi_n(t), t) = (eval'(\xi_1(t)), ..., eval'(\xi_n(t)), 0) \in \overline{P'} \cap \mathbf{R}^{n+1} = \overline{P},$$

since closure is definable over $\mathbf{A}$. We conclude then $(eval'(z), 0) \in \overline{P}$.
On the other hand, if $(x_1, ..., x_n, 0) \in \overline{P}$, then the Nash Curve Selection Lemma ([BCR] Prop.8.1.17) implies that there exist $n+1$ Nash functions $\xi_1, ..., \xi_{n+1}$ all defined over an interval $[0, \delta) \subset \mathbf{R}$ such that $(\xi_1(t), ..., \xi_{n+1}(t)) \in P$ for all $t \in (0, \delta)$ and $(\xi_1(0), ..., \xi_{n+1}(0)) = (x_1, ..., x_n, 0) \in \overline{P}$. In particular, the set $\{v \in (0, \delta) \mid \frac{d\xi_{n+1}}{dt}(v) = 0\}$, which is a semi-algebraic subset of $\mathbf{R}$, does not contain any interval of the type $(0, \delta')$, otherwise ξ_{n+1} would

be constant near 0, thus null for all t small enough, since $\xi_{n+1}(0) = 0$, but $(\xi_1(t), ..., \xi_n(t)) \in P$ implies that $\xi_{n+1}(t) > 0$). Hence there exists δ_1 such that $sign(\frac{d\xi_{n+1}}{dt})(t)$ is constant for all $t \in (0, \delta_1)$.
Let δ_2 be such that the inverse function ξ_{n+1}^{-1} is defined , continuous and semi-algebraic over the interval $(0, \delta_2)$. We define

$$u := \xi_{n+1}(t), \quad t \in (0, \delta_1)$$
$$\gamma_i(u) := (\xi_i \circ \xi_{n+1}^{-1})(u) \quad u \in (0, \delta_2)$$

The functions γ_i, $1 \leq i \leq n$, just defined are continuous and semi-algebraic. Applying [BCR] Ex.7.3.6., $\gamma_1(\varepsilon), ..., \gamma_n(\varepsilon)$ can be considered as elements of the field $\mathbf{R}'$, and thus, since $\varepsilon < \delta_2$, we obtain $(\gamma_1(\varepsilon), ..., \gamma_n(\varepsilon)) \in S(\varepsilon)$. Moreover:

$$(eval'(\gamma_1), ..., eval'(\gamma_n), 0) = \lim_{u \to 0}(\gamma_1(u), ..., \gamma_n(u), u)$$
$$= \lim_{t \to 0}(\xi_1(t), ..., \xi_n(t), \xi_{n+1}(t)) = (x_1, ..., x_n, 0)$$

The quantifier elimination algorithms described in [HRS 2,3] or [R] allow one to compute in admissible time the closure of P, and hence also $eval'(S(\varepsilon))$. □

28.2.2 Connected components of semi-algebraic curves

Proposition 2: Let $\Gamma \subset \mathbf{R}^n$ be a closed semi- algebraic set defined by a boolean combination of polynomials the sum of whose degrees is less than D and such that $dim(\Gamma) \leq 1$. There exists an admissible algorithm describing the semi-algebraic connected components of Γ. Moreover, if Γ is bounded, then the algorithm computes a family of semi-algebraic continuous functions $\gamma_i : [0,1] \to \Gamma$, $(1 \leq i \leq N)$ such that $\Gamma = \cup_{1 \leq i \leq N} im(\gamma_i)$ and $N = D^{n^{O(1)}}$.

Proof: One first computes a linear combination $a_1x_1 + ... + a_nx_n$ such that if $g : \Gamma \to \mathbf{R}$ is the function sending $x = (x_1, ..., x_n)$ to $a_1x_1 + ... + a_nx_n$, then for all t in $\mathbf{R}$, $dim(g^{-1}(t)) = 0$ or -1. This linear combination can be obtained by eliminating quantifiers in a formula with 3 blocks of quantifiers, $2n + 3$ variables and degree equal to $D + 4$, describing the fact that $dim(g^{-1}(t)) < 0$ (cf.[HKRS] Prop.12). By performing a change of variables one may suppose that $g(x) = x_1$. Corollary 2 of [HRS 4] proves that there exists a semi-algebraic partition Γ_i, $i = 1, ..., N$ of Γ such that each Γ_i is a closed connected semi-algebraic set and $N = D^{n^{O(1)}}$. One obtains a description of the semi-algebraically connected components of Γ by computing the transitive closure of the relation $R = \{(i,j) \in \{1, ..., N\}^2 \mid \Gamma_i \cap \Gamma_j \neq \emptyset\}$. Again, Corollary 2 of [HRS 4] shows that each Γ_i is the image of a continuous semi-algebraic map defined over a closed interval. □

Observation 3: If we assume in the previous proposition that Γ is bounded and semi-algebraically connected, we obtain that for all $1 \leq i, j \leq N$ there exist $i_1, ..., i_k$, with $i = i_1, j = i_k$ such that $im(\gamma_{i_\ell}) \cap im(\gamma_{i_{\ell+1}}) \neq \emptyset \quad (1 \leq \ell \leq k-1)$.

28.2.3 *Extrema of semi-algebraic functions*

The following result will be needed for the computation of the distance between semi-algebraically connected components.

Lemma 4: Let S be a semi-algebraic set and $f : S \to \mathbf{R}$ a semi-algebraic function. Let S' be the subset of elements of S which are local extrema of f. Then $E := f(S')$, the set of local extremal values of f, is a finite subset of $\mathbf{R}$..

Proof: $f(S')$ can be defined by a formula of the language of real closed fields and hence is semi-algebraic. Let us suppose that $f(S')$ is not finite. Then there exist points a, b of $\mathbf{R}$, definable over $\mathbf{A}$, with $a < b$ and such that $[a, b] \subset f(S')$. Let us consider the semi-algebraic set $T := S' \cap f^{-1}([a, b])$. According to [BCR] Prop.8.1.2, there exists a semi-algebraic subdivision of T into a finite family (T_i) (eventually a subdivision of the graph of f is necessary too) such that T_i is semi-algebraically connected and $f|_{T_i}$ is a Nash function. By definition of T there exists an index i such that $f|_{T_i}$ is not constant. By Sard's theorem, $f|_{T_i}$ has only a finite number of critical values, hence of local extremal values. Since local extremal values of f are local extremal values of $f|_{T_i}$, this leads to a contradiction. □

28.3 Proof of the Main Theorem

We subdivide the proof of our theorem into several steps; we begin with the case of a bounded and regular algebraic hypersurface, then we consider an arbitrary closed semi-algebraic set and finally we prove the general case. The following definitions allow us to simplify the statements and the proofs.

Definitions:

Let $S \subset \mathbf{R}^n$ be a semi-algebraic set.

A *roadmap* $\Re$ *for* S (defined over $\mathbf{A}$) is a pair $\Re := (\Gamma, (\Gamma_x)_{x \in S_d})$, where:
a) $\Gamma \subset S$ is a closed and bounded semi-algebraic set of dimension not greater than 1, such that for all semi-algebraically connected component S' of S, $S' \cap \Gamma$ is non-empty and semi-algebraically connected.
b) for all $x \in S_d := J\{y \in S;\ y \text{ is definable over } \mathbf{A}\}$, $\Gamma_x \subset S$ is a bounded and semi-algebraically connected path in S connecting x to Γ.

If the semi-algebraic set S is given by means of a quantifier free formula $\Phi \in \mathcal{L}(\mathbf{A})$, we say that the roadmap $\Re$ is *obtained in admissible time* if and only if:
a) Γ can be constructed from the formula Φ in admissible time,

b) for all $x \in S_d$ which is described by a formula Ψ_x we areable to construct Γ_x in admissible time from Φ and Ψ_x.

This notion of roadmap generalizes the one given in [HRS 4] ¶.4 for regular bounded hypersurfaces.

Observation: Suppose that it is possible to construct in admissible time a roadmap $\Re$ for the semi-algebraic set S, then it is clear that S verifies the Theorem, since, given $x, x' \in S_d$, one may construct (in admissible time) Γ_x and $\Gamma_{x'}$ and then apply Proposition 2 to determine whether $z \in \Gamma \cap \Gamma_x$ and $z' \in \Gamma \cap \Gamma_{x'}$ lie in the same semi-algebraically connected component of Γ. The construction of the path then follows from the constructions of Γ,Γ_x and $\Gamma_{x'}$.

Proposition 5: Let $V \subset \mathbf{R}^n$ be a closed and bounded semi-algebraic set and *suppose that it is possible to construct in admissible time a roadmap for* V. Let $V = \cup_{1 \leq i \leq N} V_i$ the decomposition of V in semi-algebraically connected components. Set $\nu_{ij} := d(V_i, V_j)$ (the distance between V_i and V_j). For all $1 \leq i, j \leq N$ it is possible to compute in admissible time ν_{ij}, and also points x_{ij}, y_{ij} such that $x_{ij} \in V_i$, $y_{ij} \in V_j$ and $\nu_{ij} = d(V_i, V_j) = d(x_{ij}, y_{ij})$.

Proof: Let $S := V \times V \subset \mathbf{R}^{2n}$, and consider the function $f : S \to \mathbf{R}$ defined by $f(x,y) = \mid x - y \mid^2$. Following the notations of Lemma 4 we have:

$$\{\nu_{ij}^2; 1 \leq i, j \leq N\} \subset E,$$

where E was the set of extremal values of f. For each $t \in E$, one computes in admissible time a system of representatives $\mathcal{R}_t$ of the semi-algebraically connected components of $f^{-1}(t)$ (computing first t and $f^{-1}(t)$ by means of the quantifier elimination algorithm, and afterwards applying [HRS 1] Th.2).

Thus, for all $i \neq j$:

$$\nu_{ij}^2 = \min\{t > 0, t \in E \mid \exists (x,y) \in \mathcal{R}_t \text{ such that } x \in V_i \text{ and } y \in V_j\}.$$

We conclude the proof by observing that by means of the previous observation one can decide whether $x \in V_i$ and $y \in V_j$. □

One notes that the proof of Proposition 5 only uses the fact that it is possible to decide whether two given points lie in the same semi- algebraically connected component of V and not the existence of the roadmap $\Re$.

We are now able to begin the proof of the Main Theorem.

Lemma 6: (The case of a bounded and regular algebraic hypersurface).

Let $F \in A[X_1, ..., X_n]$ and let $V := \{F = 0\} \subset \mathbf{R}^n$ be bounded and regular, *i.e.*, $\nabla F(x) \neq 0, \forall x \in V$. Then one can construct in admissible time a roadmap for V.

Proof: See [HRS 4] ¶4,5, Th.12. □

From now on, we denote by $\mathbf{K}$ the fraction field of the integral domain.

Lemma 7: (The case of a bounded algebraic set)
Let $F_1, ..., F_s \in \mathbf{A}[X_1, ..., X_n]$ and let $V := \{F_1 = 0, ..., F_s = 0\} \in \mathbf{R}^n$. Then, it is possible to compute in admissible time a roadmap for V.

Proof: Set $F := F_1^2 + \cdots + F_s^2$; hence $V = \{F = 0\}$.

(1) We first construct a bounded reguar algebraic hypersurface "close to" V.
By means of quantifier elimination, one computes $\rho \in \mathbf{K}$, $\rho > 0$ such that $V \subset B(0, \rho)$. Let $\pi : \mathbf{R}^{n+1} \to \mathbf{R}^n$ be defined by $\pi(x_1, ..., x_{n+1}) = (x_1, ..., x_n)$. Set

$$G := F + (\rho^2 - X_1^2 - \cdots - X_n^2 - X_{n+1}^2)^2 \in \mathbf{K}[X_1, ..., X_{n+1}]$$

and let V_1 be the zero set of G. Clearly $G(x) \geq 0$, for all $x \in \mathbf{R}^{n+1}$Jand $\pi(V_1) = V$. As before, let ε be an infinitesimally small positive element over $\mathbf{R}$, and let:

$$V_2 := \{x \in (\mathbf{R}(\varepsilon))^{n+1} \mid G(x) = \varepsilon\}$$

Sard's Theorem implies that V_2 is regular; moreover, it is also bounded (in fact, $V_2 \subset B(0, \rho + 1)$) and $eval(V_2) = V_1$ (see [GV 1] Lemma 3).

(2) We now construct a roadmap for V_2 containing curves passing through points which realize the minimal distances between semi-algebraicaly connected components.
For that, apply Lemma 6 to compute a roadmap for V_2, $\Re_2 := (\Gamma_2, (\Gamma_{2,x})_{x \in V_{2d}})$ (where as before, $V_2 - d$ is the set of points of V_2 definable over $\mathbf{A}[\varepsilon]$). Let$(x_{ij}, y_{ij}) \in V_2 \times V_2$ be the pairs of points which realize the distances between the semi-algebraically connected components of V_2 (Prop.5). Adding to Γ_2 the curves $(\Gamma_{2,x_{ij}}), (\Gamma_{2,y_{ij}})$ for all (x_{ij}, y_{ij}) one can suppose that the points x_{ij}, y_{ij} belong to Γ_2.

(3) We now define a roadmap for V_1, denoted by$\Re_1 := (\Gamma_1, (\Gamma_{1,x})_{x \in V_{1d}})$ for V_1 as follows:
a) Set $\Gamma_1 := eval(\Gamma_2)$. Since Γ_2 is computable in admissible time, applying Proposition 1 we also obtain Γ_1 in admissible time. Let us now show the connectedness of Γ_1. Let T be a semi- algebraically connected component of V; by [GV 1] Lemma 1, we get $T = eval(V_2^1 \cup ... \cup V_2^r)$, where V_2^j are the semi-algebraically connected components of V_2, and for all j there exists j' $(1 \leq j \neq j' \leq r)$ verifying $eval(d(V_2^j, V_2^{j'})) = 0$. Since the points $x_{jj'}, y_{jj'}$ which realize the distance between V_2^j and $V_2^{j'}$ $(1 \leq j \neq j' \leq r)$ are included in Γ_2, these are "glued" by the evaluation *eval*, and hence, $eval(\Gamma_2) \cap T$ is connected.
b) Let $x \in V_1$. We compute a point $y \in V_2$ such that $d(x, y) = d(x, V_2)$, and we also compute $\Gamma_{2,y}$. We define $\Gamma_{1,x} := eval(\Gamma_{2,y})$ (observing $eval(V_2) = V_1$ implies that $eval(y) = x$). By Proposition 1 the construction of $\Gamma_{1,x}$ can be done in admissible time.

(4) We now construct a roadmap for V:

Observing that $V_1 = (V_1 \cap \{X_{n+1} > 0\}) \cup (V_1 \cap \{X_{n+1} < 0\})$ is symmetric with respect to the hyperplane $\{X_{n+1} = 0\}$, the decomposition of V_1 in semi-algebraically connected components is of the form:

$$V_1 = (V_1^1 \cup ... \cup V_1^t) \cup (V_1^{t+1} \cup ... \cup V_1^{2t})$$

where for all $1 \leq j \leq t$, V_1^j are the semi-algebraically connected components of $V_1 \cap \{X_{n+1} > 0\}$, and the others are the ones for $V_1 \cap \{X_{n+1} < 0\}$; moreover

$$(x_1, ..., x_{n+1}) \in V_1^j \iff (x_1, ..., -x_{n+1}) \in V_1^{t+j} \quad (1 \leq j \leq t).$$

Let $\{p_1, ..., p_t\}$ be a system of representatives of $V_1^1, ..., V_1^t$ and $\{\bar{p}_1, ..., \bar{p}_t\}$ the corresponding points in $V_1^{t+1}, ..., V_1^{2t}$. As before, adding to Γ_1 the curves $(\Gamma_{1,p_j}), (\Gamma_{1,\bar{p}_j})$ for all j, one can suppose that all the points $p_j, \bar{p}_j$ belong to the curve Γ_1.
We finally define $\Gamma := \pi(\Gamma_1)$ (which is constructible in admissible time).
Let T be a semi-algebraically connected component of V. Then there exists $j \leq t$ such that $\pi(V_1^j) = \pi(V_1^{t+j}) = T$ (since $\pi : V_1 \cap \{X_{n+1} > 0\} \to V$ is an homeomorphism).
Clearly $\pi(\Gamma_1) \cap T$ is connected since $\pi(p_j) = \pi(\bar{p}_j)$ connects the projection.
Now let $x \in V$, then $x = \pi(y)$ where $y := (x_1, ..., x_n, \sqrt{\rho^2 - |x|^2})$, and define $\Gamma_x := \pi(\Gamma_{1,y})$. □

Lemma 8: (The case of a basic closed and bounded semi- algebraic set)
Let $F_1, ..., F_s \in \mathbf{A}[X_1, ..., X_n]$ and let $S := \{F_1 \geq 0, ..., F_s \geq 0\} \subset \mathbf{R}^n$ be a bounded semi-algebraic set. Then it is possible to compute in admissible time a roadmap for S.

Proof: We proceed by steps.

(1) By means of quantifier elimination algorithm, we compute $\rho \in \mathbf{K}$, $\rho > 0$ such that $S \subset B(0, \rho)$. We add the equation $\rho^2 - |x|^2 \geq 0$, which doesn't modify the set S. We can also suppose that $S = \partial S$ (for this we add the inequality $-Y^2 \geq 0$ and embed S in $\mathbf{R}^{n+1}$ by setting the last coordinate to 0). Let

$$S^{(\varepsilon)} := \{z \in (\mathbf{R}\langle\varepsilon\rangle)^n \mid F_1(z) + \varepsilon > 0, ..., F_s(z) + \varepsilon > 0\}.$$

Clearly $S^{(\varepsilon)} \subset B(0, \rho + 1)$ and *eval*$(S^{(\varepsilon)}) = S$.

(2) Let $F := (F_1 + \varepsilon)....(F_s + \varepsilon) - \varepsilon^s$ and $W := \{F = 0\} \subset (\mathbf{R}\langle\varepsilon\rangle)^n$. By [GV 2] Lemma 2(d), one obtains that $S = \partial S \subset$ *eval*$(S^{(\varepsilon)} \cap W) \subset$ *eval*$(S^{(\varepsilon)}) = S$, *i.e.* $S =$ *eval*$(S^{(\varepsilon)} \cap W)$. Set

$$G := F^2 + ((\rho + 1)^2 - X_1^2 - ... - X_n^2 - X_{n+1}^2)$$

and set

$$V := \{G = 0\} \subset \mathbf{R}\langle\varepsilon\rangle^{n+1}.$$

If $\pi : (\mathbf{R}(\varepsilon))^{n+1} \to (\mathbf{R}(\varepsilon))^{n}$ is the projection which forgets the last coordinate, we have $\pi(V) = W \cap \bar{B}(0, \rho+1)$. Since V is a bounded algebraic set, we apply Lemma 7 in order to compute a roadmap $\Re_1 = (\Gamma_1, (\Gamma_{1,z})_{z \in V_d})$ for V. We can also suppose (as in Lemma 7 (2)) that for all pairs of semi-algebraically connected components of V, there are points in Γ_1 realizing the distance between those pairs. Let $p_1, ..., p_N$ be a system of representatives of V. If $p_i := (x_1, ..., x_{n+1})$, we define $\bar{p}_i := (x_1, ..., -x_{n+1})$. We can again suppose that $\{p_1, ..., p_N, \bar{p}_1, ..., \bar{p}_N\} \subset \Gamma_1$.

(3) Let U be a semi-algebraically connected component of W. Applying [GV 1] Lemma 2(c), we obtain that $U \cap S^{(\varepsilon)} = \emptyset$ or $U \subset S^{(\varepsilon)}$ (*i.e.* $S^{(\varepsilon)} \cap W = \bigcup_k U_k$; where we take the union over all the semi-algebraically connected components U_k of W, such that $U_k \cap S^{(\varepsilon)} \neq \emptyset$). Let $\Gamma_1 = \Gamma_1^1 \cup ... \cup \Gamma_1^N$ be the decomposition of $\Gamma_1 \subset V$ into semi-algebraically connected components. One observes that either $\pi(\Gamma_1^j) \cap S^{(\varepsilon)} = \emptyset$ or $\pi(\Gamma_1^j) \subset S^{(\varepsilon)}$. Set $J := \{j,\ 1 \leq j \leq N$, such that $\pi(\Gamma_1^j) \subset S^{(\varepsilon)}\}$, and $\Gamma_1' = \bigcup_{j \in J} \Gamma_1^j$.

We now construct a roadmap $\Re_2 = (\Gamma_2, (\Gamma_{2,x})_{x \in S^{(\varepsilon)} \cap W_d})$ for $S^{(\varepsilon)} \cap W$ as follows:
a)$\Gamma_2 := \pi(\Gamma_1')$. Let U be a semi- algebraically connected component of $S^{(\varepsilon)} \cap W$. Thus U is a semi- algebraically connected component of W lying in $B(0, \rho+1)$ and corresponds to a semi-algebraically connected component T of V such that $\pi(T) = U$. Therefore, $\pi(\Gamma_1 \cap T) = \pi(\Gamma_1' \cap T) \subset \Gamma_2 \cap U$. As in the proof of Lemma 7 (4), the construction of p_i and $\bar{p}_i$ guarantees that $\Gamma_2 \cap U$ is semi-algebraically connected.

b) Let $x := (x_1, ..., x_n) \in S^{(\varepsilon)} \cap W_d$. We set $z := (x_1, ..., x_n, \sqrt{(\rho+1)^2 - |x|^2})$, and define $\Gamma_{2,x} := \pi(\Gamma_{1,z})$.

(4) We define, finally, a roadmap $\Re = (\Gamma, (\Gamma_x)_{x \in S_d})$ for S in the following way:
a) $\Gamma := eval(\Gamma_2)$. From [GV 1] Lemma 1, the definition of Γ_1 and the fact that $d(x,y) \geq d(\pi(x), \pi(y))$ $\quad (x, y \in (\mathbf{R}(\varepsilon))^{n+1})$, one shows that Γ verifies the desired conditions.
b)Let $x \in S$; since $S^{(\varepsilon)} \cap W$ is closed and bounded (from (3)), there exists $y \in S^{(\varepsilon)} \cap W$ such that $d(x,y) = d(x, S^{(\varepsilon)} \cap W)$. We define $\Gamma_x = eval(\Gamma_{2,y})$.
□

Lemma 9: (The case of an arbitrary closed and bounded semi-algebraic set)

Let $S \subset R^n$ be a closed and bounded semi-algebraic set described by a quantifier free formula $\Phi \in \mathcal{L}(\mathbf{A})$. We are able to construct a roadmap for S in admissible time.

Proof : Set $d := deg(\Phi)$. From the effective finiteness theorem (see [S] or [HKRS]) one computes in admissible time a decomposition $S = \cup_{1 \leq i \leq N} S_i$, with $N = d^{n^{O(1)}}$, where each S_i is a basic closed and bounded semi-algebraic set of the type $S_i = \{F_1 \geq 0, ..., F_{s_i} \geq 0\}$, and the sum of the degrees of all the polynomials F_{j_i} $(1 \leq j_i \leq s_i,\ 1 \leq i \leq N)$ is also of order

$d^{n^{O(1)}}$. Applying Lemma 8, we construct in admissible time a roadmap $\Re_i := (\Gamma_i, (\Gamma_{i,x})_{x \in S_{id}})$ for each S_i.

For all $1 \leq i, j \leq N$, we construct a system of representatives $\mathcal{P}_{ij}$ for the semi-algebraically connected components of $S_i \cap S_j$, and for all $p_{ij} \in \mathcal{P}_{ij}$ we join p_{ij} to Γ_i in S_i by the curve $\Gamma_{i,p_{ij}}$, and p_{ij} to Γ_j in S_j by the curve $\Gamma_{j,p_{ij}}$.

We define, finally, the roadmap $\Re := (\Gamma, (\Gamma_x)_{x \in S_d})$ for S as follows:

a)

$$\Gamma := \bigcup_{1 \leq i \leq N} \Gamma_i \ \cup \ \bigcup_{1 \leq i,j \leq N} \Gamma_{i,p_{ij}} \ \cup \ \bigcup_{1 \leq i,j \leq N} \Gamma_{j,p_{ij}}$$

(taking the union over all $p_{ij} \in \mathcal{P}_{ij}$ for all $1 \leq i, j \leq N$)

b) For $x \in S_d$, let i be such that $x \in S_{id}$; define $\Gamma_x := \Gamma_{i,x}$. □

In order to handle the unbounded case, we need the following result :

Proposition 10: Let $S \subset \mathbf{R}^n$ be a closed semi-algebraic set described by a quantifier free formula $\Phi \in \mathcal{L}(\mathbf{A})$. It is possible to compute in admissible time two positive numbers $\alpha_1, \alpha_2 \in \mathbf{A}$ such that for all $\rho \in \mathbf{R}$ with $\rho \geq \frac{\alpha_1}{\alpha_2}$, and for any semi-algebraically connected component C of S, the set $C \cap \bar{B}(0, \rho)$ is non-empty and semi- algebraically connected.

Proof: Consider the continuous map $f : S \to \mathbf{R}$ defined by $f(x) = | x |^2$. According to Lemma 4, one can find $\rho_1 \in \mathbf{R}$ greater than all local minima of f. In fact, one can choose $\beta_1, \beta_2 \in \mathbf{A}$ such that $\rho_1 = \frac{\beta_1}{\beta_2}$. Let $S_1 \cup ... \cup S_t$ be the decomposition of $S \cap \bar{B}(0, \rho_1)$ into semi-algebraically connected components, and choose $x_i \in S_i$ $(1 \leq i \leq t)$. If two of these points, say x_i and x_j, are in the same semi-algebraically connected components of S, there exists a continuous and semi-algebraic path $\xi_{ij} : [0, 1] \to S$ such that $\xi_{ij}(0) = x_i$ and $\xi_{ij}(1) = x_j$.
Let $\rho_2 := max | \xi_{ij}(u) |^2$, where $u \in [0, 1]$, $1 \leq i \neq j \leq t$, for x_i and x_j in the same semi-algebraically connected component of S. Let ρ_0 be the max of ρ_1 and ρ_2. Clearly, in order to finish the proof, it is enough to compute α_1 and α_2, both in $\mathbf{A}$, such that $\frac{\alpha_1}{\alpha_2} \geq \rho_0$. It is sufficient to construct the paths ξ_{ij} and afterwards to compute their maximum. For this, let Ω be a positive infinitely large element over $\mathbf{R}$, and set:

$$S^{(\Omega)} := \{x \in (\mathbf{R}(\Omega)\widetilde{)}^{\,n} \mid \Phi(x) \wedge (\Omega^2 - | x |^2) \geq 0\}$$

Therefore, $S \subset S^{(\Omega)}$. It is easy to show, using well-known properties of semi-algebraically connected components (see [BCR] ¶2.4), that two points of S lie in the same semi-algebraically connected component of S if and only if they are in the same semi-algebraical- ly connected component of $S^{(\Omega)}$. The set $S^{(\Omega)}$ is closed and bounded semi-algebraic subset of $(\mathbf{R}(\Omega)\widetilde{)}^{\,n}$, defined over $\mathbf{A}[\Omega]$, so it is possible to construct in admissible time a roadmap for $S^{(\Omega)}$ defined over $\mathbf{A}[\Omega]$. Denote by Γ the semi-algebraic curve in $S^{(\Omega)}$ defined by the roadmap.

Let $\Gamma := \cup_{1\le\ell\le N} im(\gamma_\ell)$ be the decomposition described in Proposition 2 and Observation 3. Each $\gamma_\ell : [0,1] \to \Gamma$ is a semi-algebraic and continuous map, and the graph of γ_ℓ is defined over $\mathbf{A}[\Omega]$. Moreover, we can suppose that $\gamma_1(0) = x_i$ and $\gamma_N(1) = x_j$. Denote by $\Psi_\ell(U, X, \Omega)$ the formula which defines the graph of γ_ℓ $(1 \le \ell \le N)$ (where $X := (X_1, ..., X_n)$) and consider the following formulas:

$$\Theta_1^{(\ell)}(\Omega) : \quad (\forall U \in [0,1]) \; (\exists X) \; (\Psi_\ell(U, X, \Omega)) \;\wedge\; (\forall X)(\forall X')(\forall U) \; (\Psi_\ell(U, X, \Omega) \;\wedge\; \Psi_\ell(U, X', \Omega) \Rightarrow X = X').$$

If $\ell = 1$ we add to $\Theta_1^{(\ell)}$ the condition $\Psi_1(0, x_i, \Omega)$ and if $\ell = N$ the condition $\Psi_N(1, x_j, \Omega)$.

$$\Theta_2^{(\ell)}(\Omega) : \quad (\forall X)(\forall U)(\forall \varepsilon > 0) \; (\exists \delta > 0) \; (\forall X')(\forall U') \; (| U - U' |^2 < \delta) \;\wedge\; \wedge\, \Psi_\ell(U, X, \Omega) \;\wedge\; \Psi_\ell(U', X', \Omega) \Rightarrow | X - X' |^2 < \varepsilon)$$

$$\Theta_3^{(rs)}(\Omega) : \quad (\exists X)(\exists U)(\exists U') \; (\Psi_r(U, X, \Omega) \wedge \Psi_s(U', X, \Omega)) \qquad (1 \le r, s \le N).$$

The formula $\Theta_1^{(\ell)}$ expresses that γ_ℓ is a function, defined over the whole interval $[0,1] \subset \mathbf{R}(\hat{\Omega})$, whereas the formula $\Theta_2^{(\ell)}$ expresses the continuity of γ_ℓ. Finally, $\Theta_3^{(rs)}$ says that $im(\gamma_r) \;\cap\; im(\gamma_s) \;\neq\; \emptyset$. We consider the formulas $\Theta_1^{(\ell)}(Z)$, $\Theta_2^{(\ell)}(Z)$ and $\Theta_3^{(rs)}(Z)$ $\;(1 \le \ell, r, s \le N)$ as formulas in the language $\mathcal{LJ}(\mathbf{A})$, with only one free variable Z. If we denote by $\tilde{\Theta}_1^{(\ell)}(Z)$, $\tilde{\Theta}_2^{(\ell)}(Z)$, $\tilde{\Theta}_3^{(rs)}(Z)$ the formulas in one variable Z equivalent to $\Theta_1^{(\ell)}(Z)$, $\Theta_2^{(\ell)}(Z)$ and $\Theta_3^{(rs)}(Z)$ respectively, we conclude that there exist two positive elements in $\mathbf{A}$, say a_1 and a_2, such that these formulas are either true either false for all elements $z \in \mathbf{R}(\hat{\Omega})$ greater than $\frac{a_1}{a_2}$. Moreover it is possible to compute a_1 and a_2 using effective quantifier elimination (see [HRS 2],[HRS 3] or [R]). Then the fact that Ω verifies the three formulas implies that every $z \in \mathbf{R}$ such that $z \ge \frac{a_1}{a_2}$, verifies the formulas too.
Finally, we define the set

$$\hat{\Gamma} := \bigcup_{1\le\ell\le N} \Gamma_\ell$$

where $\quad \Gamma_\ell := \{x \in \mathbf{R}^n \text{ mid exists } u \in [0,1] \text{ such that } \Psi_\ell(u, x, \frac{a_1}{a_2}) \;\}$.

Each Γ_ℓ is a semi-algebraically connected subset of S, of dimension not greater than 1, since Γ_ℓ is the image of a semi-algebraic and continuous map whose graph is described by the formula $\Psi_\ell(U, X, \frac{a_1}{a_2})$. The facts that $x_i \in \Gamma_1$, $x_j \in \Gamma_N$ and that $\hat{\Gamma}$ is connected (since the formula $\tilde{\Theta}_3^{(rs)}(\frac{a_1}{a_2})$ is verified if $im(\gamma_r) \;\cap\; im(\gamma_s) \neq \emptyset$) imply that $\hat{\Gamma}$ contains x_i and x_j. This means that $\hat{\Gamma}$ is a semi-algebraically connected path joining x_i and x_j, which concludes the proof of Proposition 10. $\square$

Lemma 11: (The case of an arbitrary closed semi-algebraic set)

Let $S \subset \mathbf{R}^n$ be a closed semi-algebraic set described by a quantifier free formula $\Phi \in \mathcal{L}(\mathbf{A})$. It is possible to construct a roadmap for S in admissible time.

Proof: Let $\frac{\alpha_1}{\alpha_2}$ be chosen as in the previous proposition, *i.e.* such that each semi-algebraically connected components of S cuts the ball $\bar{B}(0, \frac{\alpha_1}{\alpha_2})$ in a non- empty semi-algebraically connected set. By Lemma 9, we are able to compute a roadmap for the bounded and closed semi-algebraic set $S \cap \bar{B}(0, \frac{\alpha_1}{\alpha_2})$.
We denote by $\Re := (\Gamma, (\Gamma_x)_{x \in S \cap \bar{B}(0, \frac{\alpha_1}{\alpha_2})_d})$ this roadmap.

We now define a roadmap for S, $\Re_1 := (\Gamma_1, (\Gamma_{1,x})_{x \in S_d})$, as follows:
a) $\Gamma_1 := \Gamma$.
b) For $x \in S \cap \bar{B}(0, \frac{\alpha_1}{\alpha_2})_d$, $\Gamma_{1,x} := \Gamma_x$. Now let $x \in S \setminus \bar{B}(0, \frac{\alpha_1}{\alpha_2})_d$. We define $\lambda :=\mid x \mid^2$ (observe that $\lambda \geq \frac{\alpha_1}{\alpha_2}$) and we denote by $\{x_1, ..., x_M\}$ a system of representatives for the semi-algebraically connected components of Γ (which is computable, from [HRS 1]). Applying Lemma 9 to $S \cap \bar{B}(0, \lambda + 1)$, we construct a path connecting x to the unique x_j $(1 \leq j \leq M)$ lying in the same semi-algebraically connected component of $S \cap B(0, \lambda + 1)$; finally we define $\Gamma_{1,x}$ as this path. It is clear that $\Re_1$ is a roadmap for S. □

Let us make the following observation before considering the proof of the most general case.

Observation 12: Let $S \subset \mathbf{R}^n$ be a semi-algebraic set, and let $S = \cup_{1 \leq i \leq N} S^{(i)}$ be a decomposition of S where each $S^{(i)}$ is also semi-algebraic. Suppose, moreover, that each $S^{(i)}$ itself admits the following decomposition into semi- algebraically connected components $S^{(i)} = \cup_{1 \leq k \leq N_i} S_k^{(i)}$. Then each semi-algebraically connected component C of S is uniquely determined by a formula $C = \cup_{i,k_i} S_{k_i}^{(i)}$, where the indices i, k_i (which are taken respectively from the sets $\{1, ..., N\}$ and $\{1, ..., N_i\}$) satisfy the following relation: (i, k_i) and (j, k_j) appear in the description of C if and only if there exist two finite sequences $1 \leq j_1, ..., j_M \leq N$ and $k_{j_1}, ..., k_{j_M}$, with $1 \leq k_{j_\ell} \leq N_{j_\ell}$ $(1 \leq \ell \leq M)$, such that $(j_1, k_{j_1}) = (i, k_i)$, $(j_M, k_{j_M}) = (j, k_j)$ and for all $1 \leq \ell \leq M - 1$:

$$S_{k_{j_\ell}}^{(j_\ell)} \cap \overline{S_{k_{j_{\ell+1}}}^{(j_{\ell+1})}} \neq \emptyset \quad \text{or} \quad \overline{S_{k_{j_\ell}}^{(j_\ell)}} \cap S_{k_{j_{\ell+1}}}^{(j_{\ell+1})} \neq \emptyset$$

(cf [BCR], Chapter 2, 2.4.4.).

We are now ready to finish the proof of the Main Theorem.

Proof of the Main Theorem : Suppose first that S is a basic set, *i.e.* that

$$S := \{F_1 = 0 \ \wedge \ F_2 > 0 \ \wedge ... \wedge \ F_s > 0\}$$

with $s \geq 2$ (the algebraic case having already been considered). Adding a new variable Y we construct the set $T \subset \mathbf{R}^{n+1}$ as follows:

$$T := \{F_1 = 0 \ \wedge \ F_2 \geq 0 \ \wedge ... \wedge \ F_s \geq 0 \ \wedge \ 1 - \left(Y \cdot \prod_{2 \leq i \leq s} F_i \right) = 0\}.$$

Thus T is a closed semi-algebraic set and the projection $\pi : T \to S$ which forgets the coordinate Y is a semi- algebraic homeomorphism between S and T. Since T is closed, Lemma 11 allows one to compute a roadmap for it, namely $\Re := (\Gamma, (\Gamma_y)_{y \in T_d})$. We then construct a roadmap $\Re'$ for S in the following canonical way: $\Re' := (\Gamma', (\Gamma'_x)_{x \in S_d})$, where $\Gamma' = \pi(\Gamma)$ and for each $x := (x_1, ..., x_n) \in S_d$, $\Gamma'_x := \pi(\Gamma_y)$ with

$$y := (x_1, ..., x_n, \frac{1}{\prod F_i(x)}).$$

Let us now consider the case of an arbitrary semi-algebraic set S. The set S can be written in admissible time as $S = \cup_{1 \leq i \leq N} S^{(i)}$, where each $S^{(i)}$ is a basic semi-algebraic set for which we know a "basic" formula describing it.
For all $1 \leq i \leq N$ we compute a roadmap for $S^{(i)}$, $\Re^{(i)} := (\Gamma^{(i)}, (\Gamma^{(i)}_x)_{x \in S^{(i)}{}_d})$.
For all pairs (i, j), $1 \leq i, j \leq N$ we compute a system of representatives for the semi-algebraically connected components of $S^{(i)} \cap \overline{S^{(j)}}$; let $p_r^{(ij)}$ be the elements of this system $(1 \leq r \leq N_{ij})$. For each $p_r^{(ij)}$ we have a semi- algebraically connected path $\Gamma^{(i)}_{p_r^{(ij)}}$ which connects $p_r^{(ij)}$ with the path $\Gamma^{(i)}$. Applying the effective Curve Selection Lemma (see [HRS 4] or [HKRS]), it is possible to compute a semi-algebraic continuous function $\gamma_r^{(ij)} : [0,1] \to \overline{S^{(j)}}$ which verifies $\gamma_r^{(ij)}(0) = p_r^{(ij)}$ and $\gamma_r^{(ij)}((0,1]) \subset S^{(j)}$ (since $p_r^{(ij)} \in \overline{S^{(j)}}$). Set $q_r^{(ij)} := \gamma_r^{(ij)}(1) \in S^{(j)}$. We can consider the path $\Gamma^{(j)}_{q_r^{(ij)}}$ since $q_r^{(ij)} \in S_{(j)}{}_d$. Finally, we define a roadmap $\Re := (\Gamma, (\Gamma_x)_{x \in S_d})$ for S as follows:

$$\Gamma := \bigcup_{1 \leq i \leq N} \Gamma^{(i)} \cup \bigcup_{\substack{1 \leq i,j \leq N \\ 1 \leq r \leq N_{ij}}} (\Gamma^{(i)}_{p_r^{(ij)}} \cup \Gamma^{(j)}_{q_r^{(ij)}} \cup im(\gamma_r^{(ij)})) \text{ (where} S^{(i)} \cap \overline{S^{(j)}} \neq \emptyset).$$

Therefore we choose $\Gamma_x := \Gamma^{(i)}_x$, if $x \in S^{(i)}{}_d$.

By construction, $\Re$ is computable in admissible time; Observation 12 guarantees the connectedness of Γ in each component of S. □

28.4 A note about the computation of the connected components

From the procedure followed for the proof of the lemmas and the Main Theorem (especially from the continuity of the function *eval* and Proposition 1), it is easy to show that knowing how to describe in admissible time the semi-algebraically connected components of a bounded and regular algebraic hypersurface is enough to describe in admissible time the semi-algebraically connected components of every arbitrary semi-algebraic set (see also [CGV] ¶.2 and 3).

As in the proof of the theorem, we proceed by steps:
a) The case of a bounded algebraic set V (keep the notations and the procedure of Lemma 7): one constructs a bounded regular hypersurface $V_2 \subset \hat{\mathbf{R}(\varepsilon)}^{n+1}$, such that $(\pi \circ \mathit{eval})(V_2) = V$; then the image of each semi-algebraically connected component of V_2 by the map $\pi \circ \mathit{eval}$ is a semi-algebraically connected subset of V computable in admissible time (Proposition 1 and quantifier elimination). It is enough now to glue these semi-algebraically connected subsets using Observation 12.

b) The case of a basic closed and bounded semi-algebraic set S (see Lemma 8): one first constructs a semi-algebraic set $S^{(\varepsilon)}$, afterwards an algebraic hypersurface W, both in $\hat{\mathbf{R}(\varepsilon)}^{n}$, and finally a bounded algebraic set $V \subset \hat{\mathbf{R}(\varepsilon)}^{n+1}$ with the property that $\pi(V) = W \cap \bar{B}(0, \rho + 1)$. For each semi-algebraically connected component V^j of V, either $\pi(V^j) \cap S^{(\varepsilon)} = \emptyset$ or $\pi(V^j) \subset S^{(\varepsilon)}$, and a similar relation occurs also for the connected components of W (see Lemma 8 (3)). The above construction therefore allows one to compute the semi-algebraically connected components of $S^{(\varepsilon)} \cap W$, and then, since $\mathit{eval}(S^{(\varepsilon)} \cap W) = S$, to apply Proposition 1 in order to compute the connected components of S.

c) The case of an arbitrary closed and bounded semi-algebraic set S (see Lemma 9): one computes in admissible time a decomposition $S = \cup_{1 \leq i \leq n} S_i$, where each S_i is a basic closed and bounded semi-algebraic set, and again concludes by glueing and using Observation 12.

d) The case of an arbitrary closed semi-algebraic set S: let $\Phi \in \mathcal{L}(\mathbf{A})$ be a quantifier free formula which describes $S \subset \mathbf{R}^n$, Ω be a positive infinitely big element over $\mathbf{R}$ and $S^{(\Omega)}$ be as in Proposition 10, then $S^{(\Omega)} \subset \hat{\mathbf{R}(\Omega)}^{n}$ is a closed and bounded semi-algebraic set defined over $\mathbf{A}[\Omega]$. Then one can describe the semi-algebraically connected components of $S^{(\Omega)}$ by means of formulas involving polynomials in $\mathbf{A}[\Omega][X_1, ..., X_n]$. For each of these formulas Ψ, one takes the corresponding sign conditions over the polynomials in $\mathbf{A}[X_1, ..., X_n]$ appearing as the coefficients of the polynomials in Ψ. In this way one computes semi-algebraic subsets of S, each contained in exactly one semi-algebraically connected component of S. One concludes by taking closures and glueing, using Observation 12.

e) The case of a basic semi-algebraic set S (see proof of Theorem): one constructs a closed semi-algebraic set T and a semi-algebraic homeomorphism $\pi : T \to S$. Then one computes the connected components of T by d).

f) The case of an arbitrary semi-algebraic set S: one first computes a "basic" description of S, then one computes the semi-algebraically connected components of each of these basic semi-algebraic sets, and concludes by taking closures and glueing, using Observation 12.

28.5 References

[BCR] Bochnak J., Coste M., Roy M.-F. : Géométrie algébrique réelle. Springer Verlag (1987).

[Ca 1] Canny J.: Some algebraic and geometric computations in PSPACE. ACM Symposium on the theory of computation, 460-467 (1988).

[Ca 2] Canny J.: The complexity of robot motion planning. MIT Press (1989)

[CGV] Canny J., Grigor'ev D. Yu. , Vorobjov N. N. (jr): Defining connected components of a semi-algebraic set in subexponential time. Manuscript LOMI Steklov Institute of Leningrad (1990).

[Co] Collins G. : Quantifier elimination for real closed fields by cylindric algebraic decomposition. Second GI Conference on Automata Theory and Formal Languages. Lecture Notes in Computer Sciences, vol. 33, pp. 134-183, Springer-Verlag, Berlin (1975).

[Ga] von zur Gathen J.: Paralel arithmetic computations: a survey. Proc. 13 th Conf. MFCS (1986).

[G] Grigor'ev D. Yu.: Complexity of deciding Tarski algebra. J. Symbolic Computation 5 (1988) 65-108.

[GHRSV] Grigor'ev D. Yu, Heintz J., Roy M.-F., Solernó P., Vorobjov N. N. (jr): Comptage des composantes connexes d'un ensemble semi-algé brique en temps simplement exponentiel. C.R. Acad. Sci. Paris 312 (1990) 879-882.

[GV 1] Grigor'ev D. Yu. , Vorobjov N. N. (jr): Solving systems of polynomial inequalities in subexponential time. J. Symbolic Computation 5 (1988) 37-64.

[GV 2] Grigor'ev D. Yu. , Vorobjov N. N. (jr): Counting connected components of a semi-algebraic set in subexponential time. Manuscrit LOMI (1990). Submitted to Computational Complexity.

[HKRS] Heintz J., Krick T., Roy M.-F., Solernó P. : Single exponential time algorithms for basic constructions in elementary geometry. Invited Lecture AAECC'8 Tokyo (1990).

[HRS 1] Heintz J., Roy M.-F., Solernó P. : On the complexity of semi-algebraic sets. Proc. IFIP San Francisco. North-Holland (1989) 293-298.

[HRS 2] Heintz J., Roy M.-F., Solernó P.: Complexité du principe de Tarski-Seidenberg. Compte-Rendus de l'Académie des Sciences Paris. 309 (1989) 825-830.

[HRS 3] Heintz J., Roy M.-F., Solernó P.: Sur la complexité du principe de Tarski-Seidenberg. Bull.Soc.math.France, 118 (1990) 101-126.

[HRS 4] Heintz J., Roy M.-F., Solernó P.: Single exponential path finding in semi-algebraic sets: Part I : The case of a regular bounded hypersurface.

Proc. AAECC-8 Tokyo (1990).

[HRS 5] Heintz J., Roy M.-F., Solernó P.: Description of connected components of semi-algebraic sets in single exponential time. Manuscript (1990)

[R] Renegar J.: On the computational complexity and geometry of the first order theory of the reals. Technical Report 856, Cornell University Ithaca (1989). Submitted to Journal of Symbolic Computation.

[S] Solernó P.: Effective Łojasiewicz inequalities. To appear in AAECC (1990).

[SS] Schwartz J., Sharir M. : On the "piano movers" problem II. General techniques for computing topological properties of real algebraic manifolds. Adv. Applied Math. 4, 298-351 (1983).

[W] Walker R.: Algebraic curves. Princeton University Press (1950).

29
An Improved Projection for Cylindrical Algebraic Decomposition

D. Lazard

ABSTRACT It is proved that the projection needed by the CAD algorithm needs only to compute the resultants, the discriminants, the leading coefficients and the constant coefficients of the input polynomials, provided they have be made square–free and relatively prime by GCD computations. This improves [7] first improvement by removing any dimension condition and dropping out from the projection set all coefficients of the input polynomials, other than the leading and the constant ones.

29.1 Introduction

In 1975, Collins [4] gave an algorithm called Cylindrical Algebraic Decomposition or CAD for solving the problem of quantifier elimination in real closed fields. This algorithm is far better than the previous ones ; it is doubly exponential in the number of variables, but is polynomial with respect to all other parameters of the input. Moreover [6] have provided an example showing that this doubly exponential behaviour is unavoidable, and thus that CAD algorithm is, in some sense, optimal.

This algorithm is very important because most geometrical problems are instances of it. It has to be noticed that the robot planning problem (piano movers problem) is not an instance of quantifier elimination but may be solved by CAD ([9]).

The CAD algorithm has been implemented in SAC2 by [2]. It appears that the set of polynomials which are used for the projection is very important to have an efficient implementation. In original Collins [4] algorithm, given a set of polynomials in n variables viewed as univariate polynomials in the last variable x_n, the projection set consists in all the coefficients of the powers of x_n in the product of input polynomials, all subresultants of this product and its derivative (with respect to x_n) when the leading term of this product is not zero and also when any number of leading terms vanish.

[8] provided an improved projection set consisting in all the coefficients

of the product and its discriminant, but this projection set works only up to three variables, or for finding cells of codimension less than two. [3] give an improved presentation of McCallum [8] result where the list of all coefficients is replaced by the leading one and the sum of the squares of the others ones.

In this paper we prove that a smaller projection set works in any dimension if we slightly modify the way of evaluating a polynomial over a sample point.

29.2 Working domains and basic definitions.

Let R_0 be a field. Generally, R_0 will be a real field, but for most of this paper, it is only necessary to suppose that its characteristic is zero. All computations will be done in $R_n = R_0[x_1, \ldots, x_n] = R_{n-1}[x_n]$. Unless explicitely stated, the elements of R_n will be viewed as univariate polynomials in x_n ; thus, leading and constant coefficients, degree, ... will refer to this recursive representation of the polynomials.

Especially, we shall need resultants and discriminants of polynomials in x_n. The resultant is unambigousely defined, but the discriminant is not, for polynomials which are not monic.

Definition 1 *The discriminant of a polynomial p is the quotient by its leading coefficient of the resultant of p and its derivative.*

We recall now the properties of the resultant and of the discriminant that we shall need.

Proposition 1 **(i)** *The resultant of two polynomials,* resultant (p, q, x_n) *is a polynomial in the coefficients of p and q.*

(ii) resultant $(p, q, x_n) = \pm$resultant (q, p, x_n).

(iii) resultant $(pq, r, x_n) =$ resultant (p, r, x_n) resultant (q, r, x_n).

(iv) *If* $p = a_0 x_n^m + \cdots + a_m$ *then*

$$\text{discriminant}\,(p, x_n) = \pm\, \text{resultant}\,(p_1, p', x_n)/m^{m-2}$$

where p' *is the derivative of p with respect to m and* $p_1 = a_1 x^{m-1} + 2a_2 x^{m-2} + \cdots + m a_m$.

(v) discriminant $(p_1, p_2, x_n) = c\, D_1\, D_2$ resultant $(p_1, p_2, x_n)^2$ *where* $D_i =$ discriminant (p_i, x_n) *and c is a (rational) constant depending only on the degrees of* p_1 *and* p_2.

(vi) *If* $p = a(x_n - z_1) \cdots (x_n - z_m)$ *then*

$$\text{discriminant}\,(p, x_n) = \pm\, a^{2m-2} \prod_{i<j} (z_i - z_j)^2.$$

(vii) *If p is as above and $q = b(x_n - t_1) \cdots (x_n - t_k)$ then*

$$\text{resultant}\,(p, q, x_n) = \pm\, a^k b^m \prod (z_i - t_j).$$

For using the last two assertions of the preceeding proposition, we have to work in an algebraic closure K of the fraction field of R_n. We need an explicit description of it, which we shall give now.

Let us recall that if k is a field and X an indeterminate, a Puiseux series in X over k is a formal series

$$\sum_{n \geq n_0} u_n X^{n/q}$$

where n_0 and q are given integers (q is positive, but n_0 may be negative), n is integer and the u_n are in k. The main fact about the Puiseux series is that they form an algebraic closed field if k is algebraic closed. If u_{n_0} is not null, the *valuation* of the Puiseux series is n_0/q. If a is an element of k and, if we substitute $X - a$ to X in the above series, the resulting series will be called a *Puiseux series in $X - a$ or in X at a.*

Let us now define an algebraic closure of R_n. Let K_0 be an algebraic closure of R_0 and $a = (a_1, \ldots, a_n)$ be a point of K_0^n. The field of fractions of R_n may be embedded in the field $K_n(a)$ of Puiseux series in $x_1 - a_1$ whose coefficients are Puiseux series in $x_2 - a_2$ whose coefficients ... whose coefficients are Puiseux series in $x_n - a_n$ with coefficients in K_0.

Remark 1 *It should be noticed that the main variable is the last one for the polynomials and the first one for the Puiseux series. This may seem unnatural, but it is necessary for the results in this paper.*

If s is an element of $K_n(a)$, the *valuation* $v_a(s)$ of s is defined as follows. It is is an element $(v_1, \ldots, v_n)$ of $\mathbf{Q}^n$, such that v_1 is the least exponent of $x_1 - a_1$ in s ; if s_1 is the coefficient of $(x_1 - a_1)^{v_1}$ in s, then $s_2(x_2 - a_2)^{v_2}$ is the first term of s_1 as Puiseux series in $x_2 - a_2$, and, recursively, $s_i(x_i - a_i)^{v_i}$ is the first term of s_{i-1} as Puiseux series in $x_i - a_i$.

The following easy result shows that v_a is a valuation for the usual meaning if $\mathbf{Q}^n$ is ordered by the lexicographical ordering.

Proposition 2 *If s and t are in $K_n(a)$, we have $v_a(st) = v_a(s)v_a(t)$ and $v_a(s+t) \geq \min(v_a(s), v_a(t))$, this last inequality beeing an equality if $v_a(s) \neq v_a(t)$.*

For our theory, we need to see how the valuation varies with a. For this purpose, we need a topology on K_0 ; thus we suppose from now that K_0 is the field of complex numbers.

Proposition 3 *If K_0 is the field of complex numbers and s is an element of the algebraic closure of R_0, then $v_a(s)$ is a semi-continuous function of a, that is, for any a, there is a neighborough V of a such that for any b in V we have $v_b(s) \geq v_a(s)$.*

The proof follows easily from elementary theory of power series and analytic functions. •

29.3 The projection set.

We start with a family F_n of polynomials in R_n, we will associate to it a family F_{n-1} of polynomials in R_{n-1} which well describes the projection of the zeros of the polynomials in F_n. Before doing this, we need to transform F_n in a new set F'_n of polynomials in order that 0 does not appear as an element of F_{n-1}. Here is this transformation.

Replace each polynomial in F_n by its content (GCD of the coefficients as polynomial in x_n) and its primpart (quotient by this GCD).

Replace each f in the obtained list by the polynomials appearing in its square free decomposition.

If two element f and g in the list have a non trivial GCD h, then replace them by h, f/h and g/h ; do such replacments while it is possible.

If there is an element f in the list which is a multiple of x_n (there is at most one such f by the preceeding reductions), replace it by f/x_n and x_n.

Remove constants (elements of R_0).

Optionally (Probably computationally more efficient in view of proposition 1.v), replace each element of the list by their irreducible factors over R_0.

Let F'_n be the resulting list.

Remark 2 *For the validity of the method, i.e. for not obtaining zero as a member of the projection set, we need only that the polynomials in F'_n which depends on x_n are square free, pairwise relatively prime and not multiple of x_n (except, possibly, x_n itself). The other conditions are useful for getting smaller polynomials in the projection set at a low extra cost.*

We are now ready to define F_{n-1}. We take in it the elements of the following list which are not in R_0.

The elements of F'_n not depending on x_n.

The leading coefficient of each element of F'_n depending on x_n.

The constant coefficient of each element of F'_n depending on x_n.

The discriminant of each element of F'_n depending on x_n.

The resultant of each pair of elements of F'_n depending on x_n.

29.4 The CAD algorithm.

We are now able to describe CAD algorithm, supposing now that R_0 is a subfield of the reals.

Starting from F_n, we repeat above construction of a projection set until getting a set F_1 of univariate polynomials.

The univariate CAD consists in the list of the roots of the polynomials in F_1, in the intervals which separate them and in a sample point in each interval. These roots and intervals are the cells of the univariate CAD. Note that, on each cell, the valuation $v_a(f)$ of each polynomial f in F_1 is constant. The univariate CAD is easily found by real roots isolation (see, for example [5]).

Now, starting from this univariate CAD, we recursively lift it until getting a n–dimensional CAD. Thus we have only to describe the lifting from $n-1$–dimensional CAD to n–dimensional CAD.

The $n-1$–dimensional CAD consists in a partition of the n–dimensional real affine space in cells, together with a sample point for each cell. The recursion hypothesis is that the valuation $v_a(f)$ of each polynomial f in F_{n-1} is *a constant on each cell.*

Proposition 4 *Let C be a a cell of the $n-1$-dimensional CAD and f and g be two polynomials in F_n. Then*

(i) $v_a(f)$ is constant for a in $C \times 0$.

(ii) If z is a root of f, then $v_a(z)$ is constant for a in C.

(iii) If z_1 and z_2 are distinct roots of f, then $v_a(z_1 - z_2)$ is constant for a in C.

(iv) If z_1 is a root of f and z_2 is a root of g, then $v_a(z_1 - z_2)$ is constant for a in C.

Note first that the elements of F_n are products of elements of F'_n ; thus it suffices to prove this proposition for f and g in F'_n.

If f does not depends on x_n, then assertion (i) follows immediately from the recursion hypothesis and the others are empty. If $f = x_n$, then assertions (i) to (iii) are immediate. Thus we may suppose that f depends on x_n and does not vanishes for $x_n = 0$.

Let c be the constant term (with respect to x_n) of f and

$$f = b\prod(x_n - z_i)$$

be the factorization of f in the algebraic closure of R_n. We have $c = \pm b\prod z_i$; by hypothesis, $v_a(b)$ and $v_a(c)$ are constant on C. By semi–continuity and product property of the valuation, it follows that $v_a(z_i)$ is constant on C for each i (assertion ii). We have $v_a(z_i) = (v_1, \ldots, v_{n-1}, 0) \neq$

$(0, \ldots, 0, 1) = v_a(x_n)$ on $C \times 0$; thus $v_a(x_n - z_i) = \min(v_a(x_n), v_a(z_i)$ is constant on $C \times 0$, and assertion (i) follows.

Assertions (iii) and (iv) follow immediately from product properties of the discriminant and of the resultant, by the same argument of semi–continuity. •

In classical CAD algorithm, the cells of the n–dimensional CAD are defined by a stratification of the cylinder over the cells of the $n-1$–dimensional CAD ; this stratification is defined by the roots of the polynomials in F_n, viewed as functions of a point in the $n-1$–space. The difficulty is that a polynomial in F_n may vanish identically on a cell and define only the trivial stratification, even when it is not sufficent for the recursion to be correct.

Thus, we have to define an evaluation process for the polynomials in F_n over a point of K_n.

Definition 2 *Let f be a polynomial in R_n and $a = (a_1, \ldots, a_{n-1})$ be a point of K_0^{n-1}. The evaluation f_a of f at a is defined by the following algorithm which computes also the $n-1$ first coordinates of the valuation of f at a point $(a_1, \ldots, a_{n-1}, a_n)$.*

$f_a := f$

for $i := 1$ *to* $n-1$ *do*

$v_i :=$ *the greatest integer* v *such* $(x_i - a_i)$ *divides* f_a ;

$f_a := f_a/(x_i - a_i)^{v_i}$;

$f_a := subst(a_i, x_i, f_a)$.

Proposition 5 *If $f \in F_n$, the number of complex roots of f_a, the number of real roots of f_a and their multiplicities are constant on a cell of the $n-1$–dimensional CAD. If f and g are in F_n, the equalities between real roots of f_a and g_a are also constant on such a cell.*

This follows immediately from last proposition by the following lemma.

Lemma 1 *Let z be a root of f such that $v_a(z) = (v_1, \ldots, v_{n-1})$; we said that z is* infinite *at a if some v_i is negative ; if all v_i are non negative,* the value *of z at $a = (a_1, \ldots, a_{n-1})$ is obtained by substituting a_i to x_i in the Puiseux series of z ; this value is 0 if some v_i is positive.*

With these definitions, for f in F_n, the roots of f_a are exactly the values at a of the roots of f which are finite at a. A root of f has the value 0 at a iff it is finite and has a positive (for the lexicographical ordering) valuation. Two roots z_1 and z_2 of f and g have the same value at a iff both are finite and $v_a(z_1 - z_2) > 0$ for the lexicographical ordering.

This easy proof is left to the reader. •

Now, the cells of the n–dimensional CAD, over a cell C of the $n-1$–dimensional CAD may be defined as follows : let $z_1(a), \ldots, z_k(a)$ be the real roots of the f_a for f in F_n ; the cells are

$\{(a, z_i(a));\ a \in C\}$ for each i;

$\{(a, z);\ a \in C, z_i(a) < z < z_{i+1}(a)\}$ for each $i < k$;

$\{(a, z);\ a \in C, z < z_1(a)\}$;

$\{(a, z);\ a \in C, z_k(a) < z\}$.

It is now clear that the valuation $v_a(f)$ of any f in F_n is constant on each cell C_j defined over a cell C of the $n-1$–dimensional CAD : let $v = (v_1, \ldots, v_{n-1}, 0)$ be the (constant) valuation of f on $C \times 0$. If C_j is defined by a root of f_a of multiplicity m over C, then f has valuation $(v_1, \ldots, v_{n-1}, m)$ on C_j. If f_a is not 0 on C_j, then, the valuation of f is $(v_1, \ldots, v_{n-1}, 0)$ on C_j.

The computation of the n–dimensional CAD from the $n-1$–one is now completely defined :

For each cell of the $n-1$–dimensional CAD, *with sample point a, compute* $\{f_a; f \in F_{n-1}\}$.

Isolate the real roots of the f_a and define the cells by the obtained roots and the intervals between them.

Choose the n–th coordinates of the sample point in each cell, in the interval defining the cell for the "interval cells" and take the root as n–th coordinate for the "root cells".

Remark 3 *The cells are not defined by equations, but their definition are sufficent for many practicable computation. Especially, it may be decided in which cell some given point belongs : it suffices to isolate the roots of the elements of the F_i at this point ; if the point is the j–th root or is in the j–th interval, then it is in the j–th root cell or is in the j–th interval cell.*

However, the result which follows is important from a theorical point of vue.

Proposition 6 *The cells of the above defined CAD are semi–algebraic connected sets, i.e. they may be defined by polynomial equations and inequations.*

This is proved by recursion ; for 1–dimensional CAD, there is almost nothing to prove. Consider a cell C (semi–algebraic by recursion hypothesis) of the $n-1$–dimensional CAD ; we have to prove that, for each f of F_n, the roots of f_a are algebraic (and thus continuous) functions of a on C. If f does not vanishes identically on C, there is nothing to prove. If it does, choose a in C, choose b near a, in a cell D of dimension $n-1$ (this is possible because C is of dimension less than $n-1$ if f vanishes on it), and choose a neighbourhood U of a in C such that its image V by the translation from a to b is contained in D. The roots of f are algebraic functions of V. The same is true for the roots of f_c for c in U, because they are the limit of the first ones when V tends to U by the path consisting in translations along each coordinate, one after the other. •

Corollary 1 *The sign (0, + or −) of a polynomial of F_n is constant on each cell.*

This follows from the connexity : a polynomial f is 0 at a iff $v_a(f) \neq (0, \ldots, 0)$. •

29.5 Practical improvements.

In this section we give two remarks which may improve the practical efficiency.

If we look at the proofs, we see that we need that the valuation of the roots are constant on the cells only for beeing sure that a root which is finite or infinite on some point of a cell has the same property on any point of the cell. Thus we have :

Proposition 7 *If the leading coefficient of some polynomial f of F_n does not vanishes on a cell, it is not necessary to split it by the corresponding constant coefficient.*

In this case, the product of the roots, and, thus, each root are finite on the cell. •

This result is clearly interesting if the leading coefficient does not vanishe on any cell we are interested on, for example, if the leading coefficient is in R_0 or is a positive parameter. In this cases, we need not to take the constant coefficient in F_{n-1}.

But this result is also useful in the general case : let f in F_n, and b and c its leading and constant coefficients. In constructing the $n-1$–dimensional CAD, we have to separate roots of b, c an other polynomials, over the cells of the $n-2$–dimensional CAD. Let C be such a cell. If b does not vanishes identically on C, it is zero on cells over C of the $n-1$–dimensional CAD, on which c has a constant valuation.

Thus we have the following result which strongly decreases the number of cells, even if it is not easy to implement.

Proposition 8 *Let $f \in F_n$, and b and c its leading and constant coefficients as a polynomial in x_n. If b and c are in R_i and c effectively depends on x_i, for some i, then the roots of c need to be taken in account in the root isolation process (for computing the cells of the i–dimensional CAD) only over cells on which b vanishes identically.*

The second improvement is that, in a practical implementation, we generally do not need to compute the valuation. Thus the following algorithm seems better for computing the evaluation of a polynomial at a sample point. In this algorithm, primpart(f_a) is the quotient of f_a by the GCD of its coefficients (f_a being viewed as a polynomial in x_n).

$f_a := f$

for $i := 1$ *to* $n - 1$ *do*

$g :=$ *subst*(a_i, x_i, f_a) ;
if $g = 0$
then $f_a :=$ *subst*$(a_i, x_i,$ *primpart*$(f_a))$;
else $f_a := g$.

This procedure is certainly more efficient than the one which was given in definition 2, because, in most steps, the else clause is used.

We refer only to papers which are explicitely quoted above. For a complete bibliography on the subject, see [1].

29.6 References

[1] Arnon, D. S., (1988), A Bibliography for Quantifier Elimination for Real Closed Fields, *J. Symbolic Computation* **5**, pp. 267–274.

[2] Arnon, D. S. and McCallum, S., (1982), Cylindrical Algebraic Decomposition by Quantifier Elimination, Proceedings of the European Computer Algebra Conference (EUROCAM '82), *Springer Lect. Notes Comp. Sci.* **144**, pp. 215–222.

[3] Benedetti, R. and Risler J.-J., (1990), *Real Algebraic and Semi–Algebraic Sets,* Hermann (Paris).

[4] Collins, G.E., (1975), Quantifier Elimination for the Elementary Theory of Real Closed Fields by Cylindrical Algebraic Decomposition, *Lect. Notes Comput. Sci.*, **33**, pp. 134–183.

[5] Collins, G.E. and Loos, R., (1982), Reals Zeros of Polynomials, *Symbolic and Algebraic Computations (Computing Supplementum* **4***)*, (ed. Buchberger, B., Collins, G. E. and Loos R.), Springer Verlag (Wien, New York), pp. 83–94.

[6] Davenport, J. H. and Heintz, J., (1988)., Real Quantifier Elimination is Doubly Exponential, *J. Symbolic Computation*, **5**, pp. 29–35.

[7] McCallum, S., (1984)., *An Improved Projection Operator for Cylindrical Algebraic Decomposition*, PhD Thesis, Univ. Wisconsin, Madison.

[8] McCallum, S., (1988)., An Improved Projection Operator for Cylindrical Algebraic Decomposition of Three Dimensional Space, *J. Symbolic Computation*, **5**, pp. 141–161.

[9] Schwartz, J. T. and Sharir, M., (1983)., On the "Piano Movers" Problem, II, General Techniques for Computing Topological Properties of Real Algebraic Manifolds, *Adv. Applied Math.* **4**, pp. 298–351.

30

Degree Bounds of Gröbner Bases

Wei Li

ABSTRACT We show that there is no universal bound of Gröbner bases when the coefficient ring is **Z**. A similar result where the coefficient ring is $k[z], k$ a field, was proved by [1]. While the coefficient ring is a field, such bounds always exist as proved by Weispfenning [16].

30.1 Introduction

The method of Gröbner bases is a powerful tool widely used in Computer Algebra, which provides algorithms to solve many constructive problems in Commutative Algebra and Algebraic Geometry (for instance, see [5], [6], [9]). Saying algorithms, one must consider their complexity. For Gröbner bases, we may use degrees of polynomials as a measure of the complexity. Namely, we may ask the following question:

Complexity Question *Let R be a commutative ring. Is there a constant $D = D(n, d)$ such that if I is an ideal in $R[x_1, ..., x_n]$ generated by $f_1, ..., f_l$ with* $\max\ deg(f_i) \leq d$*, then there exists a minimal Gröbner basis $\{g_1, ..., g_r\}$ (with respect to a fixed admissible order) of I such that* $\max\ deg(g_i) \leq D$*?*

The terms involved will be explained in the next section, and, throughout this paper, *degree* means *total degree*.

We call the constant $D(n, d)$, if it exists, a universal bound of Gröbner bases. We do not consider any other aspects of the theory of complexity of Gröbner bases, such as steps to construct a Gröbner basis, bounds of normal form algorithms (see [5], [13], etc.). But these bounds play an essential role in the theory of complexity of Gröbner bases. In fact, if one knows such a bound, one might be able to derive others; for instance, see [7]. We now consider the complexity question above. It turns out that its answer depends on what coefficient ring, R, you are working on. If $R = k$ is a field, Weispfenning proved in [16] the existence of universal bounds (see also [3], [10], [11], [14], etc.):

Theorem A *Let k be a field. There is a constant $D = D(n, d)$ such that if I is an ideal in $k[x_1, ..., x_n]$ generated by $f_1, ..., f_l$ with* $\max\ deg(f_i) \leq d$*, then there exists a minimal Gröbner basis $\{g_1, ..., g_r\}$ (with respect to a fixed admissible order) of I such that* $\max\ deg(g_i) \leq D$.

In [16] the bounds are stated as depending also on l, the number of generators. But this dependence, once n and d are fixed, can be eliminated by taking $l = \binom{n+d}{d}$, since the vector space of polynomials with degrees not greater than d has finite dimension $\binom{n+d}{d}$. This observation works equally for any PID coefficient ring R, such as $\mathbf{Z}$ (the ring of integers) and $k[z]$ (a polynomial ring over a field in one variable). As for general commutative rings, one probably needs to take l into account while considering the complexity question above.

On the other hand, in [1] Abhyankar and this author proved

Theorem B *There is no universal bound if $R = k[z], k$ a field.*

More precisely, in [1] examples are constructed which show the nonexistence of a universal bound of degrees of Gröbner bases in a two variable polynomial ring where the coefficient ring is a polynomial ring over a field in one variable. But when the coefficient ring is a field, for the two variable case, we have not only the existence as Theorem A above, but an explicit bound d^2, where d is the maximal degree of generators, as given by [8] and [11]. Hence it is natural to search such bounds for polynomial rings over $\mathbf{Z}$. The aim of this paper is to prove

Theorem C *There is no universal bound if $R = \mathbf{Z}$.*

The admissible order used in both Theorems B and C is the lexicographic order. It is generally believed that the lexicographic order has, among all admissible orders, the worst complexity, while the reverse lexicographic order has the best complexity (see [4]). But this would not affect the strongness of Theorems B and C, since the examples showing these theorems are for two variable polynomial rings, and we know that in the two variable case, the lexicographic order and the reverse lexicographic order are the same. We remark here that we have just given an informal discussion; in fact, the lexicographic order in [4], being degree compatible, is different from ours. At any rate, the problem remains open for an arbitrary admissible order.

Theorem C will be proved in Section 31.2. We shall in the next section recall some definitions and results which are necessary for proving Theorem C. In Section 31.3, we shall make some comments on our examples and raise several problems concerning complexity of Gröbner bases.

This paper is based on part of the author's thesis [12]. It is a great pleasure for the author to take this opportunity to thank his thesis advisor, Professor Shreeram Abhyankar. He would also like to thank Professor Chanderjit Bajaj for introducing him to the theory of computational complexity.

30.2 Preliminaries

Gröbner bases were first defined for ideals of polynomial rings over fields (see [5], [6]). In [9], [15] and some other literature Gröbner bases are defined for polynomial rings over these coefficient rings which have certain computational properties. Although in this paper we are primarily interested in the case that the coefficient ring is **Z**, which of course has nice computational properties, we need to define Gröbner bases over general commutative rings. In fact, to derive the examples showing Theorem C, we need to work on some coefficient rings other than **Z**.

Let R be any commutative ring and $R[x] = R[x_1, ..., x_n]$ be the polynomial ring over R in n variables $x_1, ..., x_n$. Denote by

$$\mathcal{P} = \{x_1^{i_1} x_2^{i_2} \cdots x_n^{i_n} | \text{all } i_j \geq 0\}$$

the set of all power products in $x_1, ..., x_n$. Let $\leq$ be a total order on $\mathcal{P}$.

Definition 1 *We say* $\leq$ *is* admissible *if*

(i) $1 \leq p$ *for all* $p \in \mathcal{P}$*;*

(ii) $p \leq q \Longrightarrow p \cdot r \leq q \cdot r$ *for all* $p, q, r \in \mathcal{P}$.

We do not need to pre-assume that $\leq$ is a total order. An order being total order is indeed a consequence of the properties (i) and (ii) in the definition (see [5], [13]). The most important admissible order used in this paper is the lexicographic order defined as in the following examples.

Example 1.1 A lexicographic order on $\mathcal{P}$ is defined by

$$x_1^{i_1} \cdots x_n^{i_n} < x_1^{j_1} \cdots x_n^{j_n} \quad \Longleftrightarrow \quad \left\{ \begin{array}{l} \text{there exists an } s \text{ such that} \\ i_n = j_n, \, ..., \, i_s = j_s, \, i_{s-1} < j_{s-1} \end{array} \right\}.$$

This order is admissible. It is clear that it is uniquely determined by the order among x_1, ..., x_n, that is, $x_1 < x_2 < \cdots < x_n$. The lexicographic order with $x_1 < \cdots < x_n$ can characterized by the property:

For each i, $1 \leq i < n$, any power product in $x_1, ..., x_i <$ any power product in $x_{i+1}, ..., x_n$.□

Example 1.2 Let $x_1, ..., x_n, y_1, ..., y_m$ be $n + m$ variables, and let $\mathcal{P}_x$ ($\mathcal{P}_y$ and $\mathcal{P}_{x,y}$ respectively) the set of power products in $x_1, ..., x_n$ (in $y_1, ..., y_m$ and in $x_1, ..., x_n$, $y_1, ..., y_m$ respectively). Assume $\leq_x$ and $\leq_y$ are admissible orders on $\mathcal{P}_x$ and $\mathcal{P}_y$ respectively. Define an order $\leq_{x,y}$ on $\mathcal{P}_{x,y}$ by

$$x_1^{i_1} \cdots x_n^{i_n} y_1^{i_{n+1}} \cdots y_n^{i_{n+m}} <_{x,y} x_1^{j_1} \cdots x_n^{j_n} y_1^{j_{n+1}} \cdots y_n^{j_{n+m}}$$

$$\Longleftrightarrow \begin{cases} \text{either } x_1^{i_1}\cdots x_n^{i_n} <_x x_1^{j_1}\cdots x_n^{j_n}, \\ \text{or } x_1^{i_1}\cdots x_n^{i_n} = x_1^{j_1}\cdots x_n^{j_n},\ y_1^{i_1}\cdots y_m^{i_m} <_y y_1^{j_1}\cdots y_m^{j_m}. \end{cases}$$

The order $\leq_{x,y}$ is admissible on $\mathcal{P}_{x,y}$ satisfying the following two conditions:

(i) the restrictions to $\mathcal{P}_x$ and $\mathcal{P}_y$ are $\leq_x$ and $\leq_y$ respectively;

(ii) any power product in $y_1, ..., y_m <$ any power product in $x_1, ..., x_n$.

For example, if $\leq_x$ is the lexicographic order with $x_1 < \cdots < x_n$, $\leq_y$ is the lexicographic order with $y_1 < \cdots < y_m$, then $\leq_{x,y}$ is the lexicographic order with $y_1 < \cdots < y_m < x_1 < \cdots < x_n$.□

We now fix an admissible order, say $\leq$, on $\mathcal{P}$ and are going to define Gröbner bases for an ideal of $R[x] = R[x_1, ..., x_n]$ with respect to $\leq$. We start with a single polynomial $f \in R[x]$. Assume $f \neq 0$, we may write

$$f = ap + \sum_{q<p} a_q q$$

where $a \neq 0$ and a_q's are elements in R, i.e., p is the greatest (with respect to $\leq$) power product which has nonzero coefficient in f and a is the coefficient of p in f. We define the *leading term* of f, denoted by $\mathrm{lt}(f)$, to be ap. By convention we set $\mathrm{lt}(0) = 0$. If F is a subset of $R[x]$, the ideal $\mathrm{Lt}(F) = (\{\mathrm{lt}(f) \mid f \in F\})R[x]$ is called the *leading term ideal* of F (with respect to $\leq$).

Definition 2

(i) A subset G of an ideal $I \subseteq R[x]$ is a Gröbner basis *of I (with respect to $\leq$) if* $\mathrm{Lt}(I) = \mathrm{Lt}(G)$.

(2) A Gröbner basis G of I is said to be a minimal Gröbner basis of I if no proper subset of G is a Gröbner basis of I.

(3) A minimal Gröbner basis G of I is said to be a reduced Gröbner basis of I if each $g \in G$ satisfies:

(i) no non-leading term of g is in $\mathrm{Lt}(G)$; and

(ii) (the coefficient of the leading term of g) $= 1$.

We will need following propositions.

Proposition 1 *If G is a Gröbner basis of the ideal $I \subseteq R[x] = R[x_1, ..., x_n]$ then $I = (G)R[x]$, i.e., G is a set of generators of the ideal I.*

Proof See [9].□

Let $J \subseteq R$ be an ideal, $\bar{R} = R/J$. Let $\alpha : R \longrightarrow \bar{R}$ be the canonical homomorphism; whose extension to $R[x_1, ..., x_n] \longrightarrow \bar{R}[x_1, ..., x_n]$ is still denoted by α.

Proposition 2 *Let $\bar{I}$ be an ideal of $\bar{R}[x]$ and let $I = \alpha^{-1}(\bar{I})$. If G is a Gröbner basis of I, then $\{\alpha(g) \mid g \in G, \alpha(g) \neq 0\}$ is a Gröbner basis of $\bar{I}$.*

Proof The proof is similar to that of Proposition 3.1 in [1]. See also ([12], Proposition 1.3.1) and ([9], Proposition 3.3).□

Proposition 3 *Assume $\{\bar{g}_1, ..., \bar{g}_r\} \subset \bar{R}[x]$ is a Gröbner basis of the ideal generated by $\bar{g}_1, ..., \bar{g}_r$. Let I be the ideal in $R[x]$ generated by* $\ker \alpha$ *and $\alpha^{-1}(g_i), i = 1, ..., r$. Then I has a Gröbner basis G such that $\{\bar{g}_1, ..., \bar{g}_r\} = \{\alpha(g) \mid g \in G, \alpha(g) \neq 0\}$. Furthermore, if $\{\bar{g}_1, ..., \bar{g}_r\}$ is a minimal Gröbner basis, we can choose G to be a minimal Gröbner basis of I, provided that J has a minimal generating set.*

Proof The proof is similar to that of Proposition 3.2 in [1] (see also [12], Proposition 1.3.2). Since we will need not only the conclusion, but also the proof itself, we give a proof here.

For convenience we assume that $J = \ker(\alpha) \cap R$ is finitely generated. Note that to say a Gröbner basis of J is the same to say a set of generators of J. First let us assume that $\{\bar{g}_1, ..., \bar{g}_r\}$ is a Gröbner basis. Let $\{h_1, ..., h_l\}$ is a Gröbner basis of J. For each $i, 1 \leq i \leq r$, we can find $h_{l+i} \in \alpha^{-1}(\bar{g}_i)$ such that every coefficient of h_{l+i} is not in J. We claim that $G = \{h_1, h_2, ..., h_{l+r}\}$ is a Gröbner basis of I. Clearly for this G we have $\{\bar{g}_1, ..., \bar{g}_r\} = \{\alpha(h) \mid h \in G, \alpha(h) \neq 0\}$.

Let $f = f(x) \in I, f \neq 0$. We need to prove that $\text{lt}(f) \in \text{Lt}(G)$. Write f as

$$f(x) = a_p p + \sum_{q<p} a_q q$$

where $a_p \neq 0$. We have two cases: $a_p \in J$ and $a_p \notin J$. The first case is simple: if $a_p \in J = (h_1, ..., h_l)R$, then $\text{lt}(f) = a_p p \in (h_1, ..., h_l)R[x] \subseteq \text{Lt}(G)$ (note that $\text{lt}(h_i) = h_i, 1 \leq i \leq l$). If $a_p \notin J$, then $\alpha(a_p) \neq 0$. Hence $\text{lt}(\alpha(f)) = \alpha(a_p)p$. This implies that $\alpha(a_p)p \in \text{Lt}(\{\bar{g}_1, ..., \bar{g}_r\})$, that is,

$$\alpha(a_p)p = \sum_{i=1}^{r} \bar{b}_i(x)\text{lt}(\bar{g}_i) \tag{1.1}$$

where $\bar{b}_i(x) \in \bar{R}[x]$. Let

$$h_{l+i} = c_{p_i}^{(i)} p_i + \sum_{q<p_i} c_q^{(i)} q$$

where $c_{p_i}^{(i)}, c_q^{(i)} \in R, c_{p_i}^{(i)} \neq 0, i = 1, ..., r$. Then (1.1) becomes

$$\alpha(a_p)p = \sum_{i=1}^{r} \bar{b}_i(x)\alpha(c_{p_i}^{(i)})p_i.$$

For each i, take $b_i(x) \in \alpha^{-1}(\bar{b}_i(x))$. Then $a_p p - \sum_{i=1}^{r} b_i(x)c_{p_i}^{(i)}p_i \in \ker(\alpha)$. Hence

$$\begin{aligned} a_p p &= \sum_{i=1}^{r} b_i(x)c_{p_i}^{(i)}p_i + h(x) \\ &= \sum_{i=1}^{r} b_i(x)\mathrm{lt}(h_{l+i}) + h(x) \end{aligned}$$

where all coefficients of $h(x)$ are in J. Now

$$h(x) \in JR[x] = (h_1, ..., h_l)R[x] \subset \mathrm{Lt}(G).$$

So we obtain that $a_p p \in \mathrm{Lt}(G)$, i.e., $\mathrm{lt}(f) \in \mathrm{Lt}(G)$. This proves that G is a Gröbner basis of I.

Assume now that $\{\bar{g}_1, ..., \bar{g}_r\}$ is a minimal Gröbner basis. Let $G = \{h_1, ..., h_l, h_{l+1}, ..., h_{l+r}\}$ be as above, but in addition, let $\{h_1, ..., h_l\}$ be a minimal generating set of the ideal J. To prove G is minimal, suppose that $G_1 = G - \{h_i\}$ is still a Gröbner basis for some i. If $i \leq l$, then $\{h_1, ..., h_{i-1}, h_{i+1}, ..., h_l\} = G \cap R$ is a Gröbner basis of $I \cap R = J$. This contradicts the assumption that $\{h_1, ..., h_l\}$ is a minimal generating set. If $i > l$, then, by Proposition 2, $\{\bar{g}_1, ..., \bar{g}_{i-l-1}, \bar{g}_{i-l+1}, ..., \bar{g}_r\} = \{\alpha(h) \mid h \in G_1, \alpha(h) \neq 0\}$ is a Gröbner basis of the ideal $(\bar{g}_1, ..., \bar{g}_r)\bar{R}[x]$. This contradicts the assumption that $\{\bar{g}_1, ..., \bar{g}_r\}$ is a minimal Gröbner basis. Therefore G is a minimal Gröbner basis.□

We will also need the following Gröbner Basis Criterion for a polynomial map to be polynomially invertible; whose proof is referred to [1] or [12].

Gröbner Basis Criterion *Let $\leq$ be an admissible order on*

$$\mathcal{P} = \{x_1^{i_1} \cdots x_n^{i_n} y_1^{i_{n+1}} \cdots y_n^{i_{2n}} \mid \text{all } i_j \geq 0\}$$

satisfying:

any power product in $y_1, ..., y_n <$ any power product in $x_1, ..., x_n$.

(For example, the lexicographic order with $y_1 < \cdots < y_n < x_1 < \cdots < x_n$.) Assume $f = (f_1, ..., f_n) : R[x_1, ..., x_n] \longrightarrow R[x_1, ..., x_n]$ is an R-homomorphism. Then f is automorphism if and only if the ideal in $R[x_1, ..., x_n, y_1, ..., y_n]$ generated by $y_1 - f_1(x_1, ..., x_n), ..., y_n - f(x_1, ..., x_n)$ has a generating set

$$\{x_1 - g_1(y_1, ..., y_n), ..., x_n - g_n(y_1, ..., y_n)\}$$

which is the reduced Gröbner basis of the ideal. In the case f is automorphism, $f^{-1} = (g_1, ..., g_n)$.

Remark The statement of our Complexity Question is quite similar to a Bass' reformulation, in terms of inverse degrees of automorphisms, of the Jacobian conjecture in [2]. The criterion above, accompanied by the Bass' result, enables us to reformulate the Jacobian conjecture in terms of Gröbner bases. This and other applications to commutative algebra are referred to [1] and [12]. On the other hand, the criterion enables us to find the Gröbner bases of certain ideals, as we shall do in the next section.

30.3 Nonexistence of Bounds Over $\mathbf{Z}[\mathbf{x}, \mathbf{y}]$

Throughout this section we consider polynomial rings in two variables x, y and always use the lexicographic order with $y < x$. We shall construct examples demonstrating Theorem C announced in the introduction. In the other words, we are going to construct an example with the following properties:

Let d be a fixed positive integer. For any $t > 0$, we want to find polynomials $f_1, ..., f_r$ in $\mathbf{Z}[\mathbf{x}, \mathbf{y}]$, with $\max\deg(f_i) < d$, such that for any minimal Gröbner basis $\{g_1, ..., g_s\}$ of the ideal $(f_1, ..., f_r)\mathbf{Z}[x, y]$, $\max\deg(g_i) \geq t$.

We shall take $d = 2$. If the coefficient ring is a polynomial ring in one variable over a field, say $R = \mathbf{Q}[\mathbf{z}]$, we proved the nonexistence of a universal bound in [1] and [12] (i.e., Theorem B). Let us recall the example (Example 5.1 in [1], taking $k = \mathbf{Q}$) as follows:

Let $f_1 = g_1 = z^{t+1}$, $f_2 = y - x + zx^2$, $g_2 = x - y - \sum_{i=2}^{t+1} \dfrac{\binom{2i-2}{i} z^{i-1}}{i-1} y_i$. Then

(i) $\max\deg(f_i) = 2$;

(ii) $\max\deg(g_i) = \deg(g_2) = t + 1$;

(iii) $\{g_1, g_2\}$ is a minimal Gröbner basis of the ideal $(f_1, f_2)R[x, y]$; and

(iv) if $\{h_1, h_2\}$ is a minimal Gröbner basis of the same ideal $(f_1, f_2)R[x, y]$, then

$$\max\deg(h_i) \geq \max\deg(g_i) = t + 1.$$

To get a similar result for the case where the coefficient ring is $\mathbf{Z}$, a naive idea is to specialize f_1, f_2 to polynomials in $\mathbf{Z}[\mathbf{x}, \mathbf{y}]$ by assigning z a suitable numerical value. (This is the reason we are taking $k = \mathbf{Q}$.) But it does not work directly. The difficulty involved is that we want every thing over $\mathbf{Z}$, rather than $\mathbf{Q}$, the coefficient field of R. Also what we want are all properties (i) to (iv). Therefore we have to specialize not only f_1, f_2, but all these four properties. For example, we do not want that the degree of g_2 drops after specialization (at least we do not want it drops too much, if it does). We now proceed to construct the needed example.

We first recall more notations used in [1]. We always consider $R = \mathbf{Q}[z]$ and $A = \mathbf{Q}[z]/(z^{t+1})$ as coefficient rings. We have

$$F = x - xz^2 \in R[x];$$

$$G(x) = x + \sum_{i=2}^{t+1} \left(\frac{\binom{2i-2}{i} z^{i-1}}{i-1} \right) x_i \in R[x];$$

$\alpha : R = \mathbf{Q}[z] \longrightarrow A = \mathbf{Q}[z]/(z^{t+1})$ is canonical homomorphism, whose extension to $R[x] \longrightarrow A[x]$ is still denoted by α;

$$a = \alpha(z) \in A;$$

$$f = \alpha(F) = x - ax^2 \in A[x];$$

$$g = \alpha(G) = x + \sum_{i=2}^{t+1} \left(\frac{\binom{2i-2}{i} a^{i-1}}{i-1} \right) x_i \in A[x].$$

We have proved that $f : A[x] \longrightarrow A[x]$ is an automorphism with the inverse $g : A[x] \longrightarrow A[x]$. This means

$$\begin{cases} f(g(x)) &= x \\ g(f(x)) &= x \end{cases} \qquad \text{in } A[x].$$

Therefore there exist $U(z,x)$ and $V(z,x)$ in $R[x] = \mathbf{Q}[z,x]$ such that

$$\begin{cases} F(G(x)) &= x + z^{t+1}U(z,x) \\ G(F(x)) &= x + z^{t+1}V(z,x) \end{cases} \qquad \text{in } \mathbf{Q}[z,x]. \tag{2.1}$$

We now are able to "specialize" z to an integer σ. Take σ to be a positive integer which satisfies

(a) $\sigma U(z,x) \in \mathbf{Z}[z,x], \sigma V(z,x) \in \mathbf{Z}[z,x]$; and

(b) $t! \mid \sigma$.

Let $F_1 = F(\sigma, x), G_1 = G(\sigma, x), U_1 = \sigma U(\sigma, x)$ and $V_1 = \sigma V(\sigma, x)$. It is clear that $F_1(x) \in \mathbf{Z}[x]$, (b) implies $G_1(x) \in \mathbf{Z}[x]$, and (a) implies that $U_1(x), V_1(x) \in \mathbf{Z}[x]$. We now consider polynomials in $\mathbf{Z}[\mathbf{x}, \mathbf{y}]$. Let

$$\begin{aligned} f_1 &= g_1 = \sigma^t, \\ f_2 &= y - F_1(x) = y - x - \sigma x^2, \\ g_2 &= x - G_1(y) = x - \textstyle\sum_{i=2}^{t+1} \left(\frac{\binom{2i-2}{i} \sigma^{i-1}}{i-1} \right) y_i. \end{aligned}$$

We need to prove

(I) $\max deg(f_i) = 2$;

(II) $\max deg(g_i) = deg(g_2) = t + 1$;

(III) $\{g_1, g_2\}$ is a minimal Gröbner basis of the ideal $(f_1, f_2)\mathbf{Z}[\mathbf{x}, \mathbf{y}]$; and

(IV) if $\{h_1, h_2\}$ is a minimal Gröbner basis of the same ideal $(f_1, f_2)\mathbf{Z}[\mathbf{x}, \mathbf{y}]$, then

$$\max deg(h_i) \geq t.$$

Now (I) and (II) are obvious. Denote $B = \mathbf{Z}/(\sigma^t)$. Let

$$\beta : \mathbf{Z} \longrightarrow B = \mathbf{Z}/(\sigma^t)$$

be the canonical homomorphism; whose extensions to $\mathbf{Z}[x] \longrightarrow B[x]$ and $\mathbf{Z}[\mathbf{x}, \mathbf{y}] \longrightarrow B[x, y]$ are still denoted by β. Write

$$\begin{aligned} \bar{f} &= \beta(F_1(x)) \in B[x], \\ \bar{g} &= \beta(G_1(x)) \in B[x]. \end{aligned}$$

Proof of (III) Upon letting $z = \sigma$ in (2.1) we have

$$\begin{cases} F_1(G_1(x)) = x + \sigma^t U_1(x) \\ G_1(F_1(x)) = x + \sigma^t V_1(x) \end{cases} \quad \text{in } \mathbf{Z}[x].$$

Passing to $B[x] = \mathbf{Z}/(\sigma^t)[x]$, we have

$$\begin{cases} \bar{f}(\bar{g}(x)) = x \\ \bar{g}(\bar{f}(x)) = x \end{cases} \quad \text{in } B[x].$$

This means that $\bar{f} : B[x] \longrightarrow B[x]$ is an automorphism with the inverse $\bar{g} : B[x] \longrightarrow B[x]$. Therefore, by the Gröbner Basis Criterion stated in Section 31.2, $\{x - \bar{g}(y)\}$ is the reduced Gröbner basis of the ideal $(y - \bar{f}(x))B[x, y]$. Then we know, using Proposition 3 (and its proof), that

$\{\sigma^t, x - G_1(y)\} = \{g_1, g_2\}$ is a Gröbner basis of the ideal $(\sigma^t, y - F_1(x))\mathbf{Z}[\mathbf{x}, \mathbf{y}] = (f_1, f_2)\mathbf{Z}[\mathbf{x}, \mathbf{y}]$. It is clear that this Gröbner basis is minimal.□

Proof of (IV) We first prove that $deg(\bar{g}) \geq t$. It suffices to prove that the coefficient of x^t in $\bar{g}(x)$ is not zero. This coefficient is $\beta(\frac{2t-2t}{t-1})\sigma^{t-1})$, so it is enough to prove that

$$\frac{\binom{2t-2}{t}}{(t-1)}\sigma^{t-1} < \sigma^t,$$

that is,

$$\frac{\binom{2t-2}{t}}{(t-1)} < \sigma.$$

The property (b) in the choice of σ being $t! \mid \sigma$, it is enough to prove that

$$\frac{\binom{2t-2}{t}}{(t-1)}) < t!.$$

But $\left(\frac{\binom{2t-2}{t}}{t-1}\right) = \frac{\binom{2t-2}{t}}{(t-1)!}$. The above inequality is equivalent to

$$\frac{(2t-2!)}{(t-1)!} < (t!)^2.$$

Now note that

$$\frac{(2t-2)!}{(t-1)!} = \prod_{i=0}^{t-2}(2t-2-i)$$

and

$$(t!)^2 = \prod_{i=0}^{t-2}[(i+2)(t-i)],$$

it suffices to prove

$$2t - 2 - i < (i+2)(t-i) \text{ for } i = 0, ..., t-2$$

which turns out to be true as easily verified.

Let now $H = \{h_1, h_2\}$ be a minimal Gröbner basis of the ideal $(f_1, f_2)\mathbf{Z}[\mathbf{x}, \mathbf{y}]$. Then $\mathrm{Lt}(H) = \mathrm{Lt}(\{g_1, g_2\}) = (\sigma^t, x)\mathbf{Z}[\mathbf{x}, \mathbf{y}]$. We may assume that $h_1 =$

$\sigma^t, h_2 = x - l(y)$. Let $\bar{l}(x) = \beta(l(x)) \in B[x]$. By Proposition 2, $\{x - \bar{l}(y)\}$ is a Gröbner basis of the ideal $(y - \bar{f}(x))B[x, y]$. Then Proposition 1 implies that

$$(x - \bar{l}(y))B[x, y] = (y - \bar{f}(x))B[x, y].$$

Using the Gröbner Basis Criterion, $\bar{l} : B[x] \longrightarrow B[x]$ is the inverse of $\bar{f} : B[x] \longrightarrow B[x]$. Hence $\bar{l}(x) = \bar{g}(x)$. Therefore,

$$deg(l(x)) \geq deg(\bar{l}(x)) = deg(\bar{g}(x)) \geq t. \square$$

30.4 Some Problems on Complexity

We now discuss the examples we have constructed and pose some problems for further research on the theory of computational complexity. We first note that the admissible order we used to show Theorems B and C is the lexicographic order. But apparently the lexicographic order has bad complexity (at least when the number of variables is larger than 2), we pose the following question.

Problem 1 *Is it true that for any admissible order there is no universal bound of Gröbner bases when the coefficient ring is* $\mathbf{Z}$*?*

On the other hand, we would like to search bounds for other classes of coefficient rings. The arguments used in this paper might be modified to solve the following question.

Problem 2 *(A generalization of Theorems B and C) Let R be a UFD with infinitely many primes. Is it true that there is no universal bound of Gröbner bases when the coefficient ring is R?*

Although Theorem C is true for only the lexicographic order, it is a serious challenge to all algorithms using Gröbner bases (over $\mathbf{Z}$). Especially, we point out that for primary decomposition algorithms of ideals in $\mathbf{Z}[x_1, ..., x_n]$, so far one must use the lexicographic order (see [9]). Therefore we would like to achieve a better understanding of the situation. We now take a close look at our examples in [1] and in the last section. We note that all these examples have a common feature: the ideal $(f_1, f_2)R[x, y]$ (where R is either $k[z]$ or $\mathbf{Z}$) contains a non-zero element in R, namely,

$$f_1 = \begin{cases} z^{t+1} & \text{if } R = k[z] \\ \sigma^t & \text{if } R = \mathbf{Z}. \end{cases}$$

This might be one of reasons for the nonexistence of a universal bound of Gröbner bases, since it cannot happen when R is a field (otherwise the ideal is the unit ideal) where such a bound does exist. We also note that the complexity we considered so far is degrees of polynomials, and have not considered the "size" of coefficients. In fact our f_1 has very large "size" (when t goes to large): if $R = k[z], deg(f_1) = t + 1$; if $R = \mathbf{Z}, |\sigma^t| = \sigma^t$ is a large number since $t! \mid \sigma$. Finally, if we consider $R = k[z]$ and take "size" (i.e., degrees) of elements of R into account, then there is a universal bound of Gröbner bases; this is indeed reduced to the case where the coefficient ring is k, a field, hence Weispfenning's result (Theorem A) applies. All above observations support the following conjecture.

Problem 3 (Conjecture) *Let n, d be positive integers and let a be a non-negative integer. There exists a constant, $D = D(n, d, a)$, depending on n, d and a, such that if an ideal $I \subset \mathbf{Z}[x_1, ..., x_n]$ is generated by $f_1, ..., f_s$ satisfying*

(i) $I \cap \mathbf{Z} = \mathbf{Z} f_1$; and

(ii) $|f_1| \leq a$, $\max deg(f_i) \leq d$,

then I has a minimal Gröbner basis $\{g_1, ..., g_t\}$ such that $\max deg(g_i) \leq D$.

In the course of constructing the examples in [A-L] showing Theorem B we had the following Example 3.1; similarly in the last section we had the following Example 3.2.

Example 3.1 Let $A = k[z]/(z^{t+1})$ where k is a field of characteristic zero, a the residue class of z in A. Recall that

$$f(x) = x - ax^2,$$

$$g(x) = x + \sum_{i=2}^{t+1} \frac{\binom{2i-2}{i} a^{i-1}}{i-1} x^i.$$

Then $\{x - g(y)\}$ is the reduced Gröbner basis of the ideal $(y - f(x))A[x, y]$. We have in this case

$$deg(y - f(x)) = 2, \; deg(x - g(y)) = t + 1.$$

But we note that now the ideal does not contain a non-zero element of A, the coefficient ring.□

Example 3.2 Let $B = \mathbf{Z}/(\sigma^t)$ where σ is the integer in the last section, b the residue class of σ in B. Recall that

$$\bar{f}(x) = x - bx^2,$$

$$\bar{g}(x) = x + \sum_{i=2}^{t+1} \frac{\binom{2i-2}{i} b^{i-1}}{i-1} x^i.$$

Then $\{x - \bar{g}(y)\}$ is the reduced Gröbner basis of the ideal $(y - \bar{f}(x))B[x, y]$. We have in this case

$$deg(y - \bar{f}(x)) = 2, \; deg(x - \bar{g}(y)) \geq t.$$

We also note that the ideal does not contain a non-zero element of B, the coefficient ring.□

The above two coefficient rings have a common feature, namely, they are no longer integral domains. In fact their nilpotent indexes are $t+1$ and t respectively. Here we are back to the similarity between degree bounds of Gröbner bases and bounds of inverse degree of automorphisms. H. Bass [2] proved that there exist bounds of inverse degree of automorphisms which depend on not only the number of variables and degrees of automorphisms, but also the nilpotent indexes of coefficient rings. (Hence the Jacobian conjecture is equivalent to eliminating this dependence on the nilpotent indexes.) These observations suggest the following conjecture.

Problem 4 (Conjecture) *Let n, d, l be positive integers and let r be a non-negative integer. There exists a constant, $D = D(n, d, l, r)$, depending on n, d, l and r, such that if an ideal $I \subset R[x_1, ..., x_n]$ is generated by $f_1, ..., f_l$ satisfying*

(i) R is any commutative Noetherian ring with the nilpotent index $\leq r$;

(ii) $I \cap R = 0$; and

(iii) $\max deg(f_i) \leq d$,

then I has a minimal Gröbner basis $\{g_1, ..., g_t\}$ such that $\max deg(g_i) \leq D$.

30.5 References

[1] S. Abhyankar and W. Li, On the Jacobian Conjecture: A New Approach Via Gröbner Bases, *J. Pure Appl. Algebra*, **61**, 1989, pp. 211–222.

[2] H. Bass, The Jacobian Conjecture and Inverse Degrees, *Arithmetic and Geometry II*, (edited by M. Artin and J. Tate), Birkhauser, 1983.

[3] D. Bayer, The Division Algorithm and Hilbert Scheme, Ph. D. Thesis, Harvard University, 1982.

[4] D. Bayer and M. Stillman, A Criterion for Detecting m-Regularity, *Invent. Math.*, **87**, 1987, pp. 1–11.

[5] B. Buchberger, Gröbner Bases: An Algorithmic Method in Polynomial Ideal Theory, *Chapter 6 in: Multidimensional System Theory*, (edited by H. K. Bose) D. Reidel Pub. Comp., 1985,

[6] B. Buchberger, Applications of Gröbner Bases in Non-Linear Computational Geometry, *Mathematical Aspects of Scientific Software*, (edited by J. Rice), Springer-Verlag, 1988.

[7] B. Buchberger, A Criterion for Detecting Unnecessary Reductions in the Construction of Gröbner–Bases, *Symbolic and Algebraic Computation*, (edited by E. W. Ng), Lecture Notes in Computer Science 72, Springer-Verlag, 1979, pp.3-21.

[8] B. Buchberger, A Note on the Complexity of Constructing Gröbner–Bases, *Proc. EUROCAL'83*, (edited by J. van Hulzen), Lecture Notes in Computer Science 162, Springer-Verlag, 1983, pp.137-145,

[9] P. Gianni, B. Trager and G. Zacharias, Gröbner Bases and Primary Decomposition of Polynomial Ideals, *J. Symbolic Comput.*, **6**, 1988, pp. 149–167.

[10] M. Giusti, Some Effectivity Problems in Polynomial Ideal Theory, *Proc. EUROSAM 84*, (edited by J. Fitch), Lecture Notes in Computer Science 174, Springer-Verlag, 1984, pp.159-171.

[11] D. Lazard, Gröbner Bases, Gaussian Elimination and Resulution of Systems of Algebraic Equations, *Proc. EUROCAL'83* (edited by J. van Hulzen), Lecture Notes in Computer Science 162, Springer-Verlag, 1983, pp.146-156.

[12] W. Li, Gröbner Bases and Automorphisms of Polynomial Rings, Ph. D. Thesis, Purdue University, 1990,

[13] B. Mishra and C. Yap, Notes on Gröbner Bases, *Inform. Sci.*, **48**, 1989, pp. 219–252.

[14] H. Möller and F. Mora, Upper and Lower Bounds for the Degree of Gröbner Bases, *Proc. EUROSAM 84*, (edited by J. Fitch), Lecture Notes in Computer Science 174, Springer-Verlag, 1984, pp. 172-183.

[15] W. Trinks, Über B.Buchbergers Verfahren, Systeme algebraischer Gleichungen zu lösen, *J. Number Theory*, **10**, 1978, pp. 475–488.

[16] V. Weispfenning, Some bounds for the construction of Gröbner bases, *Proc. AAECC-4*, (edited by Th. Beth and M. Clausen), Lecture Notes in Computer Science 307, Springer-Verlag, 1988, pp.195-201.

31
Elastica and Computer Vision

David Mumford

31.1 Introduction

I want to discuss the problem from differential geometry of describing those plane curves C which minimize the integral

$$\int_C (\alpha\kappa^2 + \beta)ds. \tag{1}$$

Here α and β are constants, κ is the curvature of C, ds the arc length and, to make the fewest boundary conditions, we mean minimizing for infinitesimal variations of C on a compact set not containing the endpoints of C. Alternately, one may minimize

$$\int_C \kappa^2 ds$$

over variations of C which preserve the total length.

This problem has a very long history: it was first proposed and solved by Euler [4] in 1744 in the appendix "*De curvis elastica*" to his monumental work "*Methodus inveniendi lineas curvas maximi minimive proprietate gaudentes*". These curves have been called "elastica" ever since then, at least by those who knew of Euler's work. Curiously for such an elementary problem, Euler's solution is not even mentioned in any textbooks on variational calculus and therefore it has been rediscovered innumerable times by people like me who needed a description of these curves. I want to thank Garrett Birkhoff for steering me first to Love's treatist [11] on elasticity, thence to Born's prize-winning Ph.D. thesis on these curves and finally to Euler. I guess that the reason these beautiful curves have remained so obscure is a) because they can only be described by non-elementary functions and b) because they were developed chiefly by applied mathematicians and pure mathematicians never looked at this literature.

Among pure mathematicians, a recent but surely not the only rediscovery is due Bryant and Griffiths [3]. Among applied mathematicians, they were proposed as potential interpolating curves by Birkhoff and collaborators and called non-linear splines (cf. [2], p. 171, and [1]). This application has been pursued by Golumb and Jerome (cf. for instance [5] and the references cited there). My own interest in them was motivated by computer vision problems and I discovered later not only that I was redoing Euler but also an earlier computer vision scientist, [8] who had similar ideas. Maybe this paper will serve to forestall yet more rediscoveries of this 200 year old topic!

In this paper, in §§32.1-32.3, I want to present the problem from the point of view of Bayerian computer vision, which leads to elastica as a maximum likelihood reconstruction of occluded edges. Then, in §§32.4-32.5, I want to present what I think are new formulae for elastica using theta functions. The manipulation of elliptic functions has remained something of a black art ever since Euler and I believe that in spite of sporadic efforts for 200 years, my formulae for them are in some ways the simplest, yet don't seem to appear in the literature.

31.2 Edges in Computer Vision

From one point of view, the central problem of computer vision is this: One is given a function $I(x, y)$ representing the light intensity produced by a 3D world and striking a lens from direction x, y. One seeks to compute a second function $d(x, y)$, the distance from the lens to the nearest opaque surface in direction x, y, hence reconstructing the 3D geometry. A simplified but still useful idealization of the problem assumes that the 3D world consists of a small number of opaque objects with smooth boundaries with smoothly varying albedo illuminated by a smoothly varying light source. Then both the intensity image $I(x, y)$ and the range image $d(x, y)$ will be piecewise smooth functions, with discontinuities along "edges" Γ in their domain R. These edges are the directions where a more distant object is just visible along a ray from the lens grazing a closer object (see Figure 1)[1] Thus discontinuities in I are the first major clue to the 3D-geometry.

A second clue are the singularities of Γ. Suppose for instance that there are 3 objects: a nearest one A, a further one B and a farthest object, or background C. Suppose the edge of B, viewed against C, disappears behind A. Then the locus of visible edges Γ looks like the letter "T": the edges of A and B are smooth curves but the edge of B ends at a point on the edge of A. See Figure 2 which depicts schematically 3 blades of grass against a distant background. In fact, from a generic viewpoint, our simplified world produces an edge Γ with only 2 types of singularities: these so-called *T-junctions* and *cuspidal crack tips* which arise when a smoth object develops a crease. The latter is the "elementary catastrophe" given in suitable coorindates by

[1]One may think of Γ as the visible part of the branch locuses of the maps of surfaces $p : \partial A_i \to R$, if A_i are 3D objects and $p : U \to R$, $U \subset \mathbf{R}^3$, maps the points of the 3D-world to rays through the lens.

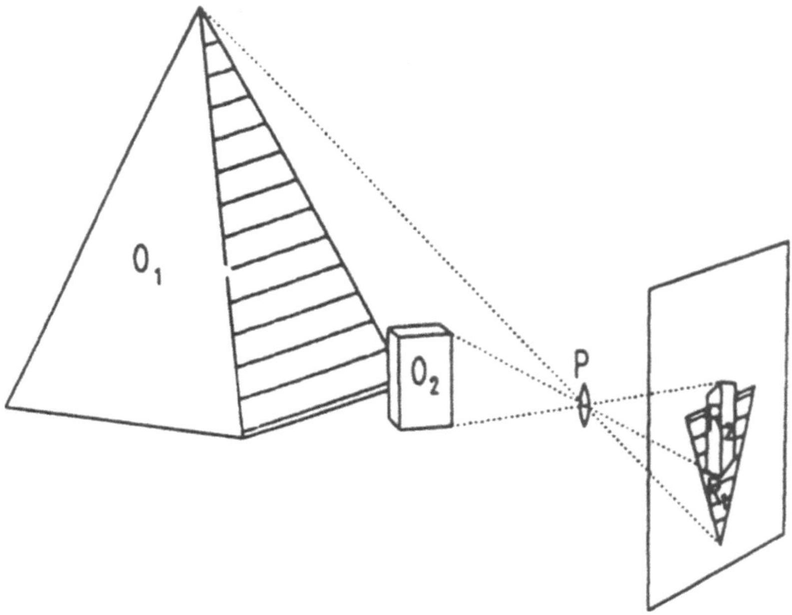

FIGURE 1. An Image of a 3D Scene

$$d(x, y) = (\text{smallest solution } z \text{ of } z^3 - xz - y = 0) + d(0, 0).$$

This $d(x, y)$ has discontinuities along the curve $27y^2 = 4x^3$, $y \geq 0$ ending in a cusp at $x = y = 0$ (see Figure 3). Note that although Γ comes to an endpoint, a crack tip, at which it has a cusp, it is the visible part of a smooth 3-dimensional curve $\Gamma^* : x = 3z^2, y = z^3$ and only gets a singularity because Γ^* turns directly away from the lens at $z = 0$ and afterwards becomes invisible in the crease.

A third clue to 3D geometry comes from the fact that an edge of an object A which disappears at a point P behind object B may often reappear further on at another point Q. This is seen in Figure 2 where the edges of the blades of grass reappear and its signature is a pair of T-junctions in Γ which nearly "match up", i.e., 2 T's as in Figure 4 where the dotted line is the inferred invisible part of the edge of object A behind B. Unlike the previous clues, this one cannot be "read off" from Γ because the 2 pieces of Γ that end at the T's are unlikely to line up exactly. What we need is to measure the relative likelihood of two disappearing edges to be matched up by an invisible edge in terms of how long and how curved an invisible edge is needed to link them. To measure this probability requires a stochastic process to model the space of all possible edges. This is an approach to computer vision that has been pioneered by Grenander and

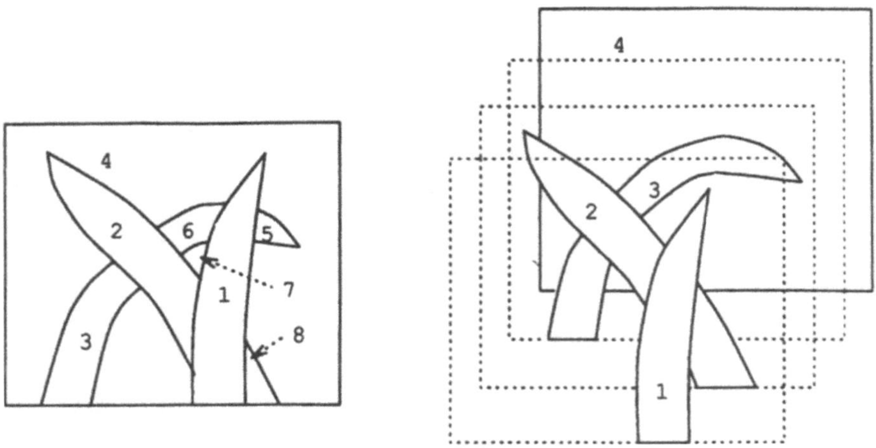

FIGURE 2. Blades of Grass

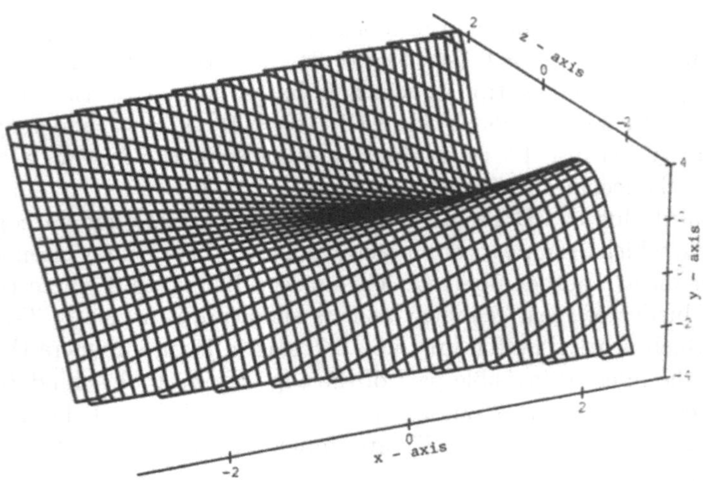

FIGURE 3. $z^3 - xz - y = 0$

FIGURE 4. An Illusion of Kanizsa

other statisticians and probabilists (see [7] and [6]).

It may be worthwhile to try to convince the skeptic of the psychological reality of these probabilistic reconstructions of 3D scenes. Some very striking "optical illusions" are due to the fact that reconstructions of the above type are made automatically at an early stage in visual processing so strongly later stages cannot reverse them if they are absurd on the basis of one's knowledge of plausible worlds. An example is the man and woman entwined in the fence in Figure 4. This and many other beautiful demonstrations are due to Kanizsa [10].

31.3 A Brownian Prior for Edges

What sort of stochastic process is a plausible candidate for modelling the relative likelihood of different edges appearing in a scene of the world? Our edges are to be continuous and almost everywhere differentiable so that, when occluded in part, they will tend to reappear with approximately the same tangent line. The simplest way to do this is to allow curvature $\kappa(s)$, as a function of arc length, to be white noise $n(s)$, so that once integrated, the tangent direction $\theta(s)$ is a Brownian motion $W(s)$. Then the positiion is given by:

$$\begin{aligned} x(s) &= \int_0^s \cos\theta(t)dt + x(0) \\ y(s) &= \int_0^s \sin\theta(t)dt + y(0). \end{aligned} \tag{2}$$

Let's also assume that the total length of the curves is exponentially distributed.

Choosing a very large N, we may approximate this by a discrete time system in which the curve Γ is a polygon of length ℓ:

$$\Gamma = \bigcup_{i=0}^{N} \overline{P_i P_{i+1}}$$

whose sides have length ℓ/N:

$$P_{i+1} = P_i + \frac{\ell}{N}(\cos\theta_i, \sin\theta_i), \quad 0 \leq i < N$$

and the θ_i are discrete Brownian motion scaled down by $\frac{\ell}{N}$:

$$\theta_{i+1} = \theta_i + \sqrt{\frac{\ell}{N}} n_{i+1}, \quad 0 \leq i < N-1,$$

n_i independent normal random variables with mean 0, standard deviation σ and the random variable ℓ has exponential distribution $\lambda \cdot e^{-\lambda\ell} d\ell$. Then

$$\begin{aligned} Pr(\Gamma) &= \frac{\lambda}{(\sqrt{2\pi}\cdot\sigma)^{N-1}} e^{-\Sigma n_i^2/2\sigma^2 - \lambda\ell} \cdot dn_1 \cdots dn_{N-1} d\ell \\ &= \text{cnst.} e^{-\Sigma \frac{\ell}{N} \cdot \left(\frac{\theta_{i+1}-\theta_i}{\ell/N}\right)^2 / 2\sigma^2 - \lambda\ell} \cdot dn_1 \cdots dn_{N-1} d\ell \end{aligned} \tag{3}$$

which is a discrete approximation to

$$e^{-\int (\alpha\kappa^2 + \beta) ds}$$

if $\alpha = \frac{1}{2\sigma^2}, \beta = \lambda$. Thus we see that elastica have the interpetation of being the *mode* of the probability distribution underlying this stochastic process restricted to curves with prescribed boundary behavior, e.g., the maximum likelihood curve with which to reconstruct hidden contours. Some typical elastica are shown in Figure 5.

For applications to computer vision, another invariant of this stochastic process is even more important. When you find some set of T-junctions in an image, you must decide which pairs are most likely to represent the disappearance and reappearance of a single contour behind a nearer object. For this we want the generating function of the stochastic process (2):

$$p(x, y, \theta, t)\Delta x \Delta y \Delta\theta = Pr\left[\begin{array}{c|c} x \leq x(t) \leq x + \Delta x & x(0) = 0 \\ y \leq y(t) \leq y + \Delta y & y(0) = 0 \\ \theta \leq \theta(t) \leq \theta + \Delta\theta & \theta(0) = 0 \end{array}\right].$$

The forwards and backwards diffusion equations for p are:

$$\begin{aligned} \frac{\partial p}{\partial t} &= \sigma \cdot \frac{\partial^2 p}{\partial \theta^2} - \cos\theta \cdot \frac{\partial p}{\partial x} - \sin\theta \cdot \frac{\partial p}{\partial y} \quad \text{and} \\ \frac{\partial p}{\partial t} &= \sigma \left(x\frac{\partial}{\partial y} - y\frac{\partial}{\partial x} - \frac{\partial}{\partial \theta}\right)^2 p - \frac{\partial p}{\partial x} \end{aligned} \tag{4}$$

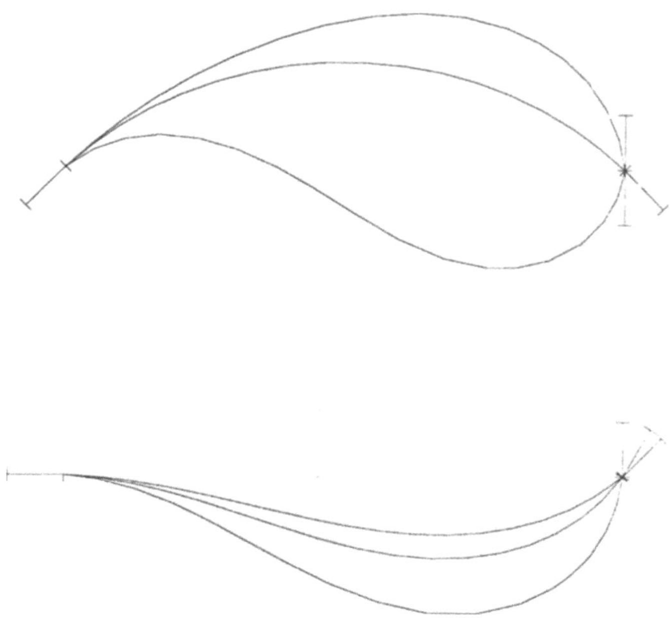

FIGURE 5. Some Elastica

where σ measures the amount of noise in the curvature. For each t, $p(x, y, \theta, t)$ is a probability distribution on $\mathbf{R}^2 \times S^1$ in the variables x, y, θ. For $t = 0$, it starts at the delta function

$$\delta(x)\delta(y)\delta(\theta)$$

and then spreads out. It spreads in each of 3 directions in totally different ways – it is transported in the $(\cos\theta\frac{\partial}{\partial x} + \sin\theta\frac{\partial}{\partial y})$-direction, it diffuses in the θ-direction and it fills x, y, θ-space only as a 2nd order effect, via the non-integrability of the $(\sin\theta \cdot dx - \cos\theta \cdot dy)$-foliation. We can also integrate the backwards equation with respect to θ and get a diffusion equation for the marginal probability $\bar{p}(x, y)$:

$$\frac{\partial \bar{p}}{\partial t} = \sigma\left(x\frac{\partial}{\partial y} - y\frac{\partial}{\partial x} - \frac{\partial}{\partial \theta}\right)^2 \bar{p} - \frac{\partial \bar{p}}{\partial x}.$$

I have looked for an explicit formula for p and $\bar{p}$ but in vain. Still, on the basis of the results of §32.2, I would conjecture that a formula exists, in terms of elliptic functions of some kind.

The function we want for computer vision is the probability that an occluded curve reappears at a particular point (x, y, θ) before dying its exponential death, i.e.,

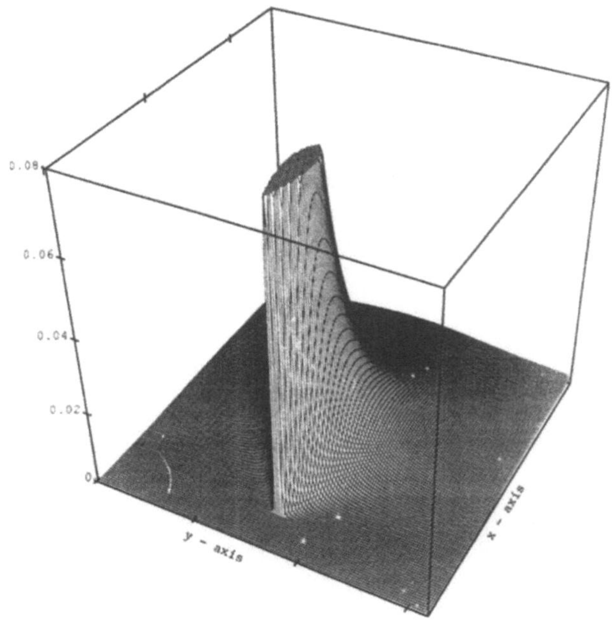

FIGURE 6. Graph of q(x,y)

$$q(x,y,\theta) = \int_0^\infty e^{-\lambda t} \cdot p(x,y,\theta,t)dt$$

which solves

$$\lambda q = \sigma \cdot \frac{\partial^2 q}{\partial \theta^2} - \cos\theta \cdot \frac{\partial q}{\partial x} - \sin\theta \cdot \frac{\partial q}{\partial y}.$$

We have integrated this equation numerically. The marginal $\bar{q}(x,y)$ is shown in Figure 6 and the conditional $q(4,y,\theta)$ is shown in Figure 7, both with $\lambda = 0.17$ (half-life of 4 units of distance) and $\sigma = \pi^2/128$ (in time 1, standard deviation of θ is $\pi/8$). Note the interesting singularity of $\bar{q}$ at $x = y = 0$ where its true value is infinite. The peak of $q(4,y,\theta)$ corresponds to a horizontal curve following the x-axis, i.e., $y = \theta = 0$, and as y increases or decreases, the corresponding most probable θ also increases or decreases.

31.4 Alternate Priors

The stochastic process (2) has a major failing as a model of edges in images: images are 2D projections of 3D scenes, so edges are 2D projections of 3D curves. In particular, this means that edges have near singularities when

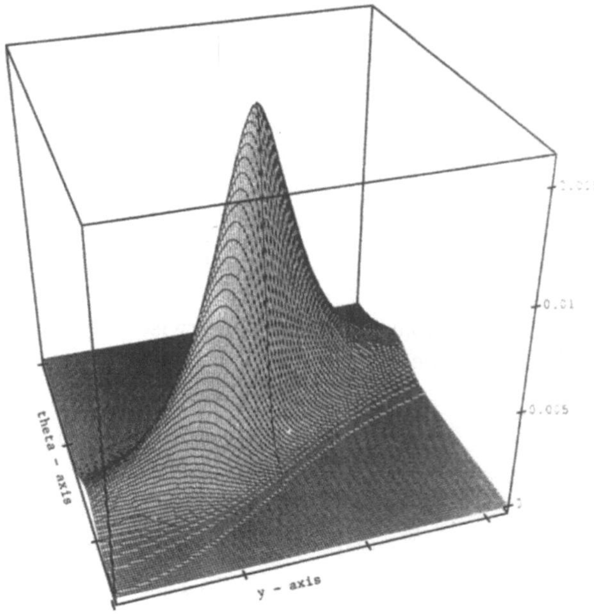

FIGURE 7. Graph of q(4,y, θ)

the 3D curve moves approximately towards or away from the imaging system. An alternate model is to generate sample paths in $\mathbf{R}^3$ by integrating Brownian motion on the sphere S^2 of 3D unit tangent vectors. This leads to:

$$\begin{array}{rcll} x(s) & = & \int_0^s \cos\theta(t)\cdot\sin\phi(t)\cdot dt & \\ y(s) & = & \int_0^s \sin\theta(s)\cdot\sin\phi(t)\cdot dt & \\ \phi(t) & = & B_1(t) + \int_0^t \frac{ds}{\tan\theta(s)}, & 0<\phi<\pi \quad \text{almost surely} \\ \theta(t) & = & B_2\left(\int_0^t \frac{ds}{\sin^2\phi(s)}\right), & 0\le\theta\le 2\pi \end{array} \tag{5}$$

where ϕ and θ are colatitude and longitude on S^2 and B_1 and B_2 are independent Brownian motions. The point is that to get Brownian motion on S^2, θ must be rescaled to go faster near the north and south poles and ϕ must have an extra drift term pushing it towards the equator (cf. [12], §4.3). This is analytically a bit of a mess, but a simple approximation to it is given by the Uhlenbeck process: this models a particle in the plane subject to white noise random forces plus friction pulling its velocity back to 0. The equations are:

$$\begin{array}{rcl} \frac{d^2x}{dt^2} & = & \alpha\cdot n_1(t) - \beta\frac{dx}{dt} \\ \frac{d^2y}{dt^2} & = & \alpha\cdot n_2(t) - \beta\frac{dy}{dt} \end{array} \tag{6}$$

where n_1 and n_2 are independent white noise. This process has the advantage of being a Gaussian process, hence its generating function $p(x, y, \dot{x}, \dot{y}, t)$ can be readily calculated and applied to image analysis. If an exponential death is included, the maximum likelihood curves which reconstruct occluded edges for this process minimize integrals of the form:

$$\int_C \left\{ \alpha \left[\left(\frac{d^2x}{dt^2}\right)^2 + \left(\frac{d^2y}{dt^2}\right)^2 \right] + \beta \left[\left(\frac{dx}{dt}\right)^2 + z \left(\frac{dy}{dt}\right)^2 \right] + \gamma \right\} dt. \quad (7)$$

If $\beta = \gamma = 0$, there are cubic splines and otherwise they are transcendental curves generalizing exponential spirals of the form

$$\begin{aligned} x(s) &= a_1 e^{\lambda_1 s} + a_2 e^{\lambda_2 s} + a_3 e^{-\lambda_2 s} + a_4 e^{-\lambda_1 s} \\ y(s) &= b_1 e^{\lambda s} + b_2 e^{\lambda_2 s} + b_3 e^{-\lambda_2 s} + b_4 e^{-\lambda_1 s}, \qquad \lambda_1 > \lambda_2 > 0 \end{aligned}$$

or

$$\begin{aligned} x(s) &= a_1 e^{\lambda s} \sin(\mu s + a_2) + a_3 e^{-\lambda s} \sin(\mu s + a_4) \\ y(s) &= b_1 e^{\lambda s} \sin(\mu s + b_2) + b_3 e^{-\lambda s} \sin(\mu s + b_4). \end{aligned}$$

Curves which minimize the above integral (7) *plus* an "external force" term $\int_C I(x, y)dt$ have been used in computer vision under the name of "snakes" [9] to reconstruct edges which are mostly visible but obscured by noise, gaps, blur, etc.

In Figure 7, we show some sample paths from all 3 priors for comparison.

31.5 The Differential Equation of Elastica

We would like to give an explicit formula for elastica in terms of theta functions. The first step is to characterize them by a differential equation for their curvature $\kappa(s)$ as a function of arc length. Using vector notation, start with the curve itself $\vec{x}(s)$ in arc-length parametrization, its tangent and normal vectors are

$$\begin{aligned} \vec{t}(s) &= \dot{\vec{x}}(s) \\ \vec{n}(s) &= \vec{t}(s)^{\perp}. \end{aligned}$$

Then

$$\dot{\vec{t}} = \kappa \cdot \vec{n}, \quad \dot{\vec{n}} = -\kappa \cdot \vec{t}.$$

Now consider an infinitesimal deformation:

$$(\text{new}\vec{x}) = \vec{x}(s) + \delta(s) \cdot \vec{n}(s).$$

Then

$$\begin{aligned} (\text{new}\vec{x})^{\cdot} &= \vec{t} + \delta' \cdot \vec{n} - \delta\kappa \cdot \vec{t} \\ &= (1 - \delta\kappa) \cdot (\vec{t} + \delta' \cdot \vec{n}) \end{aligned}$$

(since $\delta \cdot \delta'$ is effectively zero). Then

$$\begin{aligned}(\text{new}ds) &= (1-\delta\kappa)ds \\ (\text{new}\vec{t}) &= \vec{t}+\delta'\cdot\vec{n} \\ (\text{new}\vec{n}) &= \vec{n}-\delta'\cdot\vec{t}.\end{aligned}$$

We want the deformation to preserve arc length, i.e.,

$$\int(\text{new}ds) = \int ds$$

or

$$\int_C \delta\kappa \cdot ds = 0.$$

Now

$$(\text{new}\kappa)\cdot(\text{new}\vec{n}) = \frac{d(\text{new}\vec{t})}{\text{new}ds}$$

or

$$\begin{aligned}(\text{new}\kappa)(\vec{n}-\delta'\cdot\vec{t}) &= \tfrac{1}{1-\delta\kappa}\tfrac{d}{ds}(\vec{t}+\delta'\cdot\vec{n}) \\ &= (1+\delta\kappa)(\kappa\vec{n}+\delta''\cdot\vec{n}-\delta'\kappa\cdot\vec{t}) \\ &= (\kappa+\delta''+\delta\kappa^2)(\vec{n}-\delta'\vec{t}).\end{aligned}$$

Therefore

$$\text{new}\kappa = \kappa+\delta''+\delta\kappa^2.$$

Thus

$$\begin{aligned}\text{new}(\textstyle\int\kappa^2 ds) &= \textstyle\int(\kappa+\delta''+\kappa^2\delta)^2\cdot(1-\delta\kappa)ds \\ &= \textstyle\int\kappa^2 ds+\int(2\kappa\delta''+\kappa^3\delta)ds \\ &= \textstyle\int\kappa^2 ds+\int(2\kappa''+\kappa^3)\delta ds \\ &\quad \text{(integration by parts).}\end{aligned}$$

If C is an elastica, $\int\kappa^2 ds$ is a minimum for all deformations which preserve $\int ds$, hence for some constant R:

$$2\kappa''+\kappa^3 \equiv \frac{R}{2}\kappa. \tag{8}$$

Conversely, this differential equation implies that C is a critical point for $\int(\kappa^2+\frac{R}{2})ds$. We won't consider the question of which critical points are minima and which are not, but go on to write down the solutions of (8).

31.6 Solving for Elastica

Multiplying (8) by κ' and integrating, we get

$$4(\frac{d\kappa}{ds})^2+\kappa^4-R\kappa^2-S=0$$

for some constant S, so that s is an elliptic integral in κ

$$s = \int \frac{2d\kappa}{\sqrt{-\kappa^4 + R\kappa^2 + S}}.$$

Now let $\mathcal{E}$ be the real elliptic curve

$$v^2 = -u^4 + Ru^2 + S.$$

What this means is that the map

$$s \longmapsto (\kappa(s), 2\kappa'(s))$$

identifies the elastica with the real points $\mathcal{E}_{\mathbf{R}}$ of $\mathcal{E}$. Under this map, ds becomes the differential $2du/v$, the differential on $\mathcal{E}$, unique up to scalars, without zeroes or poles. If Λ is the lattice of periods of ds, then, as usual, $\mathcal{E}$ is uniformized

$$\begin{aligned} \mathcal{E} &\cong \mathbf{C}/\Lambda \\ \mathcal{E}_{\mathbf{R}} &\cong \mathbf{R}/\mathbf{R} \cap \Lambda \end{aligned}$$

and arc length s on the elastica is just the additive coordinate on the elliptic curve, the variable of the universal cover $\mathbf{C}$. The curve $\mathcal{E}$ has 2 conjugate imaginary points P and $\bar{P}$ at ∞ and the function $\kappa(s)$ on the elastica is now the restriction to $\mathcal{E}_{\mathbf{R}}$ of the meromorphic function u on $\mathcal{E}$ with 2 simple poles at P and $\bar{P}$. What we want to do is to integrate κ *twice* to get the x and y coordinates of the elastica as functions on $\mathcal{E}_{\mathbf{R}}$. To do this, combine them into the complex-valued function

$$z(s) = x(s) + \sqrt{-1}\; y(s)$$

on $\mathcal{E}_{\mathbf{R}}$. We must solve:

$$\begin{aligned} \frac{dz}{ds} &= e^{i\theta(s)} \\ \frac{d\theta}{ds} &= u(s), \end{aligned}$$

which determine z up to rotation and translation.

Note on $\mathcal{E}$, the identities

$$\begin{aligned} [v + i(u^2 - \tfrac{R}{2})] \cdot [v - i(u^2 - \tfrac{R}{2})] &= v^2 + (u^4 - Ru^2 + \tfrac{R^2}{4}) \\ &= S + \tfrac{R^2}{4}. \end{aligned}$$

This shows that $v \pm i(u^2 - \frac{R}{2})$ are functions on $\mathcal{E}$ with no zeroes or poles at finite values of (u, v). Examining their behavior at the infinite points $P, \bar{P}$, we compute their divisors (possibly interchanging $P, \bar{P}$):

$$\begin{aligned} (v + i(u^2 - \tfrac{R}{2})) &= 2P - 2\bar{P} \\ (v - i(u^2 - \tfrac{R}{2})) &= 2\bar{P} - 2P, \end{aligned}$$

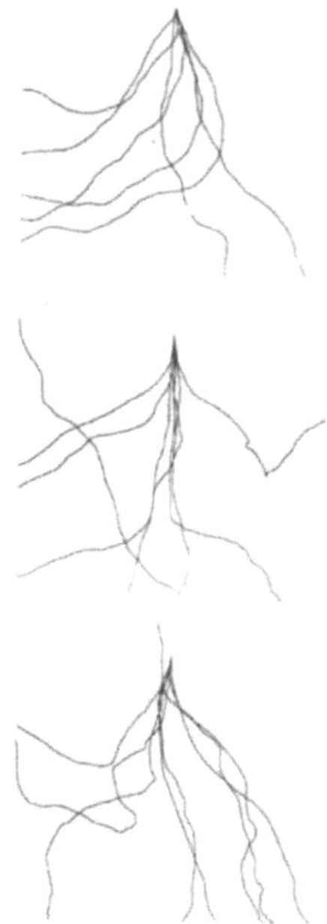

FIGURE 8. Sample Paths forom the 3 priors

i.e., $P - \bar{P}$ is a divisor class of order 2. Now let $s_0 \in \mathbf{C}$ map to P, so $\bar{s}_0$ maps to $\bar{P}$. Then $s_0 - \bar{s}_0 \in \frac{1}{2}\Lambda$ expresses this fact on $\mathbf{C}$.

Let ω_1 be a generator of $\Lambda \cap \mathbf{R}$ and $\{\omega_1, \omega_2\}$ a basis of Λ. Then the theta function:

$$\vartheta_\Lambda(s) = \sum_{n \in \mathbf{Z}} e^{\pi i n^2 (\frac{\omega_2}{\omega_1}) + 2\pi i n (\frac{s}{\omega_1})} \tag{9}$$

is a very rapidly converging holomorphic function on $\mathbf{C}$ such that:

a) $\vartheta_\Lambda(s + \omega_1) = \vartheta_\Lambda(s)$

b) $\vartheta_\Lambda(s + \omega_2) = e^{-2\pi i \frac{s}{\omega_1} - \pi i \frac{\omega_2}{\omega_1}} \vartheta_\Lambda(s)$

c) ϑ_Λ has simple zeroes at the points $\frac{1}{2}\omega_1 + \frac{1}{2}\omega_2 + \Lambda$ and no other zeroes.

(Cf. [13], §1.4).

Let

$$F(s) = c\left(\frac{d}{ds}\log\vartheta_\Lambda(\mathrm{s} - \mathrm{s}_0 + \frac{\omega_1 + \omega_2}{2})\right) - as \tag{10}$$

(a and c to be fixed shortly). The a), b), and c) imply:

a′) $F(s + \omega_1) = F(s)$

b′) $F(s+\omega_2) = F(s) - \frac{2\pi i}{\omega_1}c - a\omega_2$

c′) F is meromorphic with simple poles at the points $s_0 + \Lambda$ and no other points.

It follows that $\frac{dF}{ds}$ is periodic mod Λ, hence is a meromorphic function on $\mathcal{E}$. It has a *double* pole at P and now choose the ratio a/c so that

$$\frac{dF}{ds}(\bar{P}) = 0,$$

i.e.,

$$\frac{a}{c} = \frac{d^2}{ds^2}\log\vartheta_\Lambda(\bar{\mathrm{s}}_0 - \mathrm{s}_0 + \frac{\omega_1+\omega_2}{2}).$$

Then $\frac{\partial F}{\partial s}$ is zero at $\bar{P}$ and has one remaining zero. Since $2P - \bar{P} \equiv \bar{P}$, this last zero is $\bar{P}$ too, and

$$\frac{dF}{ds} = d\cdot(v(s) + iu(s)^2 - \frac{iR}{2}), \qquad \text{some constant } d.$$

Thus, if $s \in \mathbf{R}, u(s), v(s) \in \mathbf{R}$, so

$$\begin{aligned} \overline{\left(\frac{dF}{ds}\right)} &= d\cdot(v(s) - iu(s)^2 + \tfrac{iR}{2}) \\ \therefore \left|\frac{dF}{ds}\right|^2 &= |d|^2\cdot(S + \tfrac{R^2}{4}), \qquad \textit{a constant.} \end{aligned}$$

Finally choose a, c so that

$$\left|\frac{dF}{ds}\right| \equiv 1 \qquad \text{on } \mathbf{R}.$$

It now follows that

$$\begin{aligned} \mathbf{R} &\longrightarrow \mathbf{C} \\ s &\longmapsto F(s) \end{aligned}$$

is an arc length parametrization of a plane curve. Finally write

$$\frac{dF}{ds}(s) = e^{i\theta(s)} \qquad \text{or } \theta(s) = \frac{1}{i}\log\frac{\mathrm{dF}}{\mathrm{ds}}$$

and

$$\frac{d\theta}{ds}(s) = \kappa(s) \qquad \text{or } \kappa(s) = \frac{1}{i}\cdot\frac{d^2F/ds^2}{dF/ds}.$$

κ is again a meromorphic function on $\mathcal{E}$, now with simple poles at P and $\bar{P}$ only and using $ds = 2du/v$

$$
\begin{aligned}
\frac{dF}{ds} &= d(v + iu^2 - \tfrac{iR}{2}) \\
\frac{d^2F}{ds^2} &= d(\tfrac{dv}{ds} + 2iu\tfrac{du}{ds}) \\
&= d(\tfrac{1}{2}v\tfrac{dv}{du} + 2iu \cdot \tfrac{v}{2}) \\
&= d(\tfrac{1}{4}\tfrac{d}{du}(v^2) + iuv) \\
&= d \cdot (\tfrac{1}{4}(-4u^3 + 2Ru) + iuv) \\
&= iu \cdot d \cdot (iu^2 - \tfrac{iR}{2} + v) \\
&= iu \cdot \tfrac{dF}{ds}.
\end{aligned}
$$

Thus $\kappa(s) = u$ and this proves the final result:

Theorem *Elastica are all given in their arc length parametrization by maps:*

$$
\begin{aligned}
&\mathbf{R} \longrightarrow \mathbf{C} \\
&s \longmapsto c \cdot \tfrac{d}{ds}\log\vartheta_\Lambda(\mathrm{s} - \eta) - \mathrm{a} \cdot \mathrm{s}
\end{aligned}
$$

where $\Lambda \subset \mathbf{C}$ is a lattice such that $\bar{\Lambda} = \Lambda$, η satisfies $\eta - \bar{\eta} \in \frac{1}{2}\Lambda$, a, c are suitable constants defined above, and ϑ_Λ is the theta function defined in (9).

One can continue the analysis and classify the pairs (Λ, η) which give elastica. It turns out that 2 types suffice:

$$\Lambda = \mathbf{Z} + \mathbf{Z} \cdot it, \qquad \eta = -\frac{it}{4},$$

and

$$\Lambda = \mathbf{Z} + \mathbf{Z} \cdot \frac{it + 1}{2}, \qquad \eta = 0.$$

31.7 References

[1] G. Birkhoff, H. Burchard & D. Thomas, Non-linear interpolation by splines, pseudosplines and elastica, *General Motors Research Lab. report* 468, 1965.

[2] G. Birkhoff & C.R. De Boor, Piecewise polynomial interpolation and approximation, in *Approximation of Functions*, ed. by H. Garabedian, Elsevier, 1965.

[3] R. Bryant & P. Griffiths, Reduction for constrained variational problems and $\int \kappa^2/2\, ds$, *Am. J. Math.*, vol. 108, pp. 525-570, 1986.

[4] L. Euler, *Methodus inveniendi lineas curvas maximi minimive proprietate gaudentes*, Lausanne, 1744.

[5] M. Golumb & J. Jerome, Equilibria of the curvature functional and manifolds of non-linear interpolating spline curves, *Siam J. Math. Anal.*, vol. 13, pp. 421-458, 1982.

[6] Donald Geman, Random fields and inverse problems in imaging, *Math. Dept., U.Mass., preprint.*

[7] Ulf Grenander, *Lectures in Pattern Theory*, vol.1-3, Springer-Verlag, 1981.

[8] B.K.P. Horn, The curve of least energy, *ACM Trans. on Math. Software*, vol. 9, pp. 441-460, 1983.

[9] M. Kaas, A. Witkin & D. Terzopoulos, Snakes: Active contour models, *Proc. 1st Int. Conf. Comp. Vision*, pp.259-268, IEEE, 1987.

[10] Gaetano Kanizsa, *Organization in Vision: Essays on Gestalt Perception*, Praeger Scientific, 1979.

[11] A.E.H. Love, *A Treatise on the Mathematical Theory of Elasticity*, Camb. Univ. Press, 4th ed., 1927.

[12] H. McKean, *Stochastic integrals*, Academic Press, 1969.

[13] D. Mumford, *Tata Lectures on Theta*, vol. 1, Birkhauser-Boston, 1983.

32
Isolator Polynomials

Thomas W. Sederberg
Geng-Zhe Chang

ABSTRACT This paper explores the problem of isolating the real roots of a polynomial $p(x)$ with real coefficients, that is, of locating intervals which contain exactly one real root of p. A new solution to this problem is presented, consisting of finding a pair of auxiliary polynomials whose set of combined real roots contain at least one value in every closed interval defined by each pair of adjacent real roots in p. It is shown that any member of the polynomial remainder sequence generated by p and p' can serve as one of these auxiliary polynomials.

32.1 Introduction

Polynomial real root isolation is a classical problem with an extensive literature (see, for example, [1] and [2]). This paper presents a new approach to the root isolation problem by introducing, in Section 33.2, the notion of isolator polynomials. Section 33.3 provides examples of isolator polynomials for low degree cases and Section 33.4 shows how polynomial remainders can be used as isolator polynomials.

32.2 Isolator Polynomials

We begin by proving

Theorem 1. Given any polynomial $p(x)$ with two adjacent real roots ρ_1 and ρ_2, and given any two other polynomials $b(x)$ and $c(x)$, define

$$a(x) = b(x)p'(x) + c(x)p(x). \tag{32.1}$$

Then $a(x)$ or $b(x)$ has at least one real root in the closed interval $[\rho_1, \rho_2]$.

Proof: Since $p(\rho_1) = 0$ and $p(\rho_2) = 0$,

$$\begin{aligned} a(\rho_1)a(\rho_2) &= b(\rho_1)p'(\rho_1)b(\rho_2)p'(\rho_2) \\ [a(\rho_1)a(\rho_2)] &= [b(\rho_1)b(\rho_2)]\,[p'(\rho_1)p'(\rho_2)] \end{aligned}$$

If ρ_1 and ρ_2 are distinct adjacent roots, then $p'(\rho_1)p'(\rho_2) < 0$. Thus, either $[a(\rho_1)a(\rho_2)] \leq 0$ or $[b(\rho_1)b(\rho_2] \leq 0$.
Q.E.D.

We refer to $a(x)$ and $b(x)$ as isolator polynomials, or *IPs*. Note that $b(x)$ and $c(x)$ can be chosen freely. As shall be shown, it is always possible to find a pair of IPs whose degree sum to $degree(p) - 1$.

Note that Theorem 1 addresses distinct real roots. However, in the case of multiple roots, $p(x) = p'(x) = a(x) = 0$, and so $a(x)$ can be said to "isolate" a multiple root of $p(x)$ in the sense that a root of $a(x)$ occurs at exactly the same value as any multiple root of $p(x)$.

Section three works out in closed form some useful IPs for $p(x)$ of degree three, four, and five. Section four demonstrates that any member of the Sturm's sequence of p can serve as the a IP.

32.3 Motivating Examples

32.3.1 Degree Three Case.

For a cubic polynomial

$$p(x) = x^3 + p_2x^2 + p_1x + p_0 \tag{32.2}$$

we seek two linear isolator polynomials. Letting $b(x) = \frac{x}{3} + \frac{p_2}{9}$ and $c(x) = -1$, we have

$$a(x) = b(x)p'(x) + c(x)p(x) = \frac{2}{3}(\frac{1}{3}p_2^2 - p_1)x + (\frac{1}{9}p_2p_1 - p_0) \tag{32.3}$$

From Theorem 1, the roots of $a(x)$ and $b(x)$ isolate the roots of $p(x)$. The isolating values are

$$x_1 = -\frac{p_2}{3}; \quad x_2 = \frac{9p_0 - p_1p_2}{2p_2^2 - 6p_1}.$$

Note that from Vieta's formulas ([3], p. 154), p_{n-1} is the negative of the sum of the roots of a degree n polynomial with $p_n = 1$. Thus, x_1 is the arithmetic mean of the three roots of this cubic p. If p has three real roots, it is obvious that the mean of those roots would lie between two of them. Furthermore, if the three roots are evenly spaced, then $x_1 = x_2 =$ the middle root.

Example: To illustrate, consider the cubic polynomial with roots $1, 2, 5$:

$$p(x) = (x-1)(x-2)(x-5) = x^3 - 8x^2 + 17x - 10.$$

In this case, the isolating polynomials are

$$a(x) = \frac{26}{9}x - \frac{46}{9}; \quad b(x) = \frac{1}{3}x - \frac{8}{9}$$

whose roots are 1.769 and 2.667.

Roots of $p(x)$:	1		2		5
Roots of $a(x)$ and $b(x)$:		1.769		2.667	

32.3.2 Degree Four Case

The closed form expressions for the isolator polynomials are simplified if the polynomial is written in so-called general form:

$$p(x) = x^4 + p_2x^2 + p_1x + p_0.$$

The following linear $b(x)$ and quadratic $g(x)$ satisfy the requirements for isolator polynomials given in equation 32.1:

$$b(x) = -x,$$

$$a(x) = 2p_2x^2 + 3p_1x + 4p_0.$$

Once again, it turns out that the root of $b(x)$ (namely, $x = 0$), is the arithmetic mean of the four roots of p.

Example: Consider the polynomial with roots $-7, 1, 2, 4$:

$$p(x) = (x+7)(x-1)(x-2)(x-4) = x^4 - 35x^2 + 90x - 56.$$

(The coefficient of x^3 vanishes because we chose our roots to sum to zero.) In this case, $a(x) = -70x^2 + 270x - 224$ with roots 1.208 and 2.649. $b(x) = -x$ with root 0.

Roots of $p(x)$:	-7		1		2		4
Roots of $a(x)$ and $b(x)$:		0		1.208		2.649	

32.3.3 Degree Five Case

Once again, it simplifies matters if the polynomial is placed in general form:

$$p(x) = x^5 + p_3x^3 + p_2x^2 + p_1x + p_0.$$

The reader can verify that the quadratic $a(x)$ and quadratic $b(x)$

$$a(x) = (12p_3^3 + 45p_2^2 - 40p_3p_1)x^2 + (8p_3^2p_2 + 60p_2p_1 - 50p_3p_0)x + (4p_3^2p_1 + 75p_2p_0)$$

$$b(x) = 10p_3x^2 - 15p_2x + 4p_3^2$$

satisfy the conditions in equation 32.1 for isolator polynomials.

Example: Consider the polynomial with roots $-4, -3, 1, 2, 4$:

$$p(x) = (x+4)(x+3)(x-1)(x-2)(x-4) = x^5 - 23x^3 + 6x^2 + 112x - 96.$$

(The coefficient of x^4 vanishes because we chose our roots to sum to zero.) In this case, $a(x) = -41344x^2 - 44688x + 193792$ with roots -2.772 and 1.691. $b(x) = -230x^2 - 90x + 2116$ with roots -3.235 and 2.844.

Roots of $p(x)$:	-4		-3		1		2		4
Roots of $a(x)$ and $b(x)$:		-3.235		-2.772		1.691		2.844	

32.4 Polynomial Remainder Sequences

The examples in Section 3 presented pairs of IPs whose degrees summed to $degree(p) - 1$. It turns out that such pairs can be generated from the polynomial remainder sequence of p and p' as follows:

$$\begin{aligned}
r_{-1}(x) &= p(x) \\
r_0(x) &= p'(x) \\
r_1(x) &= r_{-1}(x) - q_1(x)r_0(x) \\
r_2(x) &= r_0(x) - q_2(x)r_1(x) \\
r_3(x) &= r_1(x) - q_3(x)r_2(x) \\
&\;\;\vdots \\
r_i(x) &= r_{i-2}(x) - q_i(x)r_{i-1}(x)
\end{aligned}$$

Note that any member r_1 of the remainder sequence fits the recipe for the *IP* $a(x)$:

$$a(x) = b(x)p'(x) + c(x)p(x). \tag{32.4}$$

For example, in the equation

$$r_1(x) = p(x) - q_1(x)p'(x), \tag{32.5}$$

$a(x) = r_1(x)$, $b(x) = -q_1(x)$, and $c(x) = 1$. Likewise, $r_2(x)$ can also be made to fit the requirements for the isolator polynomial $a(x)$:

$$\begin{aligned}
r_2(x) &= p'(x) - q_2(x)r_1(x) \\
r_2(x) &= p'(x) - q_2(x)[p(x) - q_1(x)p'(x)] \\
r_2(x) &= [1 + q_2(x)q_1(x)]p'(x) - q_2(x)p(x)
\end{aligned}$$

In this case, $b(x) = 1 + q_2(x)q_1(x)$.

In general, we have

$$\begin{aligned}
a_i(x) &= r_i(x) \\
b_i(x) &= b_{i-2}(x) - q_i(x)b_{i-1}(x)
\end{aligned}$$

with $degree(a_i) \leq degree(p) - i - 1$ and $degree(a_i) + degree(b_i) = degree(p) - 1$.

Figure 1 shows *Mathematica* [4] code for computing a_i and b_i. A sample degree eight polynomial is defined, with roots $1, 2, 3, 4, 5, 6, 7, 8$. The isolator polynomials a_3 (degree 4) and b_3 (degree 3) are shown to have roots

$p(x)$:	1		2		3		4		5		6		7		8
a_3 & b_3:		1.46		2.34		3.80		4.5		5.21		6.66		7.54	

The isolator polynomials a_4 (degree 3) and b_4 (degree 4) are shown to have roots

$p(x)$:	1		2		3		4		5		6		7		8
a_4 & b_4:		1.17		2.89		3.21		4.5		5.79		6.15		7.83	

32.5 Conclusion

This paper has presented a method through which real roots of a polynomial can be isolated. We have done enough informal testing to convince us that this method is generally *less* efficient in floating point arithmetic than heuristic methods based, for example, on Descarte's rule of signs [1].

After this material was presented at the Conference on Algebraic Geometry and its Applications at Purdue University, 1 June 1990, the speaker was approached simultaneously by John Canny, Chi Yap, and Marie-Francois Roy. They suggested that isolator polynomials may lead to a new approach to defining the computational complexity of the NC class of problems, of which the root isolation problem is a member. Research is continuing in this area.

32.6 References

[1] Collins, G. E. and Akritas, A. G., Polynomial real root isolation using Descarte's rule of signs, in SYMSAC '76, Proceedings of the 1976 ACM Symposium on Symbolic and Algebraic Computation, August 10-12, 1976, Yorktown Heights, NY, pp. 272-275.

[2] Collins, G. E. and Loos, R., Polynomial real root isolation by differentiation, in SYMSAC '76, Proceedings of the 1976 ACM Symposium on Symbolic and Algebraic Computation, August 10-12, 1976, Yorktown Heights, NY, pp. 15-25.

[3] Kurosh, A. *Higher Algebra*, MIR Publishers, Moscow, 1975.

[4] Wolfram, S. *Mathematica: A System for Doing Mathematics by Computer*, Addison-Wesley, 1988.

```
(* Define Mathematica functions a and b *)

a[n_Integer,f_] := PolynomialRemainder[a[n-2,f],a[n-1,f],t]
a[0,f] := D[f,t]
a[-1,f] := f
a[-2,f] := 0
q[n_Integer,f_] := PolynomialQuotient[a[n-2,f],a[n-1,f],t]
b[n_Integer,f_] := b[n-2,f] - q[n,f] b[n-1,f]
b[-1,f_] := 0
b[0,f_] := -1

(* Create sample polynomial with 8 evenly spaced roots *)

f = Expand[(t-1)(t-2)(t-3)(t-4)(t-5)(t-6)(t-7)(t-8)]
```

$$40320-109584t+118124t^2-67284t^3+22449t^4-4536t^5+546t^6-36t^7+t^8$$

```
(* Compute a[3,f] and b[3,f] and solve their roots *)
a3 = Expand[a[3,f]]
```

$$\frac{22851}{4}-\frac{47223}{8}t+\frac{34551}{16}t^2-\frac{2673}{8}t^3+\frac{297}{16}t^4$$

```
N[Solve[a3==0,t],4]
```

{{$t->$ 6.661}, {$t->$ 2.339}, {$t->$ 5.205}, {$t->$ 3.795}}

```
b3 = Expand[b[3,f]]
```

$$\frac{99}{64}-\frac{103}{64}t+\frac{27}{64}t^2-\frac{t^3}{32}$$

```
N[Solve[b3==0,t],4]
```

{{$t->$ 4.5}, {$t->$ 7.541}, {$t->$ 1.459}}

```
(* Compute a[4,f] and b[4,f] and solve their roots *)
a4 = Expand[a[4,f]]
```

$$-(\frac{592896}{77})+\frac{1307648}{231}t-\frac{9216}{7}t^2+\frac{2048}{21}t^3$$

```
N[Solve[a4==0,t],4]
```

{{$t->$ 4.5}, {$t->$ 6.145}, {$t->$ 2.855}}

```
b4 = Expand[b[4,f]]
```

$$-(\frac{64}{7})+\frac{3104}{231}t-\frac{12176}{2079}t^2+\frac{32}{33}t^3-\frac{16}{297}t^4$$

```
N[Solve[b4==0,t],4]
```

{{$t->$ 7.834}, {$t->$ 1.166}, {$t->$ 5.792}, {$t->$ 3.208}}

FIGURE 1. *Mathematica* program for computing isolator polynomials.

33
A Bound on the Implicit Degree of Polygonal Bézier Surfaces

Joe Warren

ABSTRACT Recent work has shown that the triangular rational Bézier surface representation can be used to create polygonal surface patches. The key to the construction is the judicious use of zero weights in creating the surface parameterization. These zero weights introduce base points into the resulting rational parameterizations. Base points also lower the degree of the implicit representation of these polygonal surface patches. This paper states and proves in a simple, constructive manner a bound on the implicit degree of these rational surfaces.

33.1 Introduction

Triangular parametric surface patches are an important tool in geometric modeling. Mathematically, a triangular surface patch S is the image of a domain triangle T under some parametric mapping. If T has vertices $\mathbf{v}_0$, $\mathbf{v}_1$, and $\mathbf{v}_2$, any point $\mathbf{q}$ in T can be expressed uniquely in the form $\mathbf{q} = \alpha_0\mathbf{v}_0 + \alpha_1\mathbf{v}_1 + \alpha_2\mathbf{v}_2$ where $\alpha_0, \alpha_1, \alpha_2 \geq 0$ and $\sum \alpha_i = 1$; in this case, $(\alpha_0, \alpha_1, \alpha_2)$ are called the *barycentric coordinates* of $\mathbf{q}$ with respect to the domain triangle T.

A *rational* Bézier surface of parametric degree d is of the form:

$$\begin{pmatrix} x \\ y \\ z \\ w \end{pmatrix} = \sum_{\substack{i_0, i_1, i_2 \geq 0 \\ i_0+i_1+i_2=d}} \begin{pmatrix} x_{i_0i_1i_2}w_{i_0i_1i_2} \\ y_{i_0i_1i_2}w_{i_0i_1i_2} \\ z_{i_0i_1i_2}w_{i_0i_1i_2} \\ w_{i_0i_1i_2} \end{pmatrix} B_{i_0i_1i_2}(\alpha_0, \alpha_1, \alpha_2),$$

where $B_{i_0i_1i_2}$ are the bivariate Bernstein basis functions of degree d. These basis functions have the form:

$$B_{i_0i_1i_2}(\alpha_0, \alpha_1, \alpha_2) = \frac{d!}{i_0!i_1!i_2!}\alpha_0^{i_0}\alpha_1^{i_1}\alpha_2^{i_2}.$$

The parameterization yields points whose homogeneous coordinates are (x, y, z, w). Since the affine coordinates of this point are of the form $(\frac{x}{w}, \frac{y}{w}, \frac{z}{w})$, the resulting affine surface patch is the image of a rational function. This representation is useful in geometric design because coefficients of these functions have geometric meaning. The points $(x_{i_0i_1i_2}, y_{i_0i_1i_2}, z_{i_0i_1i_2})$ are called *control points.* Associated with each control point is a *weight* $w_{i_0i_1i_2}$. The set of all control points and their weights forms a weighted control polyhedron that approximates the resulting surface patch. [Far88] describes some of the properties of rational Bézier surfaces.

33.2 Base points

Parameter values for which $(x, y, z, w) = (0, 0, 0, 0)$ are called *base points* of the parameterization. It is easy to introduce a base point into the parameterization of a rational Bézier surface. For example, when the weight w_{d00} is zero, the values of x, y, z, and w are all zero at the parameter value $(1, 0, 0)$. Base points have two important effects on rational surfaces.

First, the image of a base point under the rational parameterization is a set of rational curves on the resulting rational surface [Cle68, S+70]. This observation follows from the fact that at a base point, the rational parameterization evaluates to $(\frac{0}{0}, \frac{0}{0}, \frac{0}{0})$. The image of the base point depends on the direction in the parameter plane from which the base point is approached. This observation is the basis for a method to create polygonal patches [War90].

Let S be a rational Bézier surface patch of parametric degree d where S is the image of a domain triangle T with barycentric coordinates $(\alpha_0, \alpha_1, \alpha_2)$. The convex hull of the points $\{(i_0/d, i_1/d, i_2/d) \mid w_{i_0i_1i_2} \neq 0\}$ is known as the *Newton polygon* of the weight function of S [Wal50]. The Newton polygon provides a key to understanding the structure of S. The following theorem from [War90] demonstrates its importance.

Theorem 1 *Let P be the Newton polygon of the weight function for a rational Bézier surface patch S. The boundary curves of S are in $1-1$ correspondence with edges of P.*

The control points of S that lie on a single edge of P define a rational Bézier curve that is a boundary curve of S. Figure 1 illustrates the relationship between P and S and shows a hexagonal surface patch created using this technique.

The other important effect of base points is that their presence lowers the *implicit* or *algebraic* degree of S. Any rational surface patch S can be expressed implicitly as the zero set of a homogeneous polynomial function [Har77]

$$F(x, y, z, w) = 0.$$

The degree of the polynomial F is called the implicit or algebraic degree of S. The following argument, from [Sal14, Sed90], illustrates the relationship between the degree of a rational parameterization, its number of base points, and the implicit degree of resulting surface patch.

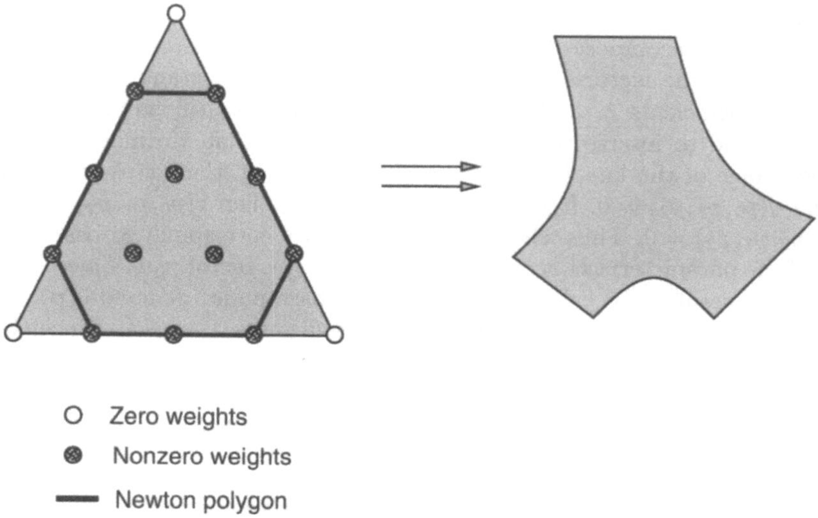

FIGURE 1. Transforming a domain triangle into a hexagonal patch

The implicit degree of S is equivalent to the number of intersections of S and a line L [Har77, Ch. I]. Define L as the intersection of the planes with equations

$$\lambda_1 x + \lambda_2 y + \lambda_3 z + \lambda_4 w = 0,$$
$$\delta_1 x + \delta_2 y + \delta_3 z + \delta_4 w = 0.$$

If $(x(\alpha_0, \alpha_1, \alpha_2), y(\alpha_0, \alpha_1, \alpha_2), z(\alpha_0, \alpha_1, \alpha_2), w(\alpha_0, \alpha_1, \alpha_2))$ is a parameterization for S, then each intersection of S with one of the planes defining L is a curve in the parameter space of S. These curves are $f = 0$ and $g = 0$ where

$$\begin{aligned} f(\alpha_0, \alpha_1, \alpha_2) &= \lambda_1 x(\alpha_0, \alpha_1, \alpha_2) + \lambda_2 y(\alpha_0, \alpha_1, \alpha_2) \\ &+ \lambda_3 z(\alpha_0, \alpha_1, \alpha_2) + \lambda_4 w(\alpha_0, \alpha_1, \alpha_2) \qquad (33.1) \\ g(\alpha_0, \alpha_1, \alpha_2) &= \delta_1 x(\alpha_0, \alpha_1, \alpha_2) + \delta_2 y(\alpha_0, \alpha_1, \alpha_2) \\ &+ \delta_3 z(\alpha_0, \alpha_1, \alpha_2) + \delta_4 w(\alpha_0, \alpha_1, \alpha_2) \qquad (33.2) \end{aligned}$$

Bézout's theorem [Ful78] guarantees that the curves $f = 0$ and $g = 0$ intersect in at most d^2 points in parameter space. However, not all of these points necessarily correspond to points of $L \cap S$. Base points of the parameterization account for some of these intersections in parameter space. Taking the u-resultant [vdW49] of

$$\begin{aligned} f(\alpha_0, \alpha_1, \alpha_2) &= 0 \\ g(\alpha_0, \alpha_1, \alpha_2) &= 0 \\ \alpha_0 u_0 + \alpha_1 u_1 + \alpha_2 u_2 &= 0 \qquad (33.3) \end{aligned}$$

yields a homogeneous polynomial of degree d^2 in u_0, u_1, and u_2. This polynomial decomposes into linear factors of the form $p_0u_0 + p_1u_1 + p_2u_2$. In some of the factors, the coefficients p_i may be constants, independent of the coefficients λ_i and δ_i. These points (p_0, p_1, p_2) in parameter space correspond to intersections of $f = 0$ and $g = 0$ that are independent of the choice of the line L. If $L = (x = 0) \cap (y = 0)$, then $x(p_0, p_1, p_2) = 0$ and $y(p_0, p_1, p_2) = 0$. If $L = (z = 0) \cap (w = 0)$, then $z(p_0, p_1, p_2) = 0$ and $w(p_0, p_1, p_2) = 0$. Thus, these parameter values correspond to base points $\mathbf{p}$ of the parameterization. The multiplicity of this factor $p_0u_0+p_1u_1+p_2u_2$ in the u-resultant is called the *order* of the base point, denoted $o(\mathbf{p})$.

The remaining factors depend on the coefficients λ_i and δ_i. Since there are only a finite number of base points, L may be chosen so that each of these factors is of the form $q_0u_0 + q_1u_1 + q_2u_2$ where (q_0, q_1, q_2) is distinct from any base point. Thus, the image of these remaining points are exactly the points of $L \cap S$.

Theorem 2 *If S is rational surface with a parameterization of degree d, then the implicit degree of S is exactly $d^2 - \sum_j o(\mathbf{p}_j)$ where $\mathbf{p}_j$ denote the base points of the parameterization.*

The main objective of this paper is to establish a tighter bound on the implicit degree of the polygonal patches used in [War90]. Such a bound is useful for several reasons. The complexity of many geometric applications for processing such surface patches depends on the implicit degree of the patch. For example, the implicit degree bounds the number of intersections of a straight line and the patch. This quantity directly affects the time required for ray tracing the surface patch. The degree of an intersection curve between two such patches is bounded by the product of the degrees of the patches. Bounding the implicit degree of S is also an important first step in the process of creating specialized resultants for converting S to its implicit form.

33.3 Intersection multiplicity and Newton polygons

The key to establishing a bound on the degree of a polygonal Bézier patch is the relationship between the orders of the base points of $(f = 0) \cap (g = 0)$ and the Newton polygons associated with f and g. Recall that the order of a base point (p_0, p_1, p_2), $o(\mathbf{p})$, is simply the power to which the linear factor $p_0u_0 + p_1u_1 + p_2u_2$ appears in the u-resultant when the coefficients λ_i and δ_i that determine the line L are treated as indeterminates. A base point $\mathbf{p}$ of order greater than one corresponds to several intersection points of $f = 0$ and $g = 0$ all lying at $\mathbf{p}$.

Intersection multiplicity is a tool for precisely measuring the number of intersection points that coincide at $\mathbf{p}$. Intersection multiplicity was originally developed to allow strict equality in Bézouts theorem. [Ful78] shows that any measure satisfying the following seven axioms is equivalent to the notion of intersection multiplicity from Bézouts theorem.

1. $I_{\mathbf{p}}(f,g) \geq 0$ and $I_{\mathbf{p}}(f,g) = \infty$ if $f = 0$ and $g = 0$ share a common curve at p.

2. $I_{\mathbf{p}}(f,g) = 0$ if and only if $f(\mathbf{p}) \neq 0$ or $g(\mathbf{p}) \neq 0$.

3. $I_{\mathbf{p}}(f,g) = 1$ if and only if $f = 0$ and $g = 0$ have distinct tangent lines at p.

4. $I_{\mathbf{p}}(f,g) = I_{\mathbf{p}}(g,f)$.

5. $I_{\mathbf{p}}(f,gh) = I_{\mathbf{p}}(f,g) + I_{\mathbf{p}}(f,h)$.

6. $I_{\mathbf{p}}(f,g) = I_{\mathbf{p}}(f,af+g)$ for all a in $\mathbf{C}[\alpha_0,\alpha_1,\alpha_2]$.

7. $I_{\mathbf{p}}(f,g)$ is invariant under an affine change of coordinates.

For a fixed choice of L, we claim that the multiplicity of a linear factor $p_0u_0 + p_1u_1 + p_2u_2$ in the u-resultant of f and g is exactly the intersection multiplicity of $f = 0$ and $g = 0$ at $\mathbf{p} = (p_0, p_1, p_2)$. With $o(\mathbf{p})$ so defined, it is a simple exercise to show that $o(\mathbf{p})$ satisfies these seven axioms.

Given this equivalence, we next show that the intersection multiplicity of two curves $f = 0$ and $g = 0$ at the origin is related to the shape of the Newton polygons for their defining polynomials. If f is written as the sum of monomials,

$$f(s,t) = \sum_{ij} f_{ij}s^it^j,$$

the Newton polygon P of f is defined as the convex hull of $\{(i,j) \mid f_{ij} \neq 0\}$. If f is not divisible by s or t, then P touches both the s and t-axes. The complement of P with respect to the quadrant $(s \geq 0) \cap (t \geq 0)$ forms two regions. The bounded region R containing the origin is called the *order* polygon of f. Figure 2 shows the an example of the Newton polygon and the order polygon for a polynomial.

Theorem 3 *Let $f_0(s,t)$ and $f_1(s,t)$ be two polynomials with generic coefficients sharing the same order polygon R. Then, the intersection multiplicity of $f_0 = 0$ and $f_1 = 0$ at the origin, $I_0(f_0, f_1)$, is exactly twice the area of R.*

Proof: We first prove a weaker, preliminary version of the theorem. Let f_0 and f_1 have an order polygon with vertices $(0,0)$, $(k,0)$, and $(0,j)$ where k and j are relatively prime. We show that $I_0(f_0, f_1) = k * j$.

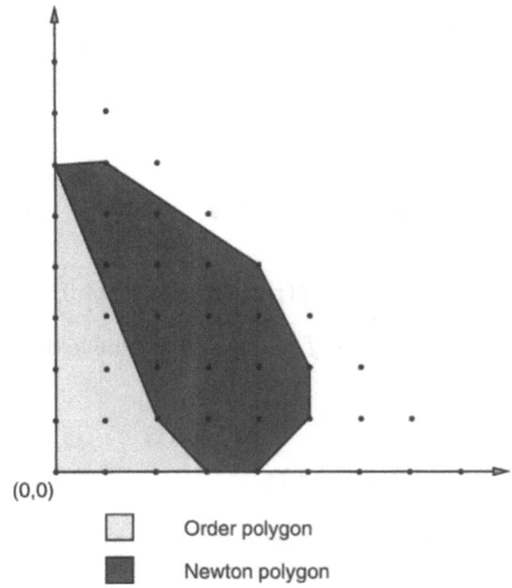

FIGURE 2. The Newton polygon and order polygon for a polynomial

The proof is by induction on $k + j$. The base case occurs when either $k = 1$ or $j = 1$. Without loss of generality, let $k = 1$. This construction involves the use of polynomial remainder sequences as outlined in [Ful78, pp. 76]. Let

$$t * f_2 = \frac{f_1(s,0)}{\gcd(f_0(s,0), f_1(s,0))} * f_0 - \frac{f_0(s,0)}{\gcd(f_0(s,0), f_1(s,0))} * f_1.$$

Since $\frac{f_1(s,0)}{\gcd(f_0(s,0),f_1(s,0))}$ does not vanish at $s = 0$, $I_0(f_1, \frac{f_1(s,0)}{\gcd(f_0(s,0),f_1(s,0))})$ must be zero by axiom 2. Therefore,

$$I_0(f_1, f_0) = I_0(f_1, \frac{f_1(s,0)}{\gcd(f_0(s,0), f_1(s,0))} * f_0)$$

by axiom 5. By axiom 6, we may add some multiple of f_1 to the right hand expression without changing the intersection multiplicity. Therefore,

$$\begin{aligned} I_0(f_1, f_0) &= I_0(f_1, \frac{f_1(s,0)}{\gcd(f_0(s,0), f_1(s,0))} * f_0 - \frac{f_0(s,0)}{\gcd(f_0(s,0), f_1(s,0))} * f_1), \\ &= I_0(f_1, t * f_2), \\ &= I_0(f_1, t) + I_0(f_1, f_2). \end{aligned}$$

Since the order polygon for f_0 and f_1 has height one, f_2 is nonzero at the origin. Therefore, by axiom 2, $I_0(f_1, f_2)$ is zero. Finally, $I_0(f_1, t)$ is equal to $I_0(f_1(s,0),t)$ by axiom 6. Since, f_1 has an order polygon with a vertex at $(k,0)$, then

$$I_0(f_0,f_1) = I_0(f_1(s,0),t) = I_0(s^k(c_0+c_1 s+...),t) = k.$$

We now complete the induction for the preliminary version. Without loss of generality, we assume that $k > j$. Let

$$\begin{aligned} t * f_{i+2} &= \frac{f_{i+1}(s,0)}{\gcd(f_i(s,0), f_{i+1}(s,0))} * f_i \\ &- \frac{f_i(s,0)}{\gcd(f_i(s,0), f_{i+1}(s,0))} * f_{i+1}. \end{aligned} \tag{33.4}$$

Given this relation, the intersection multiplicities for the f_i's are related in manner similar to that of the base case.

$$\begin{aligned} I_0(f_{i+1}, f_i) &= I_0(f_{i+1}, \frac{f_{i+1}(s,0)}{\gcd(f_i(s,0), f_{i+1}(s,0))} * f_i), \\ &= I_0(f_{i+1}, \frac{f_{i+1}(s,0)}{\gcd(f_i(s,0), f_{i+1}(s,0))} * f_i \\ &- \frac{f_i(s,0)}{\gcd(f_i(s,0), f_{i+1}(s,0))} * f_{i+1}), \\ &= I_0(f_{i+1}, t * f_{i+2}), \\ &= I_0(f_{i+1}, t) + I_0(f_{i+1}, f_{i+2}). \end{aligned}$$

Collapsing these equations yields the sum

$$I_0(f_0,f_1) = (\sum_{i=1}^{2j} I_0(f_i,t)) + I_0(f_{2j}, f_{2j+1}). \tag{33.5}$$

Since the order polygons for f_0 and f_1 have height j, f_{2j} and f_{2j+1} must be nonzero at the origin. Therefore,

$$I_0(f_0,f_1) = \sum_{i=1}^{2j} I_0(f_i,t). \tag{33.6}$$

Let g_0 and g_1 be two generic polynomials with order polygons that have vertices $(0,0)$, $(k-j,0)$ to $(0,j)$. Let

$$\begin{aligned} t * g_{i+2} &= \frac{g_{i+1}(s,0)}{\gcd(g_i(s,0), g_{i+1}(s,0))} * g_i \\ &- \frac{g_i(s,0)}{\gcd(g_i(s,0), g_{i+1}(s,0))} * g_{i+1}. \end{aligned} \tag{33.7}$$

Then, the g_i's satisfy a bound equivalent to that of equation 33.6.

$$I_0(g_0,g_1) = \sum_{i=1}^{2j} I_0(g_i,t). \tag{33.8}$$

Given that g_0 and g_1 have generic coefficients, it is straightforward to show that:

$$\begin{aligned} I_0(f_{2i}, t) &= I_0(g_{2i}, t) + j - i, \\ I_0(f_{2i+1}, t) &= I_0(g_{2i+1}, t) + j - i, \end{aligned}$$

for all $0 \leq i \leq j$.

Substituting in equation 33.6 and evaluating the resulting sum, it follows that

$$I_0(f_0, f_1) = I_0(g_0, g_1) + j^2.$$

If j and k are relatively prime, then j and $k - j$ are relatively prime. Therefore, by the inductive hypothesis, $I_0(g_0, g_1)$ is $(k - j) * j$.

$$\begin{aligned} I_0(f_0, f_1) &= (k - j) * j + j^2, \\ &= k * j. \end{aligned}$$

We now prove the general version of the theorem. The proof is by induction of the number of edges of R. Note that we treat an edge of R containing more than two points with integral coordinates as a sequence of distinct edges with the property that the interior of no edge contains a point with integral coordinates. This restriction forces an edge with vertices (i, j) and (k, l) to have the property that $i - k$ and $j - l$ are relatively prime. In the base case, if R has three edges, then the result follows immediately by the preliminary version of the theorem.

To complete the induction, let the edge of R touching the s-axis have coordinates $(l + k, 0)$ and (l, j). Again, we define the f_i as in equation 33.4. We note that equation 33.5 is still satisfied. Let g_0 and g_1 be polynomials with generic coefficients having an order polygon with vertices $(0, 0)$, $(k, 0)$, and $(0, j)$. Define the g_i as in equation 33.7. The g_i and f_i are now related by the following equation

$$I_0(f_i, t) = I_0(g_i, t) + l,$$

for $1 \leq i \leq 2j$.

By this equation and equation 33.5,

$$\begin{aligned} I_0(f_0, f_1) &= \left(\sum_{i=1}^{2j} I_0(f_i, t)\right) + I_0(f_{2j}, f_{2j+1}), \\ &= \left(\sum_{i=1}^{2j} I_0(g_i, t) + l\right) + I_0(f_{2j}, f_{2j+1}), \\ &= \left(\sum_{i=1}^{2j} I_0(g_i, t)\right) + 2l * j + I_0(f_{2j}, f_{2j+1}), \end{aligned}$$

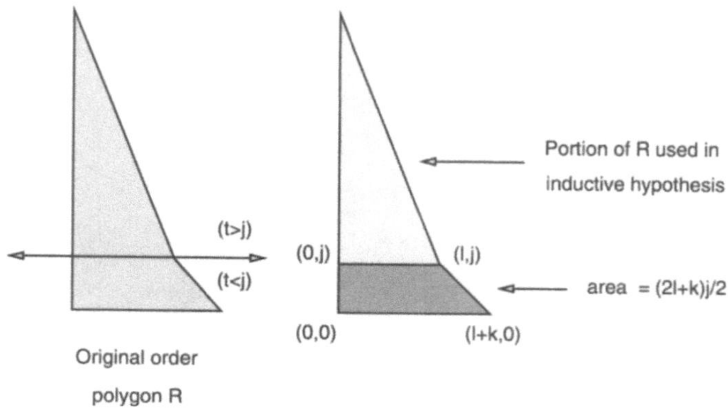

FIGURE 3. One step in the reduction of an order polygon

$$
\begin{aligned}
&= I_0(g_0, g_1) + 2l * j + I_0(f_{2j}, f_{2j+1}), \\
&= k * j + 2l * j + I_0(f_{2j}, f_{2j+1}), \\
&= (2l + k) * j + I_0(f_{2j}, f_{2j+1}).
\end{aligned}
$$

However, the order polygon for f_{2j} and f_{2j+1} is $R \cap (t \geq j)$ translated so that $(0, j)$ lies at the origin. This order polygon has one less edge than R. Therefore, by the inductive hypothesis, $I_0(f_{2j}, f_{2j+1})$ is exactly twice the area of $R \cap (t \geq j)$. But, $R \cap (t \leq j)$ has area $j * (\frac{l+(k+l)}{2})$. Therefore, $I_0(f_0, f_1)$ is exactly twice the area of R. Figure 3 illustrates this portion of the induction. □

The restriction that f_0 and f_1 have generic coefficients can be viewed as f_0 and f_1 having symbolic coefficients. During the proof of the theorem, certain specializations of these symbolic coefficients may result in the vanishing of some coefficients. The disappearance of their corresponding basis functions may increase the size of the order polygons and, thus, increase the intersection multiplicity. However, for any choice of coefficients, this theorem provides a lower bound on the intersection multiplicity.

Curves and surfaces defined by polynomials over a sparse set of basis functions are called *toric varieties* in the algebraic geometry literature [Oda85]. Results relating the number of solutions to a system of equations and the volumes of their Newton polytopes are generally known [Ber75, Kus76]. Specifically, [Oda85, pp.79] gives a bound on intersection multiplicity of several surfaces meeting at a point in terms of Newton polytopes. However, this result is derived in a very general framework that requires a deep understanding of the theory of toric varieties. The purpose of the preceding proof was to establish the theorem in a constructive manner without recourse to the general theory of toric varieties.

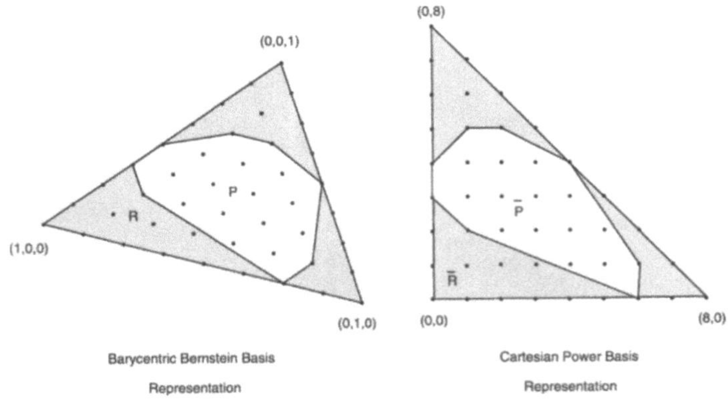

FIGURE 4. Partitioning a domain triangle

33.4 A degree bound for polygonal patches

Given the theorem of the previous section, we now prove a bound on the degree of a polygonal patch S in terms of the Newton polygon of its weight function. We assume that the Newton polygon touches each side of the of the domain triangle T. Otherwise, the parameterization of S can be trivially rewritten as a parameterization of lower degree.

Theorem 4 *Let S be a rational Bézier surface defined by a rational parameterization of degree d over a domain triangle T. Let P be the Newton polygon for the weight function of S. Then the implicit degree of of S is no greater than*

$$\frac{area(P)}{area(T)}d^2. \tag{33.9}$$

Proof: Recall from section 33.2 that the degree of S is defined as the number of intersections of S with a line L. If L is defined as the intersection of two planes, then these two planes induce two curves $f = 0$ and $g = 0$, defined by equations 33.1 and 33.2 as functions of $(\alpha_0, \alpha_1, \alpha_2)$ in the parameter plane.

For any L, the polynomials f and g share the same Newton polygon P. This polygon partitions the domain triangle T into four parts, P itself and a polygon associated with each vertex of T. Let R be the polygonal region associated with vertex $(1, 0, 0)$ (See Figure 4).

We first show that $f = 0$ and $g = 0$ intersect at $(1, 0, 0)$ with at least multiplicity

$$\frac{\text{area}(R)}{\text{area}(T)}d^2.$$

Let $\bar{f}$ and $\bar{g}$ be the power basis representations of f and g respectively, under the transformation

$$\begin{aligned} s &= \frac{\alpha_1}{\alpha_0}, \\ t &= \frac{\alpha_2}{\alpha_0}. \end{aligned}$$

The domain triangle T is transformed to a quadrant in the Cartesian plane. The Bernstein basis function $B_{i_0 i_1 i_2}$ is transformed into the monomial $\frac{s^{i_1} t^{i_2}}{1+s+t^d}$ times some constant. To complete the transformation, we multiply by $(1+s+t)^d$ to clears all denominators. This multiplication does affect the intersection multiplicity of $\bar{f}$ and $\bar{g}$ at the origin since $(1+s+t)$ does not vanish at the origin.

In terms of Newton polygons, the basis function $B_{i_0 i_1 i_2}$ with index $(i_0/d, i_1/d, i_2/d)$ is mapped to the monomial $s^{i_1} t^{i_2}$ with index (i_1, i_2). For a general polynomial of degree d, an index point $(i_0/d, i_1/d, i_2/d)$ is mapped to the point (i_1, i_2) lying in the triangle with vertices $(0,0)$, $(d,0)$ and $(0,d)$.

Let $\bar{P}$ be the image of P under this index mapping. Then $\bar{P}$ is the Newton polygon for $\bar{f}$ and $\bar{g}$. Let $\bar{R}$ be the order polygon for $\bar{f}$ and $\bar{g}$. The inverse image of $\bar{R}$ under the index mapping is exactly R.

By theorem 3, the intersection multiplicity at the origin is no less than 2 area$(\bar{R})$. However, under the index mapping,

$$\frac{\text{area}(\bar{R})}{d^2/2} = \frac{\text{area}(R)}{\text{area}(T)}.$$

If the intersection multiplicity of f and g is greater than $\frac{\text{area}R}{\text{area}T} d^2$ for all possible L, then the order of the base point at $(1,0,0)$ is greater than or equal to this quantity.

Since a similar proof holds for the remaining two vertices, the sum of the orders of the base points at the vertices of T is bounded below by

$$\frac{\text{area}(T) - \text{area}(P)}{\text{area}(T)} d^2.$$

Therefore, the degree of S is bounded above by

$$\frac{\text{area}(P)}{\text{area}(T)} d^2.$$

□

We observe that equality holds in equation 33.9 when the coefficients of S are chosen generically. Strict equality holds in theorem 3 when the nonzero coefficients are chosen to be generic. The only remaining possibility is that f and g may have base points other than those at the vertices of T. Let q be such a base point. Without loss of generality, assume the α_0, α_1

coordinates of q are nonzero. Let $(i_0, i_1, 0)$ be vertex of P on $\alpha_2 = 0$. Then, the basis function $B_{i_0 i_1 0}$ is nonzero at q. If the parametric functions (x, y, z, w) are perturbed to be $(x + \varepsilon_1 B_{i_0 i_1 0}, y + \varepsilon_2 B_{i_0 i_1 0}, z + \varepsilon_3 B_{i_0 i_1 0}, w + \varepsilon_4 B_{i_0 i_1 0})$, then q is no longer a base point of this parameterization for almost all values of $(\varepsilon_1, \varepsilon_2, \varepsilon_3, \varepsilon_4)$. Therefore, if the parameterization for S has generic coefficients, the degree of S must be exactly that of equation 33.9.

Acknowledgments: The author would like to thank Doug Moore for his assistance in proofreading this paper.

33.5 References

[Ber75] D. N. Bernshtein. The number of roots of a system of equations. *Functional Analysis and Applications*, 9(3):1–4, July–September 1975. In Russian.

[Cle68] A. Clebsch. Ueber die abbildung algebraischer flächen insbesondere der vierten and fünften ordnung. *Mathematische Annalen*, 1:253–316, 1868. In German.

[Far88] Gerald Farin. *Curves and Surfaces for Computer Aided Geometric Design: A Practical Guide.* Academic Press Inc., New York, 1988.

[Ful78] W. Fulton. *Algebraic Curves.* Benjamin/Cummings, 1978.

[Har77] R. Hartshorne. *Algebraic Geometry.* Springer-Verlag, 1977.

[Kus76] A. G. Kushnirenko. Newton polytopes and the Bézout theorem. *Functional Analysis and Applications*, 10(3):82–83, July–September 1976. In Russian.

[Oda85] T. Oda. *Convex Bodies and Algebraic Geometry.* Springer-Verlag, 1985.

[S+70] Virgil Snyder et al. *Selected Topics in Algebraic Geometry.* Chelsea Publishing Company, Bronx, New York, 1970.

[Sal14] George Salmon. *A Treatise on the Analytic Geometry of Three Dimensions.* Longmans, Green, London, 1914.

[Sed90] Thomas W. Sederberg. Techniques for cubic algebraic surfaces. *IEEE Computer Graphics and Applications*, pages 14–25, July 1990.

[vdW49] B. van der Waerden. *Modern Algebra I.* Frederick Unger, New York, 1949.

[Wal50] R. Walker. *Algebraic Curves.* Princeton University, 1950.

[War90] Joe Warren. The effect of base points on rational Bézier surfaces. Technical Report TR 90-122, Rice University, Department of Computer Science, May 1990.

Part IX

Publications of Shreeram Abhyankar

List of Publications for Shreeram S. Abhyankar

1. A common omission in textbooks on coordinate geometry, The *Mathematics Student*, 19 (1951), 61-64.

2. A note on positive polynomials, *The American Mathematical Monthly*, 61 (1954), 184-187.

3. (with O. Zariski) Splitting of valuations in extensions of local domains I, *Proceedings of the National Academy of Sciences*, U.S.A., 41 (1955), 84-90.

4. Splitting of valuations in extensions of local domains II, *Proceedings of the National Academy of Sciences*, U.S.A. 41 (1955), 220-223.

5. On the ramification of algebraic functions, *American Journal of Mathematics*, 77 (1955), 572-592.

6. On the valuations centered in a local domain, *American Journal of Mathematics*, 78 (1956), 321-348.

7. Local uniformization on algebraic surfaces over ground fields of characteristics $p \neq 0$, *Annals of Mathematics*, 63 (1956), 491-526.

8. Simultaneous resolution for algebraic surfaces, *American Journal of Mathematics*, 78 (1956), 761-790.

9. On the compositum of algebraically closed subfields, *Proceedings of the American Mathematical Society*, 7 (1956), 905-907.

10. Two notes on formal power series, *Proceedings of the American Mathematical Society*, 7 (1956), 903-905.

11. On the field of definition of a nonsingular birational transform of an algebraic surface, *Annals of Mathematics*, 65 (1957), 268-281.

12. Finite quotient groups of abelian groups of finite rational rank, *American Journal of Mathematics*, 79 (1957), 190-192.

13. Coverings of algebraic curves, *American Journal of Mathematics*, 79 (1957), 825-856.

14. On the ramification of algebraic functions Part II: Unaffected equations for characteristic two, *Transactions of American Mathematical Society*, 89 (1958), 310-324.

15. Minimal "sum of products of sums" expressions of Boolean functions, *IRE Transactions on Electronic Computers*, EC-7 (1958), 268-276.

16. Absolute minimal expressions of Boolean functions, *IRE Transactions on Electronic Computers*, EC-8 (1959), 3-8.

17. (BOOK) **Ramification Theoretic Methods in Algebraic Geometry**, Princeton University Press, Princeton, 1959.

18. Tame coverings and Fundamental groups of algebraic varieties, Part I: Branch loci with normal crossings; Applications: Theorems of Zariski and Picard, *American Journal of Mathematics*, 81 (1959), 46-94.

19. Tame coverings and fundamental groups of algebraic varieties, Part II: Branch curves with higher singularities, *American Journal of Mathematics*, 82 (1960), 120-178.

20. Tame coverings and fundamental groups of algebraic varieties, Part III: Some other sets of conditions for the fundamental group to be abelian, *American Journal of Mathematics*, 82 (1960), 179-190.

21. Tame coverings and fundamental groups of algebraic varieties, Part IV: Product theorems, *American Journal of Mathematics*, 82 (1960), 341-362.

22. Tame coverings and fundamental groups of algebraic varieties, Part V: Three cuspidal plane quartics, *American Journal of Mathematics*, 82 (1960), 365-373.

23. Tame coverings and fundamental groups of algebraic varieties, Part VI: Plane curves of order at most four, *American Journal of Mathematics*, 82 (1960), 374-388.

24. Cubic surfaces with a double line, *Memoirs of the College of Science, University of Kyoto, Series A Mathematics*, 32 (1960), 455-511.

25. Uber die endliche Erzeugung der Fundamental Gruppe einer komplex-algebraischen Mannigfaltigkeit, *Mathematische Annalen*, 139 (1960), 265-274.

26. Concepts of order and rank on a complex space and a condition for normality, *Mathematische Annalen*, 141 (1960), 171-192.

27. Remarks on Hessians and Flexes, *Nieuw Archief voor Wiskunde*, (3), 11 (1963), 110-117.

28. (Book) **Local Analytic Geometry**, Academic Press, New York, 1964.

29. Uniformization in p-cyclic extensions of algebraic surfaces over ground fields of characteristic p, *Mathematische Annalen*, 153 (1964), 81-96.

30. Reduction to multiplicity less than p in a p-cyclic extension of a two dimensional regular local ring, *Mathematische Annalen*, 154 (1964), 28-55.

31. A remark on the non-normal locus of an analytic space, *Proceedings of American Mathematical Society*, 15 (1964), 505-508.

32. Current status of the resolution problem, *Proceedings of American Mathematical Society Summer Institute on Algebraic Geometry*, 1964.

33. Nonsplitting, *Proceedings of American Mathematical Society Summer Institute on Algebraic Geometry*, 1964.

34. Uniformization of Jungian local domain, *Mathematische Annalen*, 159 (1965), 1-43.

35. Resolution of singularities of arithmetical surfaces. *Arithmetical Algebraic Geometry*, Harper and Row (1965), 111-152.

36. Ramification and resolution, *Actas del Coloquio Internacional sobre Geometrier Algebraica*, Madrid, (1965), 179-185.

37. Uniformization in a p-cyclic extension of a two dimensional regular local domain of residue field characteristic p. *Festschrift zur Gedächtnisfeier für Karl Weierstrass 1815-1895. Wissenschaftliche Abhandlungen des Landes Nordrhein-Westfallen*, 33 (1966), 243-317.

38. (Book) **Resolution of Singularities of Embedded Algebraic Surfaces**, Academic Press, New York (1966).

39. An algorithm on polynomials in one indeterminate with coefficients in a two dimensional regular local domain, *Annali di Matematica pura ed applicata-Series*, 4, 71 (1966), 25-60.

40. On the birational invariance of arithmetic genus, *Rendiconti di Matematica*, 25 (1966), 77-86.

41. Divisors on equisingular hypersurfaces, *Rendiconti della Accademia Nazionale dei Lincei-Series*, 8, 41 (1966), 49-50.

42. Nonsplitting of valuations in extensions of two dimensional regular local domains, *Mathematische Annalen*, 170 (1967), 87-144.

43. Inversion and invariance of characteristic pairs, *American Journal of Mathematics*, 89 (1967), 363-372.

44. Local rings of high embedding dimension, *American Journal of Mathematics*, 89 (1967), 1073-1077.

45. Nonprefactorial local rings, *American Journal of Mathematics*, 89 (1967), 1137-1146.

46. Remarks on equisingularity, *American Journal of Mathematics*, 90 (1968), 108-144.

47. A criterion of equisingularity, *American Journal of Mathematics*, 90 (1968), 342-345.

48. Note on coefficient fields, *American Journal of Mathematics*, 90 (1968), 346-355.

49. On the problem of resolution of singularities, *Proceedings of 1966 Moscow International Congress of Mathematicians*, Moscow (1968), 469-481.

50. Resolution of singularities of algebraic surfaces, Algebraic Geometry, *Proceedings of 1968 Bombay International Colloquium*, Oxford University Press (1969), 1-11.

51. Automorphisms of analytic local rings, *Institut des Hautes Etudes Scientifique, Publication Mathematique*, 36 (1969) 139-163.

52. (with T. T. Moh and M. van der Put) Invariants of analytic local rings, *Institut des hautes Etudes Scientifique, Publication Mathematique*, 36 (1969), 165-193.

53. Singularities of algebraic curves, *Analytic Methods in Mathematical Physics Conference Proceedings*, Gordon and Breech, (1970), 3-14.

54. (with C. Risk) Algebraic curve theory and the envelope diagrams, *Analytic Methods in Mathematical Physics Conference Proceedings Gordon and Breech*, (1970), 15-25.

55. The contributions of Ramanujan to mathematics, *Symposia on Theoretical Physics and Mathematics*, Plenum Press, 10 (1970), 1-6.

56. (with T. T. Moh) A reduction theorem for divergent power series, *Journal für die reine und angewandte Mathematik*, 241 (1970), 27-33.

57. (with M. van der Put) Homomorphisms of analytic local rings, *Journal für die reine and angewandte Mathematik*, 242 (1970), 26-60.

58. (Book) **A Glimpse of Algebraic Geometry**, Lokmanya Tilak Lectures, University of Poona, Pune, India, 1971.

59. (Book) **Algebraic Space Curves**, Les Presses de l'Universite de Montreal, Montreal, Canada, 1971.

60. *Appendix l to Chapter VIII of Zariski's Algebraic Surfaces*, 224-228, Springer-Verlag, New York, 1971.

61. Rational curves in affine space, *American Mathematical Monthly*, 78 (1971), 84-85.

62. An intersection of two surfaces, *American Mathematical Monthly*, 78 (1971), 86-88.

63. (with P. Eakin and W. Heinzer) On the uniqueness of the coefficient ring in a polynomial ring, *Journal of Algebra*, 23 (1972), 310-342.

64. On Macaulay's examples, *Conference on Commutative Algebra, Lecture Notes in Mathematics*, 311 (1973), 1-16, Springer-Verlag, New York.

65. (with T. T. Moh) Newton-Puiseux expansion and generalized Tschirnhausen transformation I, *Journal für die reine and angewandte mathematik*, 260 (1973), 47-83.

66. (with T. T. Moh) Newton-Puiseux expansion and generalized Tschirnhausen transformation II, *Journal für die reine and angewandte Mathematik*, 261 (1973), 29-54.

67. (Book) (with A. M. Sathaye) **Geometric Theory of Algebraic Space Curves**, Lecture Notes in Mathematics, No. 423, Springer-Verlag, New York, 1974.

68. High-school algebra in algebraic geometry, *Historia Mathematica*, 2 (1975), 567-572.

69. (with T. T. Moh) Embeddings of the line in the plane, *Journal für die reine and angewandte mathematik*, 276 (1975), 148-166.

70. Historical ramblings in algebraic geometry and related algebra. *American Mathematical Monthly*, 83 (1976), 409-448.

71. (with T. T. Moh) On analytic independence, *Transactions of the American Mathematical Society*, 219, (1976), 77-87.

72. (Book) **Lectures on Expansion Techniques in Algebraic Geometry**, Tata Institute of Fundamental Research, Bombay, 1977.

73. On the semigroup of a meromorphic curve, Part I, *Proceedings of the International Symposium on Algebraic Geometry*, Kyoto, 1977, 240-414.

74. (with Balwant Singh) Embeddings of certain curves in the affine plane, *American Journal of Mathematics*, 100 (1978), 99-195.

75. Generalizations of the Ancient Indian Mathematics and Applications, *Second Anniversary Souvenir of Bhaskaracharya Pratishthana*,1978, 3-13.

76. Quasirational Singularities, *American Journal of Mathematics*, 101 (1979), 276-300.

77. (with L.A. Rubel) Every difference polynomial has a connected zero-set, *Journal of the Indian Mathematical Society*, 43 (1979), 69-78.

78. (Book) **Weighted Expansions for Canonical Desingularization**, Lecture Notes in Mathematics, No. 910, Springer-Verlag, New York, (1982).

79. Desingularization of plane curves, *American Mathematical Society Proceedings of Symposia in Pure Mathematics*, Vol. 40, Part 1, 1-45, (1983).

80. On a question of Mumford, *American Journal of Mathematics*, vol. 105 (1983), 1455-1479.

81. Combinatorie des tableaux de Young, varietes determinantielles et calcul de fonctions de Hilbert, *Rendiconti del Seminario Matematico des Universita et Politecnico di Torino*, vol. 42 (1984), 65-88.

82. Characteristic Pairs, Complex Analytic Singularities, *Advanced Studies in Pure Mathematics*, vol. 8, (1986), 1-2.

83. (with C. Bajaj) Automatic parametrization of curves and surfaces I: Conics and Conicoids, *Computer-Aided Design*, vol. 19 (1987), 11-14.

84. (with C. Bajaj) Automatic parametrization of curves and surfaces II: Cubics and Cubicoids, *Computer-Aided Design*, vol. 19 (1987), 499-502.

85. Determinantal loci and enumerative combinatorics of Young tableaux, *Algebraic Geometry and Commutative Algebra in Honor of Masayoshi Nagata*, vol. I, Kinokuniya Company Ltd., Tokyo, 1987, 1-26.

86. (Book) **Enumerative Combinatorics of Young Tableaux**, Marcel Dekker, New York, 1988.

87. What is the difference between a parabola and a hyperbola?, *Mathematical Intelligencer*, vol. 10 (1988), 37-43.

88. Good points of a hypersurface, *Advances in Mathematics*, vol. 68 (1988), 87-256.

89. (with C. Bajaj) Automatic parametrization of curves and surfaces III: Algebraic plane curves, *Computer Aided Geometric Design*, vol. 5 (1988), 309-321.

90. (with D. M. Kulkarni) On hilbertian ideals, *Linear Algebra and its Applications*, vol. 116 (1989), 53-79.

91. Irreducibility Criterion for Germs of Analytic Functions of Two Complex Variables, *Advances in Mathematics*, vol. 74 (1989), 190-257.

92. (with C. Bajaj) Computations with Algebraic Curves, *Springer Lecture Notes in Computer Science*, vol. 358 (1989), 274-284.

93. (with S. B. Joshi) Generalized Codeletion and Standard Multitableaux, *Canadian Mathematical Society Conference Proceedings*, vol.10 (1989), 1-24.

94. (with S. Chandrasekar and V. Chandru) Degree Complexity Bounds on the Intersection of Algebraic Curves, *Proceedings of the Fifth Annual ACM Symposium on Computational Geometry*, vol.5 (1989), 88-93.

95. (with W. Li) On the Jacobian Conjecture: A New Approach via Gröbner Bases, *Journal of Pure and Applied Algebra*, vol.61 (1989), 211-222.

96. (with C. Bajaj) Automatic Parametrization of Rational Curves and Surfaces IV: Algebraic Space Curves, *ACM Transactions on Graphics*, vol.8 (1989), 324-333.

97. (with D. M. Kulkarni) Bijection between indexed monomials and standard bitableaux, *Discrete Mathematics*, vol. 79 (1990), 1-48.

98. (with S. Chandrasekar and V. Chandru) Improper Intersection of Algebraic Curves, *ACM Transactions on Graphics*, vol. 9 (1990), 147-159.

99. Parametrization of Curves and Surfaces,(reprinted from Curves and Surfaces in Computer Vision and Graphics), *SPIE (Society of Photo Optical Instrumentation Engineers) Proceedings Series*, vol. 1251 (1990), 122-129.

100. (Book) **Algebraic Geometry for Scientists and Engineers**, American Mathematical Society, Providence, 1990.

101. (with D. M. Kulkarni) Coinsertion and standard bitableaux, *Discrete Mathematics*, vol. 85 (1990), 115-166.

102. (with W. J. Heinzer) Derivativewise Unramified Infinite Integral Extensions, *Journal of Algebra*, vol. 136 (1991),

103. (with W. J. Heinzer and S. Wiegand) On the Compositum of Two Power Series Rings, *Proceedings of the American Mathematical Society*, vol. (1991),

104. (with T. L. Morin and T. Trafalis) Efficient Faces of Polytopes: Interior Point Algorithms, Parametrization of Algebraic Varieties, and Multiple Objective Optimization, "Mathematical Developments Arising from Linear Programming", *American Mathematical Society Contemporary Mathematics*, vol. (1991),

105. (with S. Chandrasekar and V. Chandru) Intersection of Algebraic Space Curves, *Discrete Applied Mathematics*, vol. 31 (1991),

MIX
Papier aus verantwortungsvollen Quellen
Paper from responsible sources
FSC® C105338

If you have any concerns about our products,
you can contact us on
ProductSafety@springernature.com

In case Publisher is established outside the EU,
the EU authorized representative is:
Springer Nature Customer Service Center GmbH
Europaplatz 3, 69115 Heidelberg, Germany

Printed by Libri Plureos GmbH
in Hamburg, Germany